Texts & Monographs in Symbolic Computation

A Series of the Research Institute for Symbolic Computation, Johannes Kepler University, Linz, Austria

Founding Editor

Bruno Buchberger, Research Institute for Symbolic Computation, Hagenberg, Austria

Series Editor

Peter Paule, RISC, Johannes Kepler University Linz, Austria

Mathematics is a key technology in modern society. Symbolic Computation is on its way to become a key technology in mathematics. "Texts and Monographs in Symbolic Computation" provides a platform devoted to reflect this evolution. In addition to reporting on developments in the field, the focus of the series also includes applications of computer algebra and symbolic methods in other subfields of mathematics and computer science, and, in particular, in the natural sciences. To provide a flexible frame, the series is open to texts of various kind, ranging from research compendia to textbooks for courses.

More information about this series at http://www.springer.com/series/3073

Veronika Pillwein • Carsten Schneider

Editors

Algorithmic Combinatorics: Enumerative Combinatorics, Special Functions and Computer Algebra

In Honour of Peter Paule on his 60th Birthday

 Springer

Editors
Veronika Pillwein
Research Institut for Symbolic Computation
Johannes Kepler University of Linz
Linz, Oberösterreich, Austria

Carsten Schneider
Research Institut for Symbolic Computation
Johannes Kepler University of Linz
Linz, Austria

ISSN 0943-853X   ISSN 2197-8409   (electronic)
Texts & Monographs in Symbolic Computation
ISBN 978-3-030-44561-4   ISBN 978-3-030-44559-1   (eBook)
https://doi.org/10.1007/978-3-030-44559-1

Mathematics Subject Classification: 33xx, 68W30, 05Axx, 05A17, 05A15, 05A19, Primary 11P83; Secondary 05A17

This Springer imprint is published by the registered company Springer Nature Switzerland AG.
The registered company address is: Gewerbestrasse 11, 6330 Cham, Switzerland

This volume is devoted to the 60th birthday of Peter Paule

Foreword

This volume is dedicated to Peter Paule on the occasion of his 60th birthday. It contains the proceedings of the workshop, Combinatorics, Special Functions and Computer Algebra (Paule60), held on May 17–18, 2018, at the Research Institute for Symbolic Computation (RISC) in Hagenberg, Austria. This is overwhelmingly merited in light of Peter's distinguished career. A former Humboldt Fellow, Peter has been a major player in the applications of computer algebra and has been director of RISC since 2009. He is a member of Academia Europaea and a Fellow of the American Mathematical Society.

In the early 1980s, Professor Johann Cigler gave a wonderful talk at an Oberwolfach conference on combinatorics. In the talk, he extolled the outstanding work of several of his students especially that of Peter Paule. Subsequently, at the next Oberwolfach conference on combinatorics, Peter himself gave a presentation, and I was more than pleased to make the acquaintance of this rising star. This was the beginning of a grand and lasting friendship.

I have written more papers with Peter than anyone else, 15 in all. The majority concern the computer algebra implementation of P. A. MacMahon's Partitions Analysis (often joint with Peter's student, Axel Riese). This collaboration was one of the most wonderful adventures of my career. Notable partly because this allowed us to track down a number of fascinating mathematical objects, but mostly because it is a joy to collaborate with this optimistic man who is always full of joie de vivre.

Peter is the opposite of the stereotypical mathematician. As we all know, the way to get ahead in mathematics is to sit alone in a small room for days on end concentrating intensely on esoteric abstractions. As a result, a number of us are somewhat socially challenged. Peter has completely avoided anything like this outcome. Not only do mathematicians enjoy time spent with him, but often the RISC visitor's entire family remembers him warmly. Indeed, he is often the only mathematician they do remember.

It is a great honor to be asked to prepare this foreword. Best wishes to you, Peter, from all your friends and admirers (and me especially). We look forward to your many future achievements.

University Park, PA, USA   George E. Andrews
November 2019

Preface

The book is centered around *Algorithmic Combinatorics* which covers the three research areas of *Enumerative Combinatorics, Special Functions*, and *Computer Algebra*. What these research fields share is that many of their outstanding results do not only have applications in Mathematics but also other disciplines, such as Computer Science, Physics, Chemistry, etc. A particular charm of these areas is how they interact and influence one another. For instance, combinatorial or special functions' techniques have motivated the development of new symbolic algorithms. In particular, the first proofs of challenging problems in Combinatorics and Special Functions were derived by making essential use of Computer Algebra.

This book addresses these interdisciplinary aspects with research articles and up to date reviews that are suitable for graduate students, researchers, or practitioners who are interested in solving concrete problems within mathematics and other research disciplines. Algorithmic aspects will be emphasized and the corresponding software packages for concrete problems are introduced whenever applicable.

When the Search for Solutions Can Be Terminated (Sergei A. Abramov) addresses the problem that in algorithms often the nonexistence of solutions can only be detected in the final stages, after carrying out a lot of heavy computations. In this article, it is shown how to introduce early termination checkpoints in an algorithm for finding rational solutions of differential systems.

In *Euler's Partition Theorem and Refinements Without Appeal to Infinite Products* (Krishnaswami Alladi), combinatorial arguments on 2-modular Ferrers diagrams are combined in a novel way in order to find and prove analogues of some important fundamental theorems in the theory of partitions.

In *Sequences in Partitions, Double q-Series and the Mock Theta Function $\rho_3(q)$* (George E. Andrews), a skillful mix of techniques based on q-difference equations, generating functions, series expansions, and bijective maps based on the combinatorics of integer partitions (and overpartitions) are applied to gain new insights of combinatorial aspects of a family of double sum hypergeometric q-series and their connection to many famous identities for integer partitions.

In *Refined q-Trinomial Coefficients and Two Infinite Hierarchies of q-Series Identities* (Alexander Berkovich and Ali Kemal Uncu), symbolic summation tools and classical methods from enumerative combinatorics are combined to discover and explore new doubly bounded polynomial identities in terms of refined q-trinomial coefficients.

In *Large Scale Analytic Calculations in Quantum Field Theories* (Johannes Blümlein), a general overview of computer algebra and special function tools is presented that are heavily used to solve large-scale problems in relativistic renormalizable quantum field theories. These analytic tools originate from algorithmic combinatorics or are suitable for problems coming from this field.

In *An Eigenvalue Problem for the Associated Askey–Wilson Polynomials* (Andrea Bruder, Christian Krattenthaler, and Sergei K. Suslov), an auxiliary bivariate function is introduced that links to associated ordinary Askey–Wilson polynomials. With the aid of computer algebra, from this relation an eigenvalue problem for the associated Askey–Wilson polynomials is constructed.

Context-Free Grammars and Stable Multivariate Polynomials Over Stirling Permutations (William Y.C. Chen, Robert X.J. Hao, and Harold R.L. Yang) resolves two open questions raised by Haglund and Visontai in their study of stable multivariate refinements of second-order Eulerian polynomials.

In *An Interesting Class of Hankel Determinants* (Johann Cigler and Mike Tyson), Hankel determinants $d_r(n)$ of a binomial sequence are considered for which for general integers n and r no closed-form exists. Using the methods presented here, formulas valid for all $r \geq 0$ and particular arithmetic progressions are given.

In *A Sequence of Polynomials Generated by a Kapteyn Series of the Second Kind* (Diego Dominici and Veronika Pillwein), an infinite sum involving squares of Bessel-J functions with an extra parameter n is explored. A closed-form representation is derived for this series in terms of a specific polynomial whose degree depends on n, and a recurrence relation is computed and verified with computer algebra methods that produces the coefficients of this polynomial efficiently.

In *Comparative Analysis of Random Generators* (Johannes vom Dorp, Joachim von zur Gathen, Daniel Loebenberger, Jan Lühr, and Simon Schneider), the research field of random number generators is introduced for nonexperts. A careful comparison between pseudorandom number generators and hardware controlled versions is carried out carefully in terms of their output rate.

In *Difference Equation Theory Meets Mathematical Finance* (Stefan Gerhold and Arpad Pinter), the authors make those two ends meet unexpectedly through Pringsheim's theorem and two asymptotic methods (saddle point and Hankel contour asymptotics).

In *Evaluations as L-Subsets* (Adalbert Kerber), logical systems beyond classical Boolean logic are considered by utilizing lattice-valued evaluations of statements in a novel way. In particular, examples are elaborated that demonstrate the practicality of real-world problems.

In *Exact Lower Bounds for Monochromatic Schur Triples and Generalizations* (Christoph Koutschan and Elaine Wong), exact and sharp lower bounds for the number of generalized monochromatic Schur triples subject to all 2-colorings are explored. In their challenging enterprise, they use low-dimensional polyhedral combinatorics leading to many case distinctions that could be treated successfully by the symbolic computation technique of cylindric algebraic decomposition.

In *Evaluation of Binomial Double Sums Involving Absolute Values* (Christian Krattenthaler and Carsten Schneider), double sums from a general family are considered, where the main difficulty lays in the appearance of absolute values. It is shown that these sums in general can be expressed as a linear combination of just four simple hypergeometric expressions.

In *On Two Subclasses of Motzkin Paths and Their Relation to Ternary Trees* (Helmut Prodinger, Sarah J. Selkirk, and Stephan Wagner), paths with alternating east and north-east steps are shown to give nice enumeration formulas via generalized Catalan numbers.

In *A Theorem to Reduce Certain Modular Form Relations Modulo Primes* (Cristian-Silviu Radu), a question raised by Peter Paule is settled that is of algorithmic relevance to the theory of modular forms. It is shown that the problem to decide if a certain modular form relation modulo a prime holds can be reduced to check congruences modulo p between meromorphic modular forms.

In *Trying to Solve a Linear System for Strict Partitions in "closed form"* (Volker Strehl), a challenging linear system for strict partitions in relation to Schur functions and symmetric functions is investigated that utilizes graph theory in combination with the theory of partitions in a novel way.

In *Untying the Gordian Knot via Experimental Mathematics* (Yukun Yao and Doron Zeilberger), two new applications of automated guessing are given: one related to enumerating spanning trees using transfer matrices and one about determinants of certain families of matrices. Using symbolic computation, the painful human approach is avoided.

This book is an offspring of the workshop "Combinatorics, Special Functions and Computer Algebra" at the occasion of Peter Paule's 60th birthday (https://www3.risc.jku.at/conferences/paule60/). We would like to thank Tanja Gutenbrunner and Ramona Pöchinger from RISC for all their help to organize this wonderful event. In particular, we would like to thank the Austrian FWF in the frameworks of the SFB "Algorithmic and Enumerative Combinatorics" and the Doctoral Program "Computational Mathematics" for the financial support. Furthermore, we thank Karoly Erdei for providing us with the above picture that illustrates Peter Paule's mathematical passion. Finally, we would like to thank all the authors for their stimulating contributions and the referees in the background for their valuable comments.

Linz, Austria   Veronika Pillwein
Linz, Austria   Carsten Schneider
November 2019

Contents

List of Contributors

Sergei A. Abramov Dorodnitsyn Computing Centre, Federal Research Center "Computer Science and Control" of the Russian Academy of Sciences, Moscow, Russia

Krishnaswami Alladi Department of Mathematics, University of Florida, Gainesville, FL, USA

George E. Andrews The Pennsylvania State University, University Park, PA, USA

Alexander Berkovich Department of Mathematics, University of Florida, Gainesville, FL, USA

Johannes Blümlein Deutsches Elektronen-Synchrotron, DESY, Zeuthen, Germany

Andrea Bruder Department of Mathematics and Computer Science, Colorado College, Colorado Springs, CO, USA

William Y. C. Chen Center for Applied Mathematics, Tianjin University, Tianjin, People's Republic of China

Johann Cigler Fakultät für Mathematik, Universität Wien, Vienna, Austria

Diego Dominici Johannes Kepler University Linz, Doktoratskolleg, "Computational Mathematics", Linz, Austria

Department of Mathematics State, University of New York, New Paltz, NY, USA

Stefan Gerhold TU Wien, Vienna, Austria

Robert X. J. Hao Department of Mathematics and Physics, Nanjing Institute of Technology, Nanjing, Jiangsu, People's Republic of China

Adalbert Kerber Mathematics Department, University of Bayreuth, Bayreuth, Germany

Christoph Koutschan Johann Radon Institute for Computational and Applied Mathematics, Austrian Academy of Sciences, Linz, Austria

Christian Krattenthaler Fakultät für Mathematik, Universität Wien, Vienna, Austria

Daniel Loebenberger OTH Amberg-Weiden and Fraunhofer AISEC, Weiden, Germany

Jan Lühr Bonn-Aachen International Center for Information Technology, Universität Bonn, Bonn, Germany

Veronika Pillwein Johannes Kepler University Linz, Research Institute for Symbolic Computation (RISC), Linz, Austria

Arpad Pinter TU Wien Alumnus, Vienna, Austria

Helmut Prodinger Stellenbosch University, Department of Mathematical Sciences, Stellenbosch, South Africa

Carsten Schneider Johannes Kepler University Linz, Research Institute for Symbolic Computation (RISC), Linz, Austria

Simon Schneider Bonn-Aachen International Center for Information Technology, Universität Bonn, Bonn, Germany

Sarah Selkirk Stellenbosch University, Department of Mathematical Sciences, Stellenbosch, South Africa

Cristian-Silviu Radu Johannes Kepler University Linz, Research Institute for Symbolic Computation (RISC), Linz, Austria

Volker Strehl Department Informatik, Friedrich-Alexander Universität, Erlangen, Germany

Sergei K. Suslov School of Mathematical and Statistical Sciences, Arizona State University, Tempe, AZ, USA

Mike Tyson

Ali Kemal Uncu Johannes Kepler University Linz, Research Institute for Symbolic Computation (RISC), Linz, Austria

Johannes vom Dorp Fraunhofer FKIE, Bonn, Germany

Joachim von zur Gathen Bonn-Aachen International Center for Information Technology, Universität Bonn, Bonn, Germany

Stephan Wagner Stellenbosch University, Department of Mathematical Sciences, Stellenbosch, South Africa

Elaine Wong Johann Radon Institute for Computational and Applied Mathematics (RICAM), Austrian Academy of Sciences, Linz, Austria

Harold R. L. Yang School of Science, Tianjin University of Technology and Education, Tianjin, People's Republic of China

Yukun Yao Rutgers University-New Brunswick, Piscataway, NJ, USA

Doron Zeilberger Rutgers University-New Brunswick, Piscataway, NJ, USA

When the Search for Solutions Can Be Terminated

Sergei A. Abramov

Dedicated to Peter Paule on the occasion of his 60th birthday

1 Introduction

One of actual computer algebra problems is the development of algorithms for finding solutions to differential equations and systems of such equations. Usually solutions belonging to some fixed class are discussed. Often the proposed algorithms are such that the absence of solutions of the desired form is detected only in the final stages, when many of the quantities required to construct such a (potential) solution are already computed.

However, it is possible that in the algorithm one can choose some checkpoints and, accordingly, associate with them some tests which make it possible to ascertain already at an early stage that there are no solutions of the desired type. This will save time and other computing resources. Thus there is the problem of choosing checkpoints and tests. On the one hand, one can think about this choice already in the development of the algorithm and seek the appearance in the algorithm of such points equipped with easily performable tests; on the other hand, one can take a known algorithm and insert checkpoints in it. In this case, it may be necessary to modify the algorithm in order for suitable checkpoints to be discovered and for these points to precede some resource-consuming fragments of the algorithm.

In the present paper, we consider this problem as applied to the search for rational solutions.

Let K be a field of characteristic 0. The ring of polynomials and the field of rational functions of x are conventionally denoted as $K[x]$ and $K(x)$, respectively. The ring of formal Laurent series is denoted as $K((x))$. If R is a ring (in particular, a field), then $\mathrm{Mat}_m(R)$ denotes the ring of $m \times m$-matrices with entries from R. We

S. A. Abramov (✉)
Dorodnicyn Computing Centre, Federal Research Center "Computer Science and Control"
of the Russian Academy of Sciences, Moscow, Russia

© Springer Nature Switzerland AG 2020
V. Pillwein, C. Schneider (eds.), *Algorithmic Combinatorics: Enumerative Combinatorics, Special Functions and Computer Algebra*, Texts & Monographs in Symbolic Computation, https://doi.org/10.1007/978-3-030-44559-1_1

consider systems of the form

$$A_r(x)D^r y(x) + \cdots + A_1(x)Dy(x) + A_0(x)y(x) = 0 \tag{1}$$

where $D = \frac{d}{dx}$, and $A_i(x)$, for $i = 0, 1, \ldots, r$, are matrices of size $m \times m$ with entries from $K[x]$. Here $A_r(x)$ is the *leading* matrix (we suppose that non-zero), and $y(x) = (y_1(x), y_2(x), \ldots, y_m(x))^T$ is a column of unknown functions (T denotes transposition). The number r is called the *order* of the system. The system under study is assumed to be of full rank; i.e., the equations of the system are linearly independent over the ring of operators $K(x)[D]$. In some cases, the *trailing* matrix of a system is also considered. (If $k = \min\{l \mid A_l \neq 0\}$ then A_k is the trailing matrix of (1).)

The system (1) can be written in the form

$$L(y) = 0 \tag{2}$$

where

$$L = A_r(x)D^r + \cdots + A_1(x)D + A_0(x). \tag{3}$$

A solution $y(x) = (y_1(x), y_2(x), \ldots, y_m(x))^T \in K(x)^m$ of (1) is called a *rational* solution. If $y(x) \in K[x]^m$, it is called a *polynomial* solution (a particular case of a rational solution). Algorithms for finding all rational solutions to a first-order system of the form

$$Dy(x) = A(x)y(x), \tag{4}$$

where $A(x) \in \mathrm{Mat}_m(K(x))$, are well known (see, e.g., [3, 10]). The problem of finding rational solutions for full-rank systems (1) in the case where the matrix $A_r(x)$ considered much less frequently. Nevertheless, an appropriate algorithm was suggested in [7]. This algorithm is based on finding a *universal denominator* of rational solutions to the original system (for brevity, we call it the universal denominator for the original system), i.e., a polynomial $U(x) \in K[x]$ such that, if the system has a rational solution $y(x) \in K(x)^m$, then it can be represented as $\frac{1}{U(x)}z(x)$, where $z(x) \in K[x]^m$. If a universal denominator is known, we can make the substitution

$$y(x) = \frac{1}{U(x)}z(x) \tag{5}$$

where $z(x) = (z_1, \ldots, z_m)^T$ is a vector of new unknowns, and then apply one of the algorithms for finding polynomial solutions (see, e.g., [3, 11, 15]). A *denominator bound* for the original system is a rational function $S(x)$ such that any rational solution of the original system can be represented in the form $S(x)f(x)$ with

$f(x) \in K[x]^m$. So a denominator bound can also be used for finding rational solutions by using the substitution

$$y(x) = S(x)z(x)$$

instead of (5). (If $U(x)$ is a universal denominator for (2) then $\frac{1}{U(x)}$ is obviously a denominator bound for the same system.)

Other approaches are also possible. For example, the approach presented in [2] is based on expanding a general solution of the original system (2) into a series whose coefficients linearly depend on arbitrary constants. After multiplication by a universal denominator $U(x)$ (or by $S^{-1}(x)$, where $S(x)$ is a denominator bound) the series corresponding to rational solutions turn into polynomials.

In the sequel, it will be useful to consider formal Laurent series, i.e., for example, elements of the field $K((x))$ (or the field $\bar{K}((x))$, where $\bar{K}$ is the algebraic closure of K). Recall that the *valuation* val $y(x)$ of $y(x) \in K((x))$ is the minimal integer i such that the coefficient of x^i in $y(x)$ is non-zero. If $y(x)$ is the zero series then we set val $y(x) = +\infty$. We can also consider the field $K((x - \alpha))$ of formal Laurent series in $x - \alpha$ and, correspondingly, $\mathrm{val}_{x-\alpha} t(x)$ for $t \in K((x - \alpha))$.

We consider also the formal series in terms of decreasing powers (this can also be viewed as expansion at ∞); the field of such series is denoted by $K((x^{-1}))$. Each series of this kind contains only a finite number of powers of x with nonnegative exponents and, possibly, an infinite number of powers with negative ones. The greatest exponent of x with a nonzero coefficient occurring in a series $y(x)$ is the valuation $\mathrm{val}_\infty y(x)$. If $y(x) \in K((x^{-1}))$ is the zero series, then we set $\mathrm{val}_\infty y(x) = -\infty$.

For a vector $f(x) = (f_1(x), \ldots, f_m(x))^T \in K((x))^m$ we set val $f(x) = \min_{i=1}^m \mathrm{val}\, f_i$ (similarly for $\mathrm{val}_{x-\alpha} f(x)$). For $g(x) = (g_1(x), \ldots, g_m(x))^T \in K((x^{-1}))^m$ we set $\mathrm{val}_\infty g(x) = \max_{i=1}^m \mathrm{val}_\infty g_i$. It is easy to see that $\mathrm{val}_\infty p(x) = \deg p(x)$ for a polynomial $p(x)$ and $v(\frac{f(x)}{g(x)}) = v(f(x)) - v(g(x))$ for $f(x), g(x) \in K[x]$, $v \in \{\mathrm{val}, \mathrm{val}_{x-\alpha}, \mathrm{val}_\infty\}$. It is also significant that the valuation of any type under consideration of a product is the sum of the valuations of the factors.

The checkpoints and tests mentioned at the beginning of the Introduction may help detect situations where substitutions of the series in question lead to a system that obviously has no polynomial solutions. In this case, we would like to obtain tests that do not require a complete calculation of the universal denominators or denominator bounds, but involve just some preliminary estimates. For scalar difference equations, such points and tests were found by A. Gheffar in [12, 13]. In the present paper, we generalize those ideas for linear systems of differential equations with polynomial coefficients.

2 Preliminaries: Indicial Polynomials

A rational solution of a system of the form (1) can be represented by formal Laurent series both at an arbitrary finite point α and at ∞.

It is well known (see, e.g., [8, Sect. 7.2]) that it is possible to construct for (1) a finite set of irreducible polynomials over K

$$p_1(x), \ldots, p_k(x) \tag{6}$$

such that if for some $\alpha \in \overline{K}$ there exists a solution $F \in \overline{K}((x - \alpha))^m$ such that $\mathrm{val}_{x-\alpha} F < 0$ then $p_i(\alpha) = 0$ for some $1 \leq i \leq k$, and for each $p_i(x)$ a polynomial $I_{L,p_i}(\lambda) \in K[\lambda]$ can be constructed such that for a solution $F \in K((x - \alpha))^m$, $p_i(\alpha) = 0$, one has $I_{L,p_i}(\mathrm{val}_{x-\alpha} F) = 0$ [8]. It is also possible to construct such a polynomial $I_{L,\infty}(\lambda) \in K[\lambda]$ that if a system $L(y) = 0$ has a solution $y \in K((x^{-1}))$ then $I_{L,\infty}(\mathrm{val}_\infty y(x)) = 0$. In particular, the degree of a polynomial solution is a root of $I_{L,\infty}(\lambda)$. The polynomials $I_{L,\infty}(\lambda), I_{L,p_1}(\lambda), \ldots, I_{L,p_k}(\lambda)$ are the *indicial* polynomials connected with L.

Remark In the context of this paper, by the indicial polynomial for a given operator L we mean a certain polynomial, a root of which may give useful information on solutions of the initial differential system. Absence of roots of a certain type also gives information on solutions of the initial differential system. Note that it is not necessary that every root of such a polynomial corresponds to some specific solution of the initial system, as in classical theory. To construct the needed polynomials we can use the so-called induced recurrence system and bring its leading or trailing matrix to non-singular form. Based on the determinants of those matrices, some polynomials can be obtained that play the role of the indicial polynomials [1, 8]. (The mentioned induced recurrence system is satisfied by the sequence of coefficients of any Laurent series solution of the original differential system; the elements of such a sequence belong to K^m or $\bar{K}^m$.)

3 Scheme Equipped with Control Tests

The following proposition is the main statement of the paper.

Proposition *Let L, $p_1(x), \ldots, p_k(x)$ be as in (3), (6). Let $I_{L,\infty}(\lambda), I_{L,p_1}(\lambda), \ldots, I_{L,p_k}(\lambda)$ be the corresponding indicial polynomials. In this case*

(i) *if $I_{L,\infty}(\lambda)$ has no integer root then (2) has no rational solution;*
(ii) *if at least one of the polynomials $I_{L,p_1}(\lambda), \ldots, I_{L,p_k}(\lambda)$ has no integer root then (2) has no rational solution;*
(iii) *if $b_1, \ldots, b_k \in \mathbb{Z}$ are lower bounds for integer roots of polynomials $I_{L,p_1}(\lambda), \ldots, I_{L,p_k}(\lambda)$ (e.g., $b_1, \ldots, b_k$ can be equal to the minimal integer roots of those polynomials), N is an upper bound for integer roots of the*

polynomial $I_{L,\infty}(\lambda)$ (e.g., N can be equal to the maximal integer root of that polynomial), and $N - \sum_{i=1}^{k} b_i \deg p_i < 0$, then (2) has no rational solution;

(iv) if $N - \sum_{i=1}^{k} b_i \deg p_i \geq 0$ (see (iii)) and the system (2) has a rational solution then that solution is of the form $p_1^{b_1}(x) \ldots p_k^{b_k}(x) f(x)$, where $f(x) = (f_1(x), \ldots, f_m(x))^T \in K[x]^m$ with $\deg f_j(x) \leq N - \sum_{i=1}^{k} b_i \deg p_i$, $j = 1, \ldots, m$.

Proof (i), (ii): If (2) has a rational solution $F \in K(x)^m$ then (2) has a solution in $K((x^{-1}))^m$ as well, since $F(x)$ can be represented by a series from $K((x^{-1}))$. Let $s(x)$ be a formal Laurent series over K^m for $F(x^{-1})$, then $t(x) = s(x^{-1}) \in K((x^{-1}))$ is the series for $F(x)$. So $I_{L,\infty}(\text{val}_\infty t(x)) = 0$, proving (i). Let α be such that $p_i(\alpha) = 0, 1 \leq i \leq k$, and let $s \in K((x - \alpha))^m$ be the Laurent series expansion of $F(x)$. Then $I_{L,p_i(x)}(\text{val}_{x-\alpha} s) = 0$, proving (ii).

(iii): $S(x) = p_1^{b_1}(x) \ldots p_k^{b_k}(x)$ is a denominator bound for (2) (among $b_1, \ldots, b_k$ there may be numbers of different signs). If $F(x) \in K(x)^m$ is a rational solution of (2) then $F(x) = S(x) f(x)$ for some $f(x) \in K[x]^m$. We have $0 \leq \text{val}_\infty f(x) = \text{val}_\infty F(x) - \text{val}_\infty S(x) \leq N - \text{val}_\infty S(x) = N - \sum_{i=1}^{k} b_i \deg p_i$. Thus, if there exists a rational solution then $N - \sum_{i=1}^{k} b_i \deg p_i \geq 0$.

(iv): The upper bound $N - \sum_{i=1}^{k} b_i \deg p_i \geq 0$ for $\text{val}_\infty f(x) = \max_{j=1}^{m} \deg f_j$ was obtained in the proof of (iii). □

A scheme equipped with control tests may be, for example, as follows.

1. Find $I_{L,\infty}(\lambda)$. If this polynomial does not have integer roots, then STOP. Otherwise, let N be the largest integer root of $I_{L,\infty}(\lambda)$.
2. Find $p_1(x), \ldots, p_k(x)$ and polynomials $I_{L,p_1}(\lambda), \ldots, I_{L,p_k}(\lambda)$. If at least one of $I_{L,p_1}, \ldots, I_{L,p_k}$ does not have integer roots, then STOP. Otherwise, let $e_1, \ldots, e_k$ be the smallest integer roots of these indicial polynomials and $d = e_1 \deg p_1 + \cdots + e_k \deg p_k$.
3. If $N + d < 0$ then STOP. Otherwise, perform in (1) the substitution $y = Sz$, where $S(x) = p_1(x)^{e_1} \ldots p_k(x)^{e_k}$, and z is a new unknown vector. Find all polynomial solutions of the new system $\tilde{L}(z) = 0$, using the fact that the degree of each such solution does not exceed $N + d$. If there are no such solutions, then STOP. Otherwise, rational solutions of the system $L(y) = 0$ are obtained from polynomial solutions of the system $\tilde{L}(z) = 0$ by multiplying each component of z by $S(x)$.

In this scheme, the STOP command means stopping all calculations with the message to the user: "The system has no rational solutions".

Having computed the upper bound $N - d$ for the degrees of polynomial solutions allows us to use the method of undetermined coefficients for finding polynomial solutions of the system $\tilde{L}(z) = 0$ (the problem of finding polynomial solutions is reduced to solving a system of linear algebraic equations). There exist methods which are more effective than the method of undetermined coefficients (see, for example, [15]). However, to apply the algorithm from [15], it is necessary to construct an induced recurrent system and bring its trailing matrices to non-singular form (we have mentioned induced recurrent systems in Remark 2). This preparatory work is equivalent in cost to obtaining the indicial polynomial $I_{\tilde{L},\infty}$. One can also use the approach from [2], for which one does not need the substitution $y = Sz$ into $L(y) = 0$ (we have mentioned it in Sect. 1).

4   Examples

Example For a system $L(y) = 0$ of the form

$$\begin{pmatrix} x & 1 \\ 1 & 1 \end{pmatrix} y' + \begin{pmatrix} x^2 & x \\ 1 & x \end{pmatrix} y = 0$$

we get $I_{L,\infty}(\lambda)$ as a non-zero constant. The polynomial has no integer roots and the system has no rational solutions (there is no need to look for a universal denominator and so on). If we apply the usual approach, then we would have to find the universal denominator $U(x) = x$, make the substitution (5) into the original system, then a search should be made for polynomial solutions. Finally, it would show that there are no such solutions.   □

Example If a system $L(y) = 0$ is of the form

$$\begin{pmatrix} 2 & 0 \\ 0 & x(x+1) \end{pmatrix} y' + \begin{pmatrix} -1 & 1 \\ x & 2(x+1) \end{pmatrix} y = 0$$

then $I_{L,\infty}(\lambda) = -\lambda - 3$. The only integer root is -3. We can find $U(x) = x^2$ as a universal denominator (or $S(x) = x^{-2}$ as a denominator bound). We see that $-3 + 2 = -1 < 0$. This implies that the system has no rational solutions (there is no need to produce the substitution $y = S(x)z$ and try to find polynomial solutions).   □

5   Conclusion

The present paper shows that an approach similar to the proposed in [12, 13] can be applied not only to scalar equations, but to systems of equations as well. Small changes in the scheme of the algorithm allow one to mark the points that we call the checkpoints, and write down the corresponding control tests so that without increasing the cost of the algorithm as a whole, in some cases, performing the tests on the intermediate results makes it possible to stop the algorithm as soon as these tests imply that no solutions of the desired type exist.

Apparently, this approach may be useful in the development of algorithms for finding solutions that are more complicated than rational solutions (we would emphasize that the search for many types of solutions ultimately boils down to finding rational solutions for some auxiliary systems).

This type of problem can also be posed for the case of systems of linear difference equations. In the book [14] of P. Paule and M. Kauers, in particular, the basic tools for working with scalar difference equations are described. Regarding systems, it is possible, for example, to mention publications [3–6, 8, 9, 11, 16]. The question of checkpoints and control tests for systems of linear difference equations remains a topic for future research.

Acknowledgements The author would like to thank M.Petkovšek, A.Ryabenko and D.Khmelnov for their valuable remarks to improve the paper. Supported by the Russian Foundation for Basic Research (RFBR), project no. 19-01-00032.

References

1. Abramov, S.A.: EG–eliminations. J. Diff. Equ. Appl. **5**, 393–433 (1999)
2. Abramov, S.A.: Search of rational solutions to differential and difference systems by means of formal series. Program. Comput. Softw. **41**(2), 65–73 (2015)
3. Abramov, S.A., Barkatou, M.: Rational solutions of first order linear difference systems. In: Proceedings: ISSAC'98, pp. 124–131 (1998)
4. Abramov, S., Bronstein, M.: On solutions of linear functional systems. In: ISSAC'2001 Proceedings, pp. 1–6 (2001)
5. Abramov, S., Bronstein, M.: Linear algebra for skew-polynomial matrices. Rapport de Recherche INRIA RR-4420, March 2002. http://www.inria.fr/RRRT/RR-4420.html (2002)
6. Abramov, S.A., Khmelnov, D.E.: Denominators of rational solutions of linear difference systems of arbitrary order. Program. Comput. Softw. **38**(2), 84–91 (2012)
7. Abramov, S.A., Khmelnov, D.E.: Singular points of solutions of linear ODE systems with polynomial coefficients. J. Math. Sci. **185**(3), 347–359 (2012)
8. Abramov, S.A., Khmelnov, D.E.: Linear differential and difference systems: EG_δ- and EG_σ-eliminations. Program. Comput. Softw. **39**(2), 91–109 (2013)
9. Abramov, S.A., Gheffar, A., Khmelnov, D.E.: Factorization of polynomials and gcd computations for finding universal denominators. In: Proceedings: CASC 2010. LNCS, vol. 6244, pp. 4–18 Springer, Heidelberg (2010)
10. Barkatou, M.A.: A fast algorithm to compute the rational solutions of systems of linear differential equations RR 973–M– Mars 1997. IMAG–LMC, Grenoble (1997)
11. Barkatou, M.A.: Rational solutions of matrix difference equations: problem of equivalence and factorization. In: ISSAC'99 Proceedings, pp. 277–282 (1999)
12. Gheffar, A.: Linear differential, difference and q-difference homogeneous equations having no rational solutions. ACM Commun. Comput. Algebra **44**(3), 78–83 (2010)
13. Gheffar, A.: Detecting nonexistence of rational solutions of linear difference equations in early stages of computation. ACM Commun. Comput. Algebra **48**(3), 90–97 (2014)
14. Kauers, M., Paule, P.: The concrete tetrahedron. In: Symbolic Sums, Recurrence Equations, Generating Functions, Asymptotic Estimates. Texts and Monographs in Symbolic Computation. Springer, Wien (2011)
15. Khmelnov, D.E.: Search for polynomial solutions of linear functional systems by means of induced recurrences. Program. Comput. Softw. **30**, 61–67 (2004)
16. van Hoeij, M.: Rational solutions of linear difference equations. In: ISSAC'98 Proceedings, pp. 120–123 (1998)

Euler's Partition Theorem and Refinements Without Appeal to Infinite Products

Krishnaswami Alladi

Dedicated to Peter Paule for his 60th Birthday

1 Introduction

One of the first results an entrant to the theory of partitions encounters is Euler's fundamental and beautiful theorem:

Theorem E *Let $p_d(n)$ and $p_o(n)$ denote the number of partitions of n into distinct parts and odd parts respectively. Then*

$$p_d(n) = p_o(n).$$

Euler's proof of Theorem E made use of product representations of the generating functions of $p_d(n)$ and $p_o(n)$:

$$\sum_{n=0}^{\infty} p_d(n)q^n = \prod_{m=1}^{\infty}(1+q^m) = \prod_{m=1}^{\infty}\frac{(1-q^{2m})}{(1-q^m)} = \prod_{m=1}^{\infty}\frac{1}{(1-q^{2m-1})} = \sum_{n=0}^{\infty} p_o(n)q^n.$$
$$(1.1)$$

Euler's theorem and the simple yet fundamental idea in his proof, namely the replacement of expressions of the form $1 + y$ by $(1 - y^2)/(1 - y)$ in products and the study of the resulting cancellations, plays a crucial role in the theory of partitions (see Andrews [3], for instance). Many proofs of Euler's theorem are known and a variety of important refinements of it have been obtained by Sylvester

K. Alladi (✉)
Department of Mathematics, University of Florida, Gainesville, FL, USA
e-mail: alladik@ufl.edu

© Springer Nature Switzerland AG 2020
V. Pillwein, C. Schneider (eds.), *Algorithmic Combinatorics: Enumerative Combinatorics, Special Functions and Computer Algebra*, Texts & Monographs in Symbolic Computation, https://doi.org/10.1007/978-3-030-44559-1_2

[10], Fine [7], Bessenrodt [5] and others. Our approach here (see Sect. 2) is to prove Euler's theorem by *only* considering the series generating function of $p_d(n)$ and an important (but under-utilized) *amalgamation property* of this series. We then convert the series generating function of $p_d(n)$ to the series generating function of $p_o(n)$ by a suitable *dissection* of the terms of the series. In doing so we use 2-modular Ferrers graphs to establish the equivalence. We then combine these ideas with the conjugation of the Ferrers graphs of partitions into distinct parts to improve a refinement of Euler's theorem due to Fine [7] and to obtain a dual of a refinement due to Bessenrodt [5].

Sylvester [10] improved upon many partition theorems of Euler by combinatorial methods. In particular, using a graphical representation he was led to the Theorem S below, which is a refinement of Theorem E. It is not easy to establish that the graphical representation yields a bijective proof of Theorem S; this was recently done by Kim and Yee [8].

Theorem S *The number of partitions of an integer n into odd parts of which exactly k are different is equal to the number of partitions of n into distinct parts which can be grouped into k (maximal) blocks of consecutive integers.*

Yet another refinement of Theorem E was found by Fine, namely,

Theorem F *Let $p_d(n; k)$ denote the number of partitions of n into distinct parts with largest part k. Let $p_o(n; k)$ denote the number of partitions of n into odd parts such that the largest part plus twice the number of parts is $2k+1$. Then*

$$p_d(n; k) = p_o(n; k).$$

Fine observed this in 1954 but published it only in his 1988 monograph [7]. Fine's proof of Theorem F was not combinatorial, but q-theoretic; it is sketched in [3]. Andrews [3, p. 27] provides a q-theoretic proof in detail, but prior to that in 1966 [2], noted that Theorem F also falls out from Sylvester's graphical proof of Theorem E. Theorems S and F are *refinements* of Theorem E because by summing over k we get Theorem E. Our approach to Theorem E yields a much simpler and very direct proof of Theorem F and this is given in Sect. 3.

Bessenrodt [5] obtained the following elegant reformulation of Theorem E from Sylvester's bijection for Theorem S. This refinement is also a limiting case of the deep lecture hall partition refinement of Theorem E due to Bousquet-Melou and Eriksson [6].

Theorem B *Let $p_{d,k}(n)$ denote the number of partitions of n into distinct parts such that the alternating sum starting with the largest part is k. Let $p_{o,k}(n)$ denote the number of partitions of n into odd parts with total number of parts equal to k. Then*

$$p_{d,k}(n) = p_{o,k}(n).$$

We will provide in Sect. 4 a simple direct proof of Theorem B by considering conjugates of the Ferrers graphs of partitions into distinct parts and then using the amalgamation-dissection ideas to convert the generating function of $p_{d,k}(n)$ into that of $p_{o,k}(n)$. This also yields a dual of Theorem B and an improvement of Theorem F (see Sect. 5).

As examples of some recent works pertaining to Theorem E (including its analogs and Glaisher's generalization to all odd moduli) and Theorem B, emphasizing combinatorial arguments, we mention the papers of Berkovich-Uncu [4], Straub [9], and of Xiong and Keith [11]. But our approach is quite different.

We shall use the standard notation

$$(a)_n = (a; q)_n = \prod_{j=0}^{n-1}(1 - aq^j), \tag{1.2}$$

and

$$(a)_\infty = (a; q)_\infty = \lim_{n \to \infty}(a)_n = \prod_{j=0}^{\infty}(1 - aq^j), \quad \text{when} \quad |q| < 1. \tag{1.3}$$

When the base is q, we write $(a)_n$ as in (1.2) for simplicity, but when the base is anything other than q, it will be displayed.

2 New Proof of Theorem E

The series generating function of $p_d(n)$ is

$$\sum_{n=0}^{\infty} p_d(n)q^n = \sum_{m=0}^{\infty} \frac{q^{m(m+1)/2}}{(q)_m}. \tag{2.1}$$

This is due to Euler who noted that the generating function of all partitions having exactly m parts is $q^m/(q)_m$. From such partitions we get all partitions into m distinct parts by adding 0 to the smallest part, 1 to the second smallest part, ..., and $m - 1$ to the largest part. This procedure is reversible. So the term

$$\frac{q^{m(m+1)/2}}{(q)_m} \tag{2.2}$$

on the right in (2.1) is the generating function of partitions into distinct parts with exactly m parts.

The terms in (2.2) have an interesting *amalgamation property*, namely,

$$\frac{q^{(2m-1)2m/2}}{(q)_{2m-1}} + \frac{q^{2m(2m+1)/2}}{(q)_{2m}} = \frac{q^{2m^2-m}}{(q)_{2m}}, \tag{2.3}$$

and its companion

$$\frac{q^{2m(2m+1)/2}}{(q)_{2m}} + \frac{q^{(2m+1)(2m+2)/2}}{(q)_{2m+1}} = \frac{q^{2m^2+m}}{(q)_{2m+1}}. \tag{2.4}$$

These amalgamation properties have not been fully exploited, and we shall use them here. In particular, we note from (2.1) and (2.3) that

$$\sum_{n=0}^{\infty} p_d(n)q^n = 1 + \sum_{m=1}^{\infty} \frac{q^{2m^2-m}}{(q)_{2m}}. \tag{2.5}$$

Next, having amalgamated consecutive pairs of terms in (2.1) to get (2.5), we will dissect the denominator terms in (2.5) into its odd and even components. More precisely, we rewrite (2.5) as

$$1 + \sum_{m=1}^{\infty} \frac{q^{2m^2-m}}{(q)_{2m}} = 1 + \sum_{m=1}^{\infty} \frac{q^{2m^2-m}}{(q^2; q^2)_m (q; q^2)_m}. \tag{2.6}$$

We will now show that the series on the right of (2.6) is the generating function of $p_o(n)$.

Represent a partition into odd parts as a 2-modular Ferrers graph, namely a Ferrers graph in which there is a 1 at the node on the extreme right of each row, and there is a 2 at every other node. Consider the Durfee square in this 2-modular Ferrers graph, namely the largest square of nodes starting from the upper left hand corner. Let the Durfee square be of dimension $m \times m$. Now the part below this Durfee square is a partition into odd parts the largest of which is $\leq 2m - 1$. The generating function of such partitions is

$$\frac{1}{(q; q^2)_m}. \tag{2.7}$$

The portion consisting of the Durfee square and the nodes to its right forms a partition into exactly m odd parts each $\geq 2m - 1$. If $2m - 1$ is removed from each of the m rows of this part of the graph, we remove $2m^2 - m$ in total. The remaining portion is a 2-modular Ferrer's graph with only twos in it, and which, if read columnwise, is a partition into even parts each $\leq 2m$. Thus the generating function of the Durfee square and portion to its right is

$$\frac{q^{2m^2-m}}{(q^2; q^2)_m}. \tag{2.8}$$

Thus we have shown that

$$1 + \sum_{m=1}^{\infty} \frac{q^{2m^2-m}}{(q^2;q^2)_m (q;q^2)_m} = \sum_{n=0}^{\infty} p_o(n) q^n. \tag{2.9}$$

Theorem E follows from (2.9), (2.6) and (2.5).

3   Simple Proof of Theorem F

The amalgamation-dissection idea of Sect. 2 yields the refined Theorem F as we show here. But first, we introduce the following notation for convenience: For any partition π, we let

$$\lambda(\pi) = \text{largest part of } \pi,$$

$$\nu(\pi) = \text{number of parts of } \pi,$$

and

$$\sigma(\pi) = \text{the sum of the parts of } \pi.$$

Thus $\sigma(\pi)$ is the integer being partitioned. Finally let D denote the set of partitions into distinct parts, and Ω, the set of partitions into odd parts.

By following the ideas of Euler that we described prior to (2.2), we get

$$\sum_{n=0}^{\infty} \frac{z^n t^n q^{n(n+1)/2}}{(tq)_n} = \sum_{\pi \in D} z^{\nu(\pi)} t^{\lambda(\pi)} q^{\sigma(\pi)}. \tag{3.1}$$

Even though this two parameter refined generating function of partitions into distinct parts is fundamental, it has not been given much attention because it does not have a product representation. However, if we set $t = 1$ and count only the number of parts, then we get a product representation for the expression in (3.1), namely

$$\prod_{k=0}^{\infty} (1 + zq^k),$$

and this product has been investigated in detail since the time of Euler. When we set $z = 1$ in (3.1) and keep track only of the largest part, we *do not* get a product representation for the series

$$\sum_{n=0}^{\infty} \frac{t^n q^{n(n+1)/2}}{(tq)_n} = \sum_{\pi \in D} t^{\lambda(\pi)} q^{\sigma(\pi)}. \tag{3.2}$$

Our emphasis here is on series and not infinite products. The terms of the series on the left in (3.2) amalgamate as in (2.3) and (2.4), and this is interesting and useful even though we do not have a product representation. More precisely we have,

$$\frac{t^{2m-1}q^{(2m-1)2m/2}}{(tq)_{2m-1}} + \frac{t^{2m}q^{2m(2m+1)/2}}{(tq)_{2m}} = \frac{t^{2m-1}q^{2m^2-m}}{(tq)_{2m}}. \tag{3.3}$$

and its companion

$$\frac{t^{2m}q^{2m(2m+1)/2}}{(tq)_{2m}} + \frac{t^{2m+1}q^{(2m+1)(2m+2)/2}}{(tq)_{2m+1}} = \frac{t^{2m}q^{2m^2+m}}{(tq)_{2m+1}}. \tag{3.4}$$

Thus using (3.3) we get

$$\sum_{n=0}^{\infty} \frac{t^n q^{n(n+1)/2}}{(tq)_n} = 1 + \sum_{m=1}^{\infty} \frac{t^{2m-1}q^{2m^2-m}}{(tq)_{2m}}. \tag{3.5}$$

As in (2.6), we dissect the denominator on the right in (3.5) and rewrite it as

$$\sum_{n=0}^{\infty} \frac{t^n q^{n(n+1)/2}}{(tq)_n} = 1 + \sum_{m=1}^{\infty} \frac{t^{2m-1}q^{2m^2-m}}{(tq^2; q^2)_m (tq; q^2)_m}. \tag{3.6}$$

We will now combinatorially interpret the coefficient of t^k in (3.6).
    We know already from (2.7) and (2.8) that the expression

$$\frac{t^{2m-1}q^{2m^2-m}}{(tq^2; q^2)_m (tq; q^2)_m} \tag{3.7}$$

without the parameter t, is the generating function of partitions π^* into odd parts whose 2-modular Ferrers graphs have an $m \times m$ Durfee square. How are the powers of t generated in the expression in (3.7) and what does the power of t represent in relation to the partition π^*? To understand this, we write the exponent k in the power of t, say t^k, in (3.7) as

$$k = (2m - 1) + i + j, \tag{3.8}$$

where t^i is generated from the factor $(tq^2; q^2)_m$, t^j is generated by the factor $(tq; q^2)_m$, and t^{2m-1} comes from the numerator. Thus from the arguments underlying (2.8), we see that in the 2-modular graphs under consideration, $(2m - 1) + 2i = \lambda(\pi^*)$, the largest part. Also $m + j = \nu(\pi^*)$, the number of parts, because there are j parts below the Durfee square, and the Durfee square is of size $m \times m$. Note that (3.8) yields

$$\lambda(\pi^*) + 2\nu(\pi^*) = (2m - 1) + 2i + 2(m + j) = 2k + 1. \tag{3.9}$$

On the other hand, the coefficient of t^k in (3.2) is the generating function of the number of partitions into distinct parts with largest part k. So by comparing coefficients of t^k on both sides of (3.6), we get Theorem F from (3.2) and (3.8).

Remark As mentioned in the introduction, Fine's proof of Theorem F (see [7, p. 29 and p. 46]) was not combinatorial but involved transformations of q-series. Andrews [3, p. 27] provides details of a q-theoretic proof of Theorem F using Heine's transformation and its consequences. Our proof given above is simpler and more direct because all that is needed is to amalgamate pairs of consecutive terms of the series generating function for partitions into distinct parts, and then to dissect the denominator of the resulting expression into its odd and even components. This actually leads to a refinement of Theorem F described below.

A Further Refinement Note that in the amalgamation in (2.3) and (3.3), we are adding the generating functions of partitions π into distinct parts for which the number of parts $v(\pi) = 2m - 1$ or $2m$. After the amalgamation and the dissection, we interpreted the expression in (3.7) as the generating function of partitions into odd parts π^* whose 2-modular Ferrer's graphs have an $m \times m$ Durfee square. Thus we have the the following refinement of Theorem F:

Theorem F* *Let $p_d(n; k, m)$ denote the number of partitions π of n into distinct parts with $\lambda(\pi) = k$ and $v(\pi) = 2m - 1$ or $2m$. Let $p_o(n; k, m)$ denote the number of partitions π^* of n into odd parts with $\lambda(\pi^*) + 2v(\pi^*) = 2k + 1$ and such that the 2-modular Ferrers graph of π^* has a Durfee square of dimension $m \times m$. Then*

$$p_d(n; k, m) = p_o(n; k, m).$$

4 Conjugation of Partitions into Distinct Parts

If π is a partition into distinct parts, then its conjugate π^*, namely the partition obtained by representing π as a Ferrers graph and considering the conjugate of this graph, is a partition whose set of parts is the set of consecutive integers from 1 up to $v(\pi)$. Thus letting v denote $v(\pi)$, we have

$$\sigma(\pi) = \sigma(\pi^*) = \sum_{i=1}^{v} i f_i, \tag{4.1}$$

where f_i represents the frequency with which i occurs in π^*. We may rewrite the sum on the right in (4.1) as

$$\sigma(\pi^*) = (f_1 + f_2 + \ldots + f_v) + (f_2 + f_3 + \ldots + f_v) + \ldots + (f_v), \tag{4.2}$$

where the quantities within the parenthesis represent the parts of π in decreasing order.

If we denote by $s(\pi)$ as in [5], the alternating sum of the parts of π starting with the largest part, then

$$s(\pi) = (f_1 + f_2 + \ldots + f_\nu) - (f_2 + f_3 + \ldots + f_\nu) + (f_3 + f_4 + \ldots + f_\nu) - (f_4 + f_5 + \ldots + f_\nu) + \ldots$$

$$= f_1 + f_3 + f_5 + \ldots =: \nu_o(\pi^*), \tag{4.3}$$

where

$$\nu_o(\pi^*) = \text{the number of odd parts of } \pi^*.$$

So let us reformulate Theorem B as follows:

Theorem C *Let C denote the set of partitions with the property that all integers up to the largest part occur as parts. Let $p_C^*(n; \ell)$ denote the number of partitions π^* of n, $\pi^* \in C$, with $\nu_o(\pi^*) = \ell$.*

Let $p_o^*(n; \ell)$ denote the number of partitions of n into odd parts such that the number of parts is ℓ. Then

$$p_C^*(n; \ell) = p_o^*(n; \ell).$$

A partition with the property that all integers up to the largest part occur as parts is known as a partition without gaps. We call such partitions as *chain partitions*, and C is the set of such partitions. Having reformulated Theorem B in terms of chain partitions, we give a simple proof of Theorem C.

Proof of Theorem C Using Series For $\pi^* \in C$, consider its largest part $\lambda(\pi^*)$. If the largest part is $2j - 1$, then the generating function of such chain partitions π^* is

$$\frac{q^{2j^2 - j} z^j}{(zq; q^2)_j (q^2; q^2)_{j-1}}, \tag{4.4}$$

where the power of z in (4.4) is $\nu_o(\pi^*)$. If the largest part of π^* is $2j$, then the generating function is

$$\frac{q^{2j^2 + j} z^j}{(zq; q^2)_j (q^2; q^2)_j}. \tag{4.5}$$

Now if we add the expressions in (4.4) and (4.5), they amalgamate to

$$\frac{q^{2j^2 - j} z^j}{(zq; q^2)_j (q^2; q^2)_j}.$$

So we get

$$\sum p_C(n; \ell) z^\ell t^n = \sum_{j=0}^{\infty} \frac{q^{2j^2-j} z^j}{(zq; q^2)_j (q^2; q^2)_j}. \tag{4.6}$$

Just as we showed (2.9) via Durfee squares, it follows that the series on the right in (4.6) is

$$\sum_{j=0}^{\infty} \frac{q^{2j^2-j} z^j}{(zq; q^2)_j (q^2; q^2)_j} = \sum_n \sum_\ell p_o^*(n; \ell) z^\ell q^n. \tag{4.7}$$

Theorem C follows from (4.6) and (4.7) without any appeal to infinite products.

Graphical Proof of Theorem C We now provide a bijective proof of Theorem C using 2-modular graphs.

Start by representing a partition π into odd parts as a 2-modular graph. We illustrate our bijective proof by considering the partition

$$25 + 21 + 15 + 15 + 13 + 9 + 7 + 7 + 7 + 3 + 1 + 1.$$

In this 2-modular graph, mark out the Durfee square. This is illustrated in Fig. 1.

Next delete the right most column of the Durfee square, fill them with ones, and move any twos that were in the right most column to the extreme right position on the same row. Thus the integer entries in the modified Durfee square (say of dimension j) add up to $2j^2 - j$. Group the integer entries in the modified Durfee square as indicated in Fig. 2 to see that these represent the integers $1, 2, \ldots, 2j - 1$. The rows below the modified square represent odd parts $\leq 2j - 1$. The columns to the right of the modified square represent even parts $\leq 2j$. Thus if the modified graph is viewed in this fashion, we get a chain partition π^* with largest part either $2j - 1$ or $2j$. Notice that the number of parts of π equals then number of odd parts of π^* and this proves Theorem C.

5   A Dual of Theorem C and an Improvement of Theorem F

The graphical proof of Theorem C has interesting implications.

In the graphical proof given above, we focused on the number of parts. We now see what happens if we consider the largest part.

Suppose the size of the largest part of the partition π represented in Fig. 1 is $2k + 1$. As before we let the size of the Durfee square and the modified Durfee square to be j. This means that the largest odd part $\lambda_o(\pi^*)$ of the chain partition π^* given by Fig. 2 is $2j - 1$. Consequently, $2, 4, \ldots, 2j - 2$ occur as even parts of π^*

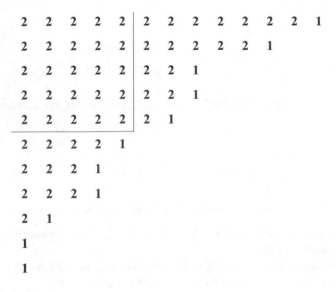

Fig. 1 5 × 5 Durfee Square

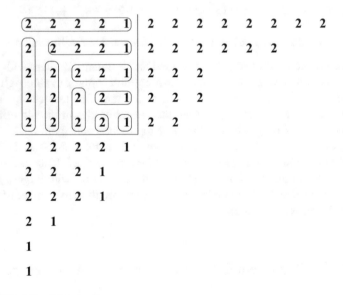

Fig. 2 Modified 5 × 5 Durfee Square

and these account for $j - 1$ even parts. The number of even parts of π^* is given by $j - 1$ plus the number of columns to the right of the modified Durfee square. We may interpret $j - 1$ as the number of twos in the first row of the modified Durfee square in Fig. 2. Thus the number of even parts of π^* is k. This leads to the following *dual* of Theorem C:

Theorem C* *Let* $p_{o,k}^*(n)$ *denote the number of partitions of n into odd parts with largest part* $2k + 1$.

Let $p_{C,k}^*(n)$ *denote the number of chain partitions of n with k even parts. Then*

$$p_{o,k}^*(n) = p_{C,k}^*(n)$$

Remarks

(i) Theorem C* is a dual of Theorem C because in Theorem C we counted the number of odd parts of π whereas in Theorem C* we count the size of the largest odd part of π; similarly in Theorem C we count the number of odd parts of π^* whereas in Theorem C* we count the number of even parts of π^*. Thus by reformulating Theorem B in terms of Theorem C, we have arrived at the dual Theorem C*.

(ii) By combining Theorems C and C* we get Fine's Theorem F. This is because with $\nu_e(\pi^*)$ representing the number of even parts of π^*, we can write Theorem C* in the form

$$\lambda(\pi) = 2\nu_e(\pi^*) + 1. \tag{5.1}$$

Similarly we may write Theorem C in the form

$$\nu(\pi) = \nu_o(\pi^*). \tag{5.2}$$

Thus (5.1) and (5.2) yield

$$\lambda(\pi) + 2\nu(\pi) = 2\nu_e(\pi^*) + 1 + \nu_o(\pi^*) = 2\nu(\pi^*) + 1. \tag{5.3}$$

By taking the conjugate of π^* the number of parts of π^* is converted to the largest part of the conjugate partition, which is a partition into distinct parts, and this is precisely Fine's theorem. Thus Theorems C and C* are improvements of Fine's theorem.

(iii) In Sect. 4 we noted that the infinite series in (4.6) and (4.7) can be interpreted in two different ways to be realized as the generating function of the two partition functions $p_C^*(n; \ell)$ and $p_o^*(n; \ell)$ in Theorem C. Similarly, the analytic version of Theorem C* is

$$\sum_n \sum_k p_{o,k}^*(n) w^k q^n = 1 + \sum_{k=0}^{\infty} \frac{w^k q^{2k+1}}{(q; q^2)_{k+1}} \tag{5.4}$$

$$= 1 + \sum_{j=1}^{\infty} \frac{w^{j-1} q^{2j^2-j}}{(q; q^2)_j (wq^2; q^2)_j} = \sum_n \sum_k p_{C,k}^*(n) w^k q^n.$$

Note that the series on the right in (5.4) is the dual of the series in (4.6) and (4.7) because in (4.6) the power of z is counting the number of odd parts, whereas on the right in (5.4) the power of w is counting the number of even parts.

Identities (4.6), (4.7), and (5.4) can be combined into a single identity as follows:

$$1 + \sum_{k=0}^{\infty} \frac{w^k z q^{2k+1}}{(zq; q^2)_{2k+1}} = 1 + \sum_{j=1}^{\infty} \frac{w^{j-1} z^j q^{2j^2-j}}{(zq; q^2)_j (wq^2; q^2)_j}. \tag{5.5}$$

Note that in (4.6) and (4.7) we did not have the series on the left in (5.5) with $w = 1$ because we did not need it. Instead we interpreted the series on the right in (5.5) with $w = 1$ in two different ways one of which relied on an amalgamation.

It turns out that the series on the right in (5.5) is a special case of a certain variant of the Rogers-Fine identity as we shall see in the next section.

6 Connection with the Rogers-Fine Identity

In the previous section we studied the 2-modular graphs of partitions into odd parts by keeping track of the largest part and the number of parts. What if we also keep track of the number of different odd parts? This leads us to a variant of the Rogers-Fine identity as we show now.

The Rogers-Fine identity in the form obtained by Fine [7, 14.1] is

$$F(\alpha, \beta, \tau; q) =: \sum_{n=0}^{\infty} \frac{(\alpha q)_n \tau^n}{(\beta q)_n} = \sum_{n=0}^{\infty} \frac{(\alpha q)_n (\alpha \tau q/\beta)_n \beta^n \tau^n q^{n^2} (1 - \alpha \tau q^{2n+1})}{(\beta q)_n (\tau)_{n+1}}. \tag{6.1}$$

Fine proved it by considering transformation properties of $F(\alpha, \beta, \tau; q)$ defined by the series on the left in (6.1).

In [1] we obtained the following variant of the Rogers-Fine identity:

$$f(a, b, c; q) =: 1 + \sum_{k=1}^{\infty} \frac{(1-a)(abq)_{k-1} bc^k q^k}{(bq)_k} \tag{6.2}$$

$$= 1 + \sum_{j=1}^{\infty} \frac{b^j c^j q^{j^2} (1-a)(abq)_{j-1}(acq)_{j-1}(1 - abcq^{2j})}{(bq)_j (cq)_j}.$$

Fine's function f and our function f are connected by the relation

$$\frac{(1-bq)}{(1-a)bcq} \{f(a, b, c; q) - 1\} = F(ab, bq, cq; q) \tag{6.3}$$

and so (6.1) and (6.2) are equivalent. The reason we investigated (6.2) was because it is combinatorially more interesting, and also can be established combinatorially in a very direct fashion.

The function $f(a, b, c; q)$ defined by the series on the left in (6.2) is the generating function of unrestricted partitions in which the power of b keeps track of the number of parts, the power of c keeps track of the largest part, and the power of $1 - a$ keeps track of the number of different parts. It is to be noted that for unrestricted partitions the generating function $f(a, b, c; q)$ has an infinite product representation only when b or c equals 1. Thus when one keeps track of all three statistics $\lambda(\pi)$, $\nu(\pi)$, and $\nu_d(\pi)$ (= the number of different parts of π), then one will NOT have a product representation but will have to deal only with a series representation. This is in line with the philosophy of this paper emphasizing series and removing dependence on infinite product representations. In order to pass from the defining series of f to the series on the right in (6.2), we studied in [1] the Ferrers graphs of unrestricted partitions using Durfee squares and the fact that under conjugation $\lambda(\pi)$ and $\nu(\pi)$ get interchanged, and $\nu_d(\pi)$ remains invariant. We needed to use the invariance of $\nu_d(\pi)$ under conjugation only on the portion of the Ferrers graph to the right of the Durfee square. This aspect will be crucial in the remark below.

The ideas in [1] can be applied to the 2-modular Ferrers graphs of partitions into odd parts. Without getting into details, we simply point out that what this means is to replace

$$q \mapsto q^2, \quad \text{and} \quad b \mapsto bq^{-1} \tag{6.4}$$

in (6.2). This yields

$$\sum_{k=1}^{\infty} \frac{(1-a)(abq; q^2)_{k-1} bc^{k-1} q^{2k-1}}{(bq; q^2)_k} \tag{6.5}$$

$$= \sum_{j=1}^{\infty} \frac{b^j c^{j-1} q^{2j^2-j} (1-a)(abq; q^2)_{j-1} (acq; q^2)_{j-1} (1 - abcq^{4j-1})}{(bq; q^2)_j (cq^2; q^2)_j}.$$

Note that if we set $a = 0$, that is if we do not count the number of different odd parts in the graph, then (6.5) reduces to (5.5) with the identifications $b = z$ and $c = w$.

We conclude by showing how Sylvester's theorem can be deduced from (6.5). For this purpose we state the dual of Sylvester's theorem by replacing partitions into distinct parts by chain partitions. Under conjugation, given a partition π into distinct parts having k maximal blocks of consecutive integers, its conjugate, namely the chain partition π^* will have $k - 1$ parts less than the largest part that repeat. Thus the dual of Sylvester's theorem is:

Theorem S* *The number of partitions of an integer into odd parts of which exactly k are different equals the number of chain partitions of that integer having $k - 1$ parts less than the largest part that repeat.*

Theorem S* can be deduced from (6.5) as follows. First decompose

$$1 - abcq^{4j-1} = cq^{2j}(1 - abq^{2j-1}) + (1 - cq^{2j}). \tag{6.6}$$

When $1 - abcq^{4j-1}$ is first replaced by $cq^{2j}(1 - abq^{2j-1})$ on the right in (6.5), the resulting series can be interpreted as the generating function of chain partitions with largest part even. Similarly when $1 - abcq^{4j-1}$ is next replaced by $(1 - cq^{2j})$ on the right in (6.5), the resulting series can be interpreted as the generating function of chain partitions with largest part odd. Thus Theorem S* will fall out of (6.5) and (6.6).

Remarks

(i) In the case of Ferrers graphs of unrestricted partitions, the number of different parts is the number of corners and this is invariant under conjugation. When one considers the 2-modular graphs of partitions into odd parts, the number of different parts is the number of corners, but under conjugation we do not have a 2-modular graph. This awkwardness is circumvented by replacing the graph in Fig. 1 by the graph in Fig. 2 so that the portion to the right of the Durfee square consists only of twos. The number of different odd parts that are at least as large as $2j - 1$ where j is the dimension of the Durfee square is given by the number of corners to the right of the Durfee square and we can keep track of this by conjugation since that portion of the graph in Fig. 2 has only twos in it.

(ii) Zeng [9] has studied combinatorially the original Rogers-Fine identity (6.1) under the dilation $q \mapsto q^2$, and with α, β chosen suitably so as to deal with partitions into odd parts. Our approach uses the variant (6.2) and so is combinatorially more direct. Also we have preferred to replace partitions into distinct parts by chain partitions. Hence there are essential differences between our approach and Zeng's.

Acknowledgements I thank the referee for a very careful reading of the manuscript and for helpful suggestions. I also thank Frank Garvan for help with the figures.

References

1. Alladi, K.: A new combinatorial study of the Rogers-Fine identity and a related partial theta series. Int. J. Number Theory **5**(7), 1311–1320 (2009)
2. Andrews, G.E.: On basic hypergeometric series, mock theta functions, and partitions. II. Q. J. Math. Oxf. Ser. (2) **17**, 132–143 (1966)
3. Andrews, G.E.: The Theory of Partitions. Cambridge Mathematical Library. Cambridge University Press, Cambridge (1998). Reprint of the 1976 original
4. Berkovich, A., Uncu, A.K.: On partitions with fixed number of even-indexed and odd-indexed odd parts. J. Number Theory **167**, 7–30 (2016)

5. Bessenrodt, C.: A bijection for Lebesgue's partition identity in the spirit of Sylvester. Discrete Math. **132**(1–3), 1–10 (1994)
6. Bousquet-Mélou, M., Eriksson, K.: Lecture hall partitions. Ramanujan J. **1**(1), 101–111 (1997)
7. Fine, N.J.: Basic Hypergeometric Series and Applications. Mathematical Surveys and Monographs, vol. 27. American Mathematical Society, Providence (1988). With a foreword by George E. Andrews
8. Kim, D., Yee, A.J.: A note on partitions into distinct parts and odd parts. Ramanujan J. **3**(2), 227–231 (1999)
9. Straub, A.: Core partitions into distinct parts and an analog of Euler's theorem. Eur. J. Combin. **57**, 40–49 (2016)
10. Sylvester, J.J.: A constructive theory of partitions, arranged in three acts, an interact and an exodion. Am. J. Math. **5**(1–4), 251–330 (1882)
11. Xiong, X., Keith, W.: Euler's partition theorem for all moduli and new companions to the Rogers-Ramanujan-Andrews-Gordon identities. Ramanujan J. (to appear)

Sequences in Partitions, Double q-Series and the Mock Theta Function $\rho_3(q)$

George E. Andrews

Dedicated to an outstanding mathematician and my good friend, Peter Paule.

1 Introduction

This paper is devoted to the partition-theoretic aspects of

$$H_{r,s}(k, a, x, q) = \sum_{n,j \geq 0} \frac{(-1)^j x^{aj+n} q^{(aj+n)^2 + k\binom{n}{2} + rn + 2asj}}{(q;q)_n (q^{2a}; q^{2a})_j}, \tag{1}$$

where $a \geq 1$ and $k \geq 0$ are integers, and

$$(A; q)_m = (1 - A)(1 - Aq) \ldots (1 - Aq^{m-1}).$$

In particular, we study two subfamilies, namely

$$f_{r,s}(a, x, q) = H_{r,s}(1, a, x, q), \tag{2}$$

and

$$g_{r,s}(a, x, q) = H_{r,s}(0, a, x, q). \tag{3}$$

There are two main theorems. In each of these theorems, the generating function in question will have the exponent of q recording the number being partitioned and the exponent of x will record the number of parts of the partitions being considered.

G. E. Andrews (✉)
The Pennsylvania State University, University Park, PA, USA
e-mail: gea1@psu.edu

© Springer Nature Switzerland AG 2020
V. Pillwein, C. Schneider (eds.), *Algorithmic Combinatorics: Enumerative Combinatorics, Special Functions and Computer Algebra*, Texts & Monographs in Symbolic Computation, https://doi.org/10.1007/978-3-030-44559-1_3

Theorem 1 $g_{0,0}(a, x, q)$ *is the generating function for partitions wherein the difference between parts is at least 2 and where maximal sequences of consecutive odd parts must be of length congruent to* $0, 1, 2, \ldots, a - 1 \pmod{2a}$. *Odd parts are consecutive if they differ by 2.*

$g_{1,1}(a, x, q)$ *is the generating function for the same partition with the added condition that 1 is not allowed as a part.*

For example,

$$g_{0,0}(2, 1, q) = 1 + q + q^2 + q^3 + q^4 + 2q^5 + 3q^6 + 3q^7 + 3q^8 + 4q^9 + \ldots$$

and the four relevant partition of 9 are 9, 8 + 1, 7 + 2, 6 + 3. Note that 5 + 3 + 1 is excluded because it is a maximal sequence of odd parts of length 3 $\not\equiv$ 0 or 1 (mod 4).

We shall also see that, as a corollary of Theorem 1 (cf. Theorem 8),

$$g_{0,0}(1, 1, q) = \prod_{\substack{n=1 \\ n \neq 0, \pm 3 (\text{mod } 7)}} (1 - q^{2n})^{-1}. \tag{4}$$

Our second central result is related to overpartitions, the subject initiated by Corteel and Lovejoy [10]. An overpartition is an ordinary integer partition with the added condition that the first appearance of any given part may be overlined. Thus the eight overpartitions of 3 are $3, \bar{3}, 2 + 1, \bar{2} + 1, 2 + \bar{1}, \bar{2} + \bar{1}, 1 + 1 + 1, \bar{1} + 1 + 1$.

Theorem 2 $f_{0,0}(a, xq^2, q)/(xq^2; q^2)_\infty$ *is the generating function for overpartitions subject to the following conditions (i) all parts are* ≥ 2 , *(ii)* $\bar{2}$ *is never a part, (iii) all odd parts are distinct and overlined, (iv) if a is odd, the following subsequence of parts is not allowed for any* $j \geq 0$:

$$\overline{(2j+3)} + \overline{(2j+6)} + (2j+6) + \overline{(2j+10)} + (2j+10) + \cdots + \overline{(2j+2a)} + (2j+2a), \tag{5}$$

(v) if a is even, the following subsequence of parts is not allowed for any $j \geq 0$:

$$\overline{(2j+4)} + (2j+4) + \overline{(2j+8)} + (2j+8) + \overline{(2j+12)} + (2j+12) + \cdots + \overline{(2j+2a)} + (2j+2a). \tag{6}$$

(vi) the difference between overlined parts is ≥ 3 .

When $a = 1$, Theorem 2 is connected to the Rogers-Ramanujan identities [2, Ch. 7] (see Sect. 9):

$$f_{0,0}(1, x, q) = \sum_{n \geq 0} \frac{x^n q^{2n^2}}{(q^2; q^2)_n}. \tag{7}$$

When $a = 3$, Theorem 2 is related to the third order mock theta function $\rho_3(q)$ [19, p. 62] (see Sect. 10), namely

$$\frac{f_{0,0}(3, q^2, q)}{(q^2; q^2)_\infty} = \frac{(q^3; q^6)_\infty}{(q; q^2)_\infty} \rho_3(q) = \frac{(q^3; q^6)_\infty}{(q; q^2)_\infty} \sum_{n=0}^{\infty} \frac{q^{2n(n+1)}(q; q^2)_{n+1}}{(q^3; q^2)_{n+1}} \quad (8)$$

$$= 1 + q^2 + q^3 + 3q^4 + 2q^5 + 5q^6 + 4q^7 + 9q^8 + \cdots,$$

and the nine relevant partitions of 8 are 8, $\bar{8}$, $6+2$, $\bar{6}+2$, $\bar{4}+4$, $4+4$, $4+2+2$, $\bar{4}+2+2$, $2+2+2+2$. $\rho_3(q)$ is the third order mock theta function[19, p. 62] defined by

$$\rho_3(q) = \sum_{n \geq 0} \frac{(q; q^2)_n q^{2n(n+1)}}{(q^3; q^6)_n}.$$

There is a somewhat scattered history of sequences in partitions dating back to Sylvester [18]. In Sect. 2, we provide a sketch of previous work including the joint paper with Bringmann and Mahlburg [6]. Section 3 provides the necessary q-difference equation satisfied by $H_{r,s}(k, a, x, q)$. The following three sections then develop the theory surrounding $g_{r,s}(a, x, q)$ including Theorem 1 and Eq. (4). Sections 7 through 10 study $f_{r,s}(a, x, q)$ including Theorem 2 and Eq. (5). We conclude with some open questions.

2 History of Sequences in Partitions

J. J. Sylvester was the first to look at sequences in partitions [18, Th. 2.12].

Sylvester's Theorem *Let $A_k(n)$ denote the number of partitions of n into odds with exactly k different parts. Let $B_k(n)$ denote the number of partitions of n into distinct parts composed exactly k noncontiguous sequences of one or more consecutive integers. Then*

$$A_k(n) = B_k(n).$$

For example $A_3(13) = 5$ enumerating $9 + 3 + 1$, $7 + 5 + 1$, $7 + 3 + 1 + 1 + 1$, $5 + 3 + 1 + 1 + 1 + 1 + 1$, and $5 + 3 + 3 + 1 + 1$; $B_3(13) = 5$ enumerating $9 + 3 + 1$, $8 + 4 + 1$, $7 + 5 + 1$, $7 + 4 + 2$, $6 + 4 + 2 + 1$.

P. A. MacMahon [15, Sec. VII, Ch. IV, pp 49–58] was the next to consider sequences in partitions. The most well-known of his theorems is the following.

MacMahon's Theorem *The number of partitions of n without sequences (i.e. no consecutive integers) and no 1's equals the number of partitions of n into parts $\not\equiv \pm 1 \pmod 6$.*

In 1978, in his unpublished Ph.D. thesis [13, Ch. 5, pp. 51–56] M. D. Hirschhorn proved that the generating function for partitions into distinct parts with all sequences of consecutive integers of length $\leq k$ and with all parts $< n$ is given by

$$\sum_{j \geq 0} x^j q^{\binom{j+1}{2}} \sum_{jl \leq j} (-1)^l q^{k\binom{l}{2}} \begin{bmatrix} n - j \\ l \end{bmatrix}_{q^k} \begin{bmatrix} n - kl - 1 \\ j - kl \end{bmatrix}_q , \tag{9}$$

where

$$\begin{bmatrix} A \\ B \end{bmatrix}_q = \begin{cases} \frac{(q;q)_A}{(q;q)_B (q;q)_{A-B}} & \text{for } 0 \leq B \leq A \\ 0 & \text{otherwise.} \end{cases} \tag{10}$$

Hirschhorn examined the case when $n \to \infty$ and related the result to the Rogers-Ramanujan identities when $k = 2$.

In 2015, Bringmann et al. [9] rediscovered some of Hirschhorn's theorems (owing to the fact that Hirschhorn's result was never published outside of his Ph.D. thesis).

In 2004, Holroyd et al. [14] looked at $p_k(n)$, the number of partitions of n that do not contain a sequence of consecutive integers of length k. They proved that if

$$G_k(q) = \sum_{n \geq 0} p_k(n) q^n, \tag{11}$$

then

$$\log G_k(q) \sim \frac{\pi^2}{6} \left(1 - \frac{2}{k(k+1)} \right) \frac{1}{1-q}, \qquad \text{as } q \to 1^- \tag{12}$$

In 2005, a double series representation of $G_k(q)$ was given [3]

$$G_k(q) = \frac{1}{(q;q)_\infty} \sum_{n,j \geq 0} \frac{(-1)^j q^{\binom{k+1}{2}(n+j)^2 + (k-1)\binom{n+1}{2}}}{(q^k; q^k)_j (q^{k+1}; q^{k+1})_n}, \tag{13}$$

and it was shown that

$$G_2(q) = \frac{(-q^3; q^3)_\infty}{(q^2; q^2)_\infty} \chi_3(q), \tag{14}$$

where $\chi_3(q)$ is one of Ramanujan's third order mock theta functions [18, p. 62] given by

$$\chi_3(q) = \sum_{n=0}^{\infty} \frac{(-q;q)_n q^{n^2}}{(-q^3;q^3)_n}.$$

An analogous identity will arise in Sect. 10.

There have been subsequent studies related to $p_k(n)$, in [8] and [7].

In 2013, Bringmann et al. [7] studied $\bar{p}_k(n)$, where the concept of sequences in partitions was extended to overpartitions.

Namely, they define *lower k-run overpartitions* to be those overpartitions in which any overlined part must occur within a run of exactly k consecutive overlined parts that terminates below with a gap. More precisely, this means that if some part $\bar{m}$ is overlined, then there is an integer j with $m \in [j+1, j+k]$ such that each of the k overlined parts $\overline{j+1}, \overline{j+2}, \ldots, \overline{j+k}$ appear (perhaps together with non-overlined versions), while no part j (overline or otherwise) appears, and no overlined part $\overline{j+k+1}$ appears.

They proved that if

$$\overline{G_k}(q) = \sum_{n \geq 0} \overline{p_k}(n) q^n, \tag{15}$$

then

$$\overline{G_k}(q) = \frac{1}{(q;q)_\infty} \sum_{n,j \geq 0} \frac{(-1)^j q^{\binom{k+1}{2}(n+j)^2 + (k+1)\binom{j+1}{2}}}{(q^k;q^k)_n (q^{k+1};q^{k+1})_j}. \tag{16}$$

In that paper, the third order mock theta function $\phi_3(q)$ arises [19, p. 62], namely

$$\overline{G_1}(q) = (q;q)_\infty \phi_3(q), \tag{17}$$

where

$$\phi_3(q) = \sum_{n=0}^{\infty} \frac{q^{n^2}}{(-q^2;q^2)_n}. \tag{18}$$

We now come to the precursor to the current paper, namely [6], a joint work with Bringmann and Mahlburg. One of the main results there (and the one that inspired this article), is, in our current notation

$$\sum_{n,j \geq 0} \frac{(-1)^j x^{n+2j} q^{(n+3j)^2 + \binom{n}{2}}}{(q;q)_n (q^6;q^6)_j} = (x;q^3)_\infty \sum_{n=0}^{\infty} \frac{(-q;q^3)_n (-q^2;q^3)_n x^n}{(q^3;q^3)_n}. \tag{19}$$

From (19), one may deduce Schur's 1926 theorem [16].

Much of [6] is devoted to special cases of

$$R(s, t, l, u, v, w) = \sum_{n,j \geq 0} \frac{(-1)^j q^{s\binom{n+uj}{2}+t(n+uj)+uv\binom{j}{2}+(w+ul)j}}{(q;q)_n (q^{uv}; q^{uv})_j}. \tag{20}$$

It is easy to see that

$$R(3, 1, 0, 2, 3, 4) = f_{0,0}(3, 1, q),$$

and this fact will lead to interesting identities is Sect. 11.

The only difference between the left side of (19) and $f_{00}(3, x, q)$ lies in the exponent on x, and, as we will see in Sect. 11, this seems to be the subtle difference between Schur's original theorem [16] and what has become known as the Alladi-Schur theorem [4–6].

3 q-Difference Equations for $H_{r,s}(k, a, x, q)$

Lemma 3 $H_{r+2,s+1}(k, a, x, q) = H_{r,s}(k, a, xq^2, q)$.

Proof

$$H_{r+2,s+1}(k, a, x, q) = \sum_{n,j \geq 0} \frac{(-1)^j x^{aj+n} q^{(aj+n)^2+k\binom{n}{2}+(r+2)n+2a(s+1)j}}{(q;q)_n (q^{2a}; q^{2a})_j}$$

$$= \sum_{n,j \geq 0} \frac{(-1)^j (xq^2)^{aj+n} q^{(aj+n)^2+k\binom{n}{2}+rn+2asj}}{(q;q)_n (q^{2a}; q^{2a})_j}$$

$$= H_{r,s}(k, a, xq^2, q).$$

Lemma 4

$$H_{r,s}(k, a, x, q) - H_{r+1,s}(k, a, x, q) = xq^{1+r} H_{r+k,s}(k, a, xq^2, q).$$

Proof

$$H_{r,s}(k, a, x, q) - H_{r+1,s}(k, a, x, q)$$

$$= \sum_{n,j \geq 0} \frac{(-1)^j x^{aj+n} q^{(aj+n)^2+k\binom{n}{2}+rn+2asj} (1 - q^n)}{(q;q)_n (q^{2a}; q^{2a})_j}$$

$$= \sum_{n,j \geq 0} \frac{(-1)^j x^{aj+n+1} q^{(aj+n+1)^2+k\binom{n+1}{2}+r(n+1)+2asj}}{(q;q)_n (q^{2a};q^{2a})_j} \qquad \text{(by shifting } n \text{ to } n+1)$$

$$= xq^{r+1} H_{r+k,s}(k,a,xq^2,q).$$

Lemma 5

$$H_{r,s}(k,a,x,q) - H_{r,s+1}(k,a,x,q) = -x^a q^{a^2+2as} H_{r,s}(k,a,xq^{2a},q).$$

Proof

$$H_{r,s}(k,a,x,q) - H_{r,s+1}(k,a,x,q)$$

$$= \sum_{n,j \geq 0} \frac{(-1)^j x^{aj+n} q^{(aj+n)^2+k\binom{n}{2}+rn+2asj} (1-q^{2aj})}{(q;q)_n (q^{2a};q^{2a})_j}$$

$$= \sum_{n,j \geq 0} \frac{(-1)^j x^{aj+n+a} q^{(aj+a+n)^2+k\binom{n}{2}+rn+2asj+2as}}{(q;q)_n (q^{2a};q^{2a})_j} \qquad \text{(by shifting } j \text{ to } j+1)$$

$$= -x^a q^{a^2+2as} H_{r,s}(k,a,xq^{2a},q).$$

Lemma 6

$$H_{r,s}(k,a,x,q) - H_{r,s}(k,a,xq^2,q)$$

$$= -x^a q^{a^2+2as} H_{r,s}(k,a,xq^{2a},q)$$

$$+ xq^{r+1} H_{r+1,s+1}(k,a,xq^2,q)$$

$$+ xq^{r+2} H_{r+k-1,s}(k,a,xq^4,q).$$

Proof

$$H_{r,s}(k,a,x,q) - H_{r,s}(k,a,xq^2,q)$$

$$= \sum_{n,j \geq 0} \frac{(-1)^j x^{aj+n} q^{(aj+n)^2+k\binom{n}{2}+rn+2asj} \left((1-q^{2aj}) + q^{2aj}(1-q^{2n})\right)}{(q;q)_n (q^{2a};q^{2a})_j}$$

$$= -x^a q^{a^2+2as} H_{r,s}(k,a,xq^{2a},q)$$

$$+ \sum_{n,j \geq 0} \frac{(-1)^j x^{n+1+aj} q^{(aj+n)^2+2(aj+n)+1+k\binom{n}{2}+kn+r(n+1)+2asj+2aj}(1+q^{n+1})}{(q;q)_n (q^{2a};q^{2a})_j}$$

$$= -x^a q^{a^2+2as} H_{r,s}(k,a,xq^{2a},q)$$

$$+ xq^{r+1} H_{r+k,s+1}(k, a, xq^2, q)$$

$$+ xq^{r+2} H_{r+k+1,s+1}(k, a, xq^2, q)$$

$$= - x^a q^{a^2+2as} H_{r,s}(k, a, xq^{2a}, q)$$

$$+ xq^{r+1} H_{r+k,s+1}(k, a, xq^2, q)$$

$$+ xq^{r+2} H_{r+k-1,s}(k, a, xq^4, q) \qquad \text{(by Lemma 3)}.$$

4   q-Difference Equation for $g_{0,0}(a, x, q)$

The lemmas of Sect. 3 allow us to obtain defining q-difference equations for $g_{0,0}(a, x, q)$. This will be the foundation for the partition-theoretic interpretation.

Theorem 7

$$g_{1,1}(a, x, q) = g_{0,0}(a, xq^2, q) + xq^2 g_{1,1}(a, xq^2, q), \tag{21}$$

$$g_{0,0}(a, x, q) = g_{1,1}(a, x, q) + xq g_{0,0}(a, xq^2, q) \tag{22}$$

$$- x^a q^{a^2} g_{0,0}(a, xq^{2a}, q) + x^{a+1} q^{(a+1)^2} g_{0,0}(a, xq^{2a+2}, q).$$

Proof Recall that

$$g_{r,s}(a, x, q) = H_{r,s}(0, a, x, q). \tag{23}$$

Hence by Lemma 4 with $r = s = 1, k = 0$

$$g_{1,1}(a, x, q) = g_{2,1}(a, x, q) + xq^2 q_{1,1}(a, xq^2, q) \tag{24}$$

$$= g_{0,0}(a, xq^2, q) + xq^2 g_{1,1}(a, xq^2, q),$$

by Lemma 3, and (21) is established.
   We now turn to (22). First, by Lemma 5, with $r = 1, s = 0, k = 0$

$$g_{1,0}(a, x, q) = g_{1,1}(a, x, q) - x^a q^{a^2} g_{1,0}(a, xq^{2a}, q). \tag{25}$$

By Lemma 4, with $r = s = k = 0$,

$$g_{0,0}(a, x, q) = g_{1,0}(a, x, q) + xq g_{0,0}(a, xq^2, q). \tag{26}$$

Utilizing (26) to eliminate g_{10} from (25), we obtain after simplification

$$g_{1,1}(a, x, q) = g_{0,0}(a, x, q) - xq g_{0,0}(a, xq^2, q) \tag{27}$$
$$+ x^a q^{a^2} \left(g_{0,0}(a, xq^{2a}, q) - xq^{2a+1} g_{0,0}(a, xq^{2a+2}, q) \right),$$

and isolating $g_{0,0}(a, x, q)$ on one side of the equation we find that (22) has been proved. $\square$

5 Proof of Theorem 1

Here is what is required, we observe from (1), that both $g_{0,0}(a, x, q)$ and $g_{1,1}(a, x, q)$ may (for integer $a \geq 1$) be expanded into double power series in x and q. We want the coefficient of $x^m q^n$ in each series to be the number of partitions of n into m parts as prescribed in Theorem 1. We also note that the initial conditions,

$$g_{r,s}(a, 0, q) = g_{r,s}(a, x, 0) = 1, \tag{28}$$

are fully consistent with the assertion that the empty partition of 0 is counted by each class of partitions.

It is also clear that the q-difference equations (21) and (22) together with (28) uniquely define both $g_{0,0}(a, x, q)$ and $g_{1,1}(a, x, q)$. So to conclude the proof of Theorem 1 we only need to show that the generating functions for the partitions described in Theorem 1 fulfill (21) and (22).

In the following, we note that the replacement of x by xq^j in any of our generating functions, in fact, adds j to each part of each partition being enumerated.

To treat (21), we split the partitions asserted to be enumerated by $g_{1,1}(a, x, q)$ into two classes: (i) those partitions that do not contain a 2, and (ii) those that do contain a 2. The partitions in (i) are clearly those enumerated by $g_{0,0}(a, xq^2, q)$. The partitions considered by (ii) must have a 2 (hence xq^2) and the remaining parts must be ≥ 4 (hence $g_{1,1}(a, xq^2, q)$). Note that the last transformation does not alter the parity of parts nor the length of subsequences of consecutive odd integers.

Equation (22) is rather more intricate. How does the right hand side of (22) account precisely for the partitions being generated by $g_{0,0}(a, x, q)$, the left hand side of (22)?

Clearly $g_{1,1}(a, x, q)$ covers the partitions that have no 1 as a part.

The term $xq g_{0,0}(a, xq^2, q)$ correctly generates the partitions that contain 1 as a part with the following exception: (A) we now have 1 in subsequences of consecutive odd parts of length $a \pmod{2a}$, and (B) we do not have 1 in any subsequences of consecutive odd parts of length $0 \pmod{2a}$.

To rectify (A) and (B) requires the final two terms on the right side of (22). The term

$$-x^a q^{a^2} g_{0,0}(a, xq^{2a}, q) = -x^a q^{1+3+4+\cdots+(2a-1)} g_{0,0}(a, xq^{2a}, q)$$

does subtract off the offending sequences from (A), but it also introduces with a minus sign sequences of odds (starting with 1) of length $a + 1, a + 2, \ldots, 2a - 1$ (mod $2a$).

To correct for this, the term

$$x^{a+1} q^{(a+1)^2} g_{0,0}(a, xq^{2(a+1)}, q)$$
$$= x^{a+1} q^{1+3+5+\cdots+(2a+1)} g_{0,0}(a, xq^{2(a+1)}, q)$$

adds back in sequences of consecutive odd parts (starting with 1) of length $a + 1, a + 2, \ldots, 2a$ (mod $2a$). I.e. it cancels the newly introduced sequences of length $a + 1, a + 2, \ldots, 2a - 1 \pmod{2a}$ and puts back in sequences of length $2a \equiv 0$ (mod $2a$).

Thus (22) has been established for $g_{0,0}(a, x, q)$ and $g_{1,1}(a, x, q)$ as generating functions for the partitions described in Theorem 1. Hence Theorem 1 is proved.

6   Rogers-Ramanujan Aspects of $g_{r,s}(1, x, q)$

We shall now reveal, both via analysis and via partitions, the relation of $g_{r,s}(1, x, q)$ to the Rogers-Ramanujan identities at modulus 14.

Theorem 8

$$g_{0,0}(1, 1, q) = \prod_{\substack{n=1 \\ n \not\equiv 0, \pm 3 (\text{mod } 7)}}^{\infty} \frac{1}{1 - q^{2n}}, \tag{29}$$

$$g_{1,1}(1, 1, q) = \prod_{\substack{n=1 \\ n \not\equiv 0, 0, \pm 2 (\text{mod } 7)}}^{\infty} \frac{1}{1 - q^{2n}}, \tag{30}$$

$$g_{0,0}(1, q^2, q) = \prod_{\substack{n=1 \\ n \not\equiv 0, \pm 1 (\text{mod } 7)}}^{\infty} \frac{1}{1 - q^{2n}}. \tag{31}$$

Analytic Proof.

$$g_{r,s}(1, x, q) = \sum_{n,j \geq 0} \frac{(-1)^j x^{n+j} q^{(j+n)^2 + rn + 2sj}}{(q; q)_n (q^2; q^2)_j}$$

$$= \sum_{n \geq 0} \frac{x^n q^{n^2 + rn} (xq^{1+2n+2s}; q^2)_\infty}{(q; q)_n} \qquad \text{(by [2, p. 19, eq. (2.2.6)])}$$

$$= (xq; q^2)_\infty \sum_{n \geq 0} \frac{x^n q^{n^2 + rn}}{(q; q)_n (xq; q^2)_{n+s}}.$$

Hence

$$g_{0,0}(1, 1, q) = (q; q^2)_\infty \sum_{n \geq 0} \frac{q^{n^2}}{(q; q)_n (q; q^2)_n}$$

$$= \frac{(q; q^2)_\infty (q^6; q^{14})_\infty (q^8; q^{14})_\infty (q^{14}; q^{14})_\infty}{(q; q)_\infty} \qquad \text{(by [17, p. 158, eq. (61)])}$$

$$= \prod_{\substack{n=1 \\ n \neq 0, \pm 3 (\text{mod } 7)}}^{\infty} \frac{1}{1 - q^{2n}}.$$

Next

$$g_{1,1}(1, 1, q) = (q; q^2)_\infty \sum_{n \geq 0} \frac{q^{n^2 + n}}{(q; q)_n (q; q^2)_{n+1}}$$

$$= \frac{(q; q^2)_\infty (q^4; q^{14})_\infty (q^{10}; q^{14})_\infty (q^{14}; q^{14})_\infty}{(q; q)_\infty} \qquad \text{(by [17, p. 158, eq, (60)])}$$

$$= \prod_{\substack{n=1 \\ n \neq 0, \pm 2 (\text{mod } 7)}}^{\infty} \frac{1}{1 - q^{2n}}.$$

Finally

$$g_{0,0}(1, q^2, q) = (q^3; q^2)_\infty \sum_{n \geq 0} \frac{q^{n^2 + 2n}}{(q; q)_n (q^3; q^2)_n}$$

$$= \frac{(q; q^2)_\infty (q^2; q^{14})_\infty (q^{12}; q^{14})_\infty (q^{14}; q^{14})_\infty}{(q; q)_\infty} \qquad \text{(by [17, p. 157, eq. (59)])}$$

$$= \prod_{\substack{n=1 \\ n \neq 0, \pm 1 (\text{mod } 7)}}^{\infty} \frac{1}{1 - q^{2n}}.$$

Proof via Partitions We recall the following special case of B. Gordon's generalization of the Rogers-Ramanujan identities [2, Ch. 7].

Theorem *The number of partitions of an even $2N$ into even parts where: (1) none appears more than twice, (2) if a part appears twice then all parts are at least 4 units away, and (3) two appears at most j times ($j = 0, 1, 2$) EQUALS the coefficient of q^{2N} in*

$$\prod_{\substack{n=1 \\ n \not\equiv 0, \pm(j+1) \pmod 7}}^{\infty} \frac{1}{1 - q^{2n}}.$$

Now we need only identify the partitions generated by $g_{0,0}(1, 1, q)$, $g_{1,1}(1, 1, q)$ and $g_{0,0}(1, q^2, q)$ respectively with the partitions given above in the special case of Gordon's Theorem.

We know that in $g_{0,0}(1, x, q)$ and $g_{1,1}(1, x, q)$ all parts differ by at least 2. Now suppose we have a sequence of consecutive odd integers as parts (note that since $a = 1$, the sequence must be of even length):

$$\big(< (2h-3)\big) + (2h-1) + (2h+1) + (2h+3) + (2h+5) + \cdots + (2i-1) + (2i+1) + (> 2i+3)$$

and we replace this by

$$\big(< (2h - 3)\big) + (2h) + (2h) + (2h + 4) + (2h + 4) + \cdots + (2i) + (2i) + (> 2i + 3).$$

Thus whenever odd parts appear they must appear in pairs as indicated, and, as we see, these directly transform into the repeated parts that are allowed in Gordon's theorem.

Finally $g_{0,0}(1, 1, q)$ would allow $1 + 3$ to appear translating into two appearing twice. So by the Theorem,

$$g_{0,0}(1, 1, q) = \prod_{\substack{n=1 \\ n \not\equiv 0, \pm 3 \pmod 7}}^{\infty} \frac{1}{1 - q^{2n}};$$

$g_{1,1}(1, 1, q)$ allows no 1's so 2 can appear at most once after translation. Hence by the theorem,

$$g_{1,1}(1, 1, q) = \prod_{\substack{n=1 \\ n \not\equiv 0, \pm 2 \pmod 7}} \frac{1}{1 - q^{2n}},$$

and lastly $g_{0,0}(1, q^2, q)$ has smallest part ≥ 3 so no 2's appear at all. Hence by the Theorem

$$g_{0,0}(1, q^2, q) = \prod_{\substack{n=1 \\ n \not\equiv 0, \pm 1 \,(\mathrm{mod}\ 7)}}^{\infty} \frac{1}{1 - q^{2n}}.$$

7 q-Difference Equations for $f_{00}(a, x, q)$

Paradoxically the q-difference equation is simpler than the one for $g_{00}(a, x, q)$ while the partition theoretic interpretation is a good deal more complicated.

Theorem 9

$$f_{0,0}(a, x, q) = f_{0,0}(a, xq^2, q) + (xq + xq^2) f_{0,0}(a, xq^4, q) \tag{32}$$

$$+ x^2 q^5 f_{0,0}(a, xq^6, q) - x^a q^{a^2} f_{0,0}(a, xq^{2a}, q).$$

Proof By Lemma 6, with $r = s = 0, k = 1$,

$$f_{0,0}(a, x, q) = f_{0,0}(a, xq^2, q) - x^a q^{a^2} f_{0,0}(a, xq^{2a}, q) \tag{33}$$

$$+ xq f_{1,1}(a, xq^2, q) + xq^2 f_{0,0}(a, xq^4, q).$$

Next by Lemma 4, with $r = s = k = 1$,

$$f_{1,1}(a, x, q) = f_{2,1}(a, x, q) + xq^2 f_{2,1}(a, xq^2, q) \tag{34}$$

$$= f_{0,0}(a, xq^2, q) + xq^2 f_{0,0}(a, xq^4, q). \qquad \text{(by Lemma 3)}$$

Using (34) to eliminate $f_{1,1}(a, xq^2, q)$ from (33), we obtain (32).   □

It would be lovely if we could use $f_{0,0}(a, x, q)$ directly as a generating function for partitions. However if $a > 1$, the expansion of $f_{0,0}(a, x, q)$ has negative terms. For example

$$f_{0,0}(2, x, q) = 1 + xq + xq^2 + xq^3 + (-x^2 + x)q^4 + \cdots$$

This problem is overcome by introducing the factor $1/(xq^2; q^2)_\infty$ in Theorem 2.

In addition, the replacement of x by xq^2 yielding $f_{0,0}(a, xq^2, q)/(xq^2; q^2)_\infty$ is done primarily to produce (8) in the case $a = 3$.

8 Proof of Theorem 2

Before we undertake this proof, a few comments are in order. First, this is a result about overpartitions, where the generating function is

$$F(a, x, q) := \frac{f_{0,0}(a, xq^2, q)}{(xq^2; q^2)_\infty}. \tag{35}$$

At first glance, it appears that the $f_{0,0}(a, xq^2, q)$ produces the overlined parts, and $1/(xq^2; q^2)_\infty$ produces the non-overlined parts; so why mix the two. Of course, the answer lies in conditions (v) and (vi) where interwoven sequences of overlined and non-overlined parts are excluded.

We observe that the xq^2 in $f_{0,0}(a, xq^2, q)$ is necessary for the proof of Theorem 2, but $f_{0,0}(a, x, q)$ would be more natural for the cases $a = 1$ and 3 treated in Sects. 9 and 10.

We note for subsequent use that the sums in (5) and (6) are both equal to $a^2 + 2a$ where $j = 0$, and each has exactly a summands.

In order to understand the intricacies of $F(a, x, q)$, we rewrite (32) in terms of $F(a, x, q)$ with x replaced by xq^2:

$$F(a, x, q) = \frac{F(a, xq^2, q)}{1 - xq^2} + \frac{(xq^{\bar{3}} + xq^{\bar{4}})}{(1 - xq^2)(1 - xq^4)} F(a, xq^4, q) \tag{36}$$

$$+ \frac{x^2 q^{\bar{3}+\bar{6}} F(a, xq^6, q)}{(1 - xq^2)(1 - xq^4)(1 - xq^6)} - \frac{x^a q^{a^2+2a} F(a, xq^{2a}, q)}{(1 - xq^2)(1 - xq^4) \cdots (1 - xq^{2a})}.$$

Now if we let $a \to \infty$, we see from (35) and (2) that

$$F(\infty, x, q) = \frac{1}{(xq^2; q^2)_\infty} \sum_{n \geq 0} \frac{x^n q^{3n(n+1)/2}}{(q; q)_n}. \tag{37}$$

Thus clearly $F(\infty, x, q)$ is the generating function for overpartitions where only even parts can avoid overlines, and the difference between overlined parts is ≥ 3, and all overlined parts are ≥ 3.

Indeed, if we let $a \to \infty$ in (36) we see that the final term vanishes and what remains is a q-difference equation that uniquely defines the generating function for the overpartitions listed in the previous paragraph.

In order to complete the proof of Theorem 2 we must determine the effect of the final term in (36). This is, indeed, accounted for by condition (iv) and (v) in Theorem 2. There are exactly a summands in each of (5) and (6). Also as noted previously, when $j = 0$, the numerical sum in both (5) and (6) is $a^2 + 2a$.

Thus final term in (36) excludes either

$$\bar{3} + \bar{6} + 6 + \overline{10} + 10 + \cdots + \overline{2a} + 2a$$

if a is odd, and

$$\bar{4} + 4 + \bar{8} + 8 + \overline{12} + 12 + \cdots + \overline{2a} + 2a$$

if a is even.

In addition, the instances of (5) and (6) with $j > 0$ are thus also excluded by the action of (36) as it generates the partitions of Theorem 2.

9 Theorem 2 for $a = 1$

We shall provide two proofs of (7).

First Proof *(Analytic)*

$$f_{0,0}(1, x, q) = \sum_{n,j \geq 0} \frac{(-1)^j x^{n+j} q^{(n+j)^2 + \binom{n}{2}}}{(q; q)_n (q^2; q^2)_j} \tag{38}$$

$$= \sum_{N \geq 0} x^N q^{N^2} \sum_{j=0}^{N} \frac{(-1)^j q^{\binom{N-j}{2}}}{(q; q)_{N-j} (q^2; q^2)_j} \tag{39}$$

$$= \sum_{N \geq 0} \frac{x^N q^{N^2}}{(q; q)_N} \sum_{j=0}^{N} \frac{(q^{-N}; q)_j q^j}{(q; q)_j (-q; q)_j} \tag{40}$$

$$= \sum_{N \geq 0} \frac{x^N q^{N^2}}{(q; q)_N} \cdot \frac{q^{N^2}}{(-q; q)_N} \qquad \text{(by [12, p. 236, eq. (II.6)])} \tag{41}$$

$$= \sum_{n \geq 0} \frac{x^N q^{2N^2}}{(q^2; q^2)_N} \tag{42}$$

which is equivalent to (7).

Second Proof *(Combinatorial)* When $a = 1$, condition (iv) and (5) required that no parts are odd. Hence the overlined parts are all even, ≥ 4, and differ by ≥ 3 and thus must differ by ≥ 4. Hence

$$\frac{f_{0,0}(1, xq^2, q)}{(xq^2; q^2)_\infty} = F(1, x, q) \tag{43}$$

$$= \frac{1}{(xq^2; q^2)_\infty} \sum_{n \geq 0} \frac{q^{2n^2 + 2n} x^n}{(q^2; q^2)_n}$$

where the product generates the non-overlined parts and the series generated the overlined parts.

Clearly (43) is equivalent to (7).

10　Theorem 2 for $a = 3$

To treat (8), we first require two lemmas:

Lemma 10

$$\rho_3(q) = \frac{(q^2; q^2)_\infty}{\displaystyle\prod_{j=1}^{\infty}(1 + q^{2j-1} + q^{4j-2})} \sum_{n \geq 0} \frac{\displaystyle\prod_{j=1}^{n}(1 + q^{2j-1} + q^{4j-2})q^{2n}}{(q^2; q^2)_n}. \tag{44}$$

Proof In [11, pg. 61, eq. (26.87)], N. J. Fine proved that if $w = e^{2\pi i/3}$

$$\rho_3(q) = \sum_{n \geq 0} \frac{(w^{-1}q)^n}{(wq; q^2)_{n+1}}. \tag{45}$$

Now

$$\sum_{n \geq 0} \frac{\displaystyle\prod_{j=1}^{n}(1 + q^{2j-1} + q^{4j-2})}{(q^2; q^2)_n} q^{2n}$$

$$= \sum_{n \geq 0} \frac{(wq; q^2)_n (w^{-1}q; q^2)_n q^{2n}}{(q^2; q^2)_n}$$

$$= \frac{(w^{-1}q; q^2)_\infty (wq^3; q^2)_\infty}{(q^2; q^2)_\infty} \sum_{n \geq 0} \frac{(w^{-1}q)^n}{(wq^3; q^2)_n} \qquad \text{(by [12, p. 241, (III.1)])}$$

$$= \frac{\displaystyle\prod_{j=1}^{\infty}(1 + q^{2j-1} + q^{4j-2})}{(q^2; q^2)_\infty} \rho_3(q),$$

by (45). □

Lemma 11

$$f_{0,0}(3, xq^2, q) = (xq^2; q^2)_\infty \sum_{n \geq 0} \frac{\prod_{j=1}^{n}(1 + q^{2j-1} + q^{4j-2})x^n q^{2n}}{(q^2; q^2)_n}. \tag{46}$$

Proof It is clear that $F(z, x, q)$ is uniquely determined by the initial conditions

$$f_{0,0}(3, 0, q) = f_{0,0}(3, x, 0) = 1,$$

and the q-difference equation (32) which simplifies to

$$f_{0,0}(3, xq^2, q) = f_{0,0}(3, xq^4, q) + (xq^3 + xq^4) f_{0,0}(3, xq^6, q) \tag{47}$$
$$+ x^2 q^9 (1 - xq^6) f_{0,0}(3, xq^8, q).$$

Now let

$$f(x) = f(x, q) := \sum_{n \geq 0} \frac{\prod_{j=1}^{n}(1 + q^{2j-1} + q^{4j-2})x^n q^{2n}}{(q^2; q^2)_n}.$$

Then clearly $f(0, q) = f(x, 0) = 1$, and

$$f(x) - f(xq^2) = \sum_{n \geq 0} \frac{\prod_{j=1}^{n}(1 + q^{2j-1} + q^{4j-2})x^n q^{2n}(1 - q^{2n})}{(q^2; q^2)_n}$$

$$= xq^2 \sum_{n \geq 0} \frac{\prod_{j=1}^{n}(1 + q^{2j-1} + q^{4j-2})x^n q^{2n}(1 + q^{2n+1} + q^{4n+2})}{(q^2; q^2)_n}$$

$$= xq^2 \left(f(x) + qf(xq^2) + q^2 f(xq^4) \right),$$

and if

$$f_1(x) := (xq^2; q^2)_\infty f(x), \tag{48}$$

then multiplying the above equation by $(xq^4; q^4)_\infty$, we obtain

$$f_1(x) = (1 + xq^3) f_1(xq^2) + xq^4(1 - xq^4) f_1(xq^4). \tag{49}$$

Iterating (49), we obtain

$$f_1(x) = f_1(xq^2) + xq^3\big((1 + xq^5)f_1(xq^4) + xq^6(1 - xq^6)f_1(xq^6)\big) \tag{50}$$
$$+ xq^4(1 - xq^4)f_1(xq^4)$$
$$= f_1(xq^2) + (xq^3 + xq^4)f_1(xq^4) + x^2q^9(1 - xq^6)f_1(xq^6).$$

Comparing (50) with (47), we see that $f_1(x)$ and $f_{00}(z, xq^2, q)$ satisfy the same q-difference equation and have the same initial value of 1 at $x = 0$ dn $q = 0$. Hence

$$f_1(x) = f_{0,0}(3, xq^2, q)$$

which is assertion (46).

First Proof of (8) Set $x = 1$ in (46) and compare with (44).

Second Proof of (8) *(Combinatorial)* We shall provide a combinatorial proof of the assertion

$$\frac{f_{0,0}(z, q^2, q)}{(q^2; q^2)_\infty} = \sum_{n=0}^{\infty} \frac{q^{2n} \prod_{j=1}^{n}(1 + q^{2j-1} + q^{5j-2})}{(q^2; q^2)_n} \tag{51}$$

which by Lemmas 10 and 11 is equivalent to (8).

It is immediate by inspection that the right-hand side of (51) is the generating function for partitions in which the largest part is even and odd parts appear at most twice.

Theorem 2 tells us that the left-hand side is the generating function for overpartitions where all parts are >1; 2 is never overlined, odd parts appear at most once and are overlined; overlined parts differ by at least 3, and there is never a sequence of parts of the form $\overline{(2j + 3)} + \overline{(2j + 6)} + (2j + 6)$.

We provide a bijection between these two classes of partitions as follows.

We begin with the overpartitions, and we consider a modified Ferrers graph as follows. Each odd part $2j + 1$ is represented by the row

$$\underbrace{2\,2\,2 \cdots 2}_{j}\, 1$$

Each even, nonoverlined part $2j$ is given a row of j 2's:

$$\underbrace{2\,2 \ldots 2}_{j \text{ times}}.$$

Each overlined $\overline{2j}$ is given a row of $j-1$ 2's and two 1's

$$\underbrace{2\,2\cdots2}_{j-1\ \text{times}}\,1\,1.$$

However if both $2j$ and $\overline{2j}$ are parts the two rows are to be:

$$2\,2\,2\cdots2\,2\,1$$

$$\underbrace{2\,2\,2\cdots2}_{j-1\ \text{times}}\,1$$

This procedure produces from the given set of overpartitions a unique set of the modified 2-modular Ferrers graphs. The uniqueness is guaranteed by exclusion of part sequences of the form $\overline{(2j+3)}+\overline{(2j+6)}+(2j+6)$ because this sequence would yield

$$2\,2\,2\cdots2\,2\,2\,1$$

$$2\,2\,2\cdots2\,2\,1$$

$$\underbrace{2\,2\,2\cdots2}_{(j+1)\ \text{times}}\,1$$

but $\overline{(2j+4)}+(2j+4)+\overline{(2j+7)}$ yields exactly the same component of the modified 2-modular Ferrers graph.

Now to complete the bijection we read these Ferrers graphs via columns instead of rows, and the resulting partitions are those generated by right-hand side of (51).

11   The Alladi-Schur Theorem

We remarked at the end of Sect. 2, that our work here was inspired by the discoveries in [6]. In particular, the identity (Eq. (19) restated):

$$\sum_{n,j\geq0}\frac{(-1)^j x^{n+2j} q^{(3j+n)^2+\binom{n}{2}}}{(q;q)_n(q^6;q^6)_j}=(x;q^3)_\infty\sum_{n\geq0}\frac{x^n(-q;q^3)_n(-q^2;q^3)_n}{(q^3;q^3)_n}\tag{52}$$

is naturally related to a proof of Schur's 1926 partition theorem. Namely, as was shown in [1, eq. (2.15)], an application of Abel's lemma reveals

$$\lim_{x\to1}(x;q^3)_\infty\sum_{n\geq0}\frac{(-q;q^3)_n(-q^2;q^3)_n x^n}{(q^3;q^3)_n}$$

$$=(-q;q^3)_\infty(-q^2;q^3)_\infty=\frac{1}{(q;q^6)_\infty(q^5;q^6)_\infty}.$$

In the current context, by Lemma 11, with x replaced by xq^{-2}

$$f_{0,0}(3, 1, q) = \lim_{x \to 1^-} (x, q^2)_\infty \sum_{n \geq 0} \frac{x^n \prod_{j=1}^{n}(1 + q^{2j-1} + q^{4j-2})}{(q^2; q^2)_n}$$

$$= \prod_{j=1}^{\infty}(1 + q^{2j-1} + q^{4j-2})$$

$$= \frac{(q^3; q^6)_\infty}{(q; q^2)_\infty}$$

$$= \frac{1}{(q; q^6)_\infty(q^5; q^6)_\infty}.$$

Identifying $f_{00}(3, 1, q)$ with the left-hand side of (44) when $x = 1$, we see that the Alladi formulation of Schur's theorem is naturally related to Lemma 11.

12 Conclusion

The most unsatisfying aspect of this paper is that we have been unable to produce a grand unified treatment of combinatorial aspects of semi-general double series such as the one given in (1). If one contrasts the theorems listed in Sect. 2 with those treated in Theorems 1 and 2, one sees the great diversity of theorems vaguely tied together by the theme of the examination of sequences of parts in partitions.

However, at this stage, one only sees the glimmer of a general theory. Nonetheless, the variety of results found to date suggest that much remains to be found.

References

1. Andrews, G.E.: On partition functions related to Schurs' second partition theorem. Proc. Am. Math. Soc. **19**, 441–444 (1968)
2. Andrews, G.E.: The Theory of Partitions. Cambridge University Press, Cambridge (1998)
3. Andrews, G.E.: Partitions with short sequences and mock theta functions. Proc. Natl. Acad. Sci. **102**, 4666–4671 (2005)
4. Andrews, G.E.: The Alladi-Schur polynomials and their factorization. In: Andrews G., Garvan F. (eds.) Analytic Number Theory, Modular Forms and q-Hypergeometric Series, pp. 25–38. Springer, Cham (2017)
5. Andrews, G.E.: A refinement of the Alladi-Schur theorem. In: Lattice Path Combinatorics and Applications, Dev. in Math. Series, pp. 17–77. Springer (2019)
6. Andrews, G.E., Bringmann, K., Mahlburg, K.: Double series representations for Schur's partition function and related identities. J. Comb. Theory (A) **132**, 102–119 (2015)

7. Bringmann, K., Holroyd, A.E., Mahlburg, K., Vlasenko, M.: k-Run overpartitions and mock theta functions. Q. J. Math. **64**, 1009–1021 (2013)
8. Bringmann, K., Dousse, J., Lovejoy, J., Mahlburg, K.: Overpartitions with restricted odd differences. Electr. J. Comb. **22**(3), #P3.17 (2015)
9. Bringmann, K., Mahlburg, K., Nataraj, K.: Distinct parts partitions without sequences. Electr. J. Comb. **22**(3), #P3.3 (2015)
10. Corteel, S., Lovejoy, J.: Overpartitions. Trans. Am. Math. Soc. **356**, 1623–1635 (2004)
11. Fine, N.J.: Basic Hypergeometric Series and Applications. American Mathematical Society, Providence (1988)
12. Gasper, G., Rahman, M.: Basic Hypergeometric Series. Cambridge University Press, Cambridge (1990)
13. Hirschhorn, M.D.: Developments in the Theory of Partitions. Ph.D. Thesis, University of New South Wales (1980)
14. Holroyd, A.E., Liggert, T.M., Romik, D.: Integrals, partitions and cellular automata. Trans. Am. Math. Soc. **356**, 3349–3368 (2004)
15. MacMahon, P.A.: Combinatory Analysis, vol. 2. A.M.S. Chelsea Publishing, Providence (1984)
16. Schur, I.: Zur additiven Zahlentheorie, Sitzungsber., Preuss. Akad. Wiss. Phys-Math. Kl. (1926)
17. Slater, L.J.: Further identities of Rogers-Ramanujan type. Proc. Lond. Math. Soc. (2) **54**, 147–167 (1952)
18. Sylvester, J.J.: A constructive theory of partitions in three acts, an interact and a exodion. Am. J. Math. **5**, 251–330 (1882)
19. Watson, G.N.: The final problem: an account of the mock theta functions. J. Lond. Math. Soc. **11**, 55–80 (1936)

Refined q-Trinomial Coefficients and Two Infinite Hierarchies of q-Series Identities

Alexander Berkovich and Ali Kemal Uncu

Peter—king of the castle—Paule on the occasion of his 60th birthday

1 Introduction

There are many important transformations for the q-binomial coefficients of the type

$$\sum_{r=0}^{L} \frac{q^{r^2}(q;q)_{2L}}{(q;q)_{L-r}(q;q)_{2r}} \begin{bmatrix} 2r \\ r-j \end{bmatrix}_q = q^{j^2} \begin{bmatrix} 2L \\ L-j \end{bmatrix}_q \qquad (1)$$

in the q-series literature (see [8], and the references there). Throughout this work $|q| < 1$. Here, the q-Pochhammer symbols are defined as

$$(a;q)_n := (1-a)(1-aq)(1-aq^2)\ldots(1-aq^{n-1}), \qquad (2)$$

for any non-negative integer n. In addition, we have

$$(a;q)_\infty := \lim_{n\to\infty} (a;q)_n, \qquad (3)$$

$$(a_1, a_2, \ldots, a_k; q)_n := (a_1;q)_n(a_2;q)_n \ldots (a_k;q)_n. \qquad (4)$$

A. Berkovich
Department of Mathematics, University of Florida, Gainesville, FL, USA
e-mail: alexb@ufl.edu

A. K. Uncu (✉)
Johannes Kepler University Linz, Research Institute for Symbolic Computation, Linz, Austria
e-mail: akuncu@risc.jku.at

© Springer Nature Switzerland AG 2020
V. Pillwein, C. Schneider (eds.), *Algorithmic Combinatorics: Enumerative Combinatorics, Special Functions and Computer Algebra*, Texts & Monographs in Symbolic Computation, https://doi.org/10.1007/978-3-030-44559-1_4

We extend the definition of q-Pochhamer symbols to negative n using

$$(a; q)_n = \frac{(a; q)_\infty}{(aq^n; q)_\infty}. \tag{5}$$

Observe that (5) implies

$$\frac{1}{(q; q)_n} = 0 \text{ if } n < 0. \tag{6}$$

The q-binomial coefficients are defined as

$$\begin{bmatrix} m + n \\ m \end{bmatrix}_q := \begin{cases} \frac{(q;q)_{m+n}}{(q;q)_m (q;q)_n}, & \text{for } m, n \geq 0, \\ 0, & \text{otherwise.} \end{cases} \tag{7}$$

The salient features of (1) are that the sum over a q-binomial coefficient multiplied by a simple factor yields a q-binomial coefficient, and the dependence on the variable j is simple. Furthermore, this summation can be applied multiple times in an iterative fashion. This type of transformations were used by Bailey [5] and Slater [16], but the real value of the iterative power was first realized by Peter Paule [12, 13] and George E. Andrews [1].

For example, we start with the simple identity

$$\delta_{L,0} = \sum_{j=-L}^{L} (-1)^j q^{\binom{j}{2}} \begin{bmatrix} 2L \\ L + j \end{bmatrix}_q, \tag{8}$$

where $\delta_{i,j}$ is the Kronecker delta function. Following the change of variable $L \mapsto r$ in (8), we multiply both sides by

$$\frac{q^{r^2} (q; q)_{2L}}{(q; q)_{L-r} (q; q)_{2r}}, \tag{9}$$

and sum both sides with respect to $r \geq 0$. This yields

$$\frac{(q; q)_{2L}}{(q; q)_L} = \sum_{j=-L}^{L} (-1)^j q^{j^2 + \binom{j}{2}} \begin{bmatrix} 2L \\ L + j \end{bmatrix}_q, \tag{10}$$

using the identity (1).

It is well known that

$$\lim_{L \to \infty} \begin{bmatrix} L \\ m \end{bmatrix}_q = \frac{1}{(q; q)_m}. \tag{11}$$

For any $j \in \mathbb{Z}_{\geq 0}$ and $\nu = 0$ or 1

$$\lim_{L \to \infty} \begin{bmatrix} 2L + \nu \\ L + j \end{bmatrix}_q = \frac{1}{(q; q)_\infty}. \tag{12}$$

Using (12) to take the limit $L \to \infty$ of (10), we get

$$(q; q)_\infty = \sum_{j=-\infty}^{\infty} (-1)^j q^{\frac{3j^2-j}{2}}. \tag{13}$$

This is nothing but Euler's Pentagonal Number Theorem. Note that this can also be viewed as a special case $(q, z) \mapsto (q^{3/2}, -q^{1/2})$ of the celebrated result:

Theorem 1 (Jacobi Triple Product Identity) *For complex numbers $z \neq 0$ and $|q| < 1$, we have*

$$\sum_{j=-\infty}^{\infty} z^j q^{j^2} = \left(q^2, -zq, -\frac{q}{z}; q^2 \right)_\infty. \tag{14}$$

We can apply (1) to the identity (10) after changing the variable $L \mapsto r$ in (10), multiply both sides by (9) and sum both sides again with respect to $r \geq 0$. This yields

$$\frac{(q; q)_{2L}}{(q; q)_L} \sum_{r=0}^{L} q^{r^2} \begin{bmatrix} L \\ r \end{bmatrix}_q = \sum_{j=-L}^{L} (-1)^j q^{2j^2 + \binom{j}{2}} \begin{bmatrix} 2L \\ L + j \end{bmatrix}_q. \tag{15}$$

Letting $L \to \infty$, using (11), (12), (14) with $(q, z) \mapsto (q^{5/2}, -q^{1/2})$, and doing some simple simplifications one obtains the first Rogers–Ramanujan identity,

$$\sum_{r \geq 0} \frac{q^{r^2}}{(q; q)_r} = \frac{1}{(q, q^4; q^5)_\infty}.$$

Proceeding in the same fashion, one can keep on applying (1) iteratively $\nu + 1$ times to (8). This way one obtains the famous Andrews–Gordon infinite hierarchy of identities as $L \to \infty$,

$$\sum_{n_1, n_2, \ldots, n_\nu \geq 0} \frac{q^{N_1^2 + N_2^2 + \cdots + N_\nu^2}}{(q; q)_{n_1} (q; q)_{n_2} \cdots (q; q)_{n_\nu}} = \prod_{n \neq 0, \pm(\nu+1) \pmod{2\nu+3}} \frac{1}{1 - q^n}, \tag{16}$$

where $N_k = n_k + n_{k+1} + \cdots + n_\nu$ for $k = 1, 2, \ldots, \nu$.

In [4], Andrews and Baxter defined the q-trinomial coefficients,

$$\binom{L,\ b}{a}_2 := \sum_{n \geq 0} q^{n(n+b)} \frac{(q;q)_L}{(q;q)_n (q;q)_{n+a} (q;q)_{L-2n-a}}, \tag{17}$$

$$T_0 \binom{L}{a};q := q^{\frac{L^2-a^2}{2}} \binom{L,\ a}{a};\frac{1}{q}_2. \tag{18}$$

Following that Warnaar [17, 18] defined a refinement of these coefficients:

$$\mathcal{T}\binom{L,\ M}{a,\ b};q := \tag{19}$$

$$\sum_{\substack{n \geq 0, \\ L-a \equiv n \,(\mathrm{mod}\ 2)}} q^{\frac{n^2}{2}} \begin{bmatrix} M \\ n \end{bmatrix}_q \begin{bmatrix} M+b+\frac{L-a-n}{2} \\ M+b \end{bmatrix}_q \begin{bmatrix} M-b+\frac{L+a-n}{2} \\ M-b \end{bmatrix}_q,$$

$$\mathcal{S}\binom{L,\ M}{a,\ b};q := \sum_{n \geq 0} q^{n(n+a)} \begin{bmatrix} M+L-a-2n \\ M \end{bmatrix}_q \begin{bmatrix} M-a+b \\ n \end{bmatrix}_q \begin{bmatrix} M+a-b \\ n+a \end{bmatrix}_q. \tag{20}$$

These refined trinomials obey transformation properties somewhat similar to (1). Therefore, they can be used in an iterative fashion [18].

In this paper, we prove a new doubly bounded polynomial identity using the symbolic tools developed by the Algorithmic Combinatorics group at the Research Institute for Symbolic Computation.

Theorem 2 *For L and M being non-negative integers, we have*

$$\sum_{\substack{m \geq 0, \\ L \equiv m \,(\mathrm{mod}\ 2)}} q^{m^2} \begin{bmatrix} 3M \\ m \end{bmatrix}_{q^2} \begin{bmatrix} 2M+\frac{L-m}{2} \\ 2M \end{bmatrix}_{q^6} = \sum_{j=-\infty}^{\infty} q^{3j^2+2j} \mathcal{T}\binom{L,\ M}{j,\ j};q^6. \tag{21}$$

Then we use transformation properties for the refined trinomials defined in (19) and (20) to obtain two new infinite hierarchies of q-series identities. An unusual feature of these identities is the presence of various q-factorial bases such as in the following theorem with bases q^2, q^3, q^6 etc.

Theorem 3 *Let v be a positive integer, and let $N_k = n_k + n_{k+1} + \cdots + n_v$, for $k = 1, 2, \ldots, v$. Then,*

$$\sum_{n_1,n_2,\ldots,n_v \geq 0} \frac{q^{3(N_1^2+N_2^2+\cdots+N_v^2)}(-q;q^2)_{3n_v}}{(q^6;q^6)_{n_1}(q^6;q^6)_{n_2}\cdots(q^6;q^6)_{n_{v-1}}(q^6;q^6)_{2n_v}}$$

$$= \frac{(-q^3;q^3)_\infty}{(q^{12};q^{12})_\infty}(q^{6(v+1)},-q^{3v+1},-q^{3v+5};q^{6(v+1)})_\infty. \tag{22}$$

This paper is structured as follows. Section 2 has a short list of known identities that will be needed later. Section 3 is totally dedicated to the proof of Theorem 2. In Sect. 4, we discuss the asymptotics of q-trinomial coefficients, and present two transformation formulas of Warnaar for the refined q-trinomial coefficients (19) and (20). We also discuss an analog of the Bailey Lemma (Theorem 6). In Sect. 5, we apply (45) to (21) in an iterative fashion. This application yields a doubly bounded infinite hierarchy. The asymptotic analysis and the proof of Theorem 3 are also given in Sect. 5. In Sect. 6, we use the second transformation (46) of Theorem 6, which yields another doubly bounded hierarchy of polynomial identities, and do its asymptotic analysis. In this way we see a connection with the Capparelli partition theorem. In Sect. 7 we briefly discuss variants of Theorem 2.

2 q-Binomial Theorem and Its Corollaries

Theorem 4 (q-Binomial Theorem) *For variables a, q, and z,*

$$\sum_{n \geq 0} \frac{(a; q)_n}{(q; q)_n} t^n = \frac{(at; q)_\infty}{(t; q)_\infty}. \tag{23}$$

Note that by setting $(a, t) \mapsto (q^{-L}, -zq^L)$ in (23), and using

$$\begin{bmatrix} L \\ n \end{bmatrix}_q = \frac{(q^{-L}; q)_n}{(q; q)_n} (-1)^n q^{Ln - \binom{n}{2}},$$

we derive

$$\sum_{n \geq 0} q^{\binom{n}{2}} z^n \begin{bmatrix} L \\ n \end{bmatrix}_q = (-z; q)_L. \tag{24}$$

We remark that (24) implies

$$\sum_{\substack{n \geq 0, \\ n \equiv \sigma \,(\mathrm{mod}\, 2)}} q^{\binom{n}{2}} z^n \begin{bmatrix} L \\ n \end{bmatrix}_q = \frac{(-z; q)_L + (-1)^\sigma (z; q)_L}{2}, \tag{25}$$

where $\sigma = 0$ or 1.

Another important corollary of the q-binomial theorem (Theorem 4) is the polynomial analog of the identity (14) [3, p. 49, Ex. 1].

$$\sum_{j=-M}^{M} q^{j^2} z^j \begin{bmatrix} 2M \\ M + j \end{bmatrix}_{q^2} = \left(-zq, -\frac{q}{z}; q^2 \right)_M. \tag{26}$$

Note that (8) is a special case of (26) with $(q, z) \mapsto (q^{1/2}, -q^{1/2})$. Another special case of (26) with $(q, z) \mapsto (q^3, q^2)$ is

$$\sum_{j=-M}^{M} q^{3j^2+2j} \begin{bmatrix} 2M \\ M + j \end{bmatrix}_{q^6} = (-q, -q^5; q^6)_M. \tag{27}$$

3 Proof of Theorem 2

Let

$$\mathcal{G}(L, M, k, q) := q^{(L-2k)^2} \begin{bmatrix} 3M \\ L - 2k \end{bmatrix}_{q^2} \begin{bmatrix} 2M + k \\ k \end{bmatrix}_{q^6} \tag{28}$$

and

$$\mathcal{F}(L, M, k, j, q) := \tag{29}$$

$$q^{3j^2+2j+3(L-j-2k)^2} \begin{bmatrix} M \\ L - j - 2k \end{bmatrix}_{q^6} \begin{bmatrix} M + j + k \\ k \end{bmatrix}_{q^6} \begin{bmatrix} M + k \\ k + j \end{bmatrix}_{q^6}. \tag{30}$$

Note that

$$\sum_{k \geq 0} \mathcal{G}(L, M, k, q) \quad \text{and} \quad \sum_{j,k \geq 0} \mathcal{F}(L, M, k, j, q)$$

are the left-hand and right-hand sides of (21), respectively.

The Mathematica packages *Sigma* [15] and *qMultiSum* [14] (both implemented by the Algorithmic Combinatorics group at the Research Institute for Symbolic Computation) are both capable of finding and automatically proving recurrences for these functions. Here we start with the recurrences that *qMultiSum* finds for the summands (28) and (29):

$$q^{9+18M} \left(1 - q^{12+6L+6M}\right) \mathcal{G}(L, M, k, q)$$

$$- \left(1 - q^{12+24M}\right) \mathcal{G}(L + 3, M + 1, k, q)$$

$$+ q^{1+6M} \left(1 - q^{24+6L+18M}\right) \left(1 + q^2 + q^4\right) \mathcal{G}(L + 2, M, k, q)$$

$$- q^{4+12M} \left(1 + q^2 + q^4\right) \left(-1 + q^{18+6L+12M}\right) \mathcal{G}(L + 1, M, k, q)$$

$$+ q^{6+12M} \left(1 + q^6\right) \left(1 - q^{6+12M}\right) \mathcal{G}(L+1, M+1, k-1, q)$$

$$+ \left(1 - q^{30+6L+24M}\right) \mathcal{G}(L+3, M, k, q) = 0$$

and

$$q^{9+18M} \left(1 - q^{12+6L+6M}\right) \mathcal{F}(L, M, k-1, j-1, q)$$

$$+ q^{4+12M} \left(1 - q^{18+6L+12M}\right) \mathcal{F}(L+1, M, k-1, j, q)$$

$$+ q^{6+12M} \left(1 - q^{18+6L+12M}\right) \mathcal{F}(L+1, M, k-1, j-1, q)$$

$$- \left(1 - q^{12+24M}\right) \mathcal{F}(L+3, M+1, k, j-1, q)$$

$$+ q^{3+6M} \left(1 - q^{24+6L+18M}\right) \mathcal{F}(L+2, M, k, j-1, q)$$

$$+ q^{5+6M} \left(1 - q^{24+6L+18M}\right) \mathcal{F}(L+2, M, k, j-2, q)$$

$$+ q^{1+6M} \left(1 - q^{24+6L+18M}\right) \mathcal{F}(L+2, M, k-1, j, q)$$

$$+ \left(1 - q^{30+6L+24M}\right) \mathcal{F}(L+3, M, k, j-1, q)$$

$$+ \left(1 + q^6\right) q^{6+12M} \left(1 - q^{6+12M}\right) \mathcal{F}(L+1, M+1, k-1, j-1, q)$$

$$+ q^{8+12M} \left(1 - q^{18+6L+12M}\right) \mathcal{F}(L+1, M, k, j-2, q) = 0.$$

Once summed over the variable k for $\mathcal{G}(L, M, k, q)$, and variables k and j for $\mathcal{F}(L, M, k, j, q)$, we see that they satisfy the same recurrence,

$$q^{9+18M} \left(1 - q^{12+6L+6M}\right) \hat{S}(L, M, q)$$

$$- q^{4+12M} \left(1 + q^2 + q^4\right) \left(-1 + q^{18+6L+12M}\right) \hat{S}(L+1, M, q)$$

$$+ q^{1+6M} \left(1 - q^{24+6L+18M}\right) \left(1 + q^2 + q^4\right) \hat{S}(L+2, M, q) \qquad (31)$$

$$+ \left(1 - q^{30+6L+24M}\right) \hat{S}(L+3, M, q)$$

$$+ q^{6+12M} \left(1 + q^6\right) \left(1 - q^{6+12M}\right) \hat{S}(L+1, M+1, q)$$

$$- \left(1 - q^{12+24M}\right) \hat{S}(L+3, M+1, q) = 0.$$

This is also the same recurrence one would get from the package *Sigma*. It remains to show that the left-hand side and the right-hand side of (21) satisfy the same initial conditions. Observe that

$$\hat{S}(L, 0, q) = \frac{1 + (-1)^L}{2}, \text{ and } \hat{S}(0, M, q) = 1, \tag{32}$$

for any non-negative integer L and M. Moreover, we have

$$\hat{S}(1, M, q) = q \begin{bmatrix} 3M \\ 1 \end{bmatrix}_{q^2}, \text{ and } \hat{S}(2, M, q) = \begin{bmatrix} 2M + 1 \\ 1 \end{bmatrix}_{q^6} + q^4 \begin{bmatrix} 3M \\ 2 \end{bmatrix}_{q^2}, \tag{33}$$

for any non-negative integer M. The recurrence (31), and the boundary conditions (32) and (33) uniquely define $\hat{S}(L, M, q)$. $\qquad\square$

4 Asymptotics and Transformations of the Refined Trinomial Coefficients

For $\sigma = 0$ or 1, we have the following limits.

$$\lim_{M \to \infty} \mathcal{T}\begin{pmatrix} L, M \\ a, b \end{pmatrix} = \frac{1}{(q; q)_L} T_0 \begin{pmatrix} L \\ a \end{pmatrix}, \tag{34}$$

$$\lim_{\substack{L \to \infty, \\ L - a \equiv \sigma \pmod 2}} \mathcal{T}\begin{pmatrix} L, M \\ a, b \end{pmatrix} = \frac{(-q^{1/2}; q)_M + (-1)^\sigma (q^{1/2}; q)_M}{2(q; q)_{2M}} \begin{bmatrix} 2M \\ M - b \end{bmatrix}_q, \tag{35}$$

$$\lim_{\substack{L \to \infty, \\ L - a \equiv \sigma \pmod 2}} T_0 \begin{pmatrix} L \\ a \end{pmatrix} = \frac{(-q^{1/2}; q)_\infty + (-1)^\sigma (q^{1/2}; q)_\infty}{2(q; q)_\infty}. \tag{36}$$

Moreover,

$$\lim_{M \to \infty} S\begin{pmatrix} L, M \\ a, b \end{pmatrix} = \frac{1}{(q; q)_L} \begin{pmatrix} L, a \\ a \end{pmatrix}_2, \tag{37}$$

$$\lim_{L \to \infty} S\begin{pmatrix} L, M \\ a, b \end{pmatrix} = \frac{1}{(q; q)_M} \begin{bmatrix} 2M \\ M - b \end{bmatrix}_q, \tag{38}$$

$$\lim_{L \to \infty} \begin{pmatrix} L, a \\ a \end{pmatrix}_2 = \frac{1}{(q; q)_\infty}. \tag{39}$$

We would like to note that the limits (34), (37), and (38) can be found in Warnaar's work [18, (2.12),(2.13),(2.17)]. The limit (36) appears in Andrews–Baxter work [4, (2.55),(2.56)]. The limit (39) is also discussed in [4, (2.48)]. The authors could not find the limit (35) in the literature. This limit can be proven by using (11) followed up with (25).

Letting $M \to \infty$ in (21), and using (34), we get

$$\sum_{\substack{m \geq 0, \\ L \equiv m \pmod 2}} q^{m^2} \frac{(q^6; q^6)_L}{(q^2; q^2)_m (q^6; q^6)_{(L-m)/2}} = \sum_{j=-L}^{L} q^{3j^2+2j} T_0 \binom{L}{j}; q^6 \right). \tag{40}$$

Observe that after the change of variables $n = (L - m)/2$, this identity becomes [7, (3.9)] with $q \mapsto q^2$.

Replacing $L \mapsto 2L + \sigma$, with $\sigma = 0, 1$, letting L tend to ∞, we get the following with the aid of (35),

$$(-q; q^2)_{3M} + (-1)^\sigma (q; q^2)_{3M}$$

$$= (-q^3; q^6)_M \sum_{j=-M}^{M} q^{3j^2+2j} \begin{bmatrix} 2M \\ M+j \end{bmatrix}_{q^6} \tag{41}$$

$$+ (-1)^\sigma (q^3; q^6)_M \sum_{j=-M}^{M} (-1)^j q^{3j^2+2j} \begin{bmatrix} 2M \\ M+j \end{bmatrix}_{q^6}.$$

It is easy to check that (41) follows from the identity (27).

Theorem 5 (Warnaar [17, 18]) *For $L, M, a, b \in \mathbb{Z}$ and $ab \geq 0$*

$$\sum_{i=0}^{M} q^{\frac{i^2}{2}} \begin{bmatrix} L+M-i \\ L \end{bmatrix}_q \mathcal{T} \binom{L-i, i}{a, b}; q \right) = q^{\frac{b^2}{2}} \mathcal{T} \binom{L, M}{a+b, b}; q \right). \tag{42}$$

For $L, M, a, b \in \mathbb{Z}$ with $ab \geq 0$, and $|a| \leq M$ if $|b| \leq M$ and $|a + b| \leq L$, then

$$\sum_{i=0}^{M} q^{\frac{i^2}{2}} \begin{bmatrix} L+M-i \\ L \end{bmatrix}_q \mathcal{T} \binom{i, L-i}{b, a}; q \right) = q^{\frac{b^2}{2}} \mathcal{S} \binom{L, M}{a+b, b}; q \right). \tag{43}$$

The transformation formulas (42) and (43) directly imply the following theorem.

Theorem 6 *Let $F_{L,M}(q)$ and $\alpha_j(q)$ be sequences, and $L, M, m, n \in \mathbb{Z}_{\geq 0}$. If*

$$F_{L,M}(q) = \sum_{j=-\infty}^{\infty} \alpha_j(q) \mathcal{T} \binom{L, M}{mj, nj}; q \right) \tag{44}$$

holds, then

$$\sum_{i \geq 0} q^{\frac{i^2}{2}} \begin{bmatrix} L+M-i \\ L \end{bmatrix}_q F_{L-i,i}(q) = \sum_{j=-\infty}^{\infty} q^{\frac{(nj)^2}{2}} \alpha_j(q) \mathcal{T} \binom{L, M}{(m+n)j, nj}; q \right)$$

$$\tag{45}$$

and

$$\sum_{i\geq 0} q^{\frac{i^2}{2}} \begin{bmatrix} L+M-i \\ L \end{bmatrix}_q F_{i,L-i}(q) = \sum_{j=-\infty}^{\infty} q^{\frac{(mj)^2}{2}} \alpha_j(q) \mathcal{S} \begin{pmatrix} L, \ M \\ (m+n)j, \ mj \end{pmatrix} ; q \end{pmatrix}$$

(46)

are true.

Note that (45) can be used in combination with an appropriately chosen identity in an iterative fashion. This leads to an infinite hierarchy of identities. On the other hand, the identity (46) can only be used once.

5   The First Doubly Bounded Infinite Hierarchy and Its Asymptotics

We use (45) v times with $q \mapsto q^6$ on (21) and obtain the following infinite hierarchy.

Theorem 7 *Let v be a positive integer, and let $N_k = n_k + n_{k+1} + \cdots + n_v$, for $k = 1, 2, \ldots, v$. Then,*

$$\sum_{\substack{m,n_1,n_2,\ldots,n_v \geq 0, \\ L+m \equiv N_1+N_2+\cdots+N_v \ (mod\ 2)}} q^{m^2+3(N_1^2+N_2^2\cdots+N_v^2)} \begin{bmatrix} L+M-N_1 \\ L \end{bmatrix}_{q^6} \begin{bmatrix} 3n_v \\ m \end{bmatrix}_{q^2}$$

(47)

$$\times \begin{bmatrix} 2n_v + \dfrac{L-m-N_1-N_2-\cdots-N_v}{2} \\ 2n_v \end{bmatrix}_{q^6} \prod_{j=1}^{v-1} \begin{bmatrix} L - \sum_{l=1}^{j} N_l + n_j \\ n_j \end{bmatrix}_{q^6}$$

$$= \sum_{j=-\infty}^{\infty} q^{3(v+1)j^2+2j} \mathcal{T} \begin{pmatrix} L, \ M \\ (v+1)j, \ j \end{pmatrix} ; q^6 \end{pmatrix}.$$

We replace $L \mapsto 2L + \sigma$, with $\sigma = 0, 1$, and sum over σ, in (47). Letting $L \to \infty$ and using the (24) and (35), we get the following theorem.

Theorem 8 *Let v be a positive integer, and let $N_k = n_k + n_{k+1} + \cdots + n_v$, for $k = 1, 2, \ldots, v$. Then,*

$$\sum_{n_1,n_2,\ldots,n_v \geq 0} \frac{q^{3(N_1^2+N_2^2+\cdots+N_v^2)}(-q;q^2)_{3n_v}}{(q^6;q^6)_{M-N_1}(q^6;q^6)_{n_1}(q^6;q^6)_{n_2}\ldots(q^6;q^6)_{n_{v-1}}(q^6;q^6)_{2n_v}}$$

$$= \frac{(-q^3;q^6)_M}{(q^6;q^6)_{2M}} \sum_{j=-M}^{M} q^{3(v+1)j^2+2j} \begin{bmatrix} 2M \\ M+j \end{bmatrix}_{q^6}.$$

(48)

The $v = 1$ case of the identity (48) yields a finite analog of the identity [6, (6.7)].

Corollary 1

$$\sum_{n=0}^{M} \frac{q^{3n^2}(-q; q^2)_{3n}}{(q^6; q^6)_{M-n}(q^6; q^6)_{2n}} = \frac{(-q^3; q^6)_M}{(q^6; q^6)_{2M}} \sum_{j=-M}^{M} q^{6j^2+2j} \begin{bmatrix} 2M \\ M+j \end{bmatrix}_{q^6}. \quad (49)$$

We can also take the limit $M \to \infty$ in the identity (47). Using (34) we get another infinite hierarchy.

Theorem 9 *Let v be a positive integer, and let $N_k = n_k + n_{k+1} + \cdots + n_v$, for $k = 1, 2, \ldots, v$. Then,*

$$\sum_{\substack{m,n_1,n_2,\ldots,n_v \geq 0, \\ L+m \equiv N_1+N_2+\cdots+N_v \,(mod\,2)}} q^{m^2+3(N_1^2+N_2^2+\cdots+N_v^2)} \begin{bmatrix} 3n_v \\ m \end{bmatrix}_{q^2}$$

$$\times \begin{bmatrix} 2n_v + \dfrac{L-m-N_1-N_2-\cdots-N_v}{2} \\ 2n_v \end{bmatrix}_{q^6} \prod_{j=1}^{v-1} \begin{bmatrix} L - \sum_{l=1}^{j} N_l + n_j \\ n_j \end{bmatrix}_{q^6} \quad (50)$$

$$= \sum_{j=-\infty}^{\infty} q^{3(v+1)j^2+2j} T_0 \left(\begin{matrix} L \\ (v+1)j \end{matrix}; q^6 \right).$$

Letting $M \to \infty$ in (48) and using (12) and (14) proves Theorem 3.

6 The Second Doubly Bounded Infinite Hierarchy and Its Asymptotics

Now we look at the implications of (46). First we apply this identity with $q \mapsto q^3$ to (21) with $q^2 \mapsto q$:

Theorem 10 *For L and M non-negative integers, we have*

$$\sum_{\substack{i,m \geq 0, \\ i+m \equiv 0 \,(mod\,2)}} q^{\frac{m^2+3i^2}{2}} \begin{bmatrix} L+M-i \\ L \end{bmatrix}_{q^3} \begin{bmatrix} 3(L-i) \\ m \end{bmatrix}_q \begin{bmatrix} 2(L-i) + \frac{i-m}{2} \\ 2(L-i) \end{bmatrix}_{q^3} \quad (51)$$

$$= \sum_{j=-\infty}^{\infty} q^{3j^2+j} S \left(\begin{matrix} L, & M \\ 2j, & j \end{matrix}; q^3 \right).$$

We can also apply (46) with $q \mapsto q^3$ to (47) with $q^2 \mapsto q$. This yields the following result.

Theorem 11 *Let v be a positive integer, and let $N_k = n_k + n_{k+1} + \cdots + n_v$, for $k = 1, 2, \ldots, v$. Then,*

$$\sum_{\substack{i,m,n_1,n_2,\ldots,n_v \geq 0, \\ i+m \equiv N_1+N_2+\cdots+N_v \,(\text{mod } 2)}} q^{\frac{m^2+3(i^2+N_1^2+N_2^2\cdots+N_v^2)}{2}} \begin{bmatrix} L + M - i \\ L \end{bmatrix}_{q^3} \begin{bmatrix} L - N_1 \\ i \end{bmatrix}_{q^3} \begin{bmatrix} 3n_v \\ m \end{bmatrix}_q$$

(52)

$$\times \begin{bmatrix} 2n_v + \dfrac{i-m-N_1-N_2-\cdots-N_v}{2} \\ 2n_v \end{bmatrix}_{q^3} \prod_{j=1}^{v-1} \begin{bmatrix} i - \sum_{l=1}^{j} N_l + n_j \\ n_j \end{bmatrix}_{q^3}$$

$$= \sum_{j=-\infty}^{\infty} q^{3\binom{v+2}{2}j^2+j} S\left(\begin{matrix} L, M \\ (v+2)j, \ (v+1)j \end{matrix} ; q^3 \right).$$

Taking the limit $M \to \infty$ in (51), and changing the summation variable $(i - m)/2 = n$ we get

$$\sum_{n,m \geq 0} q^{Q(m,n)} \begin{bmatrix} 3(L - 2n - m) \\ m \end{bmatrix}_q \begin{bmatrix} 2(L - 2n - m) + n \\ n \end{bmatrix}_{q^3}$$

(53)

$$= \sum_{j=-\infty}^{\infty} q^{3j^2+j} \left(\begin{matrix} L, \ 2j \\ 2j \end{matrix} ; q^3 \right)_2,$$

where $Q(m, n) := 2m^2 + 6mn + 6n^2$.

The polynomials on the right-hand side of (53) were first discussed by Andrews in [2]. The identity, on the other hand, was first proven in [6].

The limit $L \to \infty$ in (51) yields

$$\sum_{n,m \geq 0} q^{Q(m,n)} \frac{(q^3; q^3)_M}{(q; q)_m (q^3; q^3)_n (q^3; q^3)_{M-2n-m}} = \sum_{j=-\infty}^{\infty} q^{3j^2+j} \begin{bmatrix} 2M \\ M + j \end{bmatrix}_{q^3}.$$

(54)

This formula was first discussed in [6], and it is proven in a wider context in [7].

Finally, when L and M both tend to ∞, a simple change of variables together with the Jacobi Triple Product identity (14) yields

$$\sum_{m,n \geq 0} \frac{q^{Q(m,n)}}{(q; q)_m (q^3; q^3)_n} = (-q^2, -q^4; q^6)_\infty (-q^3; q^3)_\infty,$$

(55)

where $Q(m, n) = 2m^2 + 6mn + 6n^2$, after simplifications.

The identity (55) was recently proposed independently by Kanade–Russell [10] and Kurşungöz [11]. They showed that (55) is equivalent to the following partition theorem:

Theorem 12 (Capparelli's First Partition Theorem [9]) *For any integer n, the number of partitions of n into distinct parts where no part is congruent to ± 1 modulo 6 is equal to the number of partitions of n into parts, not equal to 1, where the minimal difference between consecutive parts is 2; the difference between consecutive parts is greater than or equal to 4 unless consecutive parts are 3k and $3k + 3$ (yielding a difference of 3), or $3k - 1$ and $3k + 1$ (yielding a difference of 2) for some $k \in \mathbb{N}$.*

Taking limits $M \to \infty$ and $L \to \infty$ in (52), we get Theorems 13 and 14, respectively.

Theorem 13 *Let v be a positive integer, and let $N_k = n_k + n_{k+1} + \cdots + n_v$, for $k = 1, 2, \ldots, v$. Then,*

$$
\sum_{\substack{i,m,n_1,n_2,\ldots,n_v \geq 0, \\ i+m \equiv N_1+N_2+\cdots+N_v \ (\text{mod } 2)}} q^{\frac{m^2+3(i^2+N_1^2+N_2^2+\cdots+N_v^2)}{2}} \begin{bmatrix} L - N_1 \\ i \end{bmatrix}_{q^3} \begin{bmatrix} 3n_v \\ m \end{bmatrix}_q
$$

$$
\times \begin{bmatrix} 2n_v + \dfrac{i-N_1-N_2-\cdots-N_v-m}{2} \\ 2n_v \end{bmatrix}_{q^3} \prod_{j=1}^{v-1} \begin{bmatrix} i - \sum_{k=1}^{j} N_k + n_j \\ n_j \end{bmatrix}_{q^3}
$$

(56)

$$
= \sum_{j=\infty}^{\infty} q^{3\binom{v+2}{2}j^2+j} \left(\begin{matrix} L, \ (v+2)j \\ (v+2)j \end{matrix} ; q^3 \right)_2 .
$$

Theorem 14 *Let v be a positive integer, and let $N_k = n_k + n_{k+1} + \cdots + n_v$, for $k = 1, 2, \ldots, v$. Then,*

$$
\sum_{\substack{i,m,n_1,n_2,\ldots,n_v \geq 0, \\ i+m \equiv N_1+N_2+\cdots+N_v \ (\text{mod } 2)}} q^{\frac{m^2+3(i^2+N_1^2+N_2^2+\cdots+N_v^2)}{2}} \begin{bmatrix} M \\ i \end{bmatrix}_{q^3} \begin{bmatrix} 3n_v \\ m \end{bmatrix}_q
$$

$$
\times \begin{bmatrix} 2n_v + \dfrac{i-N_1-N_2-\cdots-N_v-m}{2} \\ 2n_v \end{bmatrix}_{q^3} \prod_{j=1}^{v-1} \begin{bmatrix} i - \sum_{k=1}^{j} N_k + n_j \\ n_j \end{bmatrix}_{q^3}
$$

(57)

$$
= \sum_{j=\infty}^{\infty} q^{3\binom{v+2}{2}j^2+j} \begin{bmatrix} 2M \\ M + (v+1)j \end{bmatrix}_{q^3} .
$$

Finally, by letting $L \to \infty$ in (56), and using (39) and (14), we get the following result.

Theorem 15 *Let v be a positive integer, and let $N_k = n_k + n_{k+1} + \cdots + n_v$, for $k = 1, 2, \ldots, v$. Then,*

$$
\sum_{\substack{i,m,n_1,n_2,\ldots,n_v \geq 0, \\ i+m \equiv N_1+N_2+\cdots+N_v \ (mod\ 2)}} \frac{q^{\frac{m^2+3(i^2+N_1^2+N_2^2+\cdots+N_v^2)}{2}}}{(q^3;q^3)_i} \begin{bmatrix} 3n_v \\ m \end{bmatrix}_q
$$

$$
\times \begin{bmatrix} 2n_v + \dfrac{i-N_1-N_2-\cdots-N_v-m}{2} \\ 2n_v \end{bmatrix}_{q^3} \prod_{j=1}^{v-1} \begin{bmatrix} i - \sum_{k=1}^{j} N_k + n_j \\ n_j \end{bmatrix}_{q^3} \tag{58}
$$

$$
= \frac{(q^{6\binom{v+2}{2}}, -q^{3\binom{v+2}{2}+1}, -q^{3\binom{v+2}{2}-1}; q^{6\binom{v+2}{2}})_\infty}{(q^3;q^3)_\infty}.
$$

Note that Theorem 15 can also be proven by taking the limit $M \to \infty$ in (57), and using (12) together with (14).

7 Outlook

We would like to note that the identity (21) is not an isolated incident. This shows that there is a more complex structure behind and there is much more to discover. We would like to give two such example identities that we prove similarly to Theorem 2. Let

$$
\mathcal{T}_1 \begin{pmatrix} L, M \\ a, b \end{pmatrix}; q \end{pmatrix} := \sum_{\substack{n \geq 0, \\ L-a \equiv n \ (mod\ 2)}} q^{\binom{n}{2}} \begin{bmatrix} M \\ n \end{bmatrix}_q \begin{bmatrix} M+b+\frac{L-a-n}{2} \\ M+b \end{bmatrix}_q \begin{bmatrix} M-b+\frac{L+a-n}{2} \\ M-b \end{bmatrix}_q,
$$

and

$$
\mathcal{T}_{-1} \begin{pmatrix} L, M \\ a, b \end{pmatrix}; q \end{pmatrix} := \sum_{\substack{n \geq 0, \\ L-a \equiv n \ (mod\ 2)}} q^{\binom{n+1}{2}} \begin{bmatrix} M \\ n \end{bmatrix}_q \begin{bmatrix} M+b+\frac{L-a-n}{2} \\ M+b \end{bmatrix}_q \begin{bmatrix} M-b+\frac{L+a-n}{2} \\ M-b \end{bmatrix}_q.
$$

Then we have the following theorem.

Theorem 16 *For L and M being non-negative integers, we have*

$$
\sum_{\substack{m \geq 0 \\ L \equiv m \ (mod\ 2)}} q^{m^2 \mp m} \begin{bmatrix} 3M \\ m \end{bmatrix}_{q^2} \begin{bmatrix} 2M+\frac{L-m}{2} \\ 2M \end{bmatrix}_{q^6} = \sum_{j=-\infty}^{\infty} q^{3j^2+j} \mathcal{T}_{\pm 1} \begin{pmatrix} L, M \\ j, j \end{pmatrix}; q^6 \end{pmatrix}.
$$

$$
\tag{59}
$$

We are planning to address Theorem 16 and its implications elsewhere.

Acknowledgements We would like to thank Research Institute for Symbolic Computation for the warm hospitality.

Research of the first author is partly supported by the Simons Foundation, Award ID: 308929. Research of the second author is supported by the Austrian Science Fund FWF, SFB50-07 and SFB50-09 Projects.

We thank Chris Jennings-Shaffer for his careful reading of the manuscript.

References

1. Andrews, G.E.: Multiple series Rogers–Ramanujan type identities. Pac. J. Math. **114**, 267–283 (1984)
2. Andrews, G.E.: Schur's theorem. Capparelli's conjecture and q-trinomial coefficients. Contemp. Math. **166**, 141–154 (1994)
3. Andrews, G.E.: The Theory of Partitions. Cambridge Mathematical Library. Cambridge University Press, Cambridge (1998). Reprint of the 1976 original. MR1634067 (99c:11126)
4. Andrews, G.E., Baxter, R.J.: Lattice gas generalization of the hard hexagon model. III. q-Trinomial coefficients. J. Stat. Phys. **47**(3–4), 297–330 (1987)
5. Bailey, W.N.: Identities of the Rogers–Ramanujan type. Proc. Lond. Math. Soc. **50**(2), 1–10 (1949)
6. Berkovich, A., Uncu, A.K.: Polynomial identities implying Capparelli's partition theorems (2019). arXiv:1807.10974
7. Berkovich, A., Uncu, A.K.: Elementary polynomial identities involving q-trinomial coefficients (2019). arXiv:1810.06497
8. Berkovich, A., Warnaar, S.O.: Positivity preserving transformations for q-binomial coefficients. Trans. Am. Math. Soc. **357**(6), 2291–2351 (2005)
9. Capparelli, S.: A combinatorial proof of a partition identity related to the level 3 representation of twisted affine Lie algebra. Commun. Algebra **23**(8), 2959–2969 (1995)
10. Kanade, S., Russell, M.: Staircases to analytic sum-sides for many new integer partition identities of Rogers–Ramanujan type (2018). arXiv:1803.02515 [math.CO]
11. Kurşungöz, K.: Andrews–Gordon type series for Capparelli's and Göllnitz–Gordon identities (2019). arXiv:1807.11189
12. Paule, P.: Zwei neue Transformationen als elementare Anwendungen der q-Vandermonde Formel. Ph.D. Thesis, University of Vienna (1982)
13. Paule, P.: On identities of the Rogers-Ramanujan type. J. Math. Anal. Appl. **107**(1), 255–284 (1985)
14. Riese, A.: qMultiSum – a package for proving q-hypergeometric multiple summation identities. J. Symb. Comput. **35**, 349–376 (2003)
15. Schneider, C.: Symbolic summation assists combinatorics. Sem. Lothar. Combin. **56**, 1–36 (2007). Article B56b
16. Slater, L.J.: Further identities of the Rogers-Ramanujan type. Proc. Lond. Math. Soc. **54**(2), 147–167 (1952)
17. Warnaar, S.O.: Refined q-trinomial coefficients and character identities. Proceedings of the Baxter Revolution in Mathematical Physics (Canberra, 2000). J. Stat. Phys. **102**(3–4), 1065–1081 (2001)
18. Warnaar, S.O.: The generalized Borwein conjecture. II. Refined q-trinomial coefficients. Discrete Math. **272**(2–3), 215–258 (2003)

Large Scale Analytic Calculations in Quantum Field Theories

Johannes Blümlein

Dedicated to Peter Paule on the occasion of his 60th birthday

1 Introduction

Precise theoretical predictions within the Standard Model of elementary particles are indispensable for the concise understanding of the fundamental parameters of this physical theory and the discovery of its potential extensions. At the experimental side highly precise measurements exist at e^+e^-, ep and pp-colliders as at LEP, HERA, and the LHC. In the near future the high luminosity phase of the LHC will even provide much more precise data. Other facilities, like the ILC [1] and a possible FCC [2], are currently planned. During the last three decades enormous efforts have been made to calculate key observables measured at these colliders at higher and higher accuracy, to meet the challenge provided by the accuracy of the experiments.

For zero-scale quantities currently analytic massless calculations can be performed at the five-loop and for massive calculations at the four-loop level. Single scale calculations are performed in both cases at the three-loop level. To perform these large scale calculations very demanding efforts are needed at the side of their automation, computer-algebraic implementation, and the use of highly efficient mathematical technologies. Therefore, the present problems can only be solved within a very close interdisciplinary cooperation between experts in all these different fields and it cannot be the sole tasks for theoretical physicists anymore.

While at one-loop order the mathematical solution for many scattering processes has been known early, cf. [3–5], systematic representations at higher loop order turned out to be more difficult. The core problem concerns the analytic integration of Feynman parameter integrals. Here integration is understood as anti-differentiation. An essential question is to determine the final solution space to which the respective

J. Blümlein (✉)
Deutsches Elektronen-Synchrotron, DESY, Zeuthen, Germany
e-mail: Johannes.Bluemlein@desy.de

© Springer Nature Switzerland AG 2020
V. Pillwein, C. Schneider (eds.), *Algorithmic Combinatorics: Enumerative Combinatorics, Special Functions and Computer Algebra*, Texts & Monographs in Symbolic Computation, https://doi.org/10.1007/978-3-030-44559-1_5

integrals do belong and its mathematical structure, and to find the irreducible objects through which the corresponding integrals are represented. Furthermore, one needs efficient mathematical and computer-algebraic technologies to map the given Feynman parameter integrals into the latter quantities.

In this paper we give a survey on the main technological steps to calculate higher loop zero- and single-scale quantities in renormalizable quantum field theories, with the focus on analytic integration techniques and the occurring function spaces. The systematic theory of integration in this field is vastly developing and many more new structures are expected to be revealed in the future at higher loop levels and by considering the production of more particles in the final state of the respective scattering processes. These calculations are needed to obtain stable theoretical predictions for the experimental precision measurements at the present and future colliders, which operate at high luminosity.

The paper is organized as follows. In Sect. 2 we summarize the main steps in multi-loop perturbative calculations. Different methods used in symbolic calculations of zero- and single-scale Feynman parameter integrals are described in Sect. 3. In Sect. 4 a hierarchy of function spaces, mainly for single-scale integrals, is discussed which emerge in present multi-loop calculations. Here we consider as well the representations in Mellin-N and x-space. Section 5 contains the conclusions.[1]

2   Main Steps in Multi-Loop Perturbative Calculations

In most of the large projects, which are currently dealt with, the Feynman diagrams are generated using packages like QGRAF [12] and performing the color algebra for the gauge groups using Color [13]. Standardized algorithms to obtain Feynman parameterizations exist, cf. e.g. [14–16]. At growing complexity, to perform the Dirac- and spin-algebra will be a challenge even to FORM [17–20]. One further maps the set of the contributing Feynman integrals to master integrals using the integration-by-parts (IBP) technique [21] based on Laporta's algorithm [22], of which several implementations exist, cf. e.g. [23–27] and others. The remaining main step is then the integration of the master integrals. One possibility to inspect the problem on hand, is to analyze the associated system of first order differential equations for the master integrals. Sometimes it is also useful to consider, in addition, the related system of linear difference equations. One may decouple these systems using the algorithms implemented in the packages [28, 29], as e.g. Zürcher's algorithm [30]. This leads to a single differential equation or difference equation of large order and degree and associated determining equations for the remaining master integrals. If the former equations can be factored at first order, it is known that the master integrals can be obtained in terms of indefinitely nested sums or iterated integrals over certain alphabets, which are revealed in the solution process,

---

[1] For other recent surveys on integration methods for Feynman integrals see [6–11].

e.g. using difference field and ring theory [31–43], algorithmically implemented in the package Sigma [44, 45]. This applies to a wide class of physical cases. Most of the integration and summation methods described in Sect. 3 apply to them and allow to obtain the integrals analytically in terms of the mathematical functions described in Sect. 4. Finally, efficient numerical representations of these functions have to be provided to obtain numerical predictions of the different observables for the experiments.

3 Symbolic Integration of Feynman Parameter Integrals

In the following we summarize main aspects of the analytic integration of multi-loop Feynman parameter integrals. Of course these integrals can also be evaluated numerically, without observing their particular analytic structure, to some accuracy and methods exist to separate the different pole contributions in ε, cf. e.g. [46–56], which we will not discuss in the following. These methods play a role, however, also for testing analytic results. In calculating all the integrals required to solve a large scale problem, it is usually necessary to combine different analytic methods, at least for the sake of efficiency. This requirement finally led to the creation of these methods. In the future even more and further refined technologies will be needed to solve more enhanced problems. Finally, one ends up with sets of irreducible functions which span the solutions, see Sect. 4. The numerical representation of these functions is necessary and will be discussed in Sect. 4.3.

Non of the different techniques described in the following are of universal character. In particular the solution of the most advanced problems will need a combined and sensible use of various of them. All of them have to be handled with care to achieve a steady stepwise reduction of the problem on hand and to avoid to enlarge the complexity, given the limited time and memory resources for the corresponding computer algebraic calculations. This will also apply to future developments, since more complex calculations will require further new and advanced technologies.

Many of the formalisms described below lead to summation problems. Their solution requires dedicated and efficient algorithms in difference field theory as implemented in the packages Sigma [44, 45], EvaluateMultiSums and SumProduction [57–59], see also [60].

3.1 The PSLQ Method

The PSLQ method applies to the solution of zero dimensional quantities, i.e. physical quantities given by pure numbers. If the pool of constants is known or can be guessed over which the corresponding quantity has a polynomial representation over $\mathbb{Q}$, a highly precise numerical representation of the quantity and the individual monomials allows to determine the corresponding rational coefficients, cf. [61]. This method has been applied recently in a massive calculation of the five-loop QCD β-function [62]. Here the individual master integrals certainly contain also constants of elliptic nature and probably beyond. However, they all cancel in the final result, which is spanned by multiple zeta values (MZVs) [63, 64], more precisely by $\{\zeta_2, \zeta_3, \zeta_5, \zeta_7\}$, beyond pure rational terms. Let us illustrate the method by an example. We would like to determine the harmonic polylogarithm $H_{-1,0,0,1}(1)$, cf. Sect. 4.2, which is given by a polynomial of MZVs up to weight w=4. I.e. we have to apply the PSLQ method over all monomials up to w=4

$$\left\{ \ln(2), \zeta_2, \zeta_3, \mathrm{Li}_4\left(\frac{1}{2}\right) \right\}. \tag{1}$$

Here we defined

$$\zeta_k = \sum_{l=1}^{\infty} \frac{1}{l^k}, \quad k \in \mathbb{N}, k \geq 2 \tag{2}$$

$$\mathrm{Li}_k(x) = \sum_{l=1}^{\infty} \frac{x^k}{l^k}, \quad k \in \mathbb{N}, x \in \mathbb{R}, x \in [-1, 1]. \tag{3}$$

An approximate numerical value of $H_{-1,0,0,1}(1)$ is

0.33954546908735986959066784846086020613878153397957517913047 50

2224901374197238060826826241964431821670202556970965517522470 12

11749559277 $\qquad\qquad$ (4)

and PSLQ yields

$$H_{-1,0,0,1}(1) = -\frac{1}{12}\ln^4(2) + \frac{1}{2}\ln^2(2)\zeta_2 + \frac{3}{5}\zeta_2^2 - \frac{3}{4}\ln(2)\zeta_3 - 2\mathrm{Li}_4\left(\frac{1}{2}\right).$$

$$\tag{5}$$

In particular, monomials like $\ln(2), \ln^2(2), \ln^3(2), \zeta_2, \zeta_3$ do not contribute here.

3.2 Hypergeometric Functions and Their Generalizations

Simpler Feynman-parameter integrals have representations in terms of generalized hypergeometric functions [65–67] and their generalizations such as Appell-, Kampe-De-Feriet- and related functions [68–79]. This is due to the hyperexponential nature of the Feynman-parameter integrals, implying real exponents due to the dimensional parameter ε. These representations map multiple integrals to single series (for generalized hypergeometric functions) and double infinite series (e.g. for Appell series), which finally have to be solved by applying summation theory. The simplest function is Euler's Beta-function implying the series of $_{p+1}F_p$ functions

$$B(a_1, a_2) = \int_0^1 dt \, t^{a_1-1}(1-t)^{a_2-1} \tag{6}$$

$$_3F_2(a_1, a_2, a_3; b_1, b_2; x) = \frac{\Gamma(b_2)}{\Gamma(a_3)\Gamma(b_2-a_3)} \int_0^1 dt \, t^{a_3-1}(1-t)^{-a_3+b_2-1}$$
$$\times {}_2F_1(a_1, a_2; b_1; tx). \tag{7}$$

Representations of this kind are usually sufficient for massless and massive single-scale two-loop problems [80–83]. In the case of three-loop ladder graphs Appell-functions are appearing [84, 85]. There are some more classes of higher transcendental functions of this kind, which have been studied in the mathematical literature [71, 72, 76]. The corresponding representations allow the expansion in the dimensional parameter ε. At a given level in the calculation of Feynman diagrams one will not find corresponding known function representations and one has to invoke other methods of integration. One way to derive analytic infinite sum representations are Mellin–Barnes integrals to which we turn now.

3.3 Analytic Solutions Using Mellin–Barnes Integrals

The higher transcendental functions discussed in Sect. 3.2 have representations in terms of Pochhammer–Umlauf integrals [65, 86, 87] and related to it, by Mellin–Barnes integrals [88, 89]. They are defined by

$$\frac{1}{(a+b)^\alpha} = \frac{1}{\Gamma(\alpha)} \frac{1}{2\pi i} \int_{-i\infty}^{i\infty} dz \Gamma(\alpha+z)\Gamma(-z)\frac{b^z}{a^{\alpha+z}}, \quad \alpha \in \mathbb{R}, \alpha > 0, \tag{8}$$

cf. e.g. [90]. Here the contour integral is understood to be either closed to the left or the right surrounding the corresponding singularities. The Mellin–Barnes decomposition is analogous to the binomial (series) expansion for $\alpha < 0$. After its application, various more Feynman parameters can be integrated using the technique described in Sect. 3.2. In every application the decomposition introduces a number of infinite sums of depth one according to the residue theorem. There exist some packages for Mellin–Barnes integrals [91–94], allowing also for numerical checks. Finally all the produced sums have to be solved using multi-summation methods. Therefore one is advised to apply this method very carefully. Not all expressions generated by this method can be analytically summed using the presently know technologies, cf. [44, 45]. Sometimes Mellin-N space techniques may lead to elliptic structures, while x-space techniques do not, cf. [95], and sum-representations have to be cast back into definite integral representations first.

3.4 Hyperlogarithms

In a wide class of cases Feynman integrals can be represented by combinations of Kummer–Poincaré integrals [96–100] for (a part) of their expansion coefficients in ε. Let us assume one can isolate these terms, see [101], and forms a corresponding finite multi-integral. The method of hyperlogarithms [102] has originally intended to reorganize these integrals such that one can find a sequence of integrations being linear in the Feynman parameter on hand. If this is the case the result is given in terms of Kummer–Poincaré integrals. For a corresponding implementation see [103]. The method has first been applied to the usual massless Feynman integrals. A generalization for massive integrals also containing local operator insertions has been given in [104], with an implementation in [105]. Here also certain non-linear Feynman parameter structures, breaking multi-linearity, could be integrated.

3.5 The Method of Differential Equations

In single-scale processes systems of ordinary differential equations for the master integrals are naturally obtained by the IBP-relations differentiating for a parameter x.[2] The master integrals may then be calculated by solving these systems under given physical boundary conditions [106–109]. One considers the system

$$
\frac{d}{dx}
\begin{pmatrix} f_1 \\ \vdots \\ f_n \end{pmatrix}
=
\begin{pmatrix} A_{11} & \dots & A_{1,n} \\ \vdots & & \vdots \\ A_{n1} & \dots & A_{n,n} \end{pmatrix}
\begin{pmatrix} f_1 \\ \vdots \\ f_n \end{pmatrix}
+
\begin{pmatrix} g_1 \\ \vdots \\ g_n \end{pmatrix},
\tag{9}
$$

which may also be transformed into the scalar differential equation

$$
\sum_{k=0}^{n} p_{n-k}(x) \frac{d^{n-k}}{dx^{n-k}} f_1(x) = \overline{g}(x),
\tag{10}
$$

with $p_n \neq 0$, and $(n - 1)$ equations for the remaining solutions, which are fully determined by the solution $f_1(x)$. In setting up these systems one has to perform the expansion in ε in parallel in the decoupling.

---

[2]Correspondingly, in the case of more parameters, partial differential equation systems are obtained.

An important class of differential equations is formed by the first order factorizing systems, after applying the decoupling methods [8, 30] encoded in Oresys [28], which appear as the simplest case. Equation (9) may be transformed into Mellin space, decoupled there and solved using the efficient methods of the package Sigma, cf. Ref. [85].

The decoupled differential operator of (10) can be written in form of a combination of iterative integrals, cf. Sect. 4.2,

$$f_1(x) = \sum_{k-1}^{n+1} \gamma_k g_k(x), \ \gamma_k \in \mathbb{C}, \tag{11}$$

$$g_k(x) = h_0(x) \int_0^x dy_1 h_1(y_1) \int_0^{y_1} dy_2 h_2(y_2) \ldots \int_0^{y_{k-2}} dy_{k-1} h_{k-1}(y_{k-1})$$

$$\times \int_0^{y_{k-1}} dy_k q_k(y_k) \tag{12}$$

with $q_k(x) = 0$ for $1 \leq k \leq m$. Further, $\gamma_{m+1} = 0$ if $\bar{g}(x) = 0$ in (10), and $\gamma_{m+1} = 1$ and $q_{m+1}(x)$ being a mild variation of $\bar{g}(x)$ if $\bar{g}(x) \neq 0$. These solutions are d'Alembertian [110] since the master integrals appearing in quantum field theories obey differential equations with rational coefficients, the letters h_i, which constitute the iterative integrals, have to be hyperexponential. The solution can be computed using the package HarmonicSums [111]. More generally, also Liouvillian solutions [112] can be calculated with HarmonicSums utilizing Kovacic's algorithm [113]. This algorithm has been applied in many massive three-loop calculations so far, see also [85, 114–116].

If being transformed to the associated system of difference equations, the same holds, if this system is also first order factorizing. The solution of the remaining equations are directly obtained by the first solution.

In the multi-variate case, the ε-representation of a linear system of partial differential equations

$$\partial_m f(\varepsilon, x_n) = A_m(\varepsilon, x_n) f(\varepsilon, x_n) \tag{13}$$

is important, as has been recognized in Refs. [117, 118], see also [119]. The matrices A_n can now be transformed in the non-Abelian case by

$$A'_m = B^{-1} A_m B - B^{-1} (\partial_m B), \tag{14}$$

see also [120, 121], and one now intends to find a matrix B to transform (13) into the form

$$\partial_m f(\varepsilon, x_n) = \varepsilon A_m(x_n) f(\varepsilon, x_n), \tag{15}$$

if possible. This then allows solutions in terms of iterative integrals. A formalism for the basis change to the ε-basis has been proposed in [122] and implemented in the single-variate case in [123, 124] and in the multi-variate case in [125].

3.6   The Method of Arbitrary Large Moments

In the case of single-scale problems the corresponding class of Feynman integrals depends on a real parameter $x \in [0, 1]$, which is given e.g. as the ratio of two Lorentz invariants. For any power in ε one would like to find the corresponding function in x analytically. In a series of cases, cf. e.g. [116, 126–128], one may represent the solution in terms of a formal Taylor series in the variable x. The differential equations implied by the integration-by-parts method [21, 22, 25–27] can now be turned into recurrences using the Taylor series (resp. holonomic [129]) ansatz. In solving the corresponding system one may generate a large number of Mellin moments for the different projections on the individual color factors and multiple zeta values [64]. This is the case independently of the fact that the corresponding x- or N-space solution is given by iterative integrals or iterative–noniterative integrals. The corresponding method has been described in Ref. [130]. These moments can then be used as an input to the method described in Sect. 3.7 to find the associated difference equations. In some applications for single scale massive three-loop integrals [114] 8000 moments could be calculated. This is by far more than possible using standard methods like Mincer [131], MATAD [132] or Q2E [133, 134]. Based on this number of moments, the formal power series may be used as highly precise semi-analytic numeric representations, in case the corresponding series expansion has been performed for the physical quantity to be evaluated. If analytic continuations are still necessary, the method cannot be applied directly.

3.7   Guessing One-Dimensional Integrals

As has been described in Sect. 3.6 single-variate multiple Feynman parameter integrals can be either expanded into formal Taylor series or can be Mellin-transformed

$$G(N) = \mathbf{M}[f(x)](N) = \int_0^1 dx \, x^{N-1} f(x). \tag{16}$$

In both cases one tries now to find the associated difference equation [135] to the set of moments, e.g. $\{G(2), G(4), \ldots, G(2m)\}$, $m \in \mathbb{N}$ [136–139]. Indeed such an equation exists in many cases, as e.g. for (massive) operator matrix elements [140], but also for single-scale Wilson coefficients, Ref. [141]. If a suitably large number of moments has been calculated analytically, the associated series of rational numbers can now be used as input for the guessing algorithm [142], which is also available in Sage [143], exploiting the fast integer algorithms available there. The method finally returns the wanted difference equation, and tests it by a larger series of further moments. This method has been applied in Ref. [144] to obtain from more than 5000 moments the massless unpolarized three-loop anomalous dimensions and Wilson coefficients in deep-inelastic scattering [141, 145, 146]. Recently, the method has been applied ab initio in the calculation of three-loop splitting functions [147] and the massive two- and three-loop form factor [116, 148]. In the case of a massive operator matrix element 8000 moments [114] could be calculated and difference equations were derived for all contributing color and ζ-value structures. Analytic solutions can be found using the package Sigma [44, 45], provided the problem is solvable in difference field theory. In other cases at least the first order factorizing parts can be factored off. Other techniques are then needed to determine the remainder part of the solution.

3.8 The Almkvist–Zeilberger Algorithm

Since Feynman parameter integrals, depending on an additional parameter x, can be given as integrals over $\{x_i|_{i=1}^n\} \in [0, 1]^n$, they form the multi-integral $I(x)$, depending also on ε. The dependence on the real parameter x may be transformed into one on an integer parameter N, see Sect. 3.6. The Almkvist–Zeilberger algorithm [149, 150] is providing a method to find either an associated differential equation for $I(x)$ or a difference equation for $I(N)$, the coefficients of which are either polynomials in $\{x, \varepsilon\}$ or $\{N, \varepsilon\}$,

$$\sum_{l=0}^{m} P_l(x, \varepsilon) \frac{d^l}{dx^l} I(x, \varepsilon) = N(x, \varepsilon) \tag{17}$$

$$\sum_{l=0}^{m} R_l(N, \varepsilon) I(N + l, \varepsilon) = M(N, \varepsilon). \tag{18}$$

Both equations may be inhomogeneous, where the inhomogeneities emerge as known functions from lower order problems. An optimized and improved algorithm for the input class of Feynman integrals has been implemented in the MultiIntegrate package [85, 151]. It can either produce homogeneous equations of the form (17, 18) or equations with an inhomogeneity formed out of already known functions.

3.9   Iterative-Noniterative Integrals and Elliptic Solutions

Non-first order factorizing systems of differential or difference equations for the master integrals, cf. Sect. 3.5, occur at a certain order in massive Feynman diagram calculations. Well-known examples for this are the sun-rise integral, cf. e.g. [152–158], the kite integral [159–161], the three-loop QCD-corrections to the ρ-parameter [162–164], and the three-loop QCD corrections to the massive operator matrix element A_{Qg} [114]. After separating the first-order factorizing factors a Heun differential equation [165] remains in the case of the ρ-parameter. One may write the corresponding solution also using $_2F_1$-functions with rational argument [162, 166] and rational parameters. It is now interesting to see whether these solutions can be expressed in terms of complete elliptic integrals, which can be checked algorithmically using the triangle group [167].

In the examples mentioned one can find representations in terms of complete elliptic integrals of the first and second kind, $\mathbf{K}$ and $\mathbf{E}$, cf. [168, 169], and the question arises whether an argument translation allows for a representation through only $\mathbf{K}$. Criteria for this have been given in [170, 171]. In the case of the three-loop QCD-corrections to the ρ-parameter, however, this is not possible.

The homogeneous solution of the Heun equations are given by $_2F_1$-solutions $\psi_k^{(0)}(x), k = 1, 2$, at a specific rational argument. These integrals cannot be represented such that the variable x just appears in the boundaries of the integral. The inhomogeneous solution reads

$$\psi(x) = \psi_1^{(0)}(x)\left[C_1 - \int dx\,\psi_2^{(0)}(x)\frac{N(x)}{W(x)}\right] + \{1 \to 2\}, \qquad (19)$$

with $N(x)$ and $W(x)$ the inhomogeneity and the Wronskian. $C_{1,2}$ are the integration constants. Through partial integration the ratio $N(x)/W(x)$ can be transformed into an iterative integral. Since $\psi_k^{(0)}(x)$ cannot be written as iterative integrals, $\psi(x)$ is obtained as an *iterative non-iterative integral* [162, 172] of the type

$$\mathbb{H}_{a_1,...,a_{m-1};a_m,F_m(r(y_m)),a_{m+1},...a_q}(x) =$$

$$\int_0^x dy_1 f_{a_1}(y_1) \int_0^{y_1} dy_2 ... \int_0^{y_{m-1}} dy_m f_{a_m}(y_m) F_m[r(y_m)] \mathbb{H}_{a_{m+1},...,a_q}(y_m), \qquad (20)$$

with $r(x)$ a rational function and F_m a non-iterative integral. Usually more than one non-iterative integral will appear in (20). F_m denotes *any* non-iterative integral, implying a very general representation, cf. [162].[3] In Ref. [174] an ε-form for the Feynman diagrams of elliptic cases has been found recently. However, transcendental letters contribute here. This is in accordance with our earlier finding, Eq. (20), which, as well is an iterative integral over all objects between the individual

---

[3]This representation has been used in a more special form also in [173] later.

iterations and to which now also the non-iterative higher transcendental functions $F_m[r(y_m)]$ contribute. One may obtain fast convergent representations of $\mathbb{H}(x)$ by overlapping series expansions around $x = x_0$ outside possible singularities, see Ref. [162] for details.

Let us return to the elliptic case now. Here one may transform the kinematic variable x occurring as $\mathbf{K}(k^2) = \mathbf{K}(r(x))$ into the variable $q = \exp[i\pi\tau]$ analytically with

$$k^2 = r(x) = \frac{\vartheta_2^4(q)}{\vartheta_3^4(q)}, \tag{21}$$

by applying a third order Legendre–Jacobi transformation, where $\vartheta_l, l = 1, \ldots, 4$ denote Jacobi's ϑ-functions and $\mathrm{Im}(\tau) > 0$. In this way Eq. (19) is rewritten in terms of the new variable. The integrands are given by products of meromorphic modular forms, cf. [175–177], which can be written as a linear combination of ratios of Dedekind's η-function

$$\eta(\tau) = q^{\frac{1}{12}} \prod_{k=1}^{\infty} (1 - q^{2k}). \tag{22}$$

Depending on the largest multiplier $k \in \mathbb{N}$, k_m, of τ in the argument of the η-function, the solution transforms under the congruence subgroup $\Gamma_0(k_m)$. One can perform Fourier expansions in q around the different cusps of the problem, cf. [178, 179].

In the case that the occurring modular forms are holomorphic, one obtains representations in Eisenstein series with character, while in the meromorphic case additional η-factors in the denominators are present. In the former case the q-integrands can be written in terms of elliptic polylogarithms in the representation [156, 157]

$$\mathrm{ELi}_{n,m}(x, y) = \sum_{k=1}^{\infty} \sum_{l=1}^{\infty} \frac{x^k}{k^n} \frac{y^l}{l^m} q^{kl} \tag{23}$$

and products thereof, cf. [157]. The corresponding q-integrals can be directly performed. The solution (19) usually appears for single master integrals. Other master integrals are obtained integrating further other letters, so that finally representations by $\mathbb{H}(x)$ occur. Iterated modular forms, resp. Eisenstein series, have been also discussed recently in [180, 181]. Efficient numerical calculations of modular forms based on q-series were obtained in [182].

For systems which factorize only to third and higher order much less is known.

3.10   Iterative Integrals of Functions with More Variables

The occurrence of several masses or additional external non-factorizing scales in higher order loop- and phase-space integrals leads in general to rational and root-valued letters with real parameter letters in the contributing alphabet, cf. [95, 183, 186–188]. In the case of the loop integrals one obtains letters of the kind

$$\frac{1}{1 - x(1 - \eta)}, \quad \frac{\sqrt{x(1 - x)}}{\eta + x(1 - \eta)}, \quad \sqrt{x(1 - \eta(1 - x)}, \quad \eta \in [0, 1]. \tag{24}$$

The iterative integrals and constants which appeared in [95, 183] could finally be all integrated to harmonic polylogarithms containing complicated arguments, at least up to one remaining integration, which allows their straightforward numerical evaluation.

In the case of phase space integrals with more scales, e.g. [186, 187], also letters contribute, which may imply incomplete elliptic integrals and iterated structures thereof. Contrary to the functions obtained in Sect. 3.9 these are still iterative integrals, because the boundaries of the phase-space integrals are real parameters and not constants. The integrands could not by rationalized completely by variable transformations, see also [189]. Contributing letters are e.g.

$$\frac{x}{\sqrt{1 - x^2}\sqrt{1 - k^2 x^2}}, \quad \frac{x}{\sqrt{1 - x^2}\sqrt{1 - k^2 x^2}(k^2(1 - x^2(1 - z^2)) - z^2)}, \tag{25}$$

with $k, z \in [0, 1]$. The corresponding iterative integrals are called Kummer-elliptic integrals. They are derived using the techniques described in Refs. [190–192].

4   A Series of Function Spaces

Intermediary and final results for zero- and single-scale multi-loop calculations have representations by special functions as polynomials over $\mathbb{Q}$. In the case of zero-scale quantities these are special numbers. For single scale quantities one either uses finite nested sum representations in Mellin N-space or iterative integral representations in x-space. Here x denotes a Lorentz invariant ratio of two physical quantities. Both spaces are related to each other by the Mellin transform (16), where $f(x)$ denotes an iterative integral. The zero-scale quantities can be obtained e.g. in the limit $N \to \infty$ of these Mellin transforms or by the values $f(x = 1)$.

4.1 Classes of Nested Sums

The methods described in Sect. 3 very often lead to finite nested sum representations for which algorithms exist [44, 45] to cast these sums into indefinitely nested sums. They are given by

$$
S_{b,\mathbf{a}}(N) = \sum_{k=1}^{N} g_b(k) S_{\mathbf{a}}(n), \quad S_{\emptyset} = 1, \ g_c \in \bar{\mathfrak{A}},
\tag{26}
$$

with $\bar{\mathfrak{A}}$ the associated alphabet of functions. The sums obey quasi-shuffle relations [193, 194]. The simplest structures are the finite harmonic sums [195, 196], where $g_b(k) = (\text{sign}(b))^k / k^{|b|}$, $b \in \mathbb{N}\backslash\{0\}$. A generalization is obtained in the cyclotomic case [197]. Here the characteristic summands are $g_{a,b,c}(k) = (\pm 1)^k / (ak + b)^c$, with $a, b, c \in \mathbb{N}\backslash\{0\}$. Further, the generalized harmonic sums have letters of the type b^k / k^c, with $c \in \mathbb{N}\backslash\{0\}$, $b \neq 0$, $b \in \mathbb{R}$ [198]. Another generalization are nested finite binomial and inverse-binomial sums, containing also other sums discussed before. An example is given by

$$
\sum_{i=1}^{N} \binom{2i}{i} (-2)^i \sum_{j=1}^{i} \frac{1}{j \binom{2j}{j}} S_{1,2}\left(\frac{1}{2}, 1\right)(j) = \int_0^1 dx \frac{(-x)^N - 1}{x + 1} \sqrt{\frac{x}{8 - x}}
$$

$$
\times \left[H_{w_{12},1,0}(x) - 2H_{w_{13},1,0}(x) - \zeta_2 \left(H_{w_{12}}(x) - 2H_{w_{13}}(x) \right) \right]
$$

$$
- \frac{5\zeta_3}{8\sqrt{3}} \int_0^1 dx \frac{(-2x)^N - 1}{x + \frac{1}{2}} \sqrt{\frac{x}{4 - x}} + c_1 \int_0^1 dx \frac{(-8x)^N - 1}{x + \frac{1}{8}} \sqrt{\frac{x}{1 - x}},
\tag{27}
$$

with $c_1 \approx 0.10184720\ldots$, cf. [192]. Here the indices w_k label specific letters given in [192]. Infinite binomial and inverse binomial sums have been considered in [199, 200]. Given the general structure of (26) many more iterated sums can be envisaged and may still appear in even higher order calculations.

4.2 Classes of Iterated Integrals

Iterated integrals have the structure

$$
H_{b,\mathbf{a}}(x) = \int_0^x dy f_b(y) H_{\mathbf{a}}(y), \quad H_{\emptyset} = 1, \ f_c \in \mathfrak{A},
\tag{28}
$$

where f_c are real functions and are the letters of the alphabet $\mathfrak{A}$. Iterated integrals obey shuffle relations [194, 201] which allows to represent them over a multinomial basis of fewer terms.

The simplest iterative integrals having been considered in quantum field theory are the Nielsen integrals for the two-letter alphabets $\{1/x, 1/(1-x)\}$ or $\{1/x, 1/(1+x)\}$ [202–205], covering also the polylogarithms [205–207]. This class has later been extended to the harmonic polylogarithms [208] build over the alphabet $\{1/x, 1/(1-x), 1/(1+x)\}$. A further extension is to the real representations of the cyclotomic polylogarithms, with $\{1/x, 1/\Phi_k(x)\}$ [197], where $\Phi_k(x)$ denotes the kth cyclotomic polynomial. Another extension is given by Kummer–Poincaré iterative integrals over the alphabet $\{1/(x - a_i), \ a_i \in \mathbb{C}\}$, [96–100]. Properties of these functions have been studied in Refs. [198, 209]. In general one may have also more general denominator polynomials $P(x)$, which one can factor into

$$P(x) = \prod_{k=1}^{n}(x - a_k) \prod_{l=1}^{m}(x^2 + b_l x + c_l), \quad a_k, b_l, c_l \in \mathbb{R} \tag{29}$$

in real representations. One then performs partial fractioning for $1/P(x)$ and forms iterative integrals out of the obtained letters. Further classes are found for square-root valued letters as studied e.g. in Ref. [192]. In multi-scale problems, cf. e.g. [95, 186–188] and Sect. 3.10, further root-valued letters appear, like also the Kummer-elliptic integrals [187].

4.3   Classes of Associated Special Numbers

For the sums of Sect. 4.1 which are convergent in the limit $N \rightarrow \infty$ and the iterated integrals of Sect. 4.2 which can be evaluated at $x = 1$ one obtains two sets of special numbers. They span the solution spaces for zero-scale quantities and appear as boundary values for single-scale problems. Examples for these special numbers are the multiple zeta values [64], associated to the harmonic sums and harmonic polylogarithms, special generalized numbers [198] like $\mathrm{Li}_2(1/3)$, associated to generalized sums and to Kummer–Poincaré iterated integrals, special cyclotomic numbers [197] like Catalan's number, special binomial numbers [192], as e.g. $\mathrm{arccot}(\sqrt{7})$, and special constants in the elliptic case [162, 210]. The latter numbers are given by integrals involving complete elliptic integrals at special rational arguments and related functions. In general these numbers obey more relations than the finite sums and iterated integrals. One may use the PSLQ-method to get a first information on relations between these numbers occurring in a given problem and proof the conjectured relations afterwards.

4.4 Numerical Representations

Physical observables based on single scale quantities can either be represented in Mellin N-space or x-space. Representations in Mellin N-space allow the exact analytic solution of evolution equations [211] and scheme-invariant evolution equations can be derived in this way [212, 213]. The x-space representation is then obtained by a single numerical integral around the singularities of the respective quantity for $N \in \mathbb{C}$, cf. [211], requiring to know the complex representation of the integrand in N-space. In the case of harmonic sums semi-numerical representations were given in [214, 215]. Furthermore, it is known that basic harmonic sums, except of $S_1(N)$, which is represented by the Digamma function, and its polynomials, have a representation by factorial series [216, 217], which has been used in [218, 219] for their asymptotic representation, see also [220]. One uses then the recursion relations, which can be obtained from (26), to move $N \in \mathbb{C}$ from the asymptotic region to the desired point on the integration contour in the analyticity region of the problem. This can be done for the sums of the type being described in Refs. [192, 197, 198] as well, since also in this case asymptotic expansions can be provided, at least for certain combinations of sums occurring in the respective physical problem, cf. [85, 104]. In the case that the corresponding relations are not given in tabulated form, they can be calculated using the package HarmonicSums [111, 151, 192, 195–198, 221, 222]. Relations for harmonic sums are also implemented in summer [195], and for generalized harmonic sums in nestedsums [223], Xsummer [224], and PolyLogTools [226].

In other applications one may want to work in x-space directly. Here numerical representations are available for the Nielsen integrals [203], the harmonic polylogarithms [227–231], the Kummer–Poincaré iterative integrals [231], and the cyclotomic harmonic polylogarithms [116]. These representations are also useful to lower the number of numerical integrations for more general problems, e.g. in the multi-variate case. The relations for the corresponding quantities are implemented for the harmonic polylogarithms in [208, 228] and for all iterative integrals mentioned, including general iterative integrals, in the package HarmonicSums.

5 Conclusions

In parallel to the analytic higher-loop calculations in Quantum Field Theory the associated mathematical methods have been developed by theoretical physicists and mathematicians since the 1950s. We witness a very fast development since the late 1990s approaching difficult massive problems at two-loop and higher order and massless problems form three loops onward. The classical methods of polylogarithms and Nielsen-integrals which were standard means, turned out to be not sufficient anymore. Since then more and more special number- and function spaces have been revealed, studied and were brought to flexible practical use in very many applications. Moreover, a wide host of analytic integration and summation methods have been developed during a very short period. In this way very large physics problems could be solved analytically—a triumph of the exact sciences, also thanks to various groundbreaking methods in computer algebra. In this context the goal is to improve the accuracy of the fundamental parameters of the Standard Model of the elementary particles further. Within the present projects this concerns in particular the relative precision of the strong coupling constant $\alpha_s(M_Z^2)$ to less than 1% and of the $\overline{MS}$ mass of the charm quark to better than 1.5%.

At even higher loop order and for more separated final state legs, introducing more masses and kinematic invariants, one expects further mathematical structures to contribute. Possible structures of this kind could be Abel-integrals [232] and integrals related to K3-surfaces [233]. More inclusive methods, like the method of differential equations, can certainly determine the degree of non-factorization of a physical problem. However, one would like to know in a closer sense the respective analytic solution. Here cutting methods can be of use since the underlying integrands can be systematically related to the final integral by (various) Hilbert-transforms [234–236].[4] In this way integrand structures are revealed, which are somewhat hidden in the case of differential equations. This method has been advocated early by M. Veltman [238], see also [239].

This process to master highly complex Feynman integrals using analytic methods is of course just at the beginning and will develop further given the present and future challenges in the field. All of these results put experimental analyses in precision measurements at the high energy colliders into the position to analyze the data with much reduced theory errors and we will get far closer in our insight into the structure of the micro cosmos to reveal its ultimate laws. The interdisciplinary joined effort by mathematicians, theoretical and experimental particle physicists and experts in computer algebra makes this possible and allows to answer quite a series of fundamental scientific questions of our time.

I would like to give my warmest thanks to Peter Paule for his continuous collaboration and support to the DESY–RISC collaboration, starting with our first contacts in 2005, arranged by Bruno Buchberger. This scientific symbiosis has produced a large number of methods to tackle quite a series of difficult problems since, and is continuing to do so in the future. Physics, mathematics, and computer algebra profit from this and reach new horizons, which, not at all, could have been imagined. In this way we follow together the motto D. Hilbert has given to us:

Wir müssen wissen. Wir werden wissen.

Le but unique de la science, c'est l'honneuer de l'esprit humain.[5]

---

[4] For a recent application to the one-loop case, see e.g. [237].

[5] Jacobi to Legendre, July 2nd, 1830.

Acknowledgements I would like to thank J. Ablinger, D. Broadhurst, A. De Freitas, D. Kreimer, A. von Manteuffel, P. Marquard, S.-O. Moch, W.L. van Neerven, P. Paule, C. Schneider, K. Schönwald, J. Vermaseren and S. Weinzierl for countless fruitful discussions. This work was supported by the EU TMR network SAGEX Marie Skłodowska-Curie grant agreement No. 764850 and COST action CA16201: Unraveling new physics at the LHC through the precision frontier.

References

1. The ILC: https://en.wikipedia.org/wiki/International_Linear_Collider Aguilar-Saavedra, J.A., et al., [ECFA/DESY LC Physics Working Group]: TESLA: The Superconducting electron positron linear collider with an integrated x-ray laser laboratory. Technical design report. Part 3. Physics at an e^+e^- linear collider, hep-ph/0106315;
   Accomando, E., et al., [ECFA/DESY LC Physics Working Group]: Phys. Rept. **299**, 1–78 (1998). [hep-ph/9705442]
2. The Future Circular Collider. https://en.wikipedia.org/wikiFuture_Circular_Collider. TH FCC-ee design study. http://tlep.web.cern.ch
3. 't Hooft, G., Veltman, M.J.G.: Nucl. Phys. **B153**, 365–401 (1979)
4. 't Hooft, G., Veltman, M.J.G.: NATO Sci. Ser. B **4**, 177–322 (1974)
5. Veltman, M.J.G.: Diagrammatica: The Path to Feynman Rules. Cambridge Lecture Notes in Physics, vol. 4. Cambridge University Press, Cambridge (1994)
6. Weinzierl, S.: Introduction to Feynman Integrals (2010). arXiv:1005.1855 [hep-ph]
7. Ablinger, J., Blümlein, J., Schneider, C.: J. Phys. Conf. Ser. **523**, 012060 (2014). [arXiv:1310.5645[math-ph]]. http://www.arXiv.org/abs/[arXiv:1310.5645[math-ph]]
8. Ablinger, J., Blümlein, J.: In: Schneider C., Blümlein J. (eds.) Computer Algebra in Quantum Field Theory. Integration, Summation and Special Functions, pp. 1–32. Springer, Wien (2012). [arXiv: 1304.7071 [math-ph]]
9. Weinzierl, S.: In: Schneider C., Blümlein J. (eds.) Computer Algebra in Quantum Field Theory. Integration, Summation and Special Functions, pp. 381–406. Springer, Wien (2012). [arXiv:1301.6918 [hep-ph]]
10. Duhr, C.: In: Dixon L., Petriello F. (eds.) Journeys Through the Precision Frontier: Amplitudes for Colliders, 2014 TASI Lectures, pp. 419–476. World Scientific, Singapore (2015). [arXiv:1411.7538 [hep-ph]]
11. Blümlein, J., Schneider, C.: Int. J. Mod. Phys. A **33**(17), 1830015 (2018). [arXiv:1809.02889 [hep-ph]]
12. Nogueira, P.: J. Comput. Phys. **105**, 279–289 (1993)
13. van Ritbergen, T., Schellekens, A., Vermaseren, J.A.M.: Int. J. Mod. Phys. **A14**, 41–96 (1999). [hep-ph/9802376]. http://www.arXiv.org/abs/[hep-ph/9802376]
14. Nakanishi, N.: Graph Theory and Feynman Integrals. Gordon and Breach, New York (1971)
15. Lefschetz, S.: Applications of Algebraic Topology: Graphs and Networks, the Picard-Lefschetz Theory an Feynman Integrals. Springer, Berlin (1975)
16. Bogner, C., Weinzierl, S.: Int. J. Mod. Phys. **A25**, 2585–2618 (2010). [arXiv:1002.3458[hep-ph]]. http://www.arXiv.org/abs/[arXiv:1002.3458[hep-ph]]
17. Vermaseren, J.A.M.: New features of FORM (2000). math-ph/0010025
18. Tentyukov, M., Fliegner, D., Frank, M., Onischenko, A., Retey, A., Staudenmaier, H.M., Vermaseren, J.A.M.: AIP Conf. Proc. **583**(1), 202 (2002). [cs/0407066 [cs-sc]]
19. Tentyukov, M., Vermaseren, J.A.M.: Comput. Phys. Commun. **181**, 1419–1427 (2010). [hep-ph/0702279]. http://www.arXiv.org/abs/[hep-ph/0702279]
20. Ruijl, B., Takahiro, U., Vermaseren, J.A.M.: FORM version 4.2 (2017). arXiv: 1707.06453[hep-ph]
21. Chetyrkin, K.G., Tkachov, F.V.: Nucl. Phys. **B192**, 159–204 (1981)

22. Laporta, S.: Int. J. Mod. Phys. **A15**, 5087–5159 (2000). [hep-ph/0102033]. http://www.arXiv. org/abs/[hep-ph/0102033]
23. Smirnov, A.: JHEP **10**, 107 (2008). [arXiv:0807.3243[hep-ph]]. http://www.arXiv.org/abs/ [arXiv:0807.3243[hep-ph]]
24. Smirnov, A.V., Chuharev, F.S.: FIRE6: Feynman Integral REduction with Modular Arithmetic (2019). arXiv:1901.07808 [hep-ph]
25. Studerus, C.:   Comput. Phys. Commun. **181**, 1293–1300 (2010). [arXiv:0912.2546 [physics.comp-ph]]. http://www.arXiv.org/abs/[arXiv:0912.2546[physics.comp-ph]]
26. von Manteuffel, A., Studerus, C.: Reduze 2 – Distributed Feynman Integral Reduction (2012). arXiv:1201.4330 [hep-ph]
27. Marquard, P., Seidel, D.: The Crusher algorithm (unpublished)
28. Gerhold, S.: Uncoupling systems of linear Ore operator equations. Master's thesis, RISC, J. Kepler University, Linz (2002)
29. Bostan, A., Chyzak, F., de Panafieu, É.: Complexity estimates for two uncoupling algorithms. In: Proceedings of ISSAC'13, Boston, June (2013)
30. Zürcher, B.: Rationale Normalformen von pseudo-linearen Abbildungen. Master's thesis, Mathematik, ETH Zürich (1994)
31. Karr, M.: J. ACM **28**, 305–350 (1981)
32. Bronstein, M.: J. Symbolic Comput. **29**(6), 841–877 (2000)
33. Schneider, C.: Symbolic summation in difference fields. Ph.D. Thesis RISC, Johannes Kepler University, Linz technical report 01–17 (2001)
34. Schneider, C.: An. Univ. Timisoara Ser. Mat.-Inform. **42**, 163–179 (2004)
35. Schneider, C.: J. Differ. Equations Appl. **11**, 799–821 (2005)
36. Schneider, C.: Appl. Algebra Eng. Commun. Comput. **16**, 1–32 (2005)
37. Schneider, C.: J. Algebra Appl. **6**, 415–441 (2007)
38. Schneider, C.: Clay Math. Proc. **12**, 285–308 (2010). [arXiv:0904.2323 [cs.SC]]. [arXiv:0904.2323]
39. Schneider, C.: Ann. Comb. **14**, 533–552 (2010). [arXiv:0808.2596]
40. Schneider, C.: In: Gutierrez J., Schicho J., Weimann M. (eds.) Computer Algebra and Polynomials, Applications of Algebra and Number Theory. Lecture Notes in Computer Science (LNCS), vol. 8942, pp. 157–191 (2015). [arXiv:1307.7887 [cs.SC]]
41. Schneider, C.: J. Symbolic Comput. **43**, 611–644 (2008). [arXiv:0808.2543]
42. Schneider, C.: J. Symb. Comput. **72**, 82–127 (2016). [arXiv:1408.2776 [cs.SC]]
43. Schneider, C.: J. Symb. Comput. **80**, 616–664 (2017). [arXiv:1603.04285 [cs.SC]]
44. Schneider, C.: Sém. Lothar. Combin. **56**, 1–36 (2007). article B56b. http://www.arXiv.org/ abs/articleB56b
45. Schneider, C.: In: Schneider C., Blümlein J. (eds.) Computer Algebra in Quantum Field Theory: Integration, Summation and Special Functions. Texts and Monographs in Symbolic Computation, pp. 325–360. Springer, Wien (2013). [arXiv:1304.4134 [cs.SC]]
46. Binoth, T., Heinrich, G.: Nucl. Phys. **B585**, 741–759 (2000). [hep-ph/0004013]. http://www. arXiv.org/abs/[hep-ph/0004013]
47. Nagy, Z., Soper, D.E.: Phys. Rev. **D74**, 093006 (2006). [hep-ph/0610028]. http://www.arXiv. org/abs/[hep-ph/0610028]
48. Anastasiou, C., Beerli, S., Daleo, A.: JHEP **05**, 071 (2007). [hep-ph/0703282]. http://www. arXiv.org/abs/[hep-ph/0703282]
49. Smirnov, A.V., Tentyukov, M.N.:   Comput. Phys. Commun. **180**, 735–746 (2009). [arXiv:0807.4129[hep-ph]]. http://www.arXiv.org/abs/[arXiv:0807.4129[hep-ph]]
50. Carter, J., Heinrich, G.:   Comput. Phys. Commun. **182**, 1566–1581 (2011). [arXiv:1011.5493[hep-ph]]. http://www.arXiv.org/abs/[arXiv:1011.5493[hep-ph]]
51. Smirnov, A.V., Smirnov, V.A., Tentyukov, M.: Comput. Phys. Commun. **182**, 790–803 (2011). [arXiv:0912.0158[hep-ph]]. http://www.arXiv.org/abs/[arXiv:0912.0158[hep-ph]]
52. Becker, S., Reuschle, C., Weinzierl, S.: JHEP **12**, 013 (2010). [arXiv:1010.4187[hep-ph]]. http://www.arXiv.org/abs/[arXiv:1010.4187[hep-ph]]

53. Becker, S., Reuschle, C., Weinzierl, S.: JHEP **1207**, 090 (2012). [arXiv:1205.2096[hep-ph]]. http://www.arXiv.org/abs/[arXiv:1205.2096[hep-ph]]
54. Becker, S., Götz, D., Reuschle, C., Schwan, C., Weinzierl, S.: Phys. Rev. Lett. **108**, 032005 (2012). [arXiv:1111.1733[hel-ph]]. http://www.arXiv.org/abs/[arXiv:1111.1733[hel-ph]]
55. Smirnov, A.V.: Comput. Phys. Commun. **204**, 189–199 (2016). [arXiv:1511.03614[hep-ph]]. http://www.arXiv.org/abs/[arXiv:1511.03614[hep-ph]]
56. Borowka, S., Gehrmann, T., Hulme, D.: JHEP **1808**, 111 (2018). [arXiv:1804.06824 [hep-ph]]
57. Ablinger, J., Blümlein, J., Klein, S., Schneider, C.: Nucl. Phys. Proc. Suppl. **205–206**, 110–115 (2010). [arXiv:1006.4797[math-ph]]. http://www.arXiv.org/abs/[arXiv:1006.4797[math-ph]]
58. Blümlein, J., Hasselhuhn, A., Schneider, C.: PoS (RADCOR2011), 032 (2012). [arXiv:1202.4303[math-ph]]. http://www.arXiv.org/abs/[arXiv:1202.4303[math-ph]]
59. Schneider, C.: J. Phys. Conf. Ser. **523**, 012037 (2014). [arXiv:1310.0160[cs.Sc]]. http://www.arXiv.org/abs/[arXiv:1310.0160[cs.Sc]]
60. Krattenthaler, C., Schneider, C.: Evaluation of binomial double sums involving absolute values. In: Pillwein, V., Schneider, C. (eds) Algorithmic Combinatorics: Enumerative Combinatorics, Special Functions and Computer Algebra. Texts & Monographs in Symbolic Computation. Springer, Heidelberg (2020). https://doi.org/10.1007/978-3-030-44559-1_14
61. Ferguson, H.R.P., Bailey, D.H.: A polynomial time, numerically stable integer relation algorithm, RNR Technical Report RNR-91-032, July 14 (1992)
62. Luthe, T., Maier, A., Marquard, P., Schröder, Y.: JHEP **10**, 166 (2017). [arXiv:1709.07718[hep-ph]]. http://www.arXiv.org/abs/[arXiv:1709.07718[hep-ph]]
63. Borwein, J.M., Bradley, D.M., Broadhurst, D.J., Lisonek, P.: Trans. Am. Math. Soc. **353**, 907–941 (2001). [math/9910045]. http://www.arXiv.org/abs/[math/9910045]
64. Blümlein, J., Broadhurst, D., Vermaseren, J.A.M.: Comput. Phys. Commun. **181**, 582–625 (2010). [arXiv: 0907.2557[math-ph]]. http://www.arXiv.org/abs/[arXiv:0907.2557[math-ph]]
65. Klein, F.: Vorlesungen über die hypergeometrische Funktion. Wintersemester 1893/94, Die Grundlehren der Mathematischen Wissenschaften, vol. 39. Springer, Berlin (1933)
66. Bailey, W.N.: Generalized Hypergeometric Series. Cambridge University Press, Cambridge (1935)
67. Slater, L.J.: Generalized Hypergeometric Functions. Cambridge University Press, Cambridge (1966)
68. Appell, P., Kampé de Fériet, J.: Fonctions Hypergéométriques et Hypersphériques, Polynomes D' Hermite. Gauthier-Villars, Paris (1926)
69. Appell, P.: Les Fonctions Hypergéométriques de Plusieur Variables. Gauthier-Villars, Paris (1925)
70. Kampé de Fériet, J.: La fonction hypergéométrique. Gauthier-Villars, Paris (1937)
71. Exton, H.: Multiple Hypergeometric Functions and Applications. Ellis Horwood, Chichester (1976)
72. Exton, H.: Handbook of Hypergeometric Integrals. Ellis Horwood, Chichester (1978)
73. Schlosser, M.J.: In: Schneider C., Blümlein J. (eds.) Computer Algebra in Quantum Field Theory: Integration, Summation and Special Functions, pp. 305–324. Springer, Wien (2013). [arXiv:1305.1966 [math.CA]]
74. Anastasiou, C., Glover, E.W.N., Oleari, C.: Nucl. Phys. **B572**, 307–360 (2000). [hep-ph/9907494]. http://www.arXiv.org/abs/[hep-ph/9907494]
75. Anastasiou, C., Glover, E.W.N., Oleari, C.: Nucl. Phys. **B565**, 445–467 (2000). [hep-ph/9907523]. http://www.arXiv.org/abs/[hep-ph/9907523]
76. Srivastava, H.M., Karlsson, P.W.: Multiple Gaussian Hypergeometric Series. Ellis Horwood, Chichester (1985)
77. Lauricella, G.: Rediconti del Circolo Matematico di Palermo **7**(S1), 111–158 (1893)
78. Saran, S.: Ganita **5**, 77–91 (1954)
79. Saran, S.: Acta Math. **93**, 293–312 (1955)
80. Hamberg, R., van Neerven, W.L., Matsuura, T.: Nucl. Phys. **B359**, 343–405 (1991). [Erratum: Nucl. Phys. B644 (2002) 403–404]

81. Hamberg, R.: Second order gluconic contributions to physical quantities. Ph.D. Thesis, Leiden University (1991)
82. Buza, M., Matiounine, Y., Smith, J., Migneron, R., van Neerven, W.L.: Nucl. Phys. **B472**, 611–658 (1996). [hep-ph/9601302]. http://www.arXiv.org/abs/[hep-ph/9601302]
83. Bierenbaum, I., Blümlein, J., Klein, S.: Nucl. Phys. **B780**, 40–75 (2007). [hep-ph/0703285]. http://www.arXiv.org/abs/[hep-ph/0703285]
84. Ablinger, J., Blümlein, J., Hasselhuhn, A., Klein, S., Schneider, C., Wißbrock, F.: Nucl. Phys. **B864**, 52–84 (2012). [arXiv:1206.2252[hep-ph]]. http://www.arXiv.org/abs/[arXiv: 1206.2252[hep-ph]]
85. Ablinger, J., Behring, A., Blümlein, J., De Freitas, A., von Manteuffel, A., Schneider, C.: Comput. Phys. Commun. **202**, 33–112 (2016). [arXiv:1509.08324[hep-ph]]. http://www. arXiv.org/abs/[arXiv:1509.08324[hep-ph]]
86. Pochhammer, L.: Math. Ann. **35**, 495–526 (1890)
87. Kratzer, A., Franz, W.: Transzendente Funktionen. Geest & Portig, Leipzig (1960)
88. Barnes, E.W.: Q. J. Math. **41**, 136–140 (1910)
89. Mellin, H.: Math. Ann. **68**(3), 305–337 (1910)
90. Smirnov, V.A.: Feynman Integral Calculus. Springer, Berlin (2006)
91. Czakon, M.: Comput. Phys. Commun. **175**, 559–571 (2006). [hep-ph/0511200]. http://www. arXiv.org/abs/[hep-ph/0511200]
92. Smirnov, A., Smirnov, V.: Eur. Phys. J. **C62**, 445–449 (2009). [arXiv: 0901.0386[hep-ph]]. http://www.arXiv.org/abs/[arXiv:0901.0386[hep-ph]]
93. Gluza, J., Kajda, K., Riemann, T.: Comput. Phys. Commun. **177**, 879–893 (2007). [arXiv:0704.2423[hep-ph]]. http://www.arXiv.org/abs/[arXiv:0704.2423[hep-ph]]
94. Gluza, J., Kajda, K., Riemann, T., Yundin, V.: Eur. Phys. J. **C71**, 1516 (2011). [arXiv:1010.1667[hep-ph]]. http://www.arXiv.org/abs/[arXiv:1010.1667[hep-ph]]
95. Ablinger, J., Blümlein, J., De Freitas, A., Schneider, C., Schönwald, K.: Nucl. Phys. **B927**, 339–367 (2018). [arXiv: 1711.06717[hep-th]]. http://www.arXiv.org/abs/[arXiv:1711. 06717[hep-th]]
96. Kummer, E.E.: J. Reine Angew. Math. (Crelle) **21**, 74–90; 193–225; 328–371 (1840)
97. Poincaré, H.: Acta Math. **4**, 201–312 (1884)
98. Lappo-Danilevsky, J.A.: Mémoirs sur la Théorie des Systèmes Différentielles Linéaires, Chelsea Publ. Co, New York (1953)
99. Chen, K.T: Trans. Am. Math. Soc. **156**(3), 359–379 (1971)
100. Goncharov, A.B.: Math. Res. Lett. **5**, 497–516 (1998)
101. von Manteuffel, A., Panzer, E., Schabinger, R.M.: JHEP **02**, 120 (2015). [arXiv:1411.7392[hep-ph]]. http://www.arXiv.org/abs/[arXiv:1411.7392[hep-ph]]
102. Brown, F.: Commun. Math. Phys. **287**, 925–958 (2009). [arXiv:0804.1660 [math.AG]]
103. Panzer, E.: Comput. Phys. Commun. **188**, 148–166 (2015). [arXiv:1403.3385[hep-th]]. http:// www.arXiv.org/abs/[arXiv:1403.3385[hep-th]]
104. Ablinger, J., Blümlein, J., Raab, C., Schneider, C., Wißbrock, F.: Nucl. Phys. **B885**, 409–447 (2014). [arXiv:1403.1137[hep-ph]]. http://www.arXiv.org/abs/[arXiv:1403.1137[hep-ph]]
105. Wißbrock, F.Ph.: $O(\alpha_s^3)$ contributions to the heavy flavor Wilson coefficients of the structure function $F_2(x, Q^2)$ at $Q^2 \gg m^2$. Ph.D. Thesis, TU Dortmund (2015)
106. Kotikov, A.V.: Phys. Lett. **B254**, 158–164 (1991)
107. Bern, Z., Dixon, L.J., Kosower, D.A.: Phys. Lett. **B302**, 299–308 (1993). [Erratum: Phys. Lett. B318, (1993) 649]. [hep-ph/9212308]. http://www.arXiv.org/abs/[hep-ph/9212308]
108. Remiddi, E.: Nuovo Cim. **A110**, 1435–1452 (1997). [hep-th/9711188]. http://www.arXiv. org/abs/[hep-th/9711188]
109. Gehrmann, T., Remiddi, E.: Nucl. Phys. **B580**, 485–518 (2000). [hep-ph/9912329]. http:// www.arXiv.org/abs/[hep-ph/9912329]

110. Abramov, S.A., Petkovšek, M.: D'Alembertian solutions of linear differential and difference equations. In: von zur Gathen J. (ed.) Proceedings of ISSAC'94, pp. 169–174. ACM Press, New York (1994)

111. Ablinger, J.: PoS (RADCOR2017), 001 (2017). [arXiv:1801.01039 [cs.SC]]

112. Singer, M.F.: Am. J. Math. **103**(4), 661–682 (1981)

113. Kovacic, J.J.: J. Symb. Comput. **2**, 3–43 (1986)

114. Blümlein, J., Ablinger, J., Behring, A., De Freitas, A., von Manteuffel, A., Schneider, C.: PoS (QCDEV2017), 031 (2017). [arXiv:1711.07957 [hep-ph]]

115. Ablinger, J., Behring, A., Blümlein, J., Falcioni, G., De Freitas, A., Marquard, P., Rana, N., Schneider, C.: Phys. Rev. D **97**(9), 094022 (2018). [arXiv:1712.09889 [hep-ph]]

116. Ablinger, J., Blümlein, J., Marquard, P., Rana, N., Schneider, C.: Nucl. Phys. B **939**, 253–291 (2019). [arXiv:1810.12261 [hep-ph]]

117. Kotikov, A.V.: The property of maximal transcendentality in the N=4 supersymmetric Yang-Mills. In: Diakonov D. (ed.) Subtleties in quantum field theory, pp. 150–174. [arXiv:1005.5029 [hep-th]]

118. Henn, J.M.: Phys. Rev. Lett. **110**, 251601 (2013). [arXiv:1304.1806[hep-th]]. http://www.arXiv.org/abs/[arXiv:1304.1806[hep-th]]

119. Henn, J.M.: J. Phys. **A48**, 153001 (2015). [arXiv:1412.2296[hep-ph]]. http://www.arXiv.org/abs/[arXiv:1412.2296[hep-ph]]

120. Zakharov, V.E., Manakov, S.V., Novikov, S.P., Pitaevskii, L.P.: Teoria Solitonov: metod obratnoi zadatschi. Nauka, Moskva (1980)

121. Sakovich, S.Yu.: J. Phys. A Math. Gen. **28**, 2861–2869 (1995)

122. Lee, R.N.: JHEP **04**, 108 (2015). [arXiv:1411.0911[hep-ph]]. http://www.arXiv.org/abs/[arXiv:1411.0911[hep-ph]]

123. Prausa, M.: Comput. Phys. Commun. **219**, 361–376 (2017). [arXiv:1701.00725 [hep-ph]]. http://www.arXiv.org/abs/[arXiv:1701.00725[hep-ph]]

124. Gituliar, O., Magerya, V.: Comput. Phys. Commun. **219**, 329–338 (2017). [arXiv:1701.04269[hep-ph]]. http://www.arXiv.org/abs/[arXiv:1701.04269[hep-ph]]

125. Meyer, C.: Comput. Phys. Commun. **222**, 295–312 (2018). [arXiv:1705.06252[hep-ph]]. http://www.arXiv.org/abs/[arXiv:1705.06252[hep-ph]]

126. Ablinger, J., Behring, A., Blümlein, J., De Freitas, A., Hasselhuhn, A., von Manteuffel, A., Round, M., Schneider, C., Wißbrock, F.: Nucl. Phys. **B886**, 733–823 (2014). [arXiv:1406.4654[hep-ph]]. http://www.arXiv.org/abs/[arXiv:1406.4654[hep-ph]]

127. Ablinger, J., Behring, A., Blümlein, J., De Freitas, A., von Manteuffel, A., Schneider, C.: Nucl. Phys. **B890**, 48–151 (2014). [arXiv:1409.1135[hep-ph]]. http://www.arXiv.org/abs/[arXiv:1409.1135[hep-ph]]

128. Henn, J., Smirnov, A.V., Smirnov, V.A., Steinhauser, M.: JHEP **01**, 074 (2017). [arXiv:1611.07535[hep-ph]]. http://www.arXiv.org/abs/[arXiv:1611.07535[hep-ph]]

129. Kauers, M., Paule, P.: The concrete tetrahedron. In: Texts and Monographs in Symbolic Computation. Springer, Wien (2011)

130. Blümlein, J., Schneider, C.: Phys. Lett. **B771**, 31–36 (2017). [arXiv: 1701.04614 [hep-ph]]. http://www.arXiv.org/abs/[arXiv:1701.04614[hep-ph]]

131. Gorishnii, S.G., Larin, S.A., Surguladze, L.R., Tkachov, F.V., Comput. Phys. Commun. **55**, 381–408 (1989).
Larin, S.A., Tkachov, F.V., Vermaseren, J.A.M.: The FORM version of MINCER, NIKHEF-H-91-18

132. Steinhauser, M.: Comput. Phys. Commun. **134**, 335–364 (2001). [hep-ph/ 0009029]

133. Harlander, R., Seidensticker, T., Steinhauser, M.: Phys. Lett. **B426**, 125–132 (1998). [hep-ph/9712228]. http://www.arXiv.org/abs/[hep-ph/9712228]

134. Seidensticker, T.: Automatic application of successive asymptotic expansions of Feynman diagrams. In: Proceedings of the 6th International Workshop on New Computing Techniques in Physics Research, Crete, April (1999). hep-ph/9905298

135. Nörlund, N.E.: Vorlesungen über Differenzenrechnung. Springer, Berlin (1924). Reprinted by (Chelsea Publishing Company, New York, 1954)

136. Larin, S.A., van Ritbergen, T., Vermaseren, J.A.M.: Nucl. Phys. **B427**, 41–52 (1994)
137. Larin, S.A., Nogueira, P., van Ritbergen, T., Vermaseren, J.A.M.: Nucl. Phys. **B492**, 338–378 (1997). [hep-ph/9605317]. http://www.arXiv.org/abs/[hep-ph/9605317]
138. Retey, A., Vermaseren, J.A.M.: Nucl. Phys. **B604**, 281–311 (2001). [hep-ph/0007294]. http://www.arXiv.org/abs/[hep-ph/0007294]
139. Blümlein, J., Vermaseren, J.A.M.: Phys. Lett. **B606**, 130–138 (2005). [hep-ph/0411111]. http://www.arXiv.org/abs/[hep-ph/0411111]
140. Bierenbaum, I., Blümlein, J., Klein, S.: Nucl. Phys. **B820**, 417–482 (2009). [arXiv:0904.3563[hep-ph]]. http://www.arXiv.org/abs/[arXiv:0904.3563[hep-ph]]
141. Vermaseren, J.A.M., Vogt, A., Moch, S.: Nucl. Phys. **B724**, 3–182 (2005). [hep-ph/0504242]. http://www.arXiv.org/abs/[hep-ph/0504242]
142. Kauers, M., Jaroschek, M., Johansson, F.: In: Gutierrez J., Schicho J., Weimann M. (eds.) Computer Algebra and Polynomials. Lecture Notes in Computer Science, vol. 8942, pp. 105–125. Springer, Berlin (2015). [arXiv:1306.4263 [cs.SC]]
143. Sage. http://www.sagemath.org/
144. Blümlein, J., Kauers, M., Klein, S., Schneider, C.: Comput. Phys. Commun. **180**, 2143–2165 (2009). [arXiv:0902.4091[hep-ph]]. http://www.arXiv.org/abs/[arXiv:0902.4091[hep-ph]]
145. Moch, S., Vermaseren, J.A.M., Vogt, A.: Nucl. Phys. **B688**, 101–134 (2004). [hep-ph/0403192]. http://www.arXiv.org/abs/[hep-ph/0403192]
146. Vogt, A., Moch, S., Vermaseren, J.A.M.: Nucl. Phys. **B691**, 129–181 (2004). [hep-ph/0404111]. http://www.arXiv.org/abs/[hep-ph/0404111]
147. Ablinger, J., Behring, A., Blümlein, J., De Freitas, A., von Manteuffel, A., Schneider, C.: Nucl. Phys. **B922**, 1–40 (2017). [arxiv:1705.01508[hep-ph]]. http://www.arXiv.org/abs/[arxiv:1705.01508[hep-ph]]
148. Ablinger, J., Blümlein, J., Marquard, P., Rana, N., Schneider, C.: Phys. Lett. B **782**, 528–532 (2018). [arXiv:1804.07313 [hep-ph]]
149. Almkvist, G., Zeilberger, D.: J. Symb. Comput. **10**, 571–591 (1990)
150. Apagodu, M., Zeilberger, D.: Adv. Appl. Math. (Special Regev Issue) **37**, 139–152 (2006)
151. Ablinger, J.: Computer Algebra Algorithms for Special Functions in Particle Physics. Ph.D. Thesis, Linz U. (2012). arXiv:1305.0687[math-ph]
152. Broadhurst, D.J., Fleischer, J., Tarasov, O.V.: Z. Phys. **C60**, 287–302 (1993). [hep-ph/9304303]. http://www.arXiv.org/abs/[hep-ph/9304303]
153. Bloch, S., Vanhove, P.: J. Number Theor. **148**, 328–364 (2015). [hep-th/ 1309.5865]. http://www.arXiv.org/abs/[hep-th/1309.5865]
154. Laporta, S., Remiddi, E.: Nucl. Phys. **B704**, 349–386 (2005). [hep-ph/0406160]. http://www.arXiv.org/abs/[hep-ph/0406160]
155. Adams, L., Bogner, C., Weinzierl, S.: J. Math. Phys. **54**, 052303 (2013). [hep-ph/1302.7004]. http://www.arXiv.org/abs/[hep-ph/1302.7004]
156. Adams, L., Bogner, C., Weinzierl, S.: J. Math. Phys. **55**(10), 102301 (2014). [hep-ph/1405.5640]. http://www.arXiv.org/abs/[hep-ph/1405.5640]
157. Adams, L., Bogner, C., Weinzierl, S.: J. Math. Phys. **56**(7), 072303 (2015). [hep-ph/1504.03255]. http://www.arXiv.org/abs/[hep-ph/1504.03255]
158. Adams, L., Bogner, C., Weinzierl, S.: J. Math. Phys. **57**(3), 032304 (2016). [hep-ph/1512.05630]. http://www.arXiv.org/abs/[hep-ph/1512.05630]
159. Sabry, A.: Nucl. Phys. **33**, 401–430 (1962)
160. Remiddi, E., Tancredi, L.: Nucl. Phys. **B907**, 400–444 (2016). [arXiv:1602.01481[hep-ph]]. http://www.arXiv.org/abs/[arXiv:1602.01481[hep-ph]]
161. Adams, L., Bogner, C., Schweitzer, A., Weinzierl, S.: J. Math. Phys. **57**(12), 122302 (2016). [hep-ph/1607.01571]. http://www.arXiv.org/abs/[hep-ph/1607.01571]
162. Ablinger, J., Blümlein, J., De Freitas, A., van Hoeij, M., Imamoglu, E., Raab, C.G., Radu, C.S., Schneider, C.: J. Math. Phys. **59**(6), 062305 (2018). [arXiv: 1706.01299 [hep-th]]
163. Grigo, J., Hoff, J., Marquard, P., Steinhauser, M.: Nucl. Phys. B **864**, 580–596 (2012). [arXiv:1206.3418 [hep-ph]]

164. Blümlein, J., De Freitas, A., Van Hoeij, M., Imamoglu, E., Marquard, P., Schneider, C.: PoS LL **2018**, 017 (2018). [arXiv:1807.05287 [hep-ph]]
165. Ronveaux, A. (ed.): Heun's Differential Equations. The Clarendon Press Oxford, Oxford (1995)
166. Imamoglu, E., van Hoeij, M.: J. Symbolic Comput. **83**, 245–271 (2017). [arXiv:1606.01576 [cs.SC]]
167. Takeuchi, K.: J. Fac. Sci Univ. Tokyo, Sect. 1A **24**, 201–272 (1977)
168. Tricomi, F.G.: Elliptische Funktionen. Geest & Portig, Leipzig (1948). übersetzt und bearbeitet von M. Krafft
169. Whittaker, E.T., Watson, G.N.: A course of modern analysis. Cambridge University Press, Cambridge (1996). Reprint of 4th edition (1927)
170. Herfurtner, S.: Math. Ann. **291**, 319–342 (1991)
171. Movasati, H., Reiter, S.: Bull. Braz. Math Soc. **43**, 423–442 (2012). [arXiv: 0902.0760[math.AG]]
172. Blümlein, J.: Talks at: The 5th International Congress on Mathematical Software ZIB Berlin from July 11 to July 14, 2016, Session: Symbolic computation and elementary particle physics. https://www.risc.jku.at/conferences/ICMS2016/ (2016); and QCD@LHC2016, U. Zürich, August 22 to August 26, 2016. https://indico.cern.ch/event/516210/timetable/#all.detailed.
173. Remiddi, E., Tancredi, L.: Nucl. Phys. **B925**, 212–251 (2017). [arXiv: 1709.03622[hep-ph]]. http://www.arXiv.org/abs/[arXiv:1709.03622[hep-ph]]
174. Adams, L., Weinzierl, S.: Phys. Lett. **B781**, 270–278 (2018). [arXiv: 1802. 05020[hep-ph]]. http://www.arXiv.org/abs/[arXiv:1802.05020[hep-ph]]
175. Serre, J.-P.: A Course in Arithmetic. Springer, Berlin (1973)
176. Cohen, H., Strömberg, F.: Modular Forms, A Classical Approach. Graduate Studies in Mathematics, vol. 179. AMS, Providence (2017)
177. Ono, K.: The Web of Modularity: Arithmetic of the Coefficients of Modular Forms and q-series. CBMS Regional Conference Series in Mathematics, vol. 102. AMS, Providence (2004)
178. Chan, H.H., Zudilin, W.: Mathematika **56**, 107–117 (2010)
179. Broadhurst, D.J.: Eta quotients, Eichler integrals and L-series, talk at HMI Bonn, February (2018)
180. Brödel, J., Duhr, C., Dulat, F., Penante, B., Tancredi, L.: JHEP **1808**, 014 (2018). [arXiv:1803.10256 [hep-th]]
181. Adams, L., Weinzierl, S.: Commun. Numer. Theor. Phys. **12**, 193–251 (2018). [arXiv:1704.08895 [hep-ph]]
182. Bogner, C., Schweitzer, A., Weinzierl, S.: Nucl. Phys. B **922**, 528–550 (2017). [arXiv:1705.08952 [hep-ph]]
183. Ablinger, J., Blümlein, J., De Freitas, A., Goedicke, A., Schneider, C., Schönwald, K.: Nucl. Phys. B **932**, 129–240 (2018). [arXiv:1804.02226 [hep-ph]]
184. Blümlein, J. De Freitas, A., Raab, C., Schönwald, K.: The $O(\alpha^2)$ Initial State QED Corrections to $e^+e^- \rightarrow \gamma^*/Z_0^*$, [arXiv:2003.14289 [hep-ph]].
185. Ablinger, J., Blümlein, J., De Freitas, A., Schönwald, K.: Subleading Logarithmic QED Initial State Corrections to $e^+e^- \rightarrow \gamma^*/Z^{0*}$ to $O(\alpha^6 L^5)$, [arXiv:2004.04287 [hep-ph]].
186. Blümlein, J., De Freitas, A., Raab, C.G., Schönwald, K.: Phys. Lett. B **791**, 206–209 (2019). [arXiv:1901.08018 [hep-ph]]
187. Blümlein, J., De Freitas, A., Raab, C.G., Schönwald, K.: Nucl. Phys. B **945**, 114659 (2019). arXiv:1903.06155 [hep-ph]
188. Blümlein, J., Raab, C.G., Schönwald, K.: Nucl. Phys. B **948**, 114736 (2019). arXiv:1904.08911 [hep-ph]
189. Besier, M., Van Straten, D., Weinzierl, S.: Commun. Num. Theor. Phys. **13**, 253–297 (2018). arXiv:1809.10983 [hep-th]
190. Raab, C.G.: On the arithmetic of d'Alembertian functions (in preparation)
191. Guo, L., Regensburger, G., Rosenkranz, M.: J. Pure Appl. Algebra **218**, 456–473 (2014)

192. Ablinger, J., Blümlein, J., Raab, C.G., Schneider, C.: J. Math. Phys. **55**, 112301 (2014). [arXiv: 1407.1822[hep-th]]. http://www.arXiv.org/abs/[arXiv:1407.1822[hep-th]]
193. Hoffman, M.E.: J. Algebraic Combin. **11**, 49–68 (2000). [arXiv:math/9907173 [math.QA]]
194. Blümlein, J.: Comput. Phys. Commun. **159**, 19–54 (2004). [hep-ph/0311046]. http://www.arXiv.org/abs/[hep-ph/0311046]
195. Vermaseren, J.: Int. J. Mod. Phys. **A14**, 2037–2076 (1999). [hep-ph/9806280]. http://www.arXiv.org/abs/[hep-ph/9806280]
196. Blümlein, J., Kurth, S.: Phys. Rev. **D60**, 014018 (1999). [hep-ph/9810241]. http://www.arXiv.org/abs/[hep-ph/9810241]
197. Ablinger, J., Blümlein, J., Schneider, C.: J. Math. Phys. **52**, 102301 (2011). [arXiv: 1105.6063[math-ph]]. http://www.arXiv.org/abs/[arXiv:1105.6063[math-ph]]
198. Ablinger, J., Blümlein, J., Schneider, C.: J. Math. Phys. **54**, 082301 (2013). [arXiv: 1302.0378[math-ph]]. http://www.arXiv.org/abs/[arXiv:1302.0378[math-ph]]
199. Davydychev, A.I., Kalmykov, M.Yu.: Nucl. Phys. **B699**, 3–64 (2004). [hep-th/0303162]. http://www.arXiv.org/abs/[hep-th/0303162]
200. Weinzierl, S.: J. Math. Phys. **45**, 2656–2673 (2004). [hep-ph/0402131]. http://www.arXiv.org/abs/[hep-ph/0402131]
201. Reutenauer, C.: Free Lie Algebras. Calendron Press, Oxford (1993)
202. Nielsen, N.: Nova Acta Leopold. **XC**(3), 125–211 (1909)
203. Kölbig, K.S., Mignoco, J.A., Remiddi, E.: BIT **10**, 38–74 (1970)
204. Kölbig, K.S.: SIAM J. Math. Anal. **17**, 1232–1258 (1986)
205. Devoto, A., Duke, D.W.: Riv. Nuovo Cim. **7N6**, 1–39 (1984)
206. Lewin, L.: Dilogarithms and Associated Functions. Macdonald, London (1958)
207. Lewin, L.: Polylogarithms and Associated Functions. North Holland, New York (1981)
208. Remiddi, E., Vermaseren, J.A.M.: Int. J. Mod. Phys. **A15**, 725–754 (2000). [hep-ph/9905237]. http://www.arXiv.org/abs/[hep-ph/9905237]
209. Moch, S., Uwer, P., Weinzierl, S.: J. Math. Phys. **43**, 3363–3386 (2002). [hep-ph/0110083]. http://www.arXiv.org/abs/[hep-ph/0110083]
210. Laporta, S.: Phys. Lett. **B772**, 232–238 (2017). [arXiv:1704.06996[hep-ph]]. http://www.arXiv.org/abs/[arXiv:1704.06996[hep-ph]]
211. Blümlein, J., Vogt, A.: Phys. Rev. D **58**, 014020 (1998). [hep-ph/9712546]
212. Blümlein, J., Ravindran, V., van Neerven, W.L.: Nucl. Phys. B **586**, 349–381 (2000). [hep-ph/0004172]
213. Blümlein, J., Guffanti, A.: Nucl. Phys. Proc. Suppl. **152**, 87–91 (2006). [hep-ph/0411110]
214. Blümlein, J.: Comput. Phys. Commun. **133**, 76–104 (2000). [hep-ph/0003100]. http://www.arXiv.org/abs/[hep-ph/0003100]
215. Blümlein, J., Moch, S.-O.: Phys. Lett. **B614**, 53–61 (2005). [hep-ph/0503188]. http://www.arXiv.org/abs/[hep-ph/0503188]
216. Nielsen, N.: Handbuch der Theorie der Gammafunktion. Teubner, Leipzig (1906). Reprinted by (Chelsea Publishing Company, Bronx, New York, 1965)
217. Landau, E.: Über die Grundlagen der Theorie der Fakultätenreihen, S.-Ber. math.-naturw. Kl. Bayerische Akad. Wiss. München **36**, 151–218 (1906)
218. Blümlein, J.: Clay Math. Proc. **12**, 167–188 (2010). [arXiv:0901.0837[math-ph]]. http://www.arXiv.org/abs/[arXiv:0901.0837[math-ph]]
219. Blümlein, J.: Comput. Phys. Commun. **180**, 2218–2249 (2009). [arXiv:0901.3106[hep-ph]]. http://www.arXiv.org/abs/[arXiv:0901.3106[hep-ph]]
220. Kotikov, A.V., Velizhanin, V.N.: Analytic continuation of the Mellin moments of deep inelastic structure functions (2005). hep-ph/0501274
221. Ablinger, J.: PoS, 019 (2014). [arXiv:1407.6180[cs.SC]]. http://www.arXiv.org/abs/[arXiv: 1407.6180[cs.SC]]
222. Ablinger, J.: A computer algebra toolbox for harmonic sums related to particle physics. Diploma Thesis, JKU Linz (2009). arXiv:1011.1176[math-ph]
223. Weinzierl, S.: Comput. Phys. Commun. **145**, 357–370 (2002). [math-ph/0201011]. http://www.arXiv.org/abs/[math-ph/0201011]

224. Moch, S., Uwer, P.: Comput. Phys. Commun. **174**, 759–770 (2006). [math-ph/0508008]. http://www.arXiv.org/abs/[math-ph/0508008]
225. Frellesvig, H.: Generalized Polylogarithms in Maple, arXiv:1806.02883 [hep-th].
226. Duhr, C., Dulat, F.: PolyLogTools – Polylogs for the masses, JHEP **08**, 135 (2019). arXiv:1904.07279 [hep-th] H. Frellesvig, *Generalized Polylogarithms in Maple*, arXiv:1806.02883 [hep-th].
227. Gehrmann, T., Remiddi, E.: Comput. Phys. Commun. **141**, 296–312 (2001). [hep-ph/0107173]. http://www.arXiv.org/abs/[hep-ph/0107173]
228. Maitre, D.: Comput. Phys. Commun. **174**, 222–240 (2006). [hep-ph/0507152]. http://www.arXiv.org/abs/[hep-ph/0507152]
229. Ablinger, J., Blümlein, J., Round, M., Schneider, C.: PoS (RADCOR2017) 010 (2017). [arXiv:1712.08541[hep-th]]
230. Ablinger, J. Blümlein, J., Round M. Schneider, C.: %Numerical implementation of harmonic polylogarithms to weight w = 8, Comput. Phys. Commun. **240**, 189–201 (2019). doi:10.1016/j.cpc.2019.02.005 [arXiv:1809.07084 [hep-ph]].
231. Vollinga, J., Weinzierl, S.: Comput. Phys. Commun. **167**, 177 (2005). [hep-ph/0410259]. http://www.arXiv.org/abs/[hep-ph/0410259]
232. Neumann, C.: Vorlesungen über Riemann's Theorie der Abel'schen Integrale, 2nd edn. Teubner, Leipzig (1884)
233. Brown, F., Schnetz, O.: Duke Math. J. **161**(10), 1817–1862 (2012)
234. Hilbert, D.: Grundzüge einer allgemeinen Theorie der linearen Integralgleichungen. Teubner, Leipzig (1912)
235. Kronig, R. de L.: J. Opt. Soc. Am. **12**, 547–557 (1926)
236. Kramers, H.A.: Atti Cong. Intern. Fisici, (Transactions of Volta Centenary Congress) Como **2**, 545–557 (1927)
237. Abreu, S., Britto, R., Duhr, C., Gardi, E.: JHEP **12**, 090 (2017). [arXiv:1704.07931[hep-th]]. http://www.arXiv.org/abs/[arXiv:1704.07931[hep-th]]
238. Veltman, M.J.G.: Physica **29**, 186–207 (1963)
239. Remiddi, E.: Helv. Phys. Acta **54**, 364–382 (1982).
    Remiddi, E.: Differential equations and dispersion relations for feynman amplitudes. In: Blümlein J., Schneider C., Paule P. (eds.) Elliptic Integrals, Elliptic Functions and Modular Forms in Quantum Field Theory, pp. 391–414. Springer, Wien (2019)

An Eigenvalue Problem for the Associated Askey–Wilson Polynomials

Andrea Bruder, Christian Krattenthaler, and Sergei K. Suslov

Dedicated to Peter Paule on the occasion of his 60th birthday

1 Introduction

Throughout this paper, we use the standard notation for the q-shifted factorials:

$$(a; q)_n := \prod_{j=0}^{n-1} \left(1 - aq^j\right), \qquad (a_1, a_2, \ldots, a_r; q)_n := \prod_{k=1}^{r} (a_k; q)_n,$$

$$(a; q)_\infty := \lim_{n \to \infty} (a; q)_n, \qquad (a_1, a_2, \ldots, a_r; q)_\infty := \prod_{k=1}^{r} (a_k; q)_\infty,$$

provided $|q| < 1$. The basic hypergeometric series is defined by (cf. [9])

$$_r\varphi_s \left(\begin{array}{c} a_1, a_2, \ldots, a_r \\ b_1, \ldots, b_s \end{array} ; q, z \right) := \sum_{n=0}^{\infty} \frac{(a_1, a_2, \ldots, a_r; q)_n}{(q, b_1, b_2, \ldots, b_s; q)_n} \left((-1)^n q^{n(n-1)/2}\right)^{1+s-r} z^n.$$

A. Bruder
Department of Mathematics and Computer Science, Colorado College, Tutt Science Center, Colorado Springs, CO, USA
e-mail: abruder@coloradocollege.edu

C. Krattenthaler
Fakultät für Mathematik, Universität Wien, Vienna, Austria
e-mail: Christian.Krattenthaler@univie.ac.at

S. K. Suslov (✉)
School of Mathematical and Statistical Sciences, Arizona State University, Tempe, AZ, USA
e-mail: sks@asu.edu

© Springer Nature Switzerland AG 2020
V. Pillwein, C. Schneider (eds.), *Algorithmic Combinatorics: Enumerative Combinatorics, Special Functions and Computer Algebra*, Texts & Monographs in Symbolic Computation, https://doi.org/10.1007/978-3-030-44559-1_6

If $0 < |q| < 1$, the series converges absolutely for all z if $r \leq s$, and for $|z| < 1$ if $r = s + 1$.

The Askey–Wilson polynomials are the most general extension of the classical orthogonal polynomials [1–5, 11–13, 18]. They are most conveniently given in terms of a $_4\varphi_3$-series,

$$p_n(x) = p_n(x; a, b, c, d) = p_n(x; a, b, c, d|q)$$

$$= a^{-n} (ab, ac, ad; q)_n \, _4\varphi_3 \left(\begin{matrix} q^{-n}, \ abcdq^{n-1}, \ az, \ a/z \\ ab, \ ac, \ ad \end{matrix} ; q, q \right),$$

where $x = \left(z + z^{-1}\right)/2$, and $|z| < 1$. In this normalization, the Askey–Wilson polynomials are symmetric in all four parameters due to Sears' transformation [4].

The Askey–Wilson polynomials satisfy the 3-term recurrence relation

$$2x \, p_n(x; a, b, c, d) = A_n \, p_{n+1}(x; a, b, c, d) + B_n \, p_n(x; a, b, c, d)$$

$$+ \, C_n \, p_{n-1}(x; a, b, c, d), \tag{1}$$

where

$$A_n = \frac{a^{-1}(1 - abq^n)(1 - acq^n)(1 - adq^n)(1 - abcdq^{n-1})}{(1 - abcdq^{2n-1})(1 - abcdq^{2n} - q^{2n})}, \tag{2}$$

$$C_n = \frac{a(1 - bcq^{n-1})(1 - bdq^{n-1})(1 - cdq^{n-1})(1 - q^n)}{(1 - abcdq^{2n-1})(1 - abcdq^{2n})}, \tag{3}$$

$$B_n = a + a^{-1} - A_n - C_n. \tag{4}$$

The weight function with respect to which the polynomials $p_n(x)$ are orthogonal was found by Askey and Wilson in [4]. The Askey–Wilson divided difference operator is defined by

$$L(x)u := L\,(s; a, b, c, d)\,u\,(s)$$

$$= \frac{\sigma\,(-s)\,\nabla x\,(s)\,u\,(s+1) + \sigma\,(s)\,\Delta x\,(s)\,u\,(s-1)}{\Delta x\,(s)\,\nabla x\,(s)\,\nabla x_1\,(s)}$$

$$- \frac{[\sigma\,(s)\,\Delta x\,(s) + \sigma\,(-s)\,\nabla x\,(s)]\,u\,(s)}{\Delta x\,(s)\,\nabla x\,(s)\,\nabla x_1\,(s)}, \tag{5}$$

where $\sigma\,(s) = q^{-2s}\,(q^s - a)\,(q^s - b)\,(q^s - c)\,(q^s - d)$ and, by definition,

$$x(s) = \frac{1}{2}\left(q^s + q^{-s}\right), \qquad\qquad x_1(s) = x\left(s + \frac{1}{2}\right),$$

$$\Delta f(s) = f(s+1) - f(s), \qquad\qquad \nabla f(s) = f(s) - f(s-1).$$

(We follow the notation in [7] and [8].) We will make use of an analogue of the power series expansion method, where a function is expanded in terms of generalized powers. For a positive integer m, the generalized powers are defined by

$$[x(s) - x(z)]^{(m)} = \prod_{k=0}^{m-1} [x(s) - x(z-k)], \qquad x_n(z) = x\left(z + \frac{n}{2}\right) \qquad (6)$$

(see [17, Exercises 2.9–2.11, 2.25] and [16] for more details).

2 The Associated Askey–Wilson Polynomials

The associated Askey–Wilson polynomials,

$$p_n^\alpha(x) = p_n^\alpha(x; a, b, c, d) = p_n^\alpha(x; a, b, c, d | q),$$

were introduced by Ismail and Rahman in [10]. They are solutions of the 3-term recurrence relation

$$2x \, p_n^\alpha(x; a, b, c, d) = A_{n+\alpha} \, p_{n+1}^\alpha(x; a, b, c, d) + B_{n+\alpha} \, p_n^\alpha(x; a, b, c, d)$$
$$+ C_{n+\alpha} \, p_{n-1}^\alpha(x; a, b, c, d), \qquad (7)$$

where $0 < \alpha < 1$, with initial values $p_{-1}^\alpha(x) = 0$, $p_0^\alpha(x) = 1$, and $A_{n+\alpha}$, $B_{n+\alpha}$, $C_{n+\alpha}$ are given as in (2)–(4) with n replaced by $n+\alpha$. The two linearly independent solutions to (1) found in [10] are

$$R_{n+\alpha} = \frac{(abq^{n+\alpha}, acq^{n+\alpha}, adq^{n+\alpha}, bcdq^{n+\alpha}/z; q)_\infty}{(bcq^{n+\alpha}, bdq^{n+\alpha}, cdq^{n+\alpha}, azdq^{n+\alpha}; q)_\infty} \left(\frac{a}{z}\right)^{n+\alpha}$$
$$\times \, {}_8W_7(bcd/qz; b/z, c/z, d/z, abcdq^{n+\alpha-1}, q^{-\alpha-n}; q, qz/a) \qquad (8)$$

and

$$S_{n+\alpha} = \frac{(abcdq^{2n+2\alpha}, bzq^{n+\alpha+1}, czq^{n+\alpha+1}, dzq^{n+\alpha+1}, bcdzq^{n+\alpha+1}; q)_\infty}{(bcq^{n+\alpha}, bdq^{n+\alpha}, cdq^{n+\alpha}, q^{n+\alpha+1}, bcdzq^{2n+2\alpha+1}; q)_\infty} (az)^{n+\alpha}$$
$$\times \, {}_8W_7(bcdzq^{2n+2\alpha}; bcq^{n+\alpha}, bdq^{n+\alpha}, cdq^{n+\alpha}, q^{n+\alpha+1}, zq/a; q, az). \qquad (9)$$

The weight function for the associated Askey–Wilson polynomials and an explicit polynomial representation were found by Ismail and Rahman in [10]. The latter is given by

$$p_n^\alpha(x) = p_n^\alpha(x; a, b, c, d|q)$$

$$= \sum_{k=0}^{n} \frac{(q^{-n}, abcdq^{2\alpha+n-1}, abcdq^{2\alpha-1}, ae^{i\theta}, ae^{-i\theta}; q)_k}{(q, abq^\alpha, acq^\alpha, adq^\alpha, abcdq^{\alpha-1}; q)_k} q^k$$

$$\times {}_{10}W_9(abcdq^{2\alpha+k-1}; q^\alpha, bcq^{\alpha-1}, bdq^{\alpha-1}, cdq^{\alpha-1}, q^{k+1},$$

$$abcdq^{2\alpha+n+k-1}, q^{k-n}; q, a^2). \tag{10}$$

There is another useful representation of the associated Askey–Wilson polynomials in terms of a double series due to Rahman,

$$p_n^\alpha(x) = p_n^\alpha(x; a, b, c, d|q)$$

$$= \frac{(abcdq^{2\alpha-1}, q^{\alpha+1}; q)_n}{(q, abcdq^{\alpha-1}; q)_n} q^{-\alpha n} \sum_{k=0}^{n} \frac{(q^{-n}, abcdq^{2\alpha+n-1}; q)_k}{(q^{\alpha+1}, abq^\alpha; q)_k} q^k \tag{11}$$

$$\times \frac{(aq^\alpha e^{i\theta}, aq^\alpha e^{-i\theta}; q)_k}{(acq^\alpha, acq^\alpha; q)_k} \sum_{j=0}^{k} \frac{(q^\alpha, abq^{\alpha-1}, acq^{\alpha-1}, adq^{\alpha-1}; q)_j}{(q, abcdq^{2\alpha-2}, aq^\alpha e^{i\theta}, aq^\alpha e^{-i\theta}; q)_j} q^j,$$

where $x = \cos\theta$ (see [9, Exercises 8.26–8.27] and [13–15]). This formula will be the starting point for our investigation.

3   An Overview of the Main Result

To construct an eigenvalue problem for the associated Askey–Wilson polynomials, let us consider an auxiliary function $u_n^\alpha(x, y)$ in two variables, which for $x = y$ coincides with the associated Askey–Wilson polynomials (up to a factor). We observe that the Askey–Wilson operator $L_0(x)$ (in one variable x) maps $u_n^\alpha(x, y)$ to the n-th degree ordinary Askey–Wilson polynomial (up to some factors). A similar result is obtained for the operator $L_1(y)$ applied to $u_n^\alpha(x, y)$ with respect to the second independent variable y. We will find an operator $L_2(x)$, which maps certain multiples of $(L_1(y) + \lambda) u_n^\alpha(x, y)$ to $(L_0(x) + \lambda)u_n^\alpha(x, y)$. As a result, we obtain an eigenvalue problem of the form

$$\frac{(aq^s, aq^{-s}; q)_\infty}{(aq^{\alpha+s-1}, aq^{\alpha-s-1}; q)_\infty}(L_2(x) + \lambda)\frac{(aq^{\alpha+s}, aq^{\alpha-s}; q)_\infty}{(aq^s, aq^{-s}; q)_\infty}(L_1(y) + \mu_\alpha)u_n^\alpha(x, y)$$

$$= \frac{4q^{9/2}}{(1-q)^2\gamma}(L_0(x) + \lambda_{\alpha+n})u_n^\alpha(x, y) \tag{12}$$

related to the associated Askey–Wilson polynomials of Ismail and Rahman (see Theorem 1 below for an exact statement).

We shall use the normalization

$$p_n(x; a, b, c, d) = {}_4\varphi_3\left(\begin{matrix} q^{-n}, \ abcdq^{n-1}, \ aq^s, \ aq^{-s} \\ ab, \ ac, \ ad \end{matrix}; q, q\right), \quad x = \left(q^s + q^{-s}\right)/2).$$

(13)

for the ordinary Askey–Wilson polynomials throughout this paper.

Lemma 1 *Let $u_n^\alpha(x, y)$ be the function in the two variables x and y defined by*

$$u_n^\alpha(x, y) := \frac{(aq^s, aq^{-s}, aq^{\alpha+z}, aq^{\alpha-z}; q)_\infty}{(aq^{\alpha+s}, aq^{\alpha-s}, aq^z, aq^{-z}; q)_\infty}$$

$$\times \sum_{m=0}^{n} \frac{(q^{-n}, \gamma q^{2\alpha+n-1}, aq^{\alpha+s}, aq^{\alpha-s}; q)_m}{(q^{\alpha+1}, abq^\alpha, acq^\alpha, adq^\alpha; q)_m} q^m$$

$$\times \sum_{k=0}^{m} \frac{(q^\alpha, abq^{\alpha-1}, acq^{\alpha-1}, adq^{\alpha-1}; q)_k}{(q, \gamma q^{2\alpha-2}, aq^{\alpha+z}, aq^{\alpha-z}; q)_k} q^k,$$

(14)

with $x(s) = (q^s + q^{-s})/2$ and $y(z) = (q^z + q^{-z})/2$. Then $u_n^\alpha(x, y)$ satisfies an equation of the form

$$(L_0(x) + \lambda_{\alpha+n})u_n^\alpha(x, y) = f_n^\alpha(x, y),$$

(15)

where $L_0(x) = L(s; a, b, c, d)$ is the Askey–Wilson divided difference operator in the variable x given by (5). Here,

$$f_n^\alpha(x, y) = -\frac{4q^{3/2-\alpha}}{(1-q)^2} \frac{(aq^s, aq^{-s}, aq^{\alpha+z}, aq^{\alpha-z}; q)_\infty}{(aq^{\alpha+s-1}, aq^{\alpha-s-1}, aq^z, aq^{-z}; q)_\infty}$$

$$\times (q^\alpha, abq^{\alpha-1}, acq^{\alpha-1}, adq^{\alpha-1}; q)_1$$

$$\times p_n(x; aq^{\alpha-1}, bcdq^{\alpha-1}, q^{1+z}, q^{1-z}),$$

and

$$\lambda_{\alpha+n} = \frac{4q^{3/2}}{(1-q)^2}\left(1 - q^{-\alpha-n}\right)\left(1 - \gamma q^{\alpha+n-1}\right), \qquad \gamma = abcd.$$

Note that $f_n^\alpha(x, y)$ contains the n-th degree ordinary Askey–Wilson polynomial of the form (13) in the variable x. Our function $u_n^\alpha(x, y)$ is the Askey–Wilson polynomial when $\alpha = 0$ and a constant multiple of the associated Askey–Wilson polynomial if $x = y$.

Lemma 2 *The function $u_n^\alpha(x, y)$ satisfies another equation, namely*

$$(L_1(y) + \mu_\alpha)u_n^\alpha(x, y) = g_n^\alpha(x, y),$$

where $L_1(y) := L(y; q/a, q/b, q/c, q/d)$ is the Askey–Wilson divided difference operator in y.

Here,

$$
g_n^\alpha(x, y) = -\frac{4q^{9/2-\alpha}}{(1-q)^2\gamma} \frac{(aq^s, aq^{-s}, aq^{\alpha+z+1}, aq^{\alpha-z+1}; q)_\infty}{(aq^{\alpha+s}, aq^{\alpha-s}, aq^z, aq^{-z}; q)_\infty}
$$
$$
\times (q^\alpha, abq^{\alpha-1}, acq^{\alpha-1}, adq^{\alpha-1}; q)_1
$$
$$
\times p_n(x; aq^\alpha, bcdq^{\alpha-2}, q^{1+z}, q^{1-z})
$$

and

$$
\mu_\alpha = \frac{4q^{3/2}}{(1-q)^2}\left(1 - q^\alpha\right)\left(1 - q^{3-\alpha}/\gamma\right).
$$

Note that $g_n^\alpha(x, y)$ contains another n-th degree Askey–Wilson polynomial (13) in the same variable x.

Lemma 3 *The difference differentiation formula*

$$(L(x) + \lambda)p_n(x; a, b, c, d) = \lambda p_n(x; a/q, bq, c, d) \tag{16}$$

holds for the Askey–Wilson polynomials given by (13). *Here,* $L(x) = L(s; a, a/q, c, d)$ *is the Askey–Wilson divided difference operator* (5) *and*

$$
\lambda = \frac{4q^{3/2}}{(1-q)^2}(1 - ac/q)(1 - ad/q).
$$

Lemmas 1–3 allow us to establish the eigenvalue problem (12) for the associated Askey–Wilson functions (14), see the next section.

4   Main Result

With the help of Lemmas 1–3, we now identify an operator $L_2(x)$ linking $(L_0(x) + \lambda_{\alpha+n})u_n^\alpha(x, y)$ and $(L_1(y) + \lambda_{-\alpha})u_n^\alpha(x, y)$ in such a way that an eigenvalue problem is formulated.

Theorem 1 *Let* $L_2(x) = L(s; aq^\alpha, aq^{\alpha-1}, q^{1+z}, q^{1-z})$ *be the Askey–Wilson divided difference operator defined by* (5) *with*

$$\sigma(s) = q^{-2s} \left(q^s - aq^\alpha\right) \left(q^s - aq^{\alpha-1}\right) \left(q^s - q^{1+z}\right) \left(q^s - q^{1-z}\right)$$

and

$$\lambda = \frac{4q^{3/2}}{(1-q)^2} \left(1 - aq^{\alpha-z}\right) \left(1 - aq^{\alpha+z}\right).$$

Then an eigenvalue problem for the associated Askey–Wilson functions $u_n^\alpha(x, y)$ *can be stated as*

$$\frac{\gamma}{q^3} \frac{(aq^s, aq^{-s}; q)_\infty}{(aq^{\alpha+s-1}, aq^{\alpha-s-1}; q)_\infty} (L_2(x) + \lambda) \frac{(aq^{\alpha+s}, aq^{\alpha-s}; q)_\infty}{(aq^s, aq^{-s}; q)_\infty} (L_1(y) + \mu_\alpha) u_n^\alpha(x, y)$$

$$= \frac{4q^{3/2}}{(1-q)^2} (L_0(x) + \lambda_{\alpha+n}) u_n^\alpha(x, y), \qquad (17)$$

where L_0, L_1, $\lambda_{\alpha+n}$, μ_α *and* $u_n^\alpha(x, y)$ *are defined as in Lemmas* 1–3.

Computational details are left to the reader. The explicit form of the difference operator in two variables on the left-hand side of the last equation has also been calculated, but it is too long to be displayed here.

5 Proofs

Proof of Lemma 1 Let λ_ν be an arbitrary number. We are looking for solutions of a generalization of the Eq. (15), namely,

$$(L_0(x) + \lambda_\nu)u_n^\alpha(x, y) = f_n^\alpha(x, y),$$

in terms of generalized powers (see (6) for the definition)

$$u_n^\alpha(x, y) = \sum_{m=0}^n c_m v_m [x(s) - x(\xi)]^{(\alpha+m)},$$

where

$$v_m = v_m(y) = \frac{(aq^{\alpha+z}, aq^{\alpha-z}; q)_\infty}{(aq^z, aq^{-z}; q)_\infty} \sum_{k=0}^n \frac{(q^\alpha, abq^{\alpha-1}, acq^{\alpha-1}, adq^{\alpha-1}; q)_k}{(q, \gamma q^{2\alpha-2}, aq^{\alpha+z}, aq^{\alpha-z}; q)_k} q^k,$$

and $\gamma = abcd$. (This is an analogue of the power series expansion; see [7], [16, Exercises 2.9–2.11], and [17] for properties of the generalized powers.)

Apply the Askey–Wilson operator to $u_n^\alpha(x, y)$ to obtain

$$(L_0(x) + \lambda_v)u_n^\alpha(x, y) = \lambda_v \sum_{m=0}^{n} c_m v_m [x(s) - x(\xi)]^{(\alpha+m)}$$

$$+ \sum_{m=0}^{n} c_m v_m \, L_0(x)[x(s) - x(\xi)]^{(\alpha+m)},$$

since v_m is independent of x. By [7], we have

$$L_0(x)[x(s) - x(\xi)]^{(\alpha+m)} = \gamma(\alpha + m)\gamma(\alpha + m - 1)\sigma(\xi - \alpha - m + 1)$$

$$\times [x(s) - x(\xi - 1)]^{(\alpha+m-2)}$$

$$+ \gamma(\alpha + m)\tau_{\alpha+m-1}(\xi - \alpha - m + 1)[x(s) - x(\xi - 1)]^{(\alpha+m-1)}$$

$$- \lambda_{\alpha+m}[x(s) - x(\xi)]^{(\alpha+m)}.$$

We use the same notations as in [7], [16, Exercise 2.25], or [17]. Choose $a_0 := \xi - \alpha - m + 1$ to be a root of the equation $\sigma(a_0) = 0$. Then $\xi = a_0 + \alpha + m - 1$, and one obtains

$$(L_0(x) + \lambda_v)u_n^\alpha(x, y) = \sum_{m=0}^{n} c_m v_m \gamma(\alpha + m)\tau_{\alpha+m-1}(a_0)$$

$$\times [x(s) - x(a_0 + \alpha + m - 2)]^{(\alpha+m-1)}$$

$$+ \sum_{m=0}^{n} c_m v_m (\lambda_v - \lambda_{\alpha+m})[x(s) - x(a_0 + \alpha + m - 1)]^{(\alpha+m)}$$

$$= c_0 v_0 \gamma(\alpha)\tau_{\alpha-1}(a_0)[x(s) - x(a_0 + \alpha - 2)]^{(\alpha-1)}$$

$$+ \sum_{m=1}^{n} c_m v_m \gamma(\alpha + m)\tau_{\alpha+m-1}(a_0)[x(s) - x(a_0 + \alpha + m - 2)]^{(\alpha+m-1)}$$

$$+ \sum_{m=0}^{n} c_m v_m (\lambda_v - \lambda_{\alpha+m})[x(s) - x(a_0 + \alpha + m - 1)]^{(\alpha+m)}.$$

Letting $m = k + 1$, we get

$$(L_0(x) + \lambda_v)u_n^\alpha(x, y) = c_0 v_0 \gamma(\alpha)\tau_{\alpha-1}(a_0)[x(s) - x(a_0 + \alpha - 2)]^{(\alpha-1)}$$

$$+ \sum_{k=0}^{n-1} c_{k+1} v_{k+1} \gamma(\alpha + k + 1)\tau_{\alpha+k}(a_0)[x(s) - x(a_0 + \alpha + k - 1)]^{(\alpha+k)}$$

$$+ \sum_{k=0}^{n} c_k v_k (\lambda_\nu - \lambda_{\alpha+k})[x(s) - x(a_0 + \alpha + k - 1)]^{(\alpha+k)}. \qquad (18)$$

Note that for

$$v_k = \sum_{l=0}^{k} e_l, \qquad e_l := \frac{(aq^{\alpha+z}, aq^{\alpha-z}; q)_\infty}{(aq^z, aq^{-z}; q)_\infty} \frac{(q^\alpha, abq^{\alpha-1}, acq^{\alpha-1}, adq^{\alpha-1}; q)_l}{(q, \gamma q^{2\alpha-2}, aq^{\alpha+z}, aq^{\alpha-z}; q)_l} q^l$$

one has

$$v_{k+1} = v_k + e_{k+1} \quad \text{and} \quad v_0 = e_0.$$

After choosing $\lambda_\nu = \lambda_{\alpha+n}$, Eq. (18) becomes

$$(L_0(x) + \lambda_{\alpha+n})u_n^\alpha(x, y) =$$

$$\sum_{k=-1}^{n-1} c_{k+1}e_{k+1}\gamma(\alpha + k + 1)\tau_{\alpha+k}(a_0)[x(s) - x(a_0 + \alpha + k - 1)]^{(\alpha+k)}$$

$$+ \sum_{k=0}^{n-1} c_{k+1}v_k\gamma(\alpha + k + 1)\tau_{\alpha+k}(a_0)[x(s) - x(a_0 + \alpha + k - 1)]^{(\alpha+k)}$$

$$+ \sum_{k=0}^{n-1} c_k v_k (\lambda_{\alpha+n} - \lambda_{\alpha+k})[x(s) - x(a_0 + \alpha + k - 1)]^{(\alpha+k)}.$$

The latter two sums vanish if

$$c_{k+1}\gamma(\alpha + k + 1)\tau_{\alpha+k}(a_0) = c_k(\lambda_{\alpha+n} - \lambda_{\alpha+k}).$$

Therefore,

$$(L_0(x) + \lambda_{\alpha+n})u_n^\alpha(x, y)$$

$$= \sum_{k=-1}^{n-1} c_{k+1}e_{k+1}\gamma(\alpha + k + 1)\tau_{\alpha+k}(a_0)[x(s) - x(a_0 + \alpha + k - 1)]^{(\alpha+k)}$$

$$= \sum_{m=0}^{n} c_m e_m \gamma(\alpha+m)\tau_{\alpha+m-1}(a_0)[x(s)-x(a_0+\alpha+m-2)]^{(\alpha+m-1)} =: f_n^\alpha(x, y).$$

Finally, we show that the function $f_n^\alpha(x, y)$ is, up to a factor, the n-th ordinary Askey–Wilson polynomial. The generalized powers have the property (see [16])

$$[x(s) - x(z)]^{(n+1)} = [x(s) - x(z)][x(s) - x(z - 1)]^{(n)},$$

which leads to

$$f_n^\alpha(x, y) = \sum_{m=0}^{n} c_m e_m \gamma(\alpha + m) \tau_{\alpha+m-1}(a_0) \frac{[x(s) - x(a_0 + \alpha + m - 1)]^{(\alpha+m)}}{[x(s) - x(a_0 + \alpha + m - 1)]}.$$

Moreover,

$$c_m [x(s) - x(a_0 + \alpha + m - 1)]^{(\alpha+m)}$$

$$= c_0 \frac{(q^{-n}, \gamma q^{2\alpha+n-1}; q)_m}{(q^{\alpha+1}, abq^\alpha, acq^\alpha, adq^\alpha; q)_m} q^m \, [x(s) - x(a_0 + \alpha + m - 1)]^{(\alpha+m)}$$

$$= c_0 \, \varphi_m(x) \, [x(s) - x(a_0 + \alpha - 1)]^{(\alpha)},$$

where, by definition,

$$\varphi_m(x) := \frac{(aq^s, aq^{-s}; q)_\infty}{(aq^{\alpha+s}, aq^{\alpha-s}; q)_\infty} \frac{(q^{-n}, \gamma q^{2\alpha+n-1}, aq^{\alpha+s}, aq^{\alpha-s}; q)_m}{(q^{\alpha+1}, abq^\alpha, acq^\alpha, adq^\alpha; q)_m} q^m.$$

Therefore,

$$f_n^\alpha(x, y) = \frac{(aq^s, aq^{-s}; q)_\infty}{(aq^{\alpha+s}, aq^{\alpha-s}; q)_\infty} \frac{(q^{-n}, \gamma q^{2\alpha+n-1}, aq^{\alpha+s}, aq^{\alpha-s}; q)_m}{(q^{\alpha+1}, abq^\alpha, acq^\alpha, adq^\alpha; q)_m} q^m$$

$$\times \frac{(aq^{\alpha+z}, aq^{\alpha-z}; q)_\infty}{(aq^z, aq^{-z}; q)_\infty} \frac{(q^\alpha, abq^{\alpha-1}, acq^{\alpha-1}, adq^{\alpha-1}; q)_m}{(q, \gamma q^{2\alpha-2}, aq^{\alpha+z}, aq^{\alpha-z}; q)_m} q^m$$

$$\times \frac{\gamma(\alpha + m)\tau_{\alpha+m-1}(a_0)}{[x(s) - x(a_0 + \alpha + m - 1)]}. \tag{19}$$

Recall that $a = q^{a_0}$ and

$$\gamma(\alpha + m) = q^{-\frac{\alpha+m-1}{2}} \frac{1 - q^{\alpha+m}}{1 - q},$$

$$x(s) - x(a_0 + \alpha + m - 1) = -\frac{1}{2a} q^{-\alpha-m+1}(1 - aq^{\alpha-s+m-1})(1 - aq^{\alpha+s+m-1}),$$

$$\tau_{\alpha+m-1}(a_0) = \frac{2}{a(1 - q)} q^{-2(\alpha+m-1)+\frac{\alpha+m}{2}}(1 - abq^{\alpha+m-1})(1 - acq^{\alpha+m-1})(1 - adq^{\alpha+m-1}),$$

which allows us to simplify the last term of (19) to

$$q^m \frac{\gamma(\alpha + m)\tau_{\alpha+m-1}(a_0)}{[x(s) - x(a_0 + \alpha + m - 1)]}$$

$$= -4q^{\frac{3}{2}-\alpha} \frac{1 - q^{\alpha+m}}{1-q} \frac{(1 - abq^{\alpha+m-1})(1 - acq^{\alpha+m-1})(1 - adq^{\alpha+m-1})}{(1 - aq^{\alpha-s+m-1})(1 - aq^{\alpha+s+m-1})}.$$

Thus $f_n^\alpha(x, y)$ becomes

$$f_n^\alpha(x, y) = \frac{-4q^{\frac{3}{2}-\alpha}}{(1-q)^2} \frac{(aq^s, aq^{-s}, aq^{\alpha+z}, aq^{\alpha-z}; q)_\infty}{(aq^{\alpha+s-1}, aq^{\alpha-s-1}, aq^z, aq^{-z}; q)_\infty} \tag{20}$$

$$\times (q^\alpha, abq^{\alpha-1}, acq^{\alpha-1}, adq^{\alpha-1}; q)_1$$

$$\times \sum_{m=0}^{n} \frac{(q^{-n}, \gamma q^{2\alpha+n-1}, aq^{\alpha+s-1}, aq^{\alpha-s-1}; q)_m}{(q, \gamma q^{2\alpha-2}, aq^{\alpha+z}, aq^{\alpha-z}; q)_m} q^m$$

$$= \frac{-4q^{\frac{3}{2}-\alpha}}{(1-q)^2} \frac{(aq^s, aq^{-s}, aq^{\alpha+z}, aq^{\alpha-z}; q)_\infty}{(aq^{\alpha+s-1}, aq^{\alpha-s-1}, aq^z, aq^{-z}; q)_\infty}$$

$$\times (q^\alpha, abq^{\alpha-1}, acq^{\alpha-1}, adq^{\alpha-1}; q)_1$$

$$\times p_n(x; aq^{\alpha-1}, bcdq^{\alpha-1}, q^{1+z}, q^{1-z}),$$

which completes the proof of the lemma.

Proof of Lemma 2 Consider the equation

$$(L_1(y) + \lambda_v)u_n^\alpha(x, y) = g_n^\alpha(x, y),$$

and rewrite $u_n^\alpha(x, y)$ in the form

$$u_n^\alpha(x, y) = \sum_{m=0}^{n} c_m^\alpha (aq^{\alpha+s}, aq^{\alpha-s}; q)_m \frac{(aq^s, aq^{-s}; q)_\infty}{(aq^{\alpha+s}, aq^{\alpha-s}; q)_\infty} v_m^\alpha(y),$$

where

$$c_m^\alpha = \frac{(q^{-n}, \gamma q^{2\alpha+n-1}; q)_m}{(q^{\alpha+1}, abq^\alpha, acq^\alpha, adq^\alpha; q)_m} q^m, \qquad \gamma = abcd,$$

and

$$v_m^\alpha(y) = \frac{(aq^{\alpha+z}, aq^{\alpha-z}; q)_\infty}{(aq^z, aq^{-z}; q)_\infty} \sum_{k=0}^{m} \frac{(q^\alpha, abq^{\alpha-1}, acq^{\alpha-1}, adq^{\alpha-1}; q)_k}{(q, \gamma q^{2\alpha-2}, aq^{\alpha+z}, aq^{\alpha-z}; q)_k} q^k.$$

Apply the Askey–Wilson operator $L_1(y) := L(y; q/a, q/b, q/c, q/d)$ to $u_n^\alpha(x, y)$ to obtain

$$(L_1(y) + \lambda_\nu)u_n^\alpha(x, y)$$

$$= \sum_{m=0}^{n} c_m^\alpha (aq^{\alpha+s}, aq^{\alpha-s}; q)_m \frac{(aq^s, aq^{-s}; q)_\infty}{(aq^{\alpha+s}, aq^{\alpha-s}; q)_\infty} (L_1(y) + \lambda_\nu) v_m^\alpha(y).$$

Let

$$v_m^\alpha(y) := \sum_{k=0}^{m} \frac{c_k}{[x(s) - x(\xi)]^{(\alpha+k)}}$$

in analogy with [7]. Then

$$(L_1(y) + \lambda_\nu)v_m^\alpha(y) = \lambda_\nu \sum_{k=0}^{m} \frac{c_k}{[x(s) - x(\xi)]^{(\alpha+k)}} + \sum_{k=0}^{m} c_k L_1(y) \left(\frac{1}{[x(s) - x(\xi)]^{(\alpha+k)}} \right).$$

By [7], we have

$$L_1(y) \left(\frac{1}{[x(s) - x(\xi)]^{(\alpha+k)}} \right) = \frac{\gamma(\alpha+k)\gamma(\alpha+k+1)\sigma(\xi+1)}{[x(z) - x(\xi+1)]^{(\alpha+k+2)}}$$

$$- \frac{\gamma(\alpha+k)\tau_{-\alpha-k-1}(\xi+1)}{[x(z) - x(\xi)]^{(\alpha+k+1)}} - \frac{\lambda_{-\alpha-k}}{[x(z) - x(\xi)]^{(\alpha+k)}}$$

(see also [16, Exercise 2.25]). Upon choosing $a_0 := \xi+1$ to be a root of the equation $\sigma(a_0) = 0$, we obtain

$$(L_1(y) + \lambda_\nu)v_m^\alpha(y) = \lambda_\nu \sum_{k=0}^{m} \frac{c_k}{[x(s) - x(a_0)]^{(\alpha+k)}}$$

$$- \sum_{k=0}^{m} c_k \left(\frac{\gamma(\alpha+k)\tau_{-\alpha-k-1}(a_0)}{[x(z) - x(a_0-1)]^{(\alpha+k+1)}} + \frac{\lambda_{-\alpha-k}}{[x(z) - x(a_0-1)]^{(\alpha+k)}} \right)$$

$$= \sum_{k=0}^{m} \frac{c_k (\lambda_\nu - \lambda_{-\alpha-k})}{[x(z) - x(a_0-1)]^{(\alpha+k)}} - \sum_{k=0}^{m} \frac{c_k \gamma(\alpha+k)\tau_{-\alpha-k-1}(a_0)}{[x(z) - x(a_0-1)]^{(\alpha+k+1)}}$$

$$= \frac{c_0 (\lambda_\nu - \lambda_{-\alpha})}{[x(z) - x(a_0-1)]^{(\alpha)}} + \sum_{k=1}^{m} \frac{c_k (\lambda_\nu - \lambda_{-\alpha-k})}{[x(z) - x(a_0-1)]^{(\alpha+k)}}$$

$$- \frac{c_m \gamma(\alpha+m)\tau_{-\alpha-m-1}(a_0)}{[x(z) - x(a_0-1)]^{(\alpha+m+1)}} - \sum_{k=0}^{m-1} \frac{c_k \gamma(\alpha+k)\tau_{-\alpha-k-1}(a_0)}{[x(z) - x(a_0-1)]^{(\alpha+k+1)}}.$$

Now choose $\lambda_\nu = \lambda_{-\alpha}$ and let $k = l + 1$. Then we obtain

$$(L_1(y) + \lambda_\nu)v_m^\alpha(y) = -\frac{c_m\,\gamma(\alpha + m)\tau_{-\alpha-m-1}(a_0)}{[x(z) - x(a_0 - 1)]^{(\alpha+m+1)}}$$

$$+ \sum_{l=0}^{m-1} \frac{c_{l+1}\,(\lambda_{-\alpha} - \lambda_{-\alpha-l-1})}{[x(z) - x(a_0 - 1)]^{(\alpha+l+1)}} - \sum_{l=0}^{m-1} \frac{c_l\,\gamma(\alpha + l)\tau_{-\alpha-l-1}(a_0)}{[x(z) - x(a_0 - 1)]^{(\alpha+l+1)}}.$$

The latter two sums vanish if

$$c_{l+1}\,(\lambda_{-\alpha} - \lambda_{-\alpha-l-1}) = c_l\,\gamma(\alpha + l)\tau_{-\alpha-l-1}(a_0).$$

In that case, we have

$$(L_1(y) + \lambda_\nu)v_m^\alpha(y) = -\frac{c_m\,\gamma(\alpha + m)\tau_{-\alpha-m-1}(a_0)}{[x(z) - x(a_0 - 1)]^{(\alpha+m+1)}}$$

$$= -\frac{c_{m+1}\,(\lambda_{-\alpha} - \lambda_{-\alpha-m-1})}{[x(z) - x(a_0 - 1)]^{(\alpha+m+1)}} =: h_m^\alpha(y).$$

Here,

$$\frac{c_{m+1}}{[x(z) - x(a_0 - 1)]^{(\alpha+m+1)}} = \frac{c_0}{[x(z) - x(a_0 - 1)]^{(\alpha)}}\varphi_{m+1}(z),$$

$$\varphi_{m+1}(z) = \frac{(q^\alpha, abq^{\alpha-1}, acq^{\alpha-1}, adq^{\alpha-1}; q)_{m+1}}{(q, \gamma q^{2\alpha-2}, aq^{\alpha+z}, aq^{\alpha-z}; q)_{m+1}}q^{m+1},$$

$$\frac{c_0}{[x(z) - x(a_0 - 1)]^{(\alpha)}} = \frac{(aq^{\alpha+z}, aq^{\alpha-z}; q)_\infty}{(aq^z, aq^{-z}; q)_\infty},$$

$$\lambda_{-\alpha} - \lambda_{-\alpha-m-1} = \frac{4}{(1-q)^2\gamma}q^{\frac{7}{2}-\alpha-m}(1 - q^{m+1})(1 - \gamma q^{2\alpha+m-2})$$

and

$$h_m^\alpha(y) = -\frac{4q^{\frac{9}{2}-\alpha}}{(1-q)^2\gamma}\frac{(aq^{\alpha+z}, aq^{\alpha-z}; q)_\infty}{(aq^z, aq^{-z}; q)_\infty}\frac{(q^\alpha, abq^{\alpha-1}, acq^{\alpha-1}, adq^{\alpha-1}; q)_{m+1}}{(q, \gamma q^{2\alpha-2}; q)_m(aq^{\alpha+z}, aq^{\alpha-z}; q)_{m+1}}.$$

Therefore,

$$(L_1(y) + \lambda_\nu)u_n^\alpha(x, y)$$

$$= \sum_{m=0}^{n} c_m^\alpha(aq^{\alpha+s}, aq^{\alpha-s}; q)_m\frac{(aq^s, aq^{-s}; q)_\infty}{(aq^{\alpha+s}, aq^{\alpha-s}; q)_\infty}L_1(y)v_m^\alpha(y)$$

$$= -\sum_{m=0}^{n} c_m^{\alpha}(aq^{\alpha+s}, aq^{\alpha-s}; q)_m \frac{(aq^s, aq^{-s}; q)_{\infty}}{(aq^{\alpha+s}, aq^{\alpha-s}; q)_{\infty}}$$

$$\times \frac{4q^{\frac{9}{2}-\alpha}}{(1-q)^2\gamma} \frac{(aq^{\alpha+z}, aq^{\alpha-z}; q)_{\infty}}{(aq^z, aq^{-z}; q)_{\infty}} \frac{(q^{\alpha}, abq^{\alpha-1}, acq^{\alpha-1}, adq^{\alpha-1}; q)_{m+1}}{(q, \gamma q^{2\alpha-2}; q)_m (aq^{\alpha+z}, aq^{\alpha-z}; q)_{m+1}}$$

$$= -\frac{4q^{\frac{9}{2}-\alpha}}{(1-q)^2\gamma} \frac{(aq^s, aq^{-s}, aq^{\alpha+z+1}, aq^{\alpha-z+1}; q)_{\infty}}{(aq^{\alpha+s}, aq^{\alpha-s}, aq^z, aq^{-z}; q)_{\infty}}$$

$$\times \sum_{m=0}^{n} \frac{(q^{-n}, \gamma q^{2\alpha+n-1}, aq^s, aq^{-s}; q)_m}{(q^{\alpha+1}, abq^{\alpha}, acq^{\alpha}, adq^{\alpha}; q)_m} q^m$$

$$\times \frac{(1-q^{\alpha})(1-abq^{\alpha-1})(1-acq^{\alpha-1})(1-adq^{\alpha-1})}{(q, \gamma q^{2\alpha-2}; q)_m (aq^{\alpha+z+1}, aq^{\alpha-z+1}; q)_m}$$

$$= -\frac{4q^{\frac{9}{2}-\alpha}}{(1-q)^2\gamma} \frac{(aq^s, aq^{-s}, aq^{\alpha+z+1}, aq^{\alpha-z+1}; q)_{\infty}}{(aq^{\alpha+s}, aq^{\alpha-s}, aq^z, aq^{-z}; q)_{\infty}}$$

$$\times \left(q^{\alpha}, abq^{\alpha-1}, acq^{\alpha-1}, adq^{\alpha-1}; q\right)_1$$

$$\times \;_4\varphi_3\left(\begin{matrix} q^{-n}, \gamma q^{2\alpha+n-1}, aq^{\alpha+s}, aq^{\alpha-s} \\ \gamma q^{2\alpha-2}, aq^{\alpha+z+1}, aq^{\alpha-z+1} \end{matrix}; q, q\right)$$

$$= -\frac{4q^{\frac{9}{2}-\alpha}}{(1-q)^2\gamma} \frac{(aq^s, aq^{-s}, aq^{\alpha+z+1}, aq^{\alpha-z+1}; q)_{\infty}}{(aq^{\alpha+s}, aq^{\alpha-s}, aq^z, aq^{-z}; q)_{\infty}}$$

$$\times \left(q^{\alpha}, abq^{\alpha-1}, acq^{\alpha-1}, adq^{\alpha-1}; q\right)_1 \times p_n(x; aq^{\alpha}, bcdq^{\alpha-2}, q^{1+z}, q^{1-z}).$$

This completes the proof of the lemma.

Proof of Lemma 3 The structure of the Askey–Wilson operator in (5) and the basic hypergeometric series representation (13) suggest to look for a 4-term relation of the form

$$K_1 \;_4\varphi_3\left(\begin{matrix} A, B, C, D \\ F, G, H \end{matrix}; q, q\right) + K_2 \;_4\varphi_3\left(\begin{matrix} A, B, Cq, D/q \\ F, G, H \end{matrix}; q, q\right)$$

$$+ K_3 \;_4\varphi_3\left(\begin{matrix} A, B, C/q, Dq \\ F, G, H \end{matrix}; q, q\right) + K_4 \;_4\varphi_3\left(\begin{matrix} A, B, C/q, D/q \\ F, G/q, H/q \end{matrix}; q, q\right) = 0, \quad (21)$$

for some undetermined coefficients K_1, K_2, K_3 and K_4 (up to a common factor). Doing a term-wise comparison, we may hope to find K_1, K_2, K_3, K_4 which satisfy

$$K_1(1-C)(1-D)(1-Cq^{k-1})(1-Dq^{k-1})(1-G/q)(1-H/q)$$

$$+ K_2(1-Cq^k)(1-Cq^{k-1})(1-D/q)(1-D)(1-G/q)(1-H/q)$$

$$+ K_3(1 - Dq^k)(1 - Dq^{k-1})(1 - C/q)(1 - C)(1 - G/q)(1 - H/q)$$

$$+ K_4(1 - C/q)(1 - C)(1 - D/q)(1 - D)(1 - Gq^{k-1})(1 - Hq^{k-1}) = 0.$$

If we are successful, then the above equation does indeed imply the contiguous relation (21). In the equation, we compare coefficients of powers of q^k. This yields a system of 3 linear equations in the 4 unknowns K_1, K_2, K_3, K_4. With the help of *Mathematica*, we obtain the solution

$$K_1 = \frac{(C - q)(D - q)(-GH - CDq + CGq + DGq + CHq + DHq - GHq - CDq^2)}{(G - q)(H - q)(Cq - D)(Dq - C)},$$

$$K_2 = \frac{(C - 1)(D - G)(D - H)(C - q)q}{(D - C)(G - q)(H - q)(Cq - D)},$$

$$K_3 = \frac{(D - 1)(C - G)(C - H)(D - q)q}{(C - D)(G - q)(H - q)(Dq - C)},$$

where the free parameter K_4 was chosen to be 1 (see Appendix 1 for the *Mathematica* code). The required 4-term contiguous relation is then given by

$$\frac{(C - q)(D - q)(-GH - CDq + CGq + DGq + CHq + DHq - GHq - CDq^2)}{(G - q)(H - q)(Cq - D)(Dq - C)}$$

$$\times \, {}_4\varphi_3\left(\begin{matrix} A, B, C, D \\ F, G, H \end{matrix}; q, q\right) + {}_4\varphi_3\left(\begin{matrix} A, B, C/q, D/q \\ F, G/q, H/q \end{matrix}; q, q\right)$$

$$+ \frac{(C - 1)(D - G)(D - H)(C - q)q}{(D - C)(G - q)(H - q)(Cq - D)} \, {}_4\varphi_3\left(\begin{matrix} A, B, Cq, D/q \\ F, G, H \end{matrix}; q, q\right)$$

$$+ \frac{(D - 1)(C - G)(C - H)(D - q)q}{(C - D)(G - q)(H - q)(Dq - C)} \, {}_4\varphi_3\left(\begin{matrix} A, B, C/q, Dq \\ F, G, H \end{matrix}; q, q\right) = 0.$$

$$(22)$$

(This 4-term contiguous relation for the ${}_4\varphi_3$-functions can be extended to an arbitrary ${}_r\psi_s$-function, see Appendix 1 for more details.)

When $qABCD = FGH$, in view of the structure of the Askey–Wilson operator in (5), Eq. (22) should become

$$(L(x) + \lambda) \, {}_4\varphi_3\left(\begin{matrix} A, B, Cq, D/q \\ F, G, H \end{matrix}; q, q\right)$$

$$= \frac{\sigma(-s)}{\Delta x(s)\nabla x_1(s)} \, {}_4\varphi_3\left(\begin{matrix} A, B, Cq, D/q \\ F, G, H \end{matrix}; q, q\right) + \frac{\sigma(s)}{\nabla x(s)\nabla x_1(s)} \, {}_4\varphi_3\left(\begin{matrix} A, B, C/q, Dq \\ F, G, H \end{matrix}; q, q\right)$$

$$+ \frac{\lambda\Delta x(s)\nabla x(s)\nabla x_1(s) - \sigma(s)\Delta x(s) - \sigma(-s)\nabla x(s)}{\Delta x(s)\nabla x(s)\nabla x_1(s)} \, {}_4\varphi_3\left(\begin{matrix} A, B, C, D \\ F, G, H \end{matrix}; q, q\right).$$

Equating coefficients, one obtains

$$(1-C/q)(1-D/q)(D-C)(GH+CDq-CGq-DGq-CHq-DHq+GHq+CDq^2)$$

$$= \frac{2qa^3}{1-q}\left(\sigma(-s)\nabla x(s) + \sigma(s)\Delta x(s) - \lambda\Delta x(s)\nabla x(s)\nabla x_1(s)\right)$$

and

$$(D-C)(D-C/q)(-C+D/q) = \frac{2aq^{1/2}}{1-q}\nabla x_1(s)\frac{2a}{1-q}\Delta x(s)\frac{2a}{1-q}\nabla x(s),$$

$$(C-1)(G-D)(H-D)(q-C) = -qa^2\sigma(-s),$$

$$(D-C)(-C+D/q) = \frac{2aq^{1/2}}{1-q}\nabla x_1(s)\frac{2a}{1-q}\Delta x(s),$$

$$(D-1)(G-C)(H-C)(q-D) = -qa^2\sigma(s),$$

$$(D-C)(D-C/q) = \frac{2aq^{1/2}}{1-q}\nabla x_1(s)\frac{2a}{1-q}\nabla x(s),$$

$$(G-q)(H-q) = q^2\frac{(1-q)^2}{4q^{3/2}}\lambda.$$

This gives the required formula (16) for the Askey–Wilson operator with

$$\sigma(s) = q^{-2s}\left(q^s - a\right)\left(q^s - a/q\right)\left(q^s - c\right)\left(q^s - d\right),$$

$$\lambda = \frac{4q^{3/2}}{(1-q)^2}\left(1 - ac/q\right)\left(1 - ad/q\right).$$

The proof of the lemma is complete.

Acknowledgements The contents of the paper were discussed with Mizan Rahman. We are very grateful to the referee for her/his valuable suggestions which helped us to improve the presentation. This research was partially supported by the Austrian Science Foundation FWF, grants Z130-N13 and S9607-N13, the latter in the framework of the National Research Network "Analytic Combinatorics and Probabilistic Number Theory."

Appendix 1: Four-Term Contiguous Relations

In order to derive the contiguous relation (22), one can use the following *Mathematica* program[1]:

```
In[1]:= X1 = K1*(1 - C) (1 - D) (1 - C*K/q) (1 - D*K/q) (1 - G/q)
(1 - H/q)
        + K2*(1 - C*K) (1 - C*K/q) (1 - D/q) (1 - D) (1 - G/q)
(1 - H/q)
        + K3*(1 - D*K) (1 - D*K/q) (1 - C/q) (1 - C) (1 - G/q)
(1 - H/q)
        + K4*(1 - C/q) (1 - C) (1 - D/q) (1 - D) (1 - G*K/q) (1
- H*K/q);
       X1 = Table[Coefficient[X1, K, i] == 0, i, 0, 2];
       X1 = Solve[X1, K1, K2, K3, K4];
       X1 = {K1 -> Factor[K1/.X1[[1]]], K2 -> Factor[K2/.X1[[1]]],
         K3 -> Factor[K3/.X1[[1]]], K4 -> Factor[K4/.X1[[1]]]}

Out[1]= {K1 -> (K4 (C - q) (D - q)
>         (G H + C D q - C G q - D G q - C H q - D H q + G H q + C D
q^2)) /
>            ((G - q) (H - q) (-D + C q) (C - D q)),
>       K2 -> - ((-1 + C) (D - G) (D - H) K4 (C - q) q) /
>            ((C - D) (G - q) (H - q) (-D + C q)),
>       K3 -> - ((-1 + D) (C - G) (C - H) K4 (D - q) q) /
>            ((C - D) (G - q) (H - q) (C - D q)),
>       K4 -> K4}
```

It is evident from the proof of (22) that, actually, an extension for bilateral series (see [9, equation (5.1.1)] for the definition) with an arbitrary number of parameters holds, namely:

$$\frac{(c - q)(d - q)\left(-gh - cdq + cgq + dgq + chq + dhq - ghq - cdq^2\right)}{(g - q)(h - q)(cq - d)(dq - c)}$$

$$\times \; _r\psi_s\left(\begin{matrix} a_1, \ldots, a_i, \; c, d \\ b_0, \ldots, b_k, \; g, h \end{matrix}; q, t\right)$$

---

[1] A corresponding *Mathematica* notebook is available on the article's website
http://www.mat.univie.ac.at/~kratt/artikel/AssAWPols.html.

$$+ \frac{(c-1)(d-g)(d-h)(c-q)q}{(d-c)(g-q)(h-q)(cq-d)} \; {}_r\psi_s \begin{pmatrix} a_1, \ldots, a_i, \; cq, d/q \\ b_0, \ldots, b_k, \; g, h \end{pmatrix}; q, t$$

$$+ \frac{(d-1)(c-g)(c-h)(d-q)q}{(c-d)(g-q)(h-q)(dq-c)} \; {}_r\psi_s \begin{pmatrix} a_1, \ldots, a_i, \; c/q, dq \\ b_0, \ldots, b_k, \; g, h \end{pmatrix}; q, t$$

$$+ \; {}_r\psi_s \begin{pmatrix} a_1, \ldots, a_i, \; c/q, d/q \\ b_0, \ldots, b_k, \; g/q, h/q \end{pmatrix}; q, t = 0. \tag{23}$$

Furthermore, in the same way, the following variation can be obtained[2]:

$$\frac{(g-1)(h-1)\left(-gh - cdq + cgq + dgq + chq + dhq - ghq - cdq^2\right)}{(c-1)(d-1)(gq-h)(hq-g)}$$

$$\times \; {}_r\psi_s \begin{pmatrix} a_1, \ldots, a_i, \; c, d \\ b_0, \ldots, b_k, \; g, h \end{pmatrix}; q, t$$

$$+ \frac{(c-g)(d-g)(h-1)(h-q)}{(c-1)(d-1)(h-g)(gq-h)} \; {}_r\psi_s \begin{pmatrix} a_1, \ldots, a_i, \; c, d \\ b_0, \ldots, b_k, \; gq, h/q \end{pmatrix}; q, t$$

$$+ \frac{(c-h)(d-h)(g-1)(g-q)}{(c-1)(d-1)(g-h)(hq-g)} \; {}_r\psi_s \begin{pmatrix} a_1, \ldots, a_i, \; c, d \\ b_0, \ldots, b_k, \; g/q, hq \end{pmatrix}; q, t$$

$$+ \; {}_r\psi_s \begin{pmatrix} a_1, \ldots, a_i, \; cq, dq \\ b_0, \ldots, b_k, \; gq, hq \end{pmatrix}; q, t = 0. \tag{24}$$

Appendix 2: An Inverse of the Askey–Wilson Operator

The Askey–Wilson divided difference operator on the left-hand side of Eq. (16) can be inverted by the method of Ref. [6]. The end result is

$$\frac{(q, q^2; q)_\infty}{2\pi} \int_{-1}^{1} L(x, y) \; p_n(x; a, b, c, d) \, \rho(x; a, b, c, d) \; dx$$

$$= p_n(x; aq, b/q, c, d), \tag{25}$$

---

[2] Again, a corresponding *Mathematica* notebook is available on the article's website http://www.mat.univie.ac.at/\lower0.5ex\hbox~{}kratt/artikel/AssAWPols.html.

where $\rho\,(x;a,b,c,d)$ is the weight function of the Askey–Wilson polynomials (13) and the kernel is given by

$$
\begin{aligned}
L\,(x,y) = {} & \left(ac, ad, qce^{i\varphi}, qde^{-i\varphi}; q\right)_1 \\
& \times \frac{\left(be^{i\theta}, be^{-i\theta}, qde^{i\theta}, qde^{-i\theta}, qae^{i\varphi}, qae^{-i\varphi}, qce^{i\varphi}, qce^{-i\varphi}; q\right)_\infty}{\left(qe^{i\theta+i\varphi}, qe^{i\theta-i\varphi}, qe^{i\varphi-i\theta}, qe^{-i\theta-i\varphi}; q\right)_\infty} \\
& \times {}_8\varphi_7\!\left(\begin{matrix} qde^{-i\varphi}, q\sqrt{qde^{-i\varphi}}, -q\sqrt{qde^{-i\varphi}}, qe^{i\theta-i\varphi}, qe^{-i\theta-i\varphi}, qd/c, q \\ \sqrt{qde^{-i\varphi}}, \sqrt{qde^{-i\varphi}}, qde^{-i\theta}, qde^{i\theta}, q^2, qce^{-i\varphi}, qde^{-i\varphi} \end{matrix}; q,\, ce^{i\varphi}\right).
\end{aligned}
$$

Here, $x = \cos\theta$ and $y = \cos\varphi$. Computational details are left to the reader.

References

1. Andrews, G.E., Askey, R.A.: Classical orthogonal polynomials. In: Polynômes orthogonaux et applications. Lecture Notes in Mathematics, vol. 1171, pp. 36–62. Springer, Berlin (1985)
2. Andrews, G.E., Askey, R.A., Roy, R.: Special Functions. Cambridge University Press, Cambridge (1999)
3. Askey, R.A.: Orthogonal Polynomials and Special Functions. CBMS–NSF Regional Conferences Series in Applied Mathematics, SIAM, Philadelphia, PA (1975)
4. Askey, R.A., Wilson, J.A.: Some basic hypergeometric orthogonal polynomials that generalize Jacobi polynomials. Mem. Am. Math. Soc. **319**, 55pp. (1985)
5. Askey, R.A., Koornwinder, T.H., Rahman, M.: An integral of products of ultraspherical functions and a q-extension. J. Lond. Math. Soc. **33** #2, 133–148 (1986)
6. Askey, R.A., Rahman, M., Suslov, S.K.: On a general q-Fourier transformation with nonsymmetric kernels. J. Comput. Appl. Math. **68**, 25–55 (1996)
7. Atakishiyev, N.M., Suslov, S.K.: Difference hypergeometric functions. In: Gonchar, A.A., Saff, E.B. (eds.) Progress in Approximation Theory, pp. 1–35. Springer, New York (1992)
8. Atakishiyev, N.M., Suslov, S.K.: On the Askey-Wilson polynomials. Constr. Approx. **8**, 363–369 (1992)
9. Gasper, G., Rahman, M.: Basic Hypergeometric Series, 2nd edn. Encyclopedia of Mathematics and Its Applications, vol. 96. Cambridge University Press, Cambridge (2004)
10. Ismail, M.E.H., Rahman, M.: The associated Askey-Wilson polynomials. Trans. Amer. Math. Soc. **328**, 201–237 (1991)
11. Koekoek, R., Swarttouw, R.F.: The Askey scheme of hypergeometric orthogonal polynomials and its q-analogues. Report 94–05, Delft University of Technology, 1994
12. Nikiforov, A.F., Suslov, S.K., Uvarov, V.B.: Classical Orthogonal Polynomials of a Discrete Variable. Nauka, Moscow (1985) [in Russian]; English translation, Springer, Berlin (1991)
13. Rahman, M.: Askey-Wilson functions of the first and second kind: series and integral representations of $C_n^2(x; \beta|q) + D_n^2(x; \beta|q)$. J. Math. Anal. Appl. **164**, 263–284 (1992)
14. Rahman, M.: Askey-Wilson functions of the first and second kind: series and integral representations of $C(x; 6|q) + D(x, 0|q)$. J. Comput. Appl. Math. Anal. **68**, 287–296 (1996)
15. Rahman, M.: The associated classical orthogonal polynomials. In: Bustoz, J., Ismail, M.E.H., Suslov, S.K. (eds.) Special Functions 2000. Springer, Boston (2001)

16. Suslov, S.K.: The theory of difference analogues of special functions of hypergeometric type. Rus. Math. Surv. **44**, 227–278 (1989)
17. Suslov, S.K.: An Introduction to Basic Fourier Series. Spinger Series "Developments in Mathematics", vol. 9. Springer, Boston (2003)
18. Szegő, G.: Orthogonal Polynomials, 4th edn. American Mathematical Society Colloquium Publications, vol. 23. American Mathematical Society, Providence RI (1975)

Context-Free Grammars and Stable Multivariate Polynomials over Stirling Permutations

William Y. C. Chen, Robert X. J. Hao, and Harold R. L. Yang

Dedicated to Professor Peter Paule on the occasion of his 60th birthday

1 Introduction

This paper presents an approach to the construction of stable combinatorial polynomials from the perspective of context-free grammars. The framework of using context-free grammars to generate combinatorial polynomials was proposed in [9]. We find context-free grammars leading to stable multivariate polynomials over Legendre-Stirling permutations and marked Stirling permutations. These stable multivariate polynomials provide solutions to two problems raised by Haglund and Visontai [16] in their study of stable multivariate refinements of the second-order Eulerian polynomials.

Let us first give an overview of the second-order Eulerian polynomials. These polynomials were defined by Gessel and Stanley [13] as the generating functions of the descent statistic over Stirling permutations. Let $[n]_2$ denote the multiset $\{1, 1, 2, 2, \ldots, n, n\}$. A permutation $\pi = \pi_1 \pi_2 \cdots \pi_{2n-1} \pi_{2n}$ of $[n]_2$ is called a Stirling permutation if π satisfies the following condition: if $\pi_i = \pi_j$ then $\pi_k > \pi_i$

W. Y. C. Chen (✉)
Center for Applied Mathematics, Tianjin University, Tianjin, People's Republic of China
e-mail: chenyc@tju.edu.cn

R. X. J. Hao
Department of Mathematics and Physics, Nanjing Institute of Technology, Nanjing, Jiangsu, People's Republic of China
e-mail: haoxj@njit.edu.cn

H. R. L. Yang
School of Science, Tianjin University of Technology and Education, Tianjin, People's Republic of China
e-mail: yangruilong@mail.nankai.edu.cn

© Springer Nature Switzerland AG 2020
V. Pillwein, C. Schneider (eds.), *Algorithmic Combinatorics: Enumerative Combinatorics, Special Functions and Computer Algebra*, Texts & Monographs in Symbolic Computation, https://doi.org/10.1007/978-3-030-44559-1_7

whenever $i < k < j$. For $1 \leq i \leq 2n$, we say that i is a descent of π if $i = 2n$
or $1 \leq i < 2n$ and $\pi_i > \pi_{i+1}$. Analogously, i is called an ascent of π if $i = 1$ or
$1 < i \leq 2n$ and $\pi_{i-1} < \pi_i$. For the sake of consistency, we set $\pi_0 = \pi_{2n+1} = 0$.
Let Q_n denote the set of Stirling permutations on $[n]_2$. Let $C(n, k)$ be the number
of Stirling permutations of $[n]_2$ with k descents, and let

$$C_n(x) = \sum_{k=1}^{n} C(n, k)x^k.$$

Gessel and Stanley [13] showed that

$$\sum_{n=0}^{\infty} S(n + k, k)x^n = \frac{C_n(x)}{(1 - x)^{2k+1}},$$

where $S(n, k)$, as usual, denotes the Stirling number of the second kind. The numbers $C(n, k)$ are called the second-order Eulerian numbers by Graham et al. [14],
and the polynomials $C_n(x)$ are called the second-order Eulerian polynomials by
Haglund and Visontai [16]. Besides the connection with the enumeration of Stirling
permutations, the second-order Eulerian number $C(n, k)$ has other combinatorial
interpretations, such as the number of Riordan trapezoidal words of length n with k
distinct letters [23], the number of rooted plane trees on $n + 1$ nodes with k leaves
[18] and the number of matchings on $2n$ vertices with $n - k$ left-nestings [20].

The Stirling permutations were further studied by Bóna [1], Brenti [8], Janson
[18] and Janson et al. [19]. Bóna [1] introduced the notion of a plateau of a Stirling
permutation and studied the plateau statistic. Given a Stirling permutation $\pi =
\pi_1\pi_2 \ldots \pi_{2n} \in Q_n$, an index $1 < i \leq 2n$ is called a plateau of π if $\pi_{i-1} = \pi_i$.
Bóna showed that the number of ascents, the number of descents and the number
of plateaux have the same distribution over Q_n. Analogous to real-rootedness of the
classical Eulerian polynomials, Bóna [1] proved the real-rootedness of the second-
order Eulerian polynomials $C_n(x)$.

Theorem 1 (Bóna [1]) *For any positive integer n, the roots of the polynomial
$C_n(x)$ are all real, distinct, and non-positive.*

It should be noted that the real-rootedness of $C_n(x)$ is essentially equivalent to
the real-rootedness of the generating function of generalized Stirling permutations
obtained by Brenti [8]. A permutation π of the multiset $\{1^{r_1}, 2^{r_2}, \ldots, n^{r_n}\}$ is called
a generalized Stirling permutation of rank n if π satisfies the same betweenness
condition for a Stirling permutation. Let Q_n^* denote the set of generalized Stirling
permutations of rank n. In particular, if $r_1 = r_2 = \cdots = r_n = r$ for some r,
then π is called an r-Stirling permutation of order n. Let $Q_n(r)$ denote the set
of r-Stirling permutations of order n. It is clear that 1-Stirling permutations are
ordinary permutations and 2-Stirling permutations are Stirling permutations. Brenti
[8] showed that the descent generating polynomials over Q_n^* have only real roots.

Janson [18] defined the following trivariate generating function

$$C_n(x, y, z) = \sum_{\pi \in Q_n} x^{\text{des}(\pi)} y^{\text{asc}(\pi)} z^{\text{plat}(\pi)},$$

where $\text{des}(\pi)$, $\text{asc}(\pi)$, and $\text{plat}(\pi)$ denote the number of descents, the number of ascents, and the number of plateaux of π, respectively, and proved that $C_n(x, y, z)$ is symmetric in x, y, z. This implies the equidistribution of these three statistics derived by Bóna [1].

The symmetry property of $C_n(x, y, z)$ was further extended to r-Stirling permutations by Janson et al. [19]. For an r-Stirling permutation, they introduced the notion of a j-plateau. For an r-Stirling permutation $\pi = \pi_1 \pi_2 \ldots \pi_{nr}$ and an integer $1 \le j \le r - 1$, a number $1 \le i < nr$ is called a j-plateau of π if $\pi_i = \pi_{i+1}$ and there are $j - 1$ indices $l < i$ such that $\pi_l = \pi_i$, i.e., the number π_i appears j times up to the i-th position of π. Let j-plat(π) denote the number of j-plateaux of π. Define a descent and an ascent of π as in the case of ordinary permutations, and let $\text{des}(\pi)$ and $\text{asc}(\pi)$ denote the number of descents and ascents of π. Janson et al. [19] showed that the distribution of (des, 1-plat, 2-plat, ..., $(r - 1)$-plat, asc) is symmetric over the set of r-Stirling permutations.

Based on the theory of stable multivariate polynomials recently developed by Borcea and Brändén [3–5], Haglund and Visontai [16] presented a unified approach to the stability of the generating functions of Stirling permutations and r-Stirling permutations. A polynomial $f(z_1, z_2, \ldots, z_n) \in \mathbb{C}[z_1, z_2, \ldots, z_n]$ is said to be stable, if whenever the imaginary part $\text{Im}(z_i) > 0$ for all i then $f(z_1, z_2, \ldots, z_n) \ne 0$. Clearly, a univariate polynomial $f(z) \in \mathbb{R}[z]$ has only real roots if and only if it is stable.

For the case of univariate real polynomials, Pólya and Schur [22] characterized all diagonal operators preserving stability or real-rootedness. Recently, Borcea and Brändén [3–5] characterized all linear operators preserving stability of multivariate polynomials, see also the survey by Wagner [26]. This implies a characterization of linear operators preserving stability of univariate polynomials.

A multivariate polynomial is called multiaffine if the degree of each variable is at most 1. Borcea and Brändén [4] showed that each of the operators preserving stability of multiaffine polynomials has a simple form. Using this property, Haglund and Visontai [16] obtained a stable multiaffine refinement of the second-order Eulerian polynomial $C_n(x)$. Similar methods are employed for other related combinatorial structures, see [2, 6, 7, 15, 24, 25] for a few of other instances.

Given a Stirling permutation $\pi = \pi_1 \pi_2 \cdots \pi_{2n} \in Q_n$, let

$$A(\pi) = \{i \,|\, \pi_{i-1} < \pi_i, 1 \le i \le 2n\},$$

$$D(\pi) = \{i \,|\, \pi_i > \pi_{i+1}, 1 \le i \le 2n\},$$

$$P(\pi) = \{i \,|\, \pi_{i-1} = \pi_i, 1 \le i \le 2n\}$$

denote the set of ascents, the set of descents and the set of plateaux of π, respectively. We set $\pi_0 = \pi_{2n+1} = 0$. Let $X = (x_1, x_2, \ldots, x_n)$, $Y = (y_1, y_2, \ldots, y_n)$ and $Z = (z_1, z_2, \ldots, z_n)$. Define

$$C_n(X, Y, Z) = \sum_{\pi \in Q_n} \prod_{i \in D(\pi)} x_{\pi_i} \prod_{i \in A(\pi)} y_{\pi_i} \prod_{i \in P(\pi)} z_{\pi_i}.$$

Haglund and Visontai [16] proved the stability of $C_n(X, Y, Z)$.

Theorem 2 (Haglund and Visontai [16]) *The polynomial $C_n(X, Y, Z)$ is stable.*

It is worth mentioning that, as observed by Haglund and Visontai [16], the recurrence relation between $C_{n-1}(X, Y, Z)$ and $C_n(X, Y, Z)$ can be used to derive the symmetry of $C_n(X, Y, Z)$, which implies the symmetry of $C_n(x, y, z)$ obtained by Janson et al. [19].

Moreover, Haglund and Visontai [16] extended the stability of $C_n(X, Y, Z)$ to generating polynomials of r-Stirling permutations by taking the j-plateau statistic into consideration. Let $P_j(\pi)$ denote the set of j-plateaux of π. For $i = 1, 2, \ldots, r - 1$, let $Z_i = (z_{i,1}, z_{i,2}, \ldots, z_{i,n})$. Haglund and Visontai [16] obtained the following stable multivariate polynomial over r-Stirling permutations

$$E_n(X, Y, Z_1, \ldots, Z_{r-1}) = \sum_{\pi \in Q_n(r)} \prod_{i \in D(\pi)} x_{\pi_i} \prod_{i \in A(\pi)} y_{\pi_i} \prod_{j=1}^{r-1} \prod_{i \in P_j(\pi)} z_{j,\pi_i}.$$

They also obtained a similar stable multivariate polynomial for generalized Stirling permutations.

Motivated by the real-rootedness of $C_n(x)$ and its stable multivariate refinement $C_n(X, Y, Z)$, Haglund and Visontai further considered the problem of finding stable multivariate polynomials as refinements of the generating polynomials of the descent statistic over Legendre-Stirling permutations. The Legendre-Stirling permutations were introduced by Egge [12] as a generalization of Stirling permutations in the study of Legendre-Stirling numbers of the second kind. For any $n \geq 1$, let M_n be the multiset $\{1, 1, \bar{1}, 2, 2, \bar{2}, \ldots, n, n, \bar{n}\}$. A permutation $\pi = \pi_1 \pi_2 \ldots \pi_{3n}$ on M_n is called a Legendre-Stirling permutation if whenever $i < j < k$ and $\pi_i = \pi_k$ are both unbarred, then $\pi_j > \pi_i$. For a Legendre-Stirling permutation π on M_n, we say that i is a descent if either $i = 3n$ or $\pi_i > \pi_{i+1}$. Let $B_{n,k}$ denote the number of Legendre-Stirling permutations of M_n with k descents. Define

$$B_n(x) = \sum_{k=1}^{2n-1} B_{n,k} x^k.$$

Egge [12] proved the real-rootedness of $B_n(x)$.

Theorem 3 (Egge [12]) *For $n > 0$, $B_n(x)$ has distinct, real, non-positive roots.*

In order to derive a stable multivariate refinement of $B_n(x)$, we introduce an approach of generating stable polynomials by a sequence of grammars. Based on the Stirling grammar given by Chen and Fu [10], we find a sequence $G_1, G_2, \ldots$ of context-free grammars to generate Legendre-Stirling permutations. Let D_n denote the differential operator associated with the grammar G_n, which leads to a stable multivariate refinement $B_n(X, Y, Z, U, V)$ of $B_n(x)$, that is,

$$B_n(X, Y, Z, U, V) = D_{2n} D_{2n-1} \ldots D_2 D_1(x_0),$$

where $U = (u_1, u_2, \ldots, u_n)$ and $V = (v_1, v_2, \ldots, v_n)$, respectively. Then by applying Borcea and Brändén's characterization of linear operators and the grammatical interpretation of $B_n(X, Y, Z, U, V)$, we prove the stability of $B_n(X, Y, Z, U, V)$. On the other hand, according to the grammars, we obtain the following combinatorial interpretation

$$B_n(X, Y, Z, U, V) = \sum_{\pi} \prod_{i \in X(\pi)} x_{\pi_i} \prod_{i \in Y(\pi)} y_{\pi_i} \prod_{i \in Z(\pi)} z_{\pi_i} \prod_{i \in U(\pi)} u_{\pi_i} \prod_{i \in V(\pi)} v_{\pi_i},$$

where π runs over all Legendre-Stirling permutations on M_n. Here $X(\pi)$, $Y(\pi)$, $Z(\pi)$, $U(\pi)$ and $V(\pi)$ are defined as follows: For a Legendre-Stirling permutation π on M_n,

$X(\pi) = \{i \mid \pi_{i-1} \leq \pi_i, \pi_i \text{ is unbarred and appears for the first time}\}$,

$Y(\pi) = \{i \mid \pi_i > \pi_{i+1} \text{ and } \pi_i \text{ is unbarred}\}$,

$Z(\pi) = \{i \mid \pi_{i-1} \leq \pi_i, \pi_i \text{ is unbarred and appears for the second time}\}$,

$U(\pi) = \{i \mid \pi_{i-1} \leq \pi_i \text{ and } \pi_i \text{ is barred}\}$,

$V(\pi) = \{i \mid \pi_i > \pi_{i+1} \text{ and } \pi_i \text{ is barred}\}$.

Here we set $\pi_0 = \pi_{3n+1} = 0$. Then the real-rootedness of $B_n(x)$ is a consequence of the stability of $B_n(X, Y, Z, U, V)$ by setting $v_i = y_i = y$ and $x_i = z_i = u_i = 1$ for $0 \leq i \leq n$.

Haglund and Visontai [16] also raised the question of finding stable multivariate refinements of the polynomials $T_n(x)$, which are given by

$$T_n(x) = 2^n C_n \left(\frac{x}{2}\right) = \sum_k 2^{n-k} C(n, k) x^k, \tag{1}$$

where $C(n, k)$ and $C_n(x)$, as before, denote the second-order Eulerian numbers and the second-order Eulerian polynomials respectively. The polynomials $T_n(x)$ were introduced by Riordan [23].

In view of the relation (1) between $T_n(x)$ and $C_n(x)$, we mark the Stirling permutations by some rule. We consider the following multivariate polynomials

$$T_n(X, Y, Z) = \sum_\pi \prod_{i \in D(\pi)} x_{\pi_i} \prod_{i \in A(\pi)} y_{\pi_i} \prod_{i \in P(\pi)} z_{\pi_i},$$

where π ranges over marked Stirling permutations of $[n]_2$. We shall show that the polynomials $T_n(X, Y, Z)$ are stable. The polynomial $T_n(x)$ becomes the specialization of $T_n(X, Y, Z)$ by setting $x_i = z_i = 1$ and $y_i = x$ for $0 \le i \le n$. This implies that $T_n(x)$ is real-rooted.

This paper is organized as follows. In Sect. 2, we give an overview of differential operators associated with context-free grammars and find context-free grammars to generate the polynomials $C_n(X, Y, Z)$. In Sect. 3, we give context-free grammars to generate the multivariate polynomials $T_n(X, Y, Z)$. In Sect. 4, we obtain context-free grammars that lead to the multivariate generating polynomials $B_n(X, Y, Z, U, V)$. In Sect. 5, based on Borcea and Brändén's characterization of linear operators preserving stability, we prove that the formal derivative with respect to the grammar that generates $T_n(X, Y, Z)$ preserves stability of multiaffine polynomials. This leads to the stability of $T_n(X, Y, Z)$. In Sect. 6, we provide an approach to find a new stability preserving operator when a grammar is not suitable to prove the stability of polynomials. In particular, we prove the stability of the multivariate polynomials $B_n(X, Y, Z, U, V)$.

2 Context-Free Grammars

In this section, we give an overview of the idea of using context-free grammars to generate combinatorial polynomials and combinatorial structures as developed in [9]. A context-free grammar G over an alphabet A is defined to be a set of production rules. A production rule means to substitute a letter in the alphabet A by a polynomial in A over a field. Given a context-free grammar, one may define a formal derivative D as a linear operator on polynomials in A, where the action of D on a letter is defined by the substitution rule of the grammar, the action of D on a sum of two polynomials u and v is defined by linear extension:

$$D(u + v) = D(u) + D(v),$$

and the action of D on the product of u and v is defined by the Leibniz rule, that is,

$$D(uv) = D(u)v + uD(v).$$

Many combinatorial polynomials can be generated by context-free grammars. Context-free grammars can also be used to generate combinatorial structures. More precisely, one may use a word on an alphabet to label a combinatorial structure such that the context-free grammar serves as the procedure to recursively generate the combinatorial structures. Such a labeling of a combinatorial structure is called a grammatical labeling in [10].

For example, we consider the Eulerian grammar

$$G = \{x \rightarrow xy, \ y \rightarrow xy\}$$

introduced by Dumont [11].

For a permutation $\pi = \pi_1 \pi_2 \cdots \pi_n$ of $[n]$, let

$$A(\pi) = \{i \mid \pi_{i-1} < \pi_i\},$$

$$D(\pi) = \{i \mid \pi_i > \pi_{i+1}\}$$

denote the set of ascents and the set of descents of π, respectively. Here, as usual, we set $\pi_0 = \pi_{n+1} = 0$. Let $A(n, k)$ denote the Eulerian number, that is, the number of permutations on $[n]$ with k descents.

In order to show how to use the Eulerian grammar to generate permutations, Chen and Fu [10] introduced a grammatical labeling of a permutation π on $[n]$: If i is an ascent of π, then π_{i-1} is labeled by x; if i is a descent, then π_i is labeled by y. The weight of π is defined as the product of labels of elements in π, that is,

$$w(\pi) = x^{|A(\pi)|} y^{|D(\pi)|}.$$

For example, the grammatical labeling of the permutation $\pi = 325641$ is as follows:

$$
\begin{array}{ccccccc}
3 & 2 & 5 & 6 & 4 & 1 \\
x & y & x & x & y & y & y
\end{array}.
$$

Thus the weight of π equals $w(\pi) = x^3 y^4$. This grammatical labeling leads to the following expression of the Eulerian polynomials. Dumont [11] obtained an equivalent form in terms of cyclic permutations and gave an inductive proof.

Theorem 4 (Dumont [11]) *Let D denote the formal derivative with respect to the Eulerian grammar. For $n \geq 1$, we have*

$$D^n(x) = \sum_{m=1}^{n} A(n, m) y^m x^{n+1-m}.$$

Let us now consider the grammar to generate Stirling permutations. Chen and Fu [10] introduced the grammar

$$G = \{x \rightarrow x^2 y, \ y \rightarrow x^2 y\}.$$

They defined a grammatical labeling of a Stirling permutation π in Q_n as follows: Let $1 \leq i \leq 2n$. If $i \in A(\pi)$ or $i \in P(\pi)$, the element π_{i-1} is labeled by x; if $i \in D(\pi)$, the element π_i is labeled by y. The weight of π, denoted by $w(\pi)$, is defined as the product of labels of elements in π. For example, the Stirling permutation $\pi = 233211$ has the following grammatical labeling

$$
\begin{array}{cccccc}
2 & 3 & 3 & 2 & 1 & 1 \\
x & x & x & y & y & x & y
\end{array}.
$$

Then the weight of π is $w(\pi) = x^4 y^3$.

Theorem 5 (Chen and Fu [10]) *Let D denote the formal derivative with respect to the above grammar G. For $n \geq 1$, we have*

$$
D^n(x) = \sum_{m=1}^{n} C(n, m) x^{2n+1-m} y^m.
$$

We shall give two sequences of grammars based on the Eulerian grammar and the Stirling grammar to solve the problems of Haglund and Visontai [16]. On one hand, we use these grammars to construct multivariate polynomials over Legendre-Stirling permutations and marked Stirling permutations. On the other hand, we use the grammars to construct stability preserving operators leading to the stability of the multivariate polynomials.

3 Marked Stirling Permutations

In this section, we obtain a stable multivariate refinement of the polynomial $T_n(x)$, denoted by $T_n(X, Y, Z)$, which is defined as the generating function of marked Stirling permutations on $[n]_2$. This provides a solution to the problem of Haglund and Visontai.

In order to prove the stability of $T_n(X, Y, Z)$, we find grammars $G_1, G_2, \ldots$ that can be used to generate $T_n(X, Y, Z)$. More precisely, define

$$
G_n = \{x_i, z_i \to x_n y_n z_n, \ y_i \to 2 x_n y_n z_n \mid 0 \leq i \leq n-1\}.
$$

Let D_n denote the formal derivative with respect to G_n. Using a grammatical labeling of marked Stirling permutations, we shall show that the polynomial $T_n(X, Y, Z)$ can be generated by $D_1, D_2, \ldots, D_n$. The stability of $T_n(X, Y, Z)$ can be established in Sect. 6 by using the operators $D_1, D_2, \ldots, D_n$.

A marked Stirling permutation is defined as follows. Given a Stirling permutation $\pi = \pi_1 \pi_2 \cdots \pi_{2n}$, if π_i is an element of π such that π_i occurs the second time in π and $\pi_i < \pi_{i+1}$, then we may mark the element π_i. We denote a marked element i by $\bar{i}$. A marked Stirling permutation is a Stirling permutation with some elements marked according to the above rule. Let $\bar{Q}_n$ denote the set of marked Stirling permutations on $[n]_2$. For example, there is only one marked Stirling permutation on $[1]_2$: 11, whereas there are four marked Stirling permutations on $[2]_2$:

$$2211, 1221, 1122, 1\bar{1}22.$$

Let $T(n, k)$ be the number of marked Stirling permutations on $[n]_2$ with k descents. Clearly,

$$T(n, k) = 2^{n-k} \cdot C(n, k),$$

where $C(n, k)$ denotes the second-order Eulerian number. Recall that $T_n(x)$ is defined by

$$T_n(x) = 2^n \cdot C_n \left(\frac{x}{2}\right) = \sum_{k=0}^{n} 2^{n-k} C(n, k) x^k.$$

Hence $T_n(x)$ is the generating function of marked Stirling permutations on $[n]_2$, that is,

$$T_n(x) = \sum_{k=0}^{n} T(n, k) x^k = \sum_{\pi \in \bar{Q}_n} x^{|D(\pi)|}.$$

In fact, Riordan [23] introduced the polynomials $T_n(x)$ and proved that $T_n(1)$ equals the Schröder number, namely, the number of series-reduced rooted trees with $n + 1$ labeled leaves.

We shall prove that the polynomials $T_n(x)$ can be generated by the grammar

$$G = \{x \rightarrow x^2 y, y \rightarrow 2x^2 y\}.$$

The proof relies on the following grammatical labeling of a marked Stirling permutation. Let π be a marked Stirling permutation on $[n]_2$. If $i \in D(\pi)$, we label π_i by y. If $i \in A(\pi)$ or $i \in P(\pi)$, we label π_{i-1} by x. The weight of a marked Stirling permutation π on $[n]_2$ with m descents is given by

$$w(\pi) = x^{2n+1-m} y^m.$$

Theorem 6 *Let G be the grammar $G = \{x \rightarrow x^2 y, y \rightarrow 2x^2 y\}$ and D be the formal derivative associated with G. For $n \geq 1$,*

$$D^n(x) = \sum_{k=1}^{n} T(n, k) x^{2n-k+1} y^k.$$

Setting $x = 1$, we have

$$D^n(x)|_{x=1} = T_n(y).$$

Proof We aim to show that $D^n(x)$ equals the sum of the weights of marked Stirling permutations of $[n]_2$ by induction on n, that is,

$$D^n(x) = \sum_{\pi \in \bar{Q}_n} w(\pi). \tag{2}$$

For $n = 1$, (2) follows from the fact that the weight of 11, the only marked Stirling permutation on $[1]_2$, is $x^2 y$. Assume that (2) holds for $n - 1$, that is,

$$D^{n-1}(x) = \sum_{\pi \in \bar{Q}_{n-1}} w(\pi).$$

We now use an example to demonstrate the action of D on a marked Stirling permutation of $[n - 1]_2$. Let $\pi = 12\bar{2}331$ with the following grammatical labeling

$$\begin{matrix} 1 & 2 & \bar{2} & 3 & 3 & 1 \\ x & x & x & x & x & y & y \end{matrix}.$$

If we apply the substitution rule $x \to x^2 y$ to the fourth letter x, then we insert the two elements 44 after $\bar{2}$. We keep all the old labels and assign the labels x and y to the two new letters 44 from left to right. It is not difficult to see that the generated marked Stirling permutation has a consistent grammatical labeling

$$\begin{matrix} 1 & 2 & \bar{2} & 4 & 4 & 3 & 3 & 1 \\ x & x & x & x & x & y & x & y & y \end{matrix}.$$

If we apply the substitution rule $y \to 2x^2 y$ to the first letter y, then we insert 44 after the second element 3. We change the label of the second element 3 from y to x and assign x and y to the two new elements 44 from left to right. According to the marking rule, the second element 3 may be marked or unmarked. These two choices correspond the coefficient 2 in the substitution rule $y \to 2x^2 y$. So we are led to the following two marked Stirling permutations with consistent grammatical labelings,

$$\begin{matrix} 1 & 2 & \bar{2} & 3 & 3 & 4 & 4 & 1 \\ x & x & x & x & x & x & x & y & y \end{matrix},$$

and

$$\begin{matrix} 1 & 2 & \bar{2} & 3 & \bar{3} & 4 & 4 & 1 \\ x & x & x & x & x & x & x & y & y \end{matrix}.$$

In general, it can be verified that the action of D on the weights of marked Stirling permutations in $\bar{Q}_{n-1}$ generates the weights of marked Stirling permutations in $\bar{Q}_n$. So we deduce that (2) holds for n, that is,

$$D^n(x) = D(D^{n-1}(x)) = D\left(\sum_{\pi \in \bar{Q}_{n-1}} w(\pi)\right) = \sum_{\sigma \in \bar{Q}_n} w(\sigma).$$

Hence the proof is complete by induction. □

As a multivariate refinement of $T_n(x)$, we define the following generating function of marked Stirling permutations on $[n]_2$,

$$T_n(X, Y, Z) = \sum_{\pi \in \bar{Q}_n} \prod_{i \in A(\pi)} x_{\pi_i} \prod_{i \in D(\pi)} y_{\pi_i} \prod_{i \in P(\pi)} z_{\pi_i}.$$

Let

$$G_n = \{x_i \to x_n y_n z_n, z_i \to x_n y_n z_n, y_i \to 2x_n y_n z_n \mid 0 \le i \le n - 1\}.$$

We give a grammatical labeling of a marked Stirling permutation. For a marked Stirling permutation π on $[n]_2$, if $i \in A(\pi)$, we label π_{i-1} by x_{π_i}; if $i \in D(\pi)$, we label π_i by y_{π_i}; and if $i \in P(\pi)$, we label π_{i-1} by z_{π_i}. Then the weight of π equals

$$w(\pi) = \prod_{i \in A(\pi)} x_{\pi_i} \prod_{i \in D(\pi)} y_{\pi_i} \prod_{i \in P(\pi)} z_{\pi_i}.$$

The following theorem shows that the polynomials $T_n(X, Y, Z)$ can be generated by the grammars $G_1, G_2, \ldots, G_n$.

Theorem 7 *Let D_n denote the formal derivative associated with the grammar G_n. For $n \ge 1$,*

$$T_n(X, Y, Z) = D_n D_{n-1} \cdots D_1(z_0).$$

The proof of the above theorem is analogous to that of Theorem 6. Hence the details are omitted. Here we use an example to illustrate the action of D_4 on the above marked Stirling permutation $\pi = 12\bar{2}331$ with the grammatical labeling

$$\begin{array}{cccccc} 1 & 2 & \bar{2} & 3 & 3 & 1 \\ x_1 & x_2 & z_2 & x_3 & z_3 & y_3 & y_1 \end{array}.$$

Applying the substitution rule $x_3 \to x_4 y_4 z_4$ to π, we get a marked Stirling permutation by inserting the two elements 44 after $\bar{2}$ and the consistent grammatical labeling is given below:

$$
\begin{array}{cccccccc}
1 & 2 & \bar{2} & 4 & 4 & 3 & 3 & 1 \\
x_1 & x_2 & z_2 & x_4 & z_4 & y_4 & z_3 & y_3 & y_1
\end{array}.
$$

Similarly, applying the substitution rule $y_3 \to 2x_4 y_4 z_4$ leads to two marked Stirling permutations by inserting 44 after the second element 3, since the second element 3 can be marked. The consistent grammatical labelings are

$$
\begin{array}{ccccccccc}
1 & 2 & \bar{2} & 3 & 3 & 4 & 4 & 1 \\
x_1 & x_2 & z_2 & x_3 & z_3 & x_4 & z_4 & y_4 & y_1
\end{array},
$$

and

$$
\begin{array}{ccccccccc}
1 & 2 & \bar{2} & 3 & \bar{3} & 4 & 4 & 1 \\
x_1 & x_2 & z_2 & x_3 & z_3 & x_4 & z_4 & y_4 & y_1
\end{array}.
$$

For $n = 0$, the empty permutation is labeled by z_0. We have $T_0(X, Y, Z) = z_0$. For $n = 1, 2$, we have

$$
T_1(X, Y, Z) = D_1(z_0) = x_1 \overset{1}{z_1} \overset{1}{y_1},
$$

$$
T_2(X, Y, Z) = D_2 D_1(z_0) = D_2(x_1 \overset{1}{z_1} \overset{1}{y_1})
$$

$$
= x_2 \overset{2}{z_2} \overset{2}{y_2} \overset{1}{z_1} \overset{1}{y_1} + x_1 x_2 \overset{1}{z_2} \overset{2}{y_2} \overset{2}{y_2} \overset{1}{y_1} + x_1 \overset{1}{z_1} x_2 \overset{1}{z_2} \overset{2}{y_2}
$$

$$
+ x_1 \overset{1}{z_1} x_2 \overset{\bar{1}}{z_2} \overset{2}{y_2}
$$

$$
= y_1 z_1 x_2 y_2 z_2 + x_1 y_1 x_2 y_2 z_2 + 2 x_1 z_1 x_2 y_2 z_2.
$$

4   Legendre-Stirling Permutations

In this section, we give refinements of the Stirling grammar and the Eulerian grammar, and we show that these refined grammars can be used to generate stable multivariate polynomials. For $n \geq 1$, let

$$
G_{2n-1} = \{x_i, y_i, z_i, u_i, v_i \to u_n v_n \mid 0 \leq i < n\},
$$

and let

$$G_{2n} = \{u_n \rightarrow x_n z_n u_n, v_n \rightarrow x_n y_n z_n,$$

$$x_i, y_i, z_i, u_i, v_i \rightarrow x_n y_n z_n \mid 0 \leq i < n\}.$$

Clearly, G_{2n-1} is a refinement of the Eulerian grammar, and G_{2n} is a refinement of the Stirling grammar.

Let D_n denote the formal derivative with respect to the grammar G_n. We give a grammatical labeling of Legendre-Stirling permutations, which leads to a combinatorial interpretation of the multivariate polynomial $D_{2n} D_{2n-1} \cdots D_1(x_0)$. To this end, we introduce several statistics of a Legendre-Stirling permutation. In terms of these statistics, we obtain a multivariate polynomial $B_n(X, Y, Z, U, V)$ as a refinement of $B_n(x)$, which can be generated by the operators $D_1, D_2, \ldots, D_{2n}$.

Recall that M_n denotes the multiset $\{1, 1, \bar{1}, 2, 2, \bar{2}, \ldots, n, n, \bar{n}\}$. Let L_n denote the set of Legendre-Stirling permutations on M_n. For a Legendre-Stirling permutation $\pi = \pi_1 \pi_2 \ldots \pi_{3n} \in L_n$, define

$X(\pi) = \{i \mid \pi_{i-1} \leq \pi_i, \pi_i \text{ is unbarred and appears for the first time}\},$

$Y(\pi) = \{i \mid \pi_i > \pi_{i+1} \text{ and } \pi_i \text{ is unbarred}\},$

$Z(\pi) = \{i \mid \pi_{i-1} \leq \pi_i, \pi_i \text{ is unbarred and appears for the second time}\},$

$U(\pi) = \{i \mid \pi_{i-1} \leq \pi_i \text{ and } \pi_i \text{ is barred}\},$

$V(\pi) = \{i \mid \pi_i > \pi_{i+1} \text{ and } \pi_i \text{ is barred}\}.$

As usual, we set $\pi_0 = \pi_{3n+1} = 0$.

For example, let $\pi = \bar{1} 1 \bar{2} 2 3 3 2 \bar{3} 1$. Then we have $X(\pi) = \{2, 4, 5\}$, $Y(\pi) = \{6, 9\}$, $Z(\pi) = \{6\}$, $U(\pi) = \{1, 3, 8\}$ and $V(\pi) = \{8\}$.

Define

$$B_n(X, Y, Z, U, V) = \sum_{\pi \in L_n} \prod_{i \in X(\pi)} x_{\pi_i} \prod_{i \in Y(\pi)} y_{\pi_i} \prod_{i \in Z(\pi)} z_{\pi_i} \prod_{i \in U(\pi)} u_{\pi_i} \prod_{i \in V(\pi)} v_{\pi_i}. \tag{3}$$

For example, there are only two Legendre-Stirling permutations on M_1: $11\bar{1}$ and $\bar{1}11$. So we have

$$B_1(X, Y, Z, U, V) = x_1 y_1 z_1 u_1 + x_1 z_1 u_1 v_1.$$

For $n = 2$, there are 40 Legendre-Stirling permutations on M_2 and we have

$$B_2(X, Y, Z, U, V) = 2x_2 y_2 z_2 u_2 x_1 z_1 u_1 + x_2 y_2 z_2 u_2 x_1 y_1 z_1 + x_2 y_2 z_2 u_2 x_1 y_1 u_1$$

$$+ x_2 y_2 z_2 u_2 y_1 z_1 u_1 + x_2 y_2 z_2 u_2 x_1 u_1 v_1 + x_2 y_2 z_2 u_2 z_1 u_1 v_1$$

$$+ x_2 y_2 z_2 u_2 x_1 z_1 v_1 + 2x_2 z_2 u_2 v_2 x_1 z_1 u_1 + x_2 z_2 u_2 v_2 x_1 y_1 z_1$$

$$+ x_2 z_2 u_2 v_2 x_1 y_1 u_1 + x_2 z_2 u_2 v_2 y_1 z_1 u_1 + x_2 z_2 u_2 v_2 x_1 z_1 v_1$$

$$+ x_2 z_2 u_2 v_2 x_1 u_1 v_1 + x_2 z_2 u_2 v_2 z_1 u_1 v_1 + 4 x_2 y_2 z_2 u_2 v_2 x_1 z_1$$

$$+ 4 x_2 y_2 z_2 u_2 v_2 x_1 u_1 + 4 x_2 y_2 z_2 u_2 v_2 u_1 z_1 + 2 x_2 y_2 z_2 u_2 v_2 x_1 y_1$$

$$+ 2 x_2 y_2 z_2 u_2 v_2 y_1 z_1 + 2 x_2 y_2 z_2 u_2 v_2 y_1 u_1 + 2 x_2 y_2 z_2 u_2 v_2 x_1 v_1$$

$$+ 2 x_2 y_2 z_2 u_2 v_2 u_1 v_1 + 2 x_2 y_2 z_2 u_2 v_2 z_1 v_1.$$

We now give a grammatical labeling of a Legendre-Stirling permutation. Let π be a Legendre-Stirling permutation in L_n. For $i \in X(\pi)$, $i \in Z(\pi)$ or $i \in U(\pi)$, we label π_{i-1} by x_{π_i}, z_{π_i} or u_{π_i}, respectively; for $i \in Y(\pi)$ or $i \in V(\pi)$, we label π_i by y_{π_i} or v_{π_i}, respectively. The weight of π is defined as the product of these letters labeled on entries of π and denoted by $w(\pi)$. For example, the grammatical labeling of the aforementioned Legendre-Stirling permutation $\pi = 1\bar{2}1233\bar{2}31$ is given below:

$$\begin{array}{ccccccccc} 1 & \bar{2} & \bar{1} & 2 & 3 & 3 & 2 & \bar{3} & 1 \\ x_1 & u_2 & v_2 & x_2 & x_3 & z_3 & y_3 & u_3 & v_3 & y_1 \end{array}.$$

Theorem 8 *For $n \geq 1$, let D_n denote the differential operator with respect to the grammar G_n, then we have*

$$D_{2n} D_{2n-1} \cdots D_1(x_0) = B_n(X, Y, Z, U, V). \tag{4}$$

Proof We proceed by induction on n to show that

$$D_{2n} D_{2n-1} \cdots D_1(x_0) = \sum_{\pi \in L_n} w(\pi). \tag{5}$$

It can be checked that (5) holds for $n = 1$. For $n \geq 2$, we assume that (5) holds for $n - 1$, that is,

$$D_{2n-2} D_{2n-3} \cdots D_1(x_0) = \sum_{\pi \in L_{n-1}} w(\pi).$$

Note that any Legendre-Stirling permutation on M_n can be obtained from a Legendre-Stirling permutation on M_{n-1} by inserting nn and $\bar{n}$. We use examples to illustrate that the application of the operator $D_{2n} D_{2n-1}$ reflects the insertions of nn and $\bar{n}$.

Consider the Legendre-Stirling permutation $\pi = \bar{1}1\bar{2}2233\bar{2}\bar{3}1$ with the following grammatical labeling:

$$\begin{array}{ccccccccc} \bar{1} & 1 & \bar{2} & 2 & 3 & 3 & 2 & \bar{3} & 1 \\ u_1 & x_1 & u_2 & x_2 & x_3 & z_3 & y_3 & u_3 & v_3 & y_1 \end{array}.$$

Let w be the weight of the above grammatical labeling, that is,

$$w = u_1 x_1 u_2 x_2 x_3 z_3 y_3 u_3 v_3 y_1.$$

Let us consider the action of D_7 on w. Recall that

$$G_7 = \{x_i, y_i, z_i, u_i, v_i \to u_4 v_4 \mid i = 1, 2, 3\}.$$

Consider a substitution rule that replaces a letter s by $u_4 v_4$. Assume that π_k is labeled by s, where $0 \leq k \leq 9$. This rule corresponds to an insertion of $\bar{4}$ after the entry π_k in π. Then the element π_k is relabeled by u_4, and the element $\bar{4}$ is labeled by v_4.

For example, the substitution rule $z_3 \to u_4 v_4$ corresponds to the insertion of $\bar{4}$ after the first element 3 in π. After the insertion, we obtain a Legendre-Stirling permutation with a consistent grammatical labeling:

$$
\begin{array}{cccccccccc}
\bar{1} & 1 & \bar{2} & 2 & 3 & \bar{4} & 3 & 2 & \bar{3} & 1 \\
u_1 & x_1 & u_2 & x_2 & x_3 & u_4 & v_4 & y_3 & u_3 & v_3 & y_1
\end{array}.
$$

As for the action of D_8, consider the above permutation $\sigma = \bar{1}1\bar{2}234\bar{3}2\bar{3}1$. Let w' denote the weight of σ, that is,

$$w' = u_1 x_1 u_2 x_2 x_3 u_4 v_4 y_3 u_3 v_3 y_1.$$

The two substitution rules $u_4 \to x_4 z_4 u_4$ and $v_4 \to x_4 y_4 z_4$ of G_8 correspond to the insertions of the element 44 into σ before $\bar{4}$ or after $\bar{4}$, respectively, resulting in two Legendre-Stirling permutations: $\bar{1}1\bar{2}234\bar{4}3\bar{2}31$ or $\bar{1}1\bar{2}23\bar{4}443231$.

It remains to consider the substitution rules of G_8 that are of the form $s \to x_4 y_4 z_4$, where $s \in \{x_i, y_i, z_i, u_i, v_i \mid i = 1, 2, 3\}$. Suppose that σ_i is the element in σ that is labeled by s. The substitution rule $s \to x_4 y_4 z_4$ corresponds to the insertion of 44 into σ after σ_i. Let τ denote the resulting permutation obtained from σ after the insertion. Then one can obtain a consistent grammatical labeling of τ by relabeling σ_i by x_4 and assigning the two labels z_4 and y_4 to the inserted two elements 44 from left to right. For example, by applying the substitution rule $u_2 \to x_4 y_4 z_4$, we obtain the Legendre-Stirling permutation by inserting 44 after the first element 1 with a consistent grammatical labeling:

$$
\begin{array}{cccccccccccc}
\bar{1} & 1 & 4 & 4 & \bar{2} & 2 & 3 & \bar{4} & 3 & 2 & \bar{3} & 1 \\
u_1 & x_1 & x_4 & z_4 & y_4 & x_2 & x_3 & u_4 & v_4 & y_3 & u_3 & v_3 & y_1
\end{array}.
$$

In general, it can be verified that the action of $D_{2n} D_{2n-1}$ on the weights of the Legendre-Stirling permutations in L_{n-1} generates the weights of Legendre-Stirling permutations in L_n. So we conclude that (5) holds for n, that is,

$$D_{2n}D_{2n-1}\cdots D_1(x_0) = \sum_{\pi \in L_n} w(\pi).$$

Thus (5) holds for all n. This completes the proof.   □

We note that the grammars $G_2, G_4, \ldots$ are related to the polynomials $C_n(X, Y, Z)$ introduced by Haglund and Vistonai [16], as defined by

$$C_n(X, Y, Z) = \sum_{\pi \in Q_n} \prod_{i \in D(\pi)} x_{\pi_i} \prod_{i \in A(\pi)} y_{\pi_i} \prod_{i \in P(\pi)} z_{\pi_i}.$$

Clearly, $C_1(X, Y, Z) = x_1 y_1 z_1$. Based on the combinatorial interpretation of $C_n(X, Y, Z)$, Haglund and Visontai [16] established the following recurrence relation for $n \geq 1$:

$$C_{n+1}(X, Y, Z) = x_{n+1} y_{n+1} z_{n+1} \partial C_n(X, Y, Z), \tag{6}$$

where

$$\partial = \sum_{i=1}^{n} \frac{\partial}{\partial x_i} + \sum_{i=1}^{n} \frac{\partial}{\partial y_i} + \sum_{i=1}^{n} \frac{\partial}{\partial z_i}. \tag{7}$$

The following theorem shows that the grammar D_{2n} has the same effect as the operator $x_n y_n z_n \partial$ when acting on $C_{n-1}(X, Y, Z)$.

Theorem 9 *For $n \geq 0$,*

$$D_{2n+2}(C_n(X, Y, Z)) = x_{n+1} y_{n+1} z_{n+1} \partial C_n(X, Y, Z). \tag{8}$$

The relation (8) implies that

$$D_{2n+2}D_{2n}\cdots D_4 D_2(z_0) = C_{n+1}(X, Y, Z).$$

To prove Theorem 9, we observe the following property of the formal derivative D with respect to a grammar G. The verification is straightforward.

Proposition 10 *Let X denote the set of variables of a grammar G. For a polynomial f in X, we have*

$$D(f) = \sum_{x \in X} D(x) \frac{\partial f}{\partial x}.$$

5   The Stability of $T_n(X, Y, Z)$

In this section, we prove the stability of the multivariate polynomials $T_n(X, Y, Z)$ by showing that the related formal derivatives with respect to the generating grammars are stability preserving operators. The proof relies on the characterization of stability preserving linear operators on multiaffine polynomials due to Borcea and Brändén [4].

Recall that a multivariate polynomial $f(z_1, z_2, \ldots, z_n)$ is called multiaffine if the degree of any variable in f is at most 1. An operator T is called a stability preserver of multiaffine polynomials if $T(f)$ is either stable or identically 0 for any stable multiaffine polynomial $f \in \mathbb{C}[z_1, z_2, \ldots, z_n]$.

Theorem 11 (Borcea and Brändén) *Let T denote a linear operator acting on the polynomials in $\mathbb{C}[z_1, z_2, \ldots, z_n]$. If*

$$T\left(\prod_{i=1}^{n}(z_i + w_i)\right) \in \mathbb{C}[z_1, \ldots, z_n, w_1, \ldots, w_n]$$

is stable, then T is a stability preserver of multiaffine polynomials.

To prove the stability of $T_n(X, Y, Z)$, we use the grammatical expression

$$T_n(X, Y, Z) = D_n D_{n-1} \cdots D_1(z_0)$$

in Theorem 7, where D_n is the formal derivative with respect to the grammar

$$G_n = \{x_i, z_i \to x_n y_n z_n, \ y_i \to 2x_n y_n z_n \mid 0 \le i < n\}.$$

We shall show that D_n is a stability preserver, and this proves the stability of $T_n(X, Y, Z)$.

Theorem 12 *For $n \ge 1$, $T_n(X, Y, Z)$ is stable.*

Proof Let

$$F = \prod_{i=0}^{n}(x_i + u_i)(y_i + v_i)(z_i + w_i), \tag{9}$$

and let

$$\xi = \sum_{i=0}^{n-1}\left(\frac{1}{x_i + u_i} + \frac{2}{y_i + v_i} + \frac{1}{z_i + w_i}\right). \tag{10}$$

We have

$$D_n(F) = \sum_{i=0}^{n-1} D(x_i)\frac{\partial F}{\partial x_i} + \sum_{j=0}^{n-1} D(y_j)\frac{\partial F}{\partial y_j} + \sum_{k=0}^{n-1} D(z_k)\frac{\partial F}{\partial z_k}$$

$$= \sum_{i=0}^{n-1} x_n y_n z_n \frac{F}{x_i + u_i} + \sum_{j=0}^{n-1} 2x_n y_n z_n \frac{F}{y_i + v_i} + \sum_{k=0}^{n-1} x_n y_n z_n \frac{F}{z_k + w_k}$$

$$= x_n y_n z_n \xi F.$$

To prove that D_n preserves stability of multiaffine polynomials, we assume that x_i, y_i, z_i, u_i, v_i and w_i have positive imaginary parts for all $0 \le i \le n$. We proceed to show that $D_n(F) \ne 0$.

Under the above assumptions, for $0 \le i \le n$, $x_i + u_i$, $y_i + v_i$ and $z_i + w_i$ also have positive imaginary parts. It follows that $\frac{1}{x_i+u_i}$, $\frac{2}{y_i+v_i}$ and $\frac{1}{z_i+w_i}$ have negative imaginary parts. By the definition (9), we see that $F \ne 0$. By (10), we find that $\xi \ne 0$. Hence $D_n(F) \ne 0$. Thus D_n is a stability preserver. This completes the proof.   □

6 The Stability of $B_n(X, Y, Z, U, V)$

In this section, we prove the stability of the multivariate polynomials $B_n(X, Y, Z, U, V)$. Unlike the proof for $T_n(X, Y, Z)$, the formal derivatives with respect to the grammars do not preserve stability. Fortunately, as for the multiaffine polynomials that we are concerned with, the formal derivatives in our case are equivalent to linear operators which turn out to be stability preserving.

More specifically, the idea goes as follows: Let $G_1, G_2, \ldots$ be context-free grammars, and $D_1, D_2, \ldots$ be the formal derivatives with respect to $G_1, G_2, \ldots$. Suppose that we wish to prove the stability of the multivariate polynomials

$$f_n = D_n D_{n-1} \cdots D_1(x),$$

for $n \ge 1$, where $D_1, D_2, \ldots$ may not be stability preserving. We aim to construct stability preservers $T_1, T_2, \ldots$ such that

$$T_n T_{n-1} \cdots T_1(x) = D_n D_{n-1} \cdots D_1(x).$$

Once such stability preservers $T_1, T_2, \ldots$ are found, it can be asserted that the multivariate polynomials f_n are stable. The following lemma provides a way to find such operators T_n.

Lemma 13 *Let G be a context-free grammar over the alphabet $X \cup Y$, where*

$$X = \{x_1, x_2, \ldots, x_r\}$$

and

$$Y = \{y_1, y_2, \ldots, y_s\}.$$

Let D denote the formal derivative with respect to G. Assume that $D(x_i)$ contains a factor x_i for $i = 1, 2, \ldots, r$, namely, $x_i \to x_i h_i(X, Y)$ is a substitution rule in G. Let T denote the following operator

$$T = \sum_{i=1}^{r} h_i(X, Y)I + \sum_{j=1}^{s} D(y_j) \frac{\partial}{\partial y_j},$$

where I denotes the identity operator. Let $g(Y)$ be any polynomial in Y and let $f(X, Y) = x_1 x_2 \ldots x_r g(Y)$. Then we have

$$D(f(X, Y)) = T(f(X, Y)).$$

Proof By Proposition 10, we find that

$$D(f(X, Y)) = \sum_{i=1}^{r} D(x_i) \frac{\partial f(X, Y)}{\partial x_i} + \sum_{j=1}^{s} D(y_j) \frac{\partial f(X, Y)}{\partial y_j}$$

$$= \sum_{i=1}^{r} x_i h_i(X, Y) \cdot \frac{f(X, Y)}{x_i} + \sum_{j=1}^{s} D(y_j) \frac{\partial f(X, Y)}{\partial y_j}$$

$$= \sum_{i=1}^{r} h_i(X, Y) f(X, Y) + \sum_{j=1}^{s} D(y_j) \frac{\partial f(X, Y)}{\partial y_j},$$

which equals $T(f(X, Y))$. This completes the proof. $\square$

For example, the grammar

$$G = \{a \to ax, \; x \to x\}$$

is used in [9] to generate the set of partitions of $[n]$ and the Stirling polynomials

$$S_n(x) = \sum_{i=0}^{n} S(n, k) x^k,$$

where $S(n, k)$ denotes the Stirling number of the second kind.

For $n \geq 1$, we have

$$D^n(a) = \sum_{k=1}^{n} S(n,k)ax^k = aS_n(x). \tag{11}$$

Many properties of the Stirling polynomials follow from the above expression in terms of the differential operator D with respect to the grammar G.

Let $X = \{a\}$ and $Y = \{x\}$. Then D satisfies the conditions in Lemma 13. Thus $D(af(x)) = T(af(x))$ for any polynomial $f(x)$, where the operator T is given by

$$T = x\left(I + \frac{\partial}{\partial x}\right).$$

In particular, we have

$$T(aS_n(x)) = D(aS_n(x)).$$

In fact, the above operator T corresponds to the following recurrence relation for $S_n(x)$:

$$S_n(x) = T(S_{n-1}(x)),$$

which is equivalent to the recurrence relation of $S(n,k)$:

$$S(n,k) = S(n-1,k-1) + kS(n-1,k), \tag{12}$$

where $n \geq k > 1$. Harper [17] proved that $S_n(x)$ has only real roots for $n \geq 1$. Liu and Wang [21] showed that T preserves the real-rootedness of polynomials in x.

As a generalization of the real-rootedness of $S_n(x)$, we consider the stability of the multivariate polynomials $S_n(a, x_1, x_2, \ldots, x_n)$, which can be viewed as a refinement of the Stirling polynomial $S_n(x)$. Let

$$G_n = \{a \rightarrow ax_n, x_i \rightarrow x_n \mid 1 \leq i < n\},$$

and let D_n denote the formal derivative associated with G_n. It will be shown that for $n \geq 1$, $S_n(a, x_1, x_2, \ldots, x_n)$ can be generated by $G_1, G_2, \ldots, G_n$.

The polynomial $S_n(a, x_1, x_2, \ldots, x_n)$ is defined by using the following grammatical labeling of a partition $P = \{P_1, P_2, \ldots, P_k\}$ of $[n]$. The partition itself is labeled by the letter a and a block P_i is labeled by the letter x_{m_i}, where m_i is the maximal element in P_i. The weight of P is given by the product of all labelings in P, that is,

$$w(P) = a\prod_{i=1}^{k} x_{m_i}.$$

Denote by $S_n(a, x_1, x_2, \ldots, x_n)$ the sum of weights of partitions of $[n]$. Clearly, $S_n(a, x_1, x_2, \ldots, x_n)$ is the generating function of partitions of $[n]$ involving not only the number of blocks, but also the maximal elements of the blocks.

For example, for $n = 1, 2, 3$, we have

$$S_1(a, x_1) = ax_1,$$

$$S_2(a, x_1, x_2) = ax_1x_2 + ax_2,$$

$$S_3(a, x_1, x_2, x_3) = ax_1x_2x_3 + 2ax_2x_3 + ax_1x_3 + ax_3.$$

The following theorem gives a grammatical expression of $S_n(a, x_1, x_2, \ldots, x_n)$.

Theorem 14 *For $n \geq 1$,*

$$S_n(a, x_1, x_2, \ldots, x_n) = D_n D_{n-1} \cdots D_1(a). \tag{13}$$

Let us give an example to demonstrate the action of the differential operator D_7 on a partition of $\{1, 2, 3, 4, 5, 6\}$. Recall that

$$G_7 = \{a \rightarrow ax_7, x_i \rightarrow x_7 \mid 1 \leq i \leq 6\}.$$

Consider the following partition along with its grammatical labeling:

$$\{1, 3, 6\} \{2, 5\} \{4\}$$
$$x_6 \qquad x_5 \quad x_4 \ \ a^{\textstyle.}$$

Applying the substitution rule $a \rightarrow ax_7$ to the above partition leads to a partition with a consistent grammatical labeling:

$$\{1, 3, 6\} \{2, 5\} \{4\} \{7\}$$
$$x_6 \qquad x_5 \quad x_4 \ x_7 \ \ a^{\textstyle.}$$

Similarly, applying the substitution rule $x_5 \rightarrow x_7$ to the partition, we get the following partition with a consistent grammatical labeling

$$\{1, 3, 6\} \{2, 5, 7\} \{4\}$$
$$x_6 \qquad x_7 \quad x_4 \ \ a^{\textstyle.}$$

In fact, the above arguments are sufficient to justify the expression (13).

It should be noticed that the relation (13) cannot be directly used to prove the stability of $S_n(a, x_1, x_2, \ldots, x_n)$, since the operator D_n does not preserve stability in general. Take D_2 as an example. Consider the polynomial $(a + 1)(x_1 + 1)$, which is clearly stable. But

$$D_2((a+1)(x_1+1)) = x_2(ax_1 + 2a + 1)$$

is not stable since it vanishes when $a = i$ and $x_1 = i - 2$. It follows that D_2 is not stability preserving.

Fortunately, we can find a stability preserving operator T_n for the purpose of justifying the stability of $S_n(a, x_1, x_2, \ldots, x_n)$. It is easy to see that $S_n(a, x_1, x_2, \ldots, x_n)$ can be written as $ah(X)$, where $h(X)$ is a multivariate polynomial in $x_1, x_2, \ldots, x_n$ that is independent of the variable a. Let

$$T_n = x_n I + x_n \sum_{i=1}^{n} \frac{\partial}{\partial x_i}. \tag{14}$$

According to Lemma 13, for each $n \geq 1$, we have

$$T_n(S_n(a, x_1, x_2, \ldots, x_n)) = D_n(S_n(a, x_1, x_2, \ldots, x_n)).$$

It turns out that $S_n(a, x_1, x_2, \ldots, x_n)$ can be obtained by using $T_1, T_2, \ldots, T_n$.

Theorem 15 *For $n \geq 1$, we have*

$$S_n(a, x_1, x_2, \ldots, x_n) = T_n T_{n-1} \cdots T_1(a). \tag{15}$$

The following theorem establishes the stability of $S_n(a, x_1, x_2, \ldots, x_n)$.

Theorem 16 *For $n \geq 1$, the multivariate polynomial $S_n(a, x_1, x_2, \ldots, x_n)$ is stable.*

Proof It suffices to show that the linear operator T_n preserves stability of multiaffine polynomials. By Theorem 11, it is enough to prove that $T_n(F)$ is stable, where

$$F = (a + u) \prod_{i=1}^{n} (x_i + v_i).$$

Let

$$\xi = 1 + \sum_{i=1}^{n-1} \frac{1}{x_i + v_i}.$$

Then

$$T_n(F) = x_n F + x_n \sum_{i=1}^{n-1} \frac{\partial F}{\partial x_i}$$

$$= x_n F + x_n F \sum_{i=1}^{n-1} \frac{1}{x_i + v_i}$$

$$= x_n \xi F.$$

To prove that $T_n(F)$ is stable, we assume that a, u, $x_1, x_2, \ldots, x_n$ and $v_1, v_2, \ldots, v_n$ have positive imaginary parts. It remains to show that $T_n(F) \neq 0$.

Under the above assumptions, for $1 \leq i \leq n$, $x_i + v_i$ has a positive imaginary part. It follows that $\frac{1}{x_i + v_i}$ has a negative imaginary part. Furthermore, the imaginary part of ξ is also negative. Thus we have $F \neq 0$ and $\xi \neq 0$. Consequently, $T_n(F) \neq 0$. This completes the proof. □

Next we prove the stability of $B_n(X, Y, Z, U, V)$, where $X = (x_1, x_2, \ldots, x_n)$, $Y = (y_1, y_2, \ldots, y_n)$, $Z = (z_1, z_2, \ldots, z_n)$, $U = (u_1, u_2, \ldots, u_n)$ and $V = (v_1, v_2, \ldots, v_n)$. We shall show that D_{2n-1} is stability preserving for $n \geq 1$. It should be noticed that D_{2n} is not always stability preserving for $n \geq 1$. For example, the polynomial $(u_n + 1)(v_n + 1)$ is clearly stable, but

$$D_{2n}((u_n + 1)(v_n + 1)) = x_n z_n (u_n(v_n + 1) + y_n(u_n + 1))$$

is not stable since it vanishes for $y_n = i + 2, u_n = i, v_n = i - 4$. Nevertheless, when restricted to polynomials $u_n g$, where g is a polynomial in X, Y, Z, U and V that is independent of the variable u_n, there is a stability preserving operator T_n that is equivalent to D_{2n}.

Theorem 17 *For $n \geq 1$, the multivariate polynomial $B_n(X, Y, Z, U, V)$ is stable.*

Proof For $1 \leq k \leq 2n$, let

$$f_k = D_k D_{k-1} \cdots D_1(x_0),$$

which is a polynomial in

$$A_k = \{x_i, y_i, z_i, u_i, v_i \mid 1 \leq i \leq \lfloor (k+1)/2 \rfloor\}.$$

So $f_{2n} = B_n(X, Y, Z, U, V)$. For $1 \leq k \leq 2n$, it can be seen that f_k is multiaffine. We proceed to prove the stability of f_{2n} by induction on n. The stability of z_0 is evident. For $n \geq 1$, assume that f_{2n-2} is stable. Let us consider the actions of D_{2n-1} and D_{2n}.

First, we show that D_{2n-1} preserves stability of multiaffine polynomials. Let

$$A'_k = \{x'_i, y'_i, z'_i, u'_i, v'_i \mid 1 \leq i \leq \lfloor (k+1)/2 \rfloor\}.$$

According to Theorem 11, it suffices to show that the polynomial $D_{2n-1}(F)$ is stable, where

$$F = \prod_{i=1}^n (x_i + x'_i) \prod_{i=1}^n (y_i + y'_i) \prod_{i=1}^n (z_i + z'_i) \prod_{i=1}^n (u_i + u'_i) \prod_{i=1}^n (v_i + v'_i).$$

Let

$$\xi = \sum_{i=1}^{n-1} \left(\frac{1}{x_i + x'_i} + \frac{1}{y_i + y'_i} + \frac{1}{z_i + z'_i} + \frac{1}{u_i + u'_i} + \frac{1}{v_i + v'_i} \right).$$

By Proposition 10,

$$D_{2n-1}(F) = \sum_{i=1}^{n-1} D_{2n-1}(x_i) \frac{\partial F}{\partial x_i} + \sum_{i=1}^{n-1} D_{2n-1}(y_i) \frac{\partial F}{\partial y_i} + \sum_{i=1}^{n-1} D_{2n-1}(z_i) \frac{\partial F}{\partial z_i}$$

$$+ \sum_{i=1}^{n-1} D_{2n-1}(u_i) \frac{\partial F}{\partial u_i} + \sum_{i=1}^{n-1} D_{2n-1}(v_i) \frac{\partial F}{\partial v_i}$$

$$= u_n v_n \sum_{i=1}^{n-1} \left(\frac{F}{x_i + x'_i} + \frac{F}{y_i + y'_i} + \frac{F}{z_i + z'_i} + \frac{F}{u_i + u'_i} + \frac{F}{v_i + v'_i} \right)$$

$$= u_n v_n \xi F.$$

Assume that all the variables in A_{2n} and A'_{2n} have positive imaginary parts. Then each factor in F is nonzero, and so $F \neq 0$. Similarly, each term in ξ has a negative imaginary part, which implies that $\xi \neq 0$. Hence $D_{2n-1}(F) \neq 0$. This proves that D_{2n-1} is stability preserving. By the induction hypothesis, we deduce that f_{2n-1} is stable.

Next we turn to the operator D_{2n}. Define

$$T_n = x_n z_n I + x_n y_n z_n \sum_{i=1}^{n-1} \left(\frac{\partial}{\partial x_i} + \frac{\partial}{\partial y_i} + \frac{\partial}{\partial z_i} + \frac{\partial}{\partial u_i} \right) + x_n y_n z_n \sum_{i=1}^n \frac{\partial}{\partial v_i}.$$

Since f_{2n-1} can be written in the form $u_n g$, where g is a polynomial in X, Y, Z, U and V that is independent of u_n, using Lemma 13, we find that

$$f_{2n} = D_{2n}(f_{2n-1}) = T_n(f_{2n-1}).$$

To prove that T_n preserves stability of multiaffine polynomials, let

$$F = \prod_{i=1}^{n}(x_i + x_i') \prod_{i=1}^{n}(y_i + y_i') \prod_{i=1}^{n}(z_i + z_i') \prod_{i=1}^{n}(u_i + u_i') \prod_{i=1}^{n}(v_i + v_i').$$

Then

$$T_n(F) = x_n y_n z_n F \sum_{i=1}^{n-1} \left(\frac{1}{x_i + x_i'} + \frac{1}{y_i + y_i'} + \frac{1}{z_i + z_i'} + \frac{1}{u_i + u_i'} \right)$$

$$+ x_n y_n z_n F \sum_{i=1}^{n} \frac{1}{v_i + v_i'} + x_n z_n F$$

$$= x_n y_n z_n \xi F,$$

where

$$\xi = \frac{1}{y_n} + \sum_{i=1}^{n-1} \left(\frac{1}{x_i + x_i'} + \frac{1}{y_i + y_i'} + \frac{1}{z_i + z_i'} + \frac{1}{u_i + u_i'} \right) + \sum_{i=1}^{n} \frac{1}{v_i + v_i'}.$$

Assume that all the variables in A_{2n} and A_{2n}' have positive imaginary parts. By Theorem 11, it suffices to verify that $T_n(F) \neq 0$. For $1 \leq i \leq n$, since $x_i + x_i'$, $y_i + y_i'$, $z_i + z_i'$, $u_i + u_i'$, and $v_i + v_i'$ all have positive imaginary parts, we see that

$$\frac{1}{x_i + x_i'}, \frac{1}{y_i + y_i'}, \frac{1}{z_i + z_i'}, \frac{1}{u_i + u_i'}, \quad \text{and} \quad \frac{1}{v_i + v_i'}$$

all have negative imaginary parts. Similarly, under the assumption that y_n has a positive imaginary part, it can be seen that $\frac{1}{y_n}$ has a negative imaginary part. Thus we find that $\xi \neq 0$ and $F \neq 0$. Consequently, $T_n(F) \neq 0$. This leads to the stability of $T_n(F)$. Finally, in light of Theorem 11, we conclude that f_{2n} is stable. This completes the proof. □

Acknowledgements We wish to thank the referee for valuable suggestions. This work is supported by the National Science Foundation of China, the Scientific Research Foundation of Tianjin University of Technology and Education, and the Scientific Research Foundation of Nanjing Institute of Technology.

References

1. Bóna, M.: Real zeros and normal distribution for statistics on Stirling permutations defined by Gessel and Stanley. SIAM J. Discrete Math. **23**, 401–406 (2009)
2. Borcea, J., Brändén, P.: Applications of stable polynomials to mixed determinants: Johnson's conjectures, unimodality and symmetrized Fischer products. Duke Math. J. **143**(2), 205–223 (2008)
3. Borcea, J., Brändén, P.: Pólya–Schur master theorems for circular domains and their boundaries. Ann. Math. **170**, 465–492 (2009)
4. Borcea, J., Brändén, P.: The Lee–Yang and Pólya–Schur programs I: linear operators preserving stability. Invent. Math. **177**(3), 541–569 (2009)
5. Borcea, J., Brändén, P.: The Lee–Yang and Pólya–Schur programs II: theory of stable polynomials and applications. Commun. Pure Appl. Math. **62**(12), 1595–1631 (2009)
6. Brändén, P., Leander, M.: Multivariate P-Eulerian polynomials (2016, preprint). arXiv:1604.04140
7. Brändén, P., Leander, M., Visontai, M.: Multivariate Eulerian polynomials and exclusion processes. Combin. Probab. Comput. **25**, 486–499 (2016)
8. Brenti, F.: Unimodal, Log-Concave and Pólya Frequency Sequences in Combinatorics. Memoirs of the American Mathematical Society, vol. 413. American Mathematical Society, Providence (1989)
9. Chen, W.Y.C.: Context-free grammars, differential operators and formal power series. Theor. Comput. Sci. **117**(1), 113–129 (1993)
10. Chen, W.Y.C., Fu, A.M.: Context-free grammars for permutations and increasing trees. Adv. Appl. Math. **82**, 58–82 (2017)
11. Dumont, D.: Grammaires de William Chen et dérivations dans les arbres et arborescences. Sém. Lothar. Combin. **37**, 1–21 (1996)
12. Egge, E.S.: Legendre-Stirling permutations. Eur. J. Combin. **31**(7), 1735–1750 (2010)
13. Gessel, I., Stanley, R.P.: Stirling polynomials. J. Combin. Theory Ser. A **24**(1), 24–33 (1978)
14. Graham, R.L., Knuth, D.E., Patashnik, O.: Concrete Mathematics. A Foundation for Computer Science. Addison-Wesley, Reading (1994)
15. Haglund, J., Visontai, M.: On the monotone column permanent conjecture. In: Proceedings of FPSAC 2009. Discrete Mathematics & Theoretical Computer Science (2009) pp. 443–454
16. Haglund, J., Visontai, M.: Stable multivariate Eulerian polynomials and generalized Stirling permutations. Eur. J. Combin. **33**(4), 477–487 (2012)
17. Harper, L.H.: Stirling behaviour is asymptotically normal. Ann. Math. Stat. **38**(2), 410–414 (1967)
18. Janson, S.: Plane recursive trees, Stirling permutations and an urn model. In: Proceedings of Fifth Colloquium on Mathematics and Computer Science. Discrete Mathematics & Theoretical Computer Science (2008), pp. 541–547
19. Janson, S., Kuba, M., Panholzer, A.: Generalized Stirling permutations, families of increasing trees and urn models. J. Combin. Theory Ser. A **118**(1), 94–114 (2011)
20. Levande, P.: Two new interpretations of the Fishburn numbers and their refined generating functions (2010). arXiv:1006.3013v1
21. Liu, L.L., Wang, Y.: A unified approach to polynomial sequences with only real zeros. Adv. Appl. Math. **38**(4), 542–560 (2007)
22. Pólya, G., Schur, J.: Über zwei Arten von Faktorenfolgen in der Theorie der algebraischen Gleichungen. J. Reine Angew. Math. **144**, 89–113 (1914)
23. Riordan, J.: The blossoming of Schröder's fourth problem. Acta Math. **137**(1), 1–16 (1976)
24. Sokal, A.: The multivariate Tutte polynomial (alias Potts model) for graphs and matroids. In: Surveys in Combinatorics 2005, Cambridge University Press, Cambridge (2005), pp. 173–226

25. Visontai, M., Williams, N.: Stable multivariate W-Eulerian polynomials. J. Combin. Theory Ser. A **120**(7), 1929–1945 (2013)
26. Wagner, D.G.: Multivariate stable polynomials: theory and applications. Bull. Amer. Math. Soc. (N.S.) **48**(1), 53–84 (2011)

An Interesting Class of Hankel Determinants

Johann Cigler and Mike Tyson

Dedicated to Professor Peter Paule on the occasion of his 60th birthday

1 Introduction

Let $(a_n)_{n \geq 0}$ be a sequence of real numbers with $a_0 = 1$. For each n consider the Hankel determinant

$$H_n = \det(a_{i+j})_{i,j=0}^{n-1}.$$

We are interested in the sequence $(H_n)_{n \geq 0}$ for the sequences $a_{n,r} = \binom{2n+r}{n}$ for some $r \in \mathbb{N}$. For $n = 0$ we let $H_0 = 1$.

Let

$$d_r(n) = \det\left(\binom{2i+2j+r}{i+j} \right)_{i,j=0}^{n-1}.$$

For $r = 0$ and $r = 1$ these determinants are well known and satisfy $d_0(n) = 2^{n-1}$ and $d_1(n) = 1$ for $n > 0$. Eğecioğlu et al. [3] computed $d_2(n)$ and $d_3(n)$ and stated some conjectures for $r > 3$.

J. Cigler (✉)
Fakultät für Mathematik, Universität Wien, Vienna, Austria
e-mail: johann.cigler@univie.ac.at

M. Tyson

© Springer Nature Switzerland AG 2020
V. Pillwein, C. Schneider (eds.), *Algorithmic Combinatorics: Enumerative Combinatorics, Special Functions and Computer Algebra*, Texts & Monographs in Symbolic Computation, https://doi.org/10.1007/978-3-030-44559-1_8

137

Many of these determinants are easy to guess and show an interesting modular pattern. For example

$$(d_0(n))_{n \geq 0} = (1, 1, 2, 2^2, 2^3, \dots),$$

$$(d_1(n))_{n \geq 0} = (1, 1, 1, 1, 1, \dots),$$

$$(d_2(n))_{n \geq 0} = (1, 1, -1, -1, 1, 1, -1, -1, \dots),$$

$$(d_3(n))_{n \geq 0} = (1, 1, -4, 3, 3, -8, 5, 5, -12, 7, 7, -16, \dots),$$

$$(d_4(n))_{n \geq 0} = (1, 1, -8, 8, 1, 1, -16, 16, 1, 1, -24, 24, \dots),$$

$$(d_5(n))_{n \geq 0} = (1, 1, -13, -16, 61, 9, 9, -178, -64, 370, 25, 25, -695, -144, 1127, \dots)$$

These and other computations suggest the following evaluations:

$$d_{2k+1}((2k+1)n) = d_{2k+1}((2k+1)n+1) = (2n+1)^k,$$

$$d_{2k+1}((2k+1)n+k+1) = (-1)^{\binom{k+1}{2}}4^k(n+1)^k,$$

$$d_{2k}(2kn) = d_{2k}(2kn+1) = (-1)^{kn},$$

$$d_{2k}(2kn+k) = -d_{2k}(2kn+k+1) = (-1)^{kn+\binom{k}{2}}4^{k-1}(n+1)^{k-1}.$$

The purpose of this paper is to prove these conjectures. Our methods seem to extend to the Hankel determinants of the sequences $\left(\binom{2n+r}{n-s}\right)_{n \geq 0}$, but we do not compute these here.

In Sects. 2 and 3 we review some well-known facts from the theory of Hankel determinants. In particular we compute $d_0(n)$ and $d_1(n)$. In Sect. 4 we define the matrix γ and use it to compute $d_2(n)$. In Sect. 5 we introduce the matrices α_n and β_n, which serve as the basis of our method. In Sect. 6 we write the Hankel matrices in terms of these matrices. In Sects. 7 and 8 we use this information to compute $d_r(n)$ in the aforementioned seven cases.

We would like to thank Darij Grinberg for his helpful suggestions.

2 Some Background Material

Let us first recall some well-known facts about Hankel determinants (cf. e.g. [1]). If $d_n = \det(a_{i+j})_{i,j=0}^{n-1} \neq 0$ for each n we can define the polynomials

$$p_n(x) = \frac{1}{d_n} \det \begin{pmatrix} a_0 & a_1 & \cdots & a_{n-1} & 1 \\ a_1 & a_2 & \cdots & a_n & x \\ a_2 & a_3 & \cdots & a_{n+1} & x^2 \\ \vdots & & & & \vdots \\ a_n & a_{n+1} & \cdots & a_{2n-1} & x^n \end{pmatrix}.$$

If we define a linear functional L on the polynomials by $L(x^n) = a_n$ then $L(p_n p_m) = 0$ for $n \neq m$ and $L(p_n^2) \neq 0$ (orthogonality).

By Favard's Theorem there exist complex numbers s_n and t_n such that

$$p_n(x) = (x - s_{n-1})p_{n-1}(x) - t_{n-2}p_{n-2}(x).$$

For arbitrary s_n and t_n define numbers $a_n(j)$ by

$$a_0(j) = [j = 0],$$

$$a_n(0) = s_0 a_{n-1}(0) + t_0 a_{n-1}(1), \tag{1}$$

$$a_n(j) = a_{n-1}(j-1) + s_j a_{n-1}(j) + t_j a_{n-1}(j+1).$$

These numbers satisfy

$$\sum_{j=0}^{n} a_n(j) p_j(x) = x^n. \tag{2}$$

Let $A_n = (a_i(j))_{i,j=0}^{n-1}$ and D_n be the diagonal matrix with entries $d(i,i) = \prod_{j=0}^{i-1} t_j$. Then we get

$$\left(a_{i+j}(0)\right)_{i,j=0}^{n-1} = A_n D_n A_n^\top \tag{3}$$

and

$$\det \left(a_{i+j}(0)\right)_{i,j=0}^{n-1} = \prod_{i=1}^{n-1} \prod_{j=0}^{i-1} t_j.$$

If we start with the sequence $(a_n)_{n \geq 0}$ and guess s_n and t_n and if we also can guess $a_n(j)$ and show that $a_n(0) = a_n$ then all our guesses are correct and the Hankel determinant is given by the above formula.

There is a well-known equivalence with continued fractions, so-called J-fractions:

$$\sum_{n \geq 0} a_n x^n = \cfrac{1}{1 - s_0 x - \cfrac{t_0 x^2}{1 - s_1 x - \cfrac{t_1 x^2}{1 - \cdots}}}.$$

For some sequences this gives a simpler approach to Hankel determinants.

As is well known Hankel determinants are intimately connected with the Catalan numbers $C_n = \frac{1}{n+1}\binom{2n}{n}$. Consider for example the aerated sequence of Catalan numbers $(c_n) = (1, 0, 1, 0, 2, 0, 5, 0, 14, 0, \dots)$ defined by $c_{2n} = C_n$ and $c_{2n+1} = 0$. Since the generating function of the Catalan numbers

$$C(x) = \sum_{n \geq 0} C_n x^n = \frac{1 - \sqrt{1 - 4x}}{2x}$$

satisfies

$$C(x) = 1 + xC(x)^2,$$

we get

$$C(x) = \frac{1}{1 - xC(x)}$$

and

$$C(x^2) = \frac{1}{1 - x^2 C(x^2)} = \cfrac{1}{1 - \cfrac{x^2}{1 - \cfrac{x^2}{1 - \ddots}}}$$

and therefore

$$\det(c_{i+j})_{i,j=0}^{n-1} = 1.$$

From $C(x) = 1 + xC(x)^2$ we get $C(x)^2 = 1 + 2xC(x)^2 + x^2 C(x)^4$ or

$$C(x)^2 = \frac{1}{1 - 2x - x^2 C(x)^2} = \cfrac{1}{1 - 2x - \cfrac{x^2}{1 - 2x - \cfrac{x^2}{1 - 2x - \ddots}}}. \tag{4}$$

The generating function of the central binomial coefficients $B_n = \binom{2n}{n}$ is

$$B(x) = \sum_{n \geq 0} B_n x^n = \frac{1}{\sqrt{1 - 4x}} = \frac{1}{1 - 2xC(x)} = \frac{1}{1 - 2x - 2x^2 C(x)^2}.$$

Therefore by (4) we get the J-fraction

$$B(x) = \cfrac{1}{1 - 2x - 2x^2 C(x)^2} = \cfrac{1}{1 - 2x - \cfrac{2x^2}{1 - 2x - \cfrac{x^2}{1 - 2x - \cfrac{x^2}{1 - 2x - \ddots}}}}.$$

Thus the corresponding numbers t_n are given by $t_0 = 2$ and $t_n = 1$ for $n > 0$ which implies $d_0(n) = 2^{n-1}$ for $n \geq 1$.

Let us also consider the aerated sequence (b_n) with $b_{2n} = B_n$ and $b_{2n+1} = 0$. Here we get

$$b(x) = B(x^2) = \cfrac{1}{1 - 2x^2 C(x)^2} = \cfrac{1}{1 - \cfrac{2x^2}{1 - \cfrac{x^2}{1 - \cfrac{x^2}{1 - \ddots}}}}.$$

In this case $s_n = 0$, $t_0 = 2$, and $t_n = 1$ for $n > 0$. Here we also get $\det(b_{i+j})_{i,j=0}^{n-1} = 2^{n-1}$ for $n > 0$. The corresponding orthogonal polynomials satisfy $p_0(x) = 1$, $p_1(x) = x$, $p_2(x) = xp_1(x) - 2$ and $p_n(x) = xp_{n-1}(x) - p_{n-2}(x)$ for $n > 2$. The first terms are $1, x, x^2 - 2, x^3 - 3x, \ldots$.

Now recall that the Lucas polynomials

$$L_n(x) = \sum_{k=0}^{\lfloor \frac{n}{2} \rfloor} (-1)^k \binom{n-k}{k} \frac{n}{n-k} x^{n-2k}$$

for $n > 0$ satisfy $L_n(x) = xL_{n-1}(x) - L_{n-2}(x)$ with initial values $L_0(x) = 2$ and $L_1(x) = x$. The first terms are $2, x, x^2 - 2, x^3 - 3x, \ldots$. Thus $p_n(x) = \bar{L}_n(x)$, where $\bar{L}_n(x) = L_n(x)$ for $n > 0$ and $\bar{L}_0(x) = 1$.

For the numbers $a_n(j)$ we get

$$a_{2n}(2j) = \binom{2n}{n-j},$$

$$a_{2n+1}(2j+1) = \binom{2n+1}{n-j},$$

and $a_n(j) = 0$ else. Equivalently $a_n(n-2j) = \binom{n}{j}$ and $a_n(k) = 0$ else.

For the proof it suffices to verify (1) which reduces to the trivial identities $\binom{2n}{n} = 2\binom{2n-1}{n-1}$, $\binom{2n}{n-j} = \binom{2n-1}{n-j} + \binom{2n-1}{n-1-j}$, and $\binom{2n+1}{n-j} = \binom{2n}{n-j} + \binom{2n}{n-1-j}$. Identity (2) reduces to

$$\sum_{k=0}^{\lfloor \frac{n}{2} \rfloor} \binom{n}{k} \bar{L}_{n-2k} = x^n. \tag{5}$$

3   Some Well-Known Applications of These Methods

Now let us consider

$$d_1(n) = \det \binom{2i + 2j + 1}{i + j}.$$

The generating function of the sequence $\binom{2n+1}{n}$ is

$$\sum_{n \geq 0} \binom{2n+1}{n} x^n = \frac{1}{2} \sum_{n \geq 0} \binom{2n+2}{n+1} x^n = \frac{1}{2x} \left(\frac{1}{\sqrt{1 - 4x}} - 1 \right) = \frac{C(x)}{\sqrt{1 - 4x}}.$$

Now we have

$$\sqrt{1 - 4x} = 1 - 2xC(x) = (C(x) - xC(x)^2) - 2xC(x) = C(x)(1 - 2x - xC(x))$$

$$= C(x)(1 - 2x - x(1 + xC(x)^2)) = C(x)(1 - 3x - x^2 C(x)^2).$$

Therefore

$$\frac{C(x)}{\sqrt{1 - 4x}} = \frac{1}{1 - 3x - x^2 C(x)^2} = \cfrac{1}{1 - 3x - \cfrac{x^2}{1 - 2x - \cfrac{x^2}{1 - 2x - \cfrac{x^2}{1 - 2x - \ddots}}}}.$$

The corresponding sequences s_n, t_n are $s_0 = 3$, $s_n = 2$ for $n > 0$ and $t_n = 1$. Thus $d_1(n) = 1$. The corresponding $a_i(j)$ are $a_i(j) = \binom{2i+1}{i-j}$.

To prove this one must verify (1) which reduces to

$$\binom{1}{-j} = [j = 0],$$

$$\binom{2n+1}{n} = 3\binom{2n-1}{n-1} + \binom{2n-1}{n-2},$$

$$\binom{2n+1}{n-j} = \binom{2n-1}{n-j} + 2\binom{2n-1}{n-1-j} + \binom{2n-1}{n-2-j}.$$

By (3) we see that with

$$A_n = \left(\binom{2i+1}{i-j}\right)_{i,j=0}^{n-1}$$

we get

$$A_n A_n^\top = \left(\binom{2i+2j+1}{i+j}\right)_{i,j=0}^{n-1}. \tag{6}$$

Since A_n is a triangle matrix whose diagonal elements are $\binom{2i+1}{i-i} = 1$ we get $\det(A_n A_n^\top) = 1$.

4 A New Method

Fix $k > 0$. Let us consider the determinants of the Hankel matrices $B_n(k) = \left(\binom{2i+2j+2}{i+j+1-k}\right)_{i,j=0}^{n-1}$. These have already been computed in [2], Corollary 20. There it is shown that

$$\det(B_{km}(k)) = (-1)^{\binom{m}{2}k + m\binom{k}{2}} \tag{7}$$

and $\det(B_n(k)) = 0$ else.

Definition 4.1 For $k \geq 1$, let $\gamma^{(k)}$ be the infinite matrix given by $\gamma_{ij}^{(k)} = 1$ if $|i - j| = k$ or $i + j = k - 1$ and 0 elsewhere, with rows and columns indexed by $\mathbb{Z}_{\geq 0}$. Set $\gamma^{(0)} = 2I_\infty$ and $\gamma^{(-k)} = \gamma^{(k)}$. Let us also consider the finite truncations $\gamma^{(k)}|_N$, where $A|_N$ denotes the submatrix consisting of the first N rows and columns of a matrix A. We shall also write $\gamma^{(1)} = \gamma$ and $\gamma^{(k)}|_N = \gamma_N^{(k)}$.

For example $\gamma_5^{(1)}$ and $\gamma_5^{(2)}$ are the following matrices:

$$
\gamma_5^{(1)} = \begin{pmatrix} 1&1&0&0&0 \\ 1&0&1&0&0 \\ 0&1&0&1&0 \\ 0&0&1&0&1 \\ 0&0&0&1&0 \end{pmatrix} \qquad \gamma_5^{(2)} = \begin{pmatrix} 0&1&1&0&0 \\ 1&0&0&1&0 \\ 1&0&0&0&1 \\ 0&1&0&0&0 \\ 0&0&1&0&0 \end{pmatrix}
$$

An alternative description will be useful in this section and the next. Let J_n be the n-by-n exchange matrix with 1's on its antidiagonal and 0's elsewhere. Let Q_n be the block matrix $\left(J_n \ I_n \right)$. Let σ_n be the n-by-n shift matrix with (i, j) entry equal to 1 if $j = i - 1$ and 0 otherwise. Then $\gamma_n^{(k)} = Q_n \sigma_{2n}^k Q_n^\top$.

Theorem 4.2 *For $n \geq 1$ and all integer k,*

$$
A_n \gamma_n^{(k)} A_n^\top = B_n(k). \tag{8}
$$

Proof The $k = 0$ case of (8) is (6). By symmetry, it suffices to prove the $k > 0$ case. We have $(A_n Q_n)_{ij} = \binom{2i+1}{i+j-(n-1)}$. Hence the (i, j) entry of $A_n Q_n \sigma_{2n} Q_n^\top A_n^\top$ is

$$
\sum_{0 \leq r,s \leq n-1} (A_n Q_n)_{ir} (\sigma^k)_{rs} (A_n Q_n)_{js}
$$

$$
= \sum_{0 \leq r,s \leq n-1} \binom{2i+1}{i+r-(n-1)} \delta_{r-k,s} \binom{2j+1}{j+s-(n-1)}
$$

$$
= \sum_{r=n-1-i}^{n+i} \binom{2i+1}{i+r-(n-1)} \binom{2j+1}{j+r-k-(n-1)}
$$

$$
= \sum_{r'=0}^{2i+1} \binom{2i+1}{r'} \binom{2j+1}{j-i+r'-k}
$$

$$
= \binom{2i+2j+2}{i+j+1-k}.
$$

The last identity follows from the Chu–Vandermonde formula. $\square$

Lemma 4.3

$$
\det(\gamma_{2kn}^{(k)}) = (-1)^{kn}
$$

$$
\det(\gamma_{2kn+k}^{(k)}) = (-1)^{kn+\binom{k}{2}}
$$

and all other determinants $\det(\gamma_n^{(k)})$ *vanish.*

Proof By the definition of a determinant we have

$$\det(a_{i,j})_{i,j=0}^{n-1} = \sum_\pi \operatorname{sgn}(\pi) a_{0,\pi(0)} a_{1,\pi(1)} \cdots a_{n-1,\pi(n-1)}$$

where π runs over all permutations of the set $\{0, 1, \ldots, n-1\}$. We claim that the determinants of the matrices $\gamma_n^{(k)}$ either vanish or the sum over all permutations reduces to a single term $\operatorname{sgn}\pi_n \gamma^{(k)}(0, \pi_n(0)) \gamma^{(k)}(1, \pi_n(1)) \cdots \gamma^{(k)}(n-1, \pi_n(n-1))$.

Let us first consider $k = 1$. The last row of $\gamma_n^{(1)}$ has only one non-vanishing element $\gamma^{(1)}(n-1, n-2)$. Thus each π which occurs in the determinant must satisfy $\pi(n-1) = n-2$. The next row from below contains two non-vanishing elements $\gamma^{(1)}(n-2, n-3)$ and $\gamma^{(1)}(n-2, n-1)$. The last element is the only element of the last column. Therefore we must have $\pi(n-2) = n-1$. The next row from below contains again two non-vanishing elements, $\gamma^{(1)}(n-3, n-4)$ and $\gamma^{(1)}(n-3, n-2)$. But since $n-2$ already occurs as image of π we must have $\pi(n-3) = n-4$. Thus the situation has been reduced to $\gamma_{n-2}^{(1)}$. In order to apply induction we need the two initial cases $\gamma_1^{(1)}$ and $\gamma_2^{(1)}$.

For $n = 1$ we get $\pi(0) = 0$ and for $n = 2$ $\pi(0) = 1$ and $\pi(1) = 0$ since

$$\gamma_2^{(1)} = \begin{pmatrix} 1 & 1 \\ 1 & 0 \end{pmatrix}.$$

If we write $\pi = \pi(0) \cdots \pi(n-1)$ we get in this way $\pi_1 = 0$, $\pi_2 = 10$, $\pi_3 = 021$, $\pi_4 = 1032, \ldots$ This gives $\operatorname{sgn}\pi_n = -\operatorname{sgn}\pi_{n-2}$ and thus by induction $\det \gamma_n^{(1)} = (-1)^{\binom{n}{2}}$, which agrees with (7).

For general k the situation is analogous. The last k rows and columns contain only one non-vanishing element. This implies $\pi(n-j) = n-j-k$ and $\pi(n-j-k) = n-j$ for $1 \le j \le k$ and $n \ge 2k$. Hence π restricts to a permutation of $\{0, 1, \ldots, n-2k-1\}$. Thus the determinant can be reduced to $\gamma_{n-2k}^{(k)}$ and we get $\det \gamma_n^{(k)} = (-1)^k \det \gamma_{n-2k}^{(k)}$ if $n \ge 2k$.

For $n = k$ $\gamma_k^{(k)}$ reduces to the anti-diagonal and thus $\det \gamma_k^{(k)} = (-1)^{\binom{k}{2}}$. For $0 < n < k$ the first row of $\gamma_n^{(k)}$ vanishes and thus $\det \gamma_n^{(k)} = 0$. For $k < n < 2k$ there are two identical rows because $\gamma^{(k)}(k-1, 0) = \gamma^{(k)}(k, 0) = 1$ and $\gamma^{(k)}(k-1, j) = \gamma^{(k)}(k, j) = 0$ for $0 < j < n$. Thus we see by induction that

$$\det(\gamma_{2kn}^{(k)}) = (-1)^{kn}$$

$$\det(\gamma_{2kn+k}^{(k)}) = (-1)^{kn+\binom{k}{2}}$$

and all other determinants vanish. This is the same as (7) because $(-1)^{\binom{2n}{2}k+2n\binom{k}{2}} = (-1)^{kn}$ and $(-1)^{\binom{2n+1}{2}k+(2n+1)\binom{k}{2}} = (-1)^{kn+\binom{k}{2}}$. $\square$

5 Two Useful Matrices

Recall that $\gamma = \gamma^{(1)}$. For the finite matrices $\gamma_N = \gamma|_N$ we have $\gamma_N^k \neq \gamma^k|_N$. In order to compute $\gamma^k|_N$ in the realm of N-by-N-matrices we introduce the auxiliary matrices $\alpha_N^{(k)}$ and $\beta_N^{(k)}$.

Let J_N be the exchange matrix with 1's on its antidiagonal and 0's elsewhere. Let Q_N be the block matrix $\begin{pmatrix} J_N & I_N \end{pmatrix}$. Let $\sigma_N(\varepsilon)$ be given by

$$(\sigma_N(\varepsilon))_{ij} = \begin{cases} 1 & \text{if } i = j+1 \\ \varepsilon & \text{if } (i, j) = (0, N-1) \\ 0 & \text{otherwise.} \end{cases}$$

Define $\alpha_N^{(k)}$ and $\beta_N^{(k)}$ as

$$Q_N \sigma_{2N}(\varepsilon)^k Q_N^{\top} = \begin{cases} \alpha_N^{(k)} & \text{if } \varepsilon = 1 \\ \beta_N^{(k)} & \text{if } \varepsilon = -1 \\ \gamma_N^{(k)} & \text{if } \varepsilon = 0, \end{cases}$$

and the last line has been stated before. We shall again suppress the superscripts when $k = 1$.

As a slight variation, consider the following infinite square matrices with rows and columns indexed by $\mathbb{Z} \setminus \{0\} = \{\ldots, -2, -1, 1, 2, \ldots\}$. Let $\bar{I}$ be the identity matrix and let $\bar{J}$ be the exchange matrix with $\bar{J}_{n,-n} = 1$ for all n and 0 elsewhere. Let $\bar{\sigma}$ be given by $\bar{\sigma}_{n,n-1} = 1$ and 0 elsewhere. Define also the infinite rectangular matrix $\bar{Q}$ with rows indexed by $\mathbb{Z}_+ = \{1, 2, \ldots\}$ and columns indexed by $\mathbb{Z} \setminus \{0\}$ by $\bar{Q}_{|n|,n} = 1$ for $n \in \mathbb{Z} \setminus \{0\}$ and 0 elsewhere. Note that $\gamma^{(k)} = \bar{Q} \bar{\sigma}^k \bar{Q}^T$, after shifting indices from $\mathbb{Z}_{\geq 0}$ to $\mathbb{Z}_+$.

Theorem 5.1 *When δ stands for either α_N, β_N, or γ one has $\delta^{(k)} = \delta \cdot \delta^{(k-1)} - \delta^{(k-2)}$ with initial values $\delta^{(1)} = \delta$ and $\delta^{(0)} = 2$.*

Proof For α_N and β_N, take $\sigma = \sigma_{2N}(\pm 1)$, $Q = Q_N$, $J = J_{2N}$, and $I = I_{2N}$. For γ, take $\sigma = \bar{\sigma}$, $Q = \bar{Q}$, $J = \bar{J}$, and $I = \bar{I}$. Note that in either case $Q^{\top} Q = I + J$, $\sigma J \sigma = J$, and $QJ = Q$. For $k \geq 2$,

$$\delta \cdot \delta^{(k-1)} = Q \sigma Q^{\top} Q \sigma^{k-1} Q^{\top}$$
$$= Q \sigma (I + J) \sigma^{k-1} Q^{\top}$$
$$= Q \sigma^k Q^{\top} + Q(\sigma J \sigma) \sigma^{k-2} Q^{\top}$$
$$= \delta^{(k)} + \delta^{(k-2)}.$$

$\square$

By induction we see that each $\gamma^{(k)}$ is a polynomial in γ. Therefore all $\gamma^{(k)}$ commute. Theorem 5.1 shows that the matrices $\gamma^{(k)}$ are Lucas polynomials in γ. More precisely

$$\gamma^{(k)} = L_k(\gamma). \tag{9}$$

By the same argument, $\alpha_N^{(k)} = L_k(\alpha_N)$ and $\beta_N^{(k)} = L_k(\beta_N)$.

Theorem 5.2 *For any polynomial p with $\deg p \leq 2N$, $\frac{p(\alpha_N)+p(\beta_N)}{2} = p(\gamma)|_N$.*

Proof Note that $(L_0, \ldots, L_{2N})$ is a basis of the vector space of degree at most $2N$, since $\deg(L_k) = k$. Therefore it suffices to show that $(L_k(\alpha_N) + L_k(\beta_N))/2 = L_k(\gamma)|_N$ for $k \leq 2N$. To wit,

$$\begin{aligned}
(L_k(\alpha_N) + L_k(\beta_N))/2 &= (\alpha_N^{(k)} + \beta_N^{(k)})/2 \\
&= Q_N(\sigma_{2N}(1)^k + \sigma_{2N}(-1)^k)Q_N^\top/2 \\
&= Q_N\sigma_{2N}(0)^k Q_N^\top \\
&= \gamma_N^{(k)} \\
&= L_k(\gamma)|_N.
\end{aligned}$$

$\square$

6   Relating the Determinant to the γ Matrices

Let a_n, b_n, and g_n be the characteristic polynomials of α_n, β_n, and γ_n, respectively. By cofactor expansion along the last row we get $g_n(x) = xg_{n-1}(x) - g_{n-2}(x)$, $a_n(x) = g_n(x) - g_{n-1}(x)$, and $b_n(x) = g_n(x) + g_{n-1}(x)$. This plus the initial conditions of the $n = 1$ and 2 cases gives $b_n(x) = L_n(x)$,

$$g_n(x) = \sum_{k=0}^{n}(-1)^{n-k}\bar{L}_k(x), \tag{10}$$

and

$$a_n(x) = L_n(x) + 2\sum_{k=0}^{n-1}(-1)^{n-k}\bar{L}_k(x).$$

Here $\bar{L}_n(x)$ is the Lucas polynomial $L_n(x)$ except when $n = 0$, in which case it is 1.

By Theorem 4.2, $A\phi(\gamma)A^\top$ is Hankel for all polynomials ϕ. Here A represents the infinite matrix $\left(\binom{2i+1}{i-j}\right)_{i,j\geq0}$ with finite truncations $A|_n = A_n$. This is because $\phi(x)$ can be expanded as a sum of Lucas polynomials $L_k(x)$, each of which gives a

Hankel matrix. Moreover, multiplying the polynomial by $(x + 2)$ shifts the Hankel matrix forward by 1. It suffices to show this for $L_k(x)$. Recall that

$$AL_k(\gamma)A^\top = A\gamma^{(k)}A^\top = \left(\binom{2i + 2j + 2}{i + j + 1 - k}\right)_{i,j \geq 0}.$$

Then by Theorem 5.1,

$$AL_k(\gamma)(\gamma + 2)A^\top = A(\gamma^{(k-1)} + \gamma^{(k+1)} + 2\gamma^{(k)})A^\top$$

has (i, j) entry

$$\binom{2i + 2j + 2}{i + j + 2 - k} + 2\binom{2i + 2j + 2}{i + j + 1 - k} + \binom{2i + 2j + 2}{i + j - k} = \binom{2i + 2j + 4}{i + j + 2 - k}$$

by Pascal's identity, which is the $(i + 1, j)$ entry of the original matrix.

We will now write the Hankel matrices of the sequence $\left(\binom{2n+r}{n}\right)_{n \geq 0}$ explicitly in terms of the γ matrices and A.

Theorem 6.1 *For $r \geq 1$, let $k = \lfloor \frac{r}{2} \rfloor$ and $l = \lfloor \frac{r-1}{2} \rfloor$, and define the function*

$$h_r(x) = \begin{cases} g_k(x) & \text{if } r = 2k + 1 \\ b_k(x) & \text{if } r = 2k. \end{cases}$$

For $N \geq k + l$, $d_r(N)$ equals

$$\det\left(h_r(\gamma)(\gamma + 2)^l |_N\right) = \det\left(\frac{1}{2}\left(h_r(\alpha_N)(\alpha_N + 2)^l + h_r(\beta_N)(\beta_N + 2)^l\right)\right).$$

Proof By the above results, when $i + j = n$ we have

$$(Ab_k(\gamma)A^\top)_{ij} = (AL_k(\gamma)A^\top)_{ij} = \binom{2n + 2}{n + 1 - k}$$

and

$$(Ab_k(\gamma)(\gamma + 2)^{k-1}A^\top)_{ij} = \binom{2(n + k - 1) + 2}{(n + k - 1) + 1 - k} = \binom{2n + 2k}{n}.$$

By induction on k, we will show $(Ag_k(\gamma)A^\top)_{ij} = \binom{2n+1}{n-k}$. The $k = 0$ case is (6). For $k \geq 1$,

$$(Ag_k(\gamma)A^\top)_{ij} = (A(b_k(\gamma) - g_{k-1}(\gamma))A^\top)_{ij}$$

$$= \binom{2n + 2}{n + 1 - k} - \binom{2n + 1}{n - (k - 1)} = \binom{2n + 1}{n - k}.$$

Hence

$$
(A g_k(\gamma)(\gamma + 2)^k \Lambda^\top)_{ij} = \binom{2(n+k)+1}{(n+k)-k} = \binom{2n+2k+1}{n}.
$$

The final claimed formula with α and β follows from Theorem 5.2.   □

7   Structure of the Matrices

In this section we determine the structure of the matrices $(\beta_N + 2)^{-1}$, $g_k(\alpha_N)$, $g_k(\beta_N)$, $b_k(\alpha_N)$, and $b_k(\beta_N)$, as well as the determinants of $g_k(\gamma)|_N$ and $b_k(\gamma)|_N$.

To determine $p(\alpha_N)$ and $p(\beta_N)$ for a polynomial p of degree less than N, we begin by writing $p(\gamma)$ as a sum of $\gamma^{(k)}$ matrices using the multiplicative formula of Theorem 5.1. We then apply Prop 7.2 to show that $p(\alpha_N)$ and $p(\beta_N)$ are the same as $p(\gamma)|_N$ on and above the anti-diagonal. The structure of $p(\alpha_N)$ follows from the symmetry of α_N across its anti-diagonal. The structure of $p(\beta_N)$ can be computed from $p(\alpha_N)$ and $p(\gamma)|_N$ with Theorem 5.2.

Proposition 7.1 *The determinant of a block matrix*

$$
\begin{pmatrix} A & B \\ C & D \end{pmatrix}
$$

where A and D are square and D is invertible is $\det(D)\det(A - BD^{-1}C)$.

Proof Note that

$$
\begin{pmatrix} A & B \\ C & D \end{pmatrix} \begin{pmatrix} I & 0 \\ -D^{-1}C & I \end{pmatrix} = \begin{pmatrix} A - BD^{-1}C & B \\ 0 & D \end{pmatrix},
$$

and that the determinant of a block-triangular matrix is the product of the determinants of its diagonal blocks.   □

Proposition 7.2 *Let T be an N-by-N tridiagonal matrix and let p be a polynomial of degree d. Let v be the N-by-1 column vector with a 1 in its last entry and 0 elsewhere. Then the (i, j) entries of $p(T)$ and $p(T + vv^\top)$ agree when $i + j \le 2(N-1) - d$.*

Proof It suffices to prove this for $p(x) = x^d$. Call an N-by-N matrix "k-small" if and only if its entries (i, j) with $i + j \le 2(N-1) - k$ are all 0. For instance, $vv^\top$ is 1-small.

Suppose a matrix M is k-small. For $i + j \le 2(N-1) - k - 1$, the (i, j) entry of TM is $\sum_{l=0}^{N-1} T_{il} M_{lj} = T_{i,i-1}M_{i-1,j} + T_{i,i}M_{i,j} + T_{i,i+1}M_{i+1,j}$. Since M is k-small, its $(i-1, j)$, (i, j), and $(i+1, j)$ entries are 0, which implies that TM is $(k+1)$-small. Similarly, MT, $vv^\top M$, and $Mvv^\top$ are $(k+1)$-small.

Consider $(T + vv^\top)^d - T^d$. Expanding the binomial product yields $2^d - 1$ terms, all of which are products of d T's and $vv^\top$'s and contain at least one $vv^\top$. It follows from the above that each of these terms is d-small, so $p(T + vv^\top) - p(T)$ is d-small. $\qquad\square$

Lemma 7.3 *The inverse of $(\beta_N + 2)$ is $(\frac{1}{2}(-1)^{i+j}(2\min\{i, j\} + 1))_{i,j=0}^{N-1}$. The determinant of $(\beta_N + 2)$ is 2. For example,*

$$(\beta_5 + 2)^{-1} = \frac{1}{2}\begin{pmatrix} 1 & -1 & 1 & -1 & 1 \\ -1 & 3 & -3 & 3 & -3 \\ 1 & -3 & 5 & -5 & 5 \\ -1 & 3 & -5 & 7 & -7 \\ 1 & -3 & 5 & -7 & 9 \end{pmatrix}.$$

Proof For $i \neq 0, N - 1$ the row i of $(\beta_N + 2)$ is $(2\delta_{il} + \delta_{i,l-1} + \delta_{i,l+1})_{l=0}^{N-1}$. The product of this with column j of the claimed inverse is

$$\sum_{l=0}^{N-1}(2\delta_{il} + \delta_{i,l-1} + \delta_{i,l+1})\frac{1}{2}(-1)^{l+j}(2\min\{l, j\} + 1)$$

$$= \frac{1}{2}(-1)^{i+j}(4\min\{i, j\} + 2 - 2\min\{i + 1, j\} - 1 - 2\min\{i - 1, j\} - 1)$$

$$= (-1)^{i+j}(2\min\{i, j\} - \min\{i + 1, j\} - \min\{i - 1, j\}).$$

This is 0 if $i + 1 \leq j$ or $i - 1 \geq j$ and is 1 if $i = j$.

The first row of $(\beta_N + 2)$ is $(3, 1, 0, \ldots, 0)$, and the last row is $(0, \ldots, 0, 1, 1)$. Column $j \neq 0, N - 1$ of the claimed inverse begins and ends as

$$\frac{1}{2}((-1)^j, (-1)^{j+1}3, \ldots, (-1)^{j+N-2}(2j + 1), (-1)^{j+N-1}(2j + 1)),$$

so it kills the first and last rows of $(\beta_N + 2)$. Column 0 of the claimed inverse begins and ends as $\frac{1}{2}(1, -1, \ldots, (-1)^{N-2}, (-1)^{N-1})$ while column $N - 1$ begins and ends as $\frac{1}{2}((-1)^{N-1}, (-1)^N 3, \ldots, -(2N - 3), 2N - 1)$. It is easy to verify that these columns have the correct products with rows of $(\beta_N + 2)$.

The determinant $\det(\beta + 2)$ is $(-1)^N b_N(-2)$, which can be computed with recurrence in Sect. 6 to be 2. $\qquad\square$

Lemma 7.4 *For $k < N$, the (i, j) entry of $g_k(\alpha_N)$ is $(-1)^{i+j+k}$ if $k \leq i + j \leq 2N-k-2$ and $|i-j| \leq k$ and is 0 otherwise. The (i, j) entry of $g_k(\beta_N)$ is $(-1)^{i+j+k}$ if $k \leq i + j \leq 2N - k - 2$ and $|i - j| \leq k$, is $2(-1)^{i+j+k}$ if $2N - k - 1 \leq i + j$,* *and is 0 otherwise. For example,*

$$g_2(\beta_6) = \begin{pmatrix} 0 & 0 & 1 & 0 & 0 & 0 \\ 0 & 1 & -1 & 1 & 0 & 0 \\ 1 & -1 & 1 & -1 & 1 & 0 \\ 0 & 1 & -1 & 1 & -1 & 1 \\ 0 & 0 & 1 & -1 & 1 & -2 \\ 0 & 0 & 0 & 1 & -2 & 2 \end{pmatrix}.$$

Proof Recall that $g_j(\gamma) = \gamma^{(j)} - \gamma^{(j-1)} + \cdots \pm \gamma^{(1)} \mp 1$, by (10). Therefore $\frac{1}{2}(g_k(\alpha_N) + g_k(\beta_N)) = g_k(\gamma)|_N = \gamma_N^{(k)} - \gamma_N^{(k-1)} + \cdots \pm \gamma_N^{(1)} \mp 1$. From the definition of the $\gamma_N^{(j)}$, the (i, j) entry of $g_k(\gamma)|_N$ is $(-1)^{i+j+k}$ if $k \le i + j$ and $|i - j| \le k$ and is 0 otherwise.

Note that polynomials in α_N are symmetric about their anti-diagonal. Since the degree of g_k is $k < N$, Proposition 7.2 says that $g_k(\alpha_N)$ agrees with $g_k(\gamma)|_N$ on and above its anti-diagonal. Thus, the (i, j) entry of $g_k(\alpha_N)$ is $(-1)^{i+j+k}$ if $k \le i + j \le 2N - k - 2$ and $|i - j| \le k$ and is 0 otherwise. Similarly, the (i, j) entry of $g_k(\beta_N) = 2g_k(\gamma)|_N - g_k(\alpha_N)$ is $(-1)^{i+j+k}$ if $k \le i + j \le 2N - k - 2$ and $|i - j| \le k$, $2(-1)^{i+j+k}$ if $2N - k - 1 \le i + j$, and 0 otherwise. $\square$

Lemma 7.5

$$\det g_k(\gamma)|_N = \begin{cases} 1 & \text{if } N = (2k + 1)n \\ (-1)^{\binom{k+1}{2}} & \text{if } N = (2k + 1)n + k + 1 \\ 0 & \text{otherwise.} \end{cases}$$

Proof When $N = 0$ the determinant is vacuously 1. When $0 < N < k + 1$, the first column is 0. When $N = k + 1$ the matrix is 0 above its antidiagonal and 1 on its antidiagonal, so its determinant is $(-1)^{\binom{k+1}{2}}$. When $k + 1 < N < 2k + 1$, columns $k - 1$ and $k + 1$ are equal. Thus the claim holds for all $N < 2k + 1$. We will show that for $N \ge 2k + 1$, $\det g_k(\gamma)|_N = \det g_k(\gamma)|_{N-2k-1}$.

Fix $N \ge 2k + 1$ and let $M = g_k(\gamma)|_N$. Subdivide M into a block matrix consisting of the leading principal order-$N - 1$ submatrix M_{11}, the bottom-right entry M_{22}, and the remainders of the last column and row M_{12} and M_{21}. The determinant of M is $\det(M_{22}) \det(M')$, where M' is the $N - 1$-by-$N - 1$ matrix $M_{11} - M_{12}M_{22}^{-1}M_{21}$ by Proposition 7.1.

We will perform cofactor expansion in the bottom right of M'. Since $M_{22} = (-1)^k$, the bottom right k-by-k submatrix of M' is the zero matrix. As a result, the only entry in the bottom row of M' is the 1 at $(N - 2, N - k - 2)$. After deleting its row and column, the only entry in the bottom row of M' is the 1 at $(N - 3, N - k - 3)$. This pattern continues up to the 1 at $(N - k - 1, N - 2k - 1)$. Since M' is symmetric, a similar sequence of lone 1's can be removed in the last k columns.

After the last $2k$ rows and columns have been removed, M' has been reduced to $g_k(\gamma)|_{N-2k-1}$. The $2k$ removed 1's contribute a factor of $(-1)^k$ to the determinant, which comes from the parity of the permutation $(0\ k)(1\ k+1)\cdots(k-1\ 2k)$. This cancels with the sign of M_{22}. □

Lemma 7.6 *For $k < N$, the (i, j) entry of $b_k(\alpha_N)$ is 1 if $|i - j| = k$, $i + j = k - 1$, or $i + j = 2(N - 1) - (k - 1)$ and is 0 otherwise. The (i, j) entry of $b_k(\beta_N)$ is 1 if $|i - j| = k$ or $i + j = k - 1$, is -1 if $i + j = 2(N - 1) - (k - 1)$, and is 0 otherwise. In particular $b_k(\gamma) = \gamma^{(k)}$. Moreover,*

$$
\det b_k(\gamma)|_N = \begin{cases} (-1)^{kn} & \text{if } N = 2kn \\ (-1)^{kn+\binom{k}{2}} & \text{if } N = 2kn + k \\ 0 & \text{otherwise.} \end{cases}
$$

Proof The first set of claims follow from the Lemma 7.4 and the fact that $b_k(x) = g_k(x) + g_{k-1}(x)$. The determinant of $\gamma^{(k)}$ was calculated in Lemma 4.3. □

8   Calculation of the Determinant

In this section we prove the seven formulas mentioned in the introduction. Recall Theorem 6.1 and its notation.

Let $\mu_i = \frac{1}{2}((\alpha_N + 2)^i h_r(\alpha_N) + (\beta_N + 2)^i h_r(\beta_N))$ for $0 \leq i \leq l$. From here on we will suppress the subscripts on α_N and β_N. By Theorem 6.1, we are interested in calculating $d_r(N) = \det \mu_l$. Note that

$$
\mu_{i+1} = \mu_i(\beta + 2) + (\alpha + 2)^i h_r(\alpha)vv^\top. \tag{11}
$$

The results of the previous section give us control over μ_0. We will induct on the above equation to screw the smoothing operators $\alpha + 2$ and $\beta + 2$ into place, using the matrix determinant lemma to keep track of the determinants. In the seven cases proven here, the determinant or adjugate of μ_i is multiplied by a constant factor at each step.

Proposition 8.1 (Matrix Determinant lemma) *If A is an n-by-n matrix and u and v are n-by-1 column vectors, then*

$$
\det(A + uv^\top) = \det(A) + v^\top \mathrm{adj}(A)u.
$$

Proof This is a polynomial identity in the entries of A, u, and v, so it suffices to prove it for the dense subset where A is invertible. Consider

$$
\begin{pmatrix} I & 0 \\ v^\top & 1 \end{pmatrix} \begin{pmatrix} I + A^{-1}uv^\top & u \\ 0 & 1 \end{pmatrix} \begin{pmatrix} I & 0 \\ -v^\top & 1 \end{pmatrix} = \begin{pmatrix} I & u \\ 0 & 1 + v^\top A^{-1}u \end{pmatrix},
$$

which shows that $1 \cdot \det(I + A^{-1}uv^{\top}) \cdot 1 = \det(1 + v^{\top}A^{-1}u)$. Multiplying through by $\det A$ yields $\det(A + uv^{\top}) = \det(A)(1 + v^{\top}A^{-1}u) = \det(A) + v^{\top} \mathrm{adj}(A)u$. $\quad\square$

8.1 The Case Where μ_0 Is Invertible

Lemma 8.2 *Suppose there is an N-dimensional column vector w such that $\mu_0 w = h_r(\alpha_N)v$ and that the last $l - 1$ entries of $h_r(\beta_N)w$ are 0. Then*

$$\det(\mu_l) = \det(\mu_0)2^l \left(1 + v^{\top}(\beta_N + 2)^{-1}w\right)^l.$$

Proof By Proposition 7.2, $(\alpha + 2)^i$ and $(\beta + 2)^i$ differ only in the last i columns. It follows from the second hypothesis that $(\beta + 2)^i h_r(\beta)w = (\alpha + 2)^i h_r(\beta)w$ for $0 \le i < l$. Thus

$$\mu_i w = (\alpha + 2)^i h_r(\alpha)v$$

and

$$\det(\mu_i)w = \mathrm{adj}(\mu_i)(\alpha + 2)^i h_r(\alpha)v$$

for $0 \le i < l$. By (11) and the matrix determinant lemma,

$$\det(\mu_{i+1}) = \det(\beta + 2)\left(\det(\mu_i) + v^{\top}(\beta + 2)^{-1}\mathrm{adj}(\mu_i)(\alpha + 2)^i h_r(\alpha)v\right)$$

$$= \det(\beta + 2)\left(\det(\mu_i) + v^{\top}(\beta + 2)^{-1}\det(\mu_i)w\right).$$

Hence

$$\det(\mu_{i+1}) = 2\det(\mu_i)\left(1 + v^{\top}(\beta_N + 2)^{-1}w\right).$$

$\quad\square$

Theorem 8.3 *For $n, k \ge 1$,*

$$d_{2k+1}((2k + 1)n) = (2n + 1)^k$$

$$d_{2k+1}((2k + 1)n + k + 1) = (-1)^{\binom{k+1}{2}}4^k(n + 1)^k$$

$$d_{2k}(2kn) = (-1)^{kn}$$

$$d_{2k}(2kn + k) = (-1)^{kn+\binom{k}{2}}4^{k-1}(n + 1)^{k-1}.$$

Proof Given w, it is straightforward to verify the hypotheses and evaluate the final expression of Lemma 8.2 with the lemmas of Sect. 7. For the first formula, take w to be the $(2k + 1)n$-dimensional column vector

$$w_1 = (-1)^{n-1} \left(\sum_{m=0}^{n-1} (-1)^m e_{(2k+1)m} - \sum_{m=0}^{n-1} (-1)^m e_{(2k+1)m+2k} \right) + e_{N-1},$$

where $\{e_i\}_{i=0}^{N-1}$ is the standard basis. Then $g_k(\alpha)w_1 = g_k(\beta)w_1 = e_{N-k-1}$.

For the second formula, take w to be the $(2k + 1)n + k + 1$-dimensional column vector

$$w_2 = (-1)^n \left(\sum_{m=0}^{n} (-1)^m e_{(2k+1)m+k-1} - \sum_{m=0}^{n-1} (-1)^m e_{(2k+1)m+k+1} \right) + e_{N-1},$$

which gives $g_k(\alpha)w_2 = e_{N-k-1} + e_{N-k}$ and $g_k(\beta)w_2 = e_{N-k-1} - e_{N-k}$.

For the third formula, take w to be the $2kn$-dimensional column vector

$$w_3 = (-1)^{n-1} \left(\sum_{m=0}^{n-1} (-1)^m e_{2km} - \sum_{m=0}^{n-1} (-1)^m e_{2km+2k-1} \right) + e_{N-1},$$

which gives $b_k(\alpha)w_3 = b_k(\beta)w_3 = e_{N-k-1} + e_{N-k}$.

For the fourth formula, take w to be the $2kn + k$-dimensional column vector

$$w_4 = (-1)^n \left(\sum_{m=0}^{n} (-1)^m e_{2km+k-1} - \sum_{m=0}^{n-1} (-1)^m e_{2km+k+1} \right) + e_{N-1},$$

which gives $b_k(\alpha)w_4 = e_{N-k-1} + 3e_{N-k}$ and $b_k(\beta)w_4 = e_{N-k-1} - e_{N-k}$. □

8.2 The Case Where μ_0 Is Singular

We will make use of the following fact about the adjugate matrix.

Proposition 8.4 *The rank of the adjugate* $\mathrm{adj}(M)$ *of an n-by-n matrix M satisfies*

$$\mathrm{rk}\,\mathrm{adj}(M) = \begin{cases} n & \text{if } \mathrm{rk}\,M = n \\ 1 & \text{if } \mathrm{rk}\,M = n - 1 \\ 0 & \text{otherwise.} \end{cases}$$

Proof Recall that $\mathrm{adj}(M) \cdot M = \det(M)I$. If $\mathrm{rk}\,M = n$ then M is invertible with inverse $\frac{1}{\det(M)}\,\mathrm{adj}(M)$, which also has rank n.

If rk $M = n - 1$, then $\det(M) = 0$, in which case adj(M) must send all vectors into the kernel of M, which has rank 1. In this case M also has a nonzero order-$n - 1$ minor, so adj(M) has rank 1.

If rk $M \leq n - 2$, then all order-$n - 1$ minors of M are zero, so adj$(M) = 0$.   □

Lemma 8.5 *Suppose there is a nonzero N-dimensional column vector w such that $\det(\mu_0) = 0$, $\det(\mu_0|_{N-1}) \neq 0$, $\mu_0 w = 0$, $v^\top w = 1$, $v^\top (\beta + 2)^{-1} w \neq 0$, and entries $N - k - l$ through $N - 3$ of w are 0. Then*

$$\det(\mu_l) = \det(\mu_0|_{N-1}) \left(2v^\top (\beta_N + 2)^{-1} w\right)^l \left(w^\top (\alpha + 2)^{l-1} h_r(\alpha) v\right).$$

Proof Let $c = \det(\mu_0|_{N-1})$. We will show by induction that

$$\text{adj}(\mu_i) = c \left(2v^\top (\beta_N + 2)^{-1} w\right)^i w w^\top,$$

for $0 \leq i < l$. For the base case of $i = 0$, note that the first two hypotheses imply that μ_0 has rank $N - 1$. Since w generates the kernel and μ_0 is symmetric, Proposition 8.4 implies that adj(μ_0) is a constant d times $w w^\top$. In fact $c = v^\top$ adj$(\mu_0)v = dv^\top w w^\top v = d$.

Suppose the claim holds for i. Since $\alpha + 2$ is tridiagonal, the last hypothesis combined with Lemmas 7.4 and 7.6 imply that $w^\top (\alpha + 2)^i h_r(\alpha)v = 0$. By (11) and the matrix determinant lemma,

$$\det(\mu_{i+1}) = \det(\beta + 2) \left(\det(\mu_i) + v^\top (\beta + 2)^{-1} \text{adj}(\mu_i)(\alpha + 2)^i h_r(\alpha)v\right)$$

$$= \det(\beta + 2) \left(0 + c \left(2v^\top (\beta_N + 2)^{-1} w\right)^i v^\top (\beta + 2)^{-1} w w^\top (\alpha + 2)^i h_r(\alpha)v\right)$$

$$= 0,$$

so μ_{i+1} has rank at most $n - 1$. Since $(\alpha + 2)^i h_r(\alpha)vv^\top$ does not affect the bottom-right cofactor,

$$v^\top \text{adj}(\mu_{i+1})v = v^\top \text{adj}\left(\mu_i(\beta + 2) + (\alpha + 2)^i h_r(\alpha)vv^\top\right)v$$

$$= v^\top \text{adj}\left(\mu_i(\beta + 2)\right)v$$

$$= c \det(\beta + 2)v^\top (\beta + 2)^{-1} \left(2v^\top (\beta_N + 2)^{-1} w\right)^i w w^\top v$$

$$= c(2v^\top (\beta_N + 2)^{-1} w)^{i+1}.$$

This is nonzero by assumption, so adj(μ_{i+1}) is nonzero. By Proposition 8.4, it is rank 1. The matrix μ_{i+1} is symmetric and w lies in its kernel:

$$w^\top \mu_{i+1} = w^\top \mu_i(\beta + 2) + w^\top (\alpha + 2)^i h_r(\alpha)vv^\top = 0 + 0,$$

so it is of the form $\mathrm{adj}(\mu_{i+1}) = c(2v^\top(\beta_N + 2)^{-1}w)^{i+1}ww^\top$. This completes the induction.

The final μ_l has determinant

$$\det(\mu_l) = \det(\beta + 2)\left(\det(\mu_{l-1}) + v^\top(\beta + 2)^{-1}\mathrm{adj}(\mu_{l-1})(\alpha + 2)^{l-1}h_r(\alpha)v\right)$$

$$= 2\left(0 + 2^{l-1}c(v^\top(\beta_N + 2)^{-1}w)^l w^\top(\alpha + 2)^{l-1}h_r(\alpha)v\right)$$

$$= c\left(2v^\top(\beta_N + 2)^{-1}w\right)^l\left(w^\top(\alpha + 2)^{l-1}h_r(\alpha)v\right).$$

$\square$

Theorem 8.6 *For $n, k \geq 1$,*

$$d_{2k+1}((2k + 1)n + 1) = (2n + 1)^k$$

$$d_{2k}(2kn + 1) = (-1)^{kn}$$

$$d_{2k}(2kn + k + 1) = -(-1)^{kn+\binom{k}{2}}4^{k-1}(n + 1)^{k-1}$$

Proof Given w, it is straightforward to verify the hypotheses and evaluate the final expression of Lemma 8.5 with the lemmas of Sect. 7.

For the first formula, take w to be

$$w_5 = (-1)^n\left(\sum_{m=0}^{n}(-1)^m e_{(2k+1)m} - \sum_{m=0}^{n-1}(-1)^m e_{(2k+1)m+2k}\right),$$

where $\{e_i\}_{i=0}^{N-1}$ is the standard basis.

For the second formula, take w to be

$$w_6 = (-1)^n\left(\sum_{m=0}^{n}(-1)^m e_{2km} - \sum_{m=0}^{n-1}(-1)^m e_{2km+2k-1}\right).$$

For the third formula, use

$$w_7 = (-1)^{n-1}\left(\sum_{m=0}^{n}(-1)^m e_{2km+k-1} - \sum_{m=0}^{n}(-1)^m e_{2km+k}\right).$$

$\square$

9 Conjectures

Let

$$d'_r(n) = \det\left(\frac{r}{2i+2j+r}\binom{2i+2j+r}{i+j}\right)_{i,j=0}^{n-1}.$$

These sequences are considered alongside $d_r(n)$ in [1]. Computer experiments suggest the following conjectures:

$$d'_{2k+1}((2k+1)n) = d'_{2k+1}((2k+1)n+1) = (-1)^{kn},$$

$$d'_{2k+1}((2k+1)n+k) = -d'_{2k+1}((2k+1)n+k+2)$$

$$= (-1)^{kn+\binom{k}{2}}((2k+1)(n+1))^{k-1},$$

$$d'_{2k+1}((2k+1)n+k+1) = 0,$$

$$d'_{2k}(kn) = -d'_{2k}(kn+1) = (-1)^{n\binom{k}{2}}(n+1)^{k-1}.$$

Moreover, it seems that

$$\left(\frac{r}{2i+2j+r}\binom{2i+2j+r}{i+j}\right)_{i,j\geq 0}$$

$$= \begin{cases} A(-a_k(\gamma)(\gamma+2)^{k-1})A^\top & \text{if } r = 2k \\ A((-1)^{k+1}g_k(-\gamma)(\gamma-2)(\gamma+2)^{k-1})A^\top & \text{if } r = 2k+1. \end{cases}$$

References

1. Cigler, J.: Catalan numbers, Hankel determinants and Fibonacci polynomials (2018). arXiv:1801.05608
2. Cigler, J., Krattenthaler, C.: Some determinants of path generating functions. Adv. Appl. Math. **46**(1), 144–174 (2011)
3. Eğecioğlu, Ö., Redmond, T., Ryavec, C.: A multilinear operator for almost product evaluation of Hankel determinants. J. Combin. Theory Ser. A **117**(1), 77–103 (2010)

A Sequence of Polynomials Generated by a Kapteyn Series of the Second Kind

Diego Dominici and Veronika Pillwein

Dedicated to Peter Paule, friend and mentor. Thank you for sharing your insight and enthusiasm.

1 Introduction

Series of the form

$$\sum_{k=0}^{\infty} \alpha_k^{\nu} J_{\nu+k} \left[(\nu + k) z \right], \tag{1}$$

and

$$\sum_{k=0}^{\infty} \alpha_k^{\mu,\nu} J_{\mu+k} \left[(\mu + \nu + 2k) z \right] J_{\nu+k} \left[(\mu + \nu + 2k) z \right], \tag{2}$$

where $\mu, \nu \in \mathbb{C}$ and $J_n(z)$ is the Bessel function of the first kind [32, 10.2.2]

$$J_\nu(z) = \sum_{j=0}^{\infty} \frac{(-1)^j}{\Gamma(\nu + j + 1) \, j!} \left(\frac{z}{2} \right)^{\nu + 2j}, \tag{3}$$

D. Dominici (✉)
Johannes Kepler University Linz, Doktoratskolleg, "Computational Mathematics", Linz, Austria
Department of Mathematics State University of New York at New Paltz, New Paltz, NY, USA
e-mail: diego.dominici@dk-compmath.jku.at

V. Pillwein
Johannes Kepler University Linz, RISC, Linz, Austria
e-mail: veronika.pillwein@risc.jku.at

© Springer Nature Switzerland AG 2020
V. Pillwein, C. Schneider (eds.), *Algorithmic Combinatorics: Enumerative Combinatorics, Special Functions and Computer Algebra*, Texts & Monographs in Symbolic Computation, https://doi.org/10.1007/978-3-030-44559-1_9

and $\Gamma(\cdot)$ is the Gamma function [32, Chapter 5.], are called *Kapteyn series of the first kind* and *Kapteyn series of the second kind* respectively.

Kapteyn series have a long history, going back to Lagrange's 1771 paper *Sur le Problème de Képler* [23], where he solved Kepler's equation [8]

$$M = E - \varepsilon \sin(E), \tag{4}$$

using his method for solving implicit equations [22] (now called *Lagrange inversion theorem*) and obtained [12]

$$E(M) = M + \sum_{n=1}^{\infty} \frac{\varepsilon^n}{n!} \frac{d^{n-1}}{dM^{n-1}} \sin^n(M).$$

Here M is the mean anomaly (a parameterization of time) and E is the eccentric anomaly (an angular parameter) of a body orbiting on an ellipse with eccentricity ε.

In 1819 Bessel published his paper *Analytische Auflösung der Kepler'schen Aufgabe* [3], where he approached (4) using a different method. First of all he observed that the function $g(M) = E(M) - M$ defined implicitly by $g = \varepsilon \sin(g + M)$ is $2\pi-$periodic and satisfies $g(0) = 0 = g(\pi)$. Hence, $g(M)$ can be expanded in a Fourier sine series

$$g(M) = \sum_{n=1}^{\infty} b_n \sin(nM),$$

where

$$b_n = \frac{2}{\pi} \int_0^\pi g(M) \sin(nM) dM = \frac{2}{\pi n} \int_0^\pi \cos(nE - n\varepsilon \sin E) \, dE.$$

He then introduced the functions $J_n(z)$ defined by

$$J_n(z) = \frac{1}{\pi} \int_0^\pi \cos(nE - z \sin E) \, dE, \quad n \in \mathbb{Z} \tag{5}$$

which now bear his name and obtained

$$E(M) = M + \sum_{n=1}^{\infty} \frac{2}{n} J_n(n\varepsilon) \sin(nM). \tag{6}$$

Bessel's work on (5) was continued by other researchers including Lommel [28], who defined the Bessel function of the first kind by (3).

In 1817, Francesco Carlini [4] found an expression for the true anomaly v (an angular parameter), defined in terms of E and ε by

$$\tan\left(\frac{v}{2}\right) = \sqrt{\frac{1+\varepsilon}{1-\varepsilon}} \tan\left(\frac{E}{2}\right).$$

Carlini's expression reads [7]

$$v = M + \sum_{n=1}^{\infty} B_n \sin(nM),$$

where

$$B_n = \frac{2}{n} J_n(n\varepsilon) + \sum_{k=0}^{\infty} \alpha^k \left[J_{n-k}(n\varepsilon) + J_{n+k}(n\varepsilon) \right],$$

with $\varepsilon = \frac{2\alpha}{1+\alpha^2}$. The problem considered by Carlini was to determine the asymptotic behavior of the coefficients B_n for large values of n [14]. The astronomer Johann Encke drew Jacobi's attention to the work of Carlini. In 1849, Jacobi published a paper improving and correcting Carlini's article [16] and in 1850 Jacobi published a translation from Italian into German [5], with critical comments and extensions of Carlini's investigation.

Bessel's research on series of the type (6) was continued by Ernst Meissel [34] in his papers [29, 30] and in a systematic way by Willem Kapteyn (not to be confused with his brother Jacobus Cornelius Kapteyn [1]) in the articles [17] and [18]. Most of the early work on Kapteyn series can be found in the books by Niels Nielsen [31, Chapter XXII] and Watson [38, Chapter 17]. For additional properties, see [9, 10, 36, 37], and especially the book [2].

In recent years, there has been a renewed interest on Kapteyn series, particularly from researchers in the fields of Astrophysics and Electrodynamics, see [13, 26], and [27]. In [24], Ian Lerche and Robert Tautz studied the Kapteyn series of the second kind

$$S_1(a) = \sum_{k=1}^{\infty} k^4 J_k^2(ka)$$

and derived the formula

$$S_1(a) = \frac{a^2 \left(64 + 592a^2 + 472a^4 + 27a^6\right)}{256 \left(1 - a^2\right)^{\frac{13}{2}}}.$$

They continued their investigations in [25], where they outlined a way for calculating more general Kapteyn series of the form

$$S_1(n, a) = \sum_{k=1}^{\infty} k^{2n} J_k^2(ka), \quad n \in \mathbb{N}_0, \tag{7}$$

where $\mathbb{N}$ denotes the set of natural numbers and

$$\mathbb{N}_0 = \mathbb{N} \cup \{0\} = \{0, 1, 2, \ldots\}.$$

Motivated by (7), we considered in [11] the Kapteyn series of the second kind with $\mu = \nu = 0$ and $\alpha_k^{\mu,\nu} = k^{2n}$

$$g_n(z) = \sum_{k=0}^{\infty} k^{2n} J_k^2(2kz), \quad n \in \mathbb{N}_0.$$

Using the formula [32, 10.8.3]

$$J_k^2(w) = \sum_{j=0}^{\infty} \frac{(-1)^j}{(2k+j)!j!} \binom{2k+2j}{k+j} \left(\frac{w^2}{4}\right)^{k+j}, \quad k \in \mathbb{N}_0, \tag{8}$$

it is clear that $g_n(z)$ is an even function of z and therefore we can write

$$g_n(z) = \sum_{k=0}^{\infty} b_{n,k} z^{2k}, \quad n \in \mathbb{N}_0. \tag{9}$$

In [11] we computed the first few $g_n(z)$ and obtained

$$g_0(z) = \frac{1}{2} + \frac{1}{2\sqrt{1 - 4z^2}}, \quad g_1(z) = \frac{z^2(1 + z^2)}{(1 - 4z^2)^{\frac{7}{2}}},$$

$$g_2(z) = \frac{z^2(1 + 37z^2 + 118z^4 + 27z^6)}{(1 - 4z^2)^{\frac{13}{2}}},$$

$$g_3(z) = \frac{z^2(1 + 217z^2 + 5036z^4 + 23630z^6 + 22910z^8 + 2250z^{10})}{(1 - 4z^2)^{\frac{19}{2}}},$$

which seemed to suggest that $g_n(z)$ should be of the form

$$g_n(z) = \frac{P_n(z^2)}{(1 - 4z^2)^{3n + \frac{1}{2}}} + \frac{1}{2}\delta_{n,0}, \quad n \in \mathbb{N}_0, \tag{10}$$

where $P_n(x) \in \mathbb{R}[x]$, $\deg(P_n) = 2n$, and $\delta_{n,k}$ is Kronecker's delta, defined by

$$\delta_{n,k} = \begin{cases} 1, & n = k \\ 0, & n \neq k \end{cases}.$$

The purpose of this paper is to show that this conjecture is true.

2 The Coefficients $b_{n,k}$

To begin, we find some representations of the coefficients $b_{n,k}$ appearing in the Taylor series (9).

Proposition 1 *Let $b_{n,k}$ be defined by*

$$\sum_{k=0}^{\infty} k^{2n} J_k^2(2kz) = \sum_{k=0}^{\infty} b_{n,k} z^{2k}.$$

Then,

$$b_{n,k} = \binom{2k}{k} \sum_{j=0}^{k} \frac{(-1)^{k-j}}{(k+j)!(k-j)!} j^{2k+2n}, \quad n, k \in \mathbb{N}_0. \tag{11}$$

Proof Using (8), we have

$$\sum_{k=0}^{\infty} k^{2n} J_k^2(2kz) = \sum_{l=0}^{\infty} \frac{(-1)^l}{(2k+l)!l!} \binom{2k+2l}{k+l} \left(k^2 z^2\right)^{k+l}.$$

Setting $k + l = j$, we get

$$\sum_{k=0}^{\infty} b_{n,k} z^{2k} = \sum_{k=0}^{\infty} \binom{2k}{k} z^{2k} \sum_{j=0}^{k} \frac{(-1)^{k-j}}{(k+j)!(k-j)!} j^{2(k+n)},$$

and the result follows. $\qquad \square$

Remark 1 Note that

$$b_{n,0} = 0^{2n} = \delta_{n,0}.$$

Proposition 2 *Let $b_{n,k}$ be defined by (11). Then,*

$$b_{n,k} = \frac{1}{2}\binom{2k}{k}\sum_{j=0}^{2k}\frac{(-1)^{2k-j}}{j!\,(2k-j)!}\,(k-j)^{2k+2n} + \frac{1}{2}\delta_{n+k,0}. \tag{12}$$

Proof Changing the summation variable from j to $k-j$, we have

$$\sum_{j=0}^{k}\frac{(-1)^{k-j}}{(k+j)!\,(k-j)!}\,j^{2k+2n} = \sum_{j=0}^{k}\frac{(-1)^{j}}{(2k-j)!\,j!}\,(k-j)^{2k+2n}.$$

Also, changing the summation variable from j to $k+j$, we have

$$\sum_{j=0}^{k}\frac{(-1)^{k-j}}{(k+j)!\,(k-j)!}\,j^{2k+2n} = \sum_{j=k}^{2k}\frac{(-1)^{2k-j}}{j!\,(2k-j)!}\,(j-k)^{2k+2n}.$$

Thus,

$$2\sum_{j=0}^{k}\frac{(-1)^{k-j}}{(k+j)!\,(k-j)!}\,j^{2k+2n} = \sum_{j=0}^{2k}\frac{(-1)^{j}}{(2k-j)!\,j!}\,(k-j)^{2k+2n} + \frac{(-1)^{k}}{(k!)^{2}}\delta_{n+k,0},$$

and we obtain

$$b_{n,k} = \frac{1}{2}\binom{2k}{k}\sum_{j=0}^{2k}\frac{(-1)^{j}}{(2k-j)!\,j!}\,(k-j)^{2k+2n} + \frac{1}{2}\delta_{n+k,0}.$$

$\square$

Next, we analyze the sum in the representation (12).

Lemma 1 *Let the functions $q_n(k)$ be defined by*

$$q_n(k) = \sum_{j=0}^{2k}\frac{(-1)^{2k-j}}{j!\,(2k-j)!}\,(k-j)^{2k+2n}, \quad n,k \in \mathbb{N}_0. \tag{13}$$

Then, we can write $q_n(k)$ as the forward difference of a polynomial

$$q_n(k) = \frac{1}{(2k)!}\Delta_x^{2k}\left[(x-k)^{2k+2n}\right]_{x=0}. \tag{14}$$

Proof The forward difference operator (with respect to x) Δ_x is defined by

$$\Delta_x f(x) = f(x+1) - f(x). \tag{15}$$

Iterating (15), one obtains an expression for the m-th order forward difference of a function

$$\Delta_x^m f(x) = \sum_{j=0}^{m} \binom{m}{j} (-1)^{m-j} f(x+j).$$ (16)

Comparing (13) with (16), the result follows. □

The Stirling numbers of the second kind are defined by [32, 26.8.6]

$$\left\{ {n \atop k} \right\} = \frac{1}{k!} \sum_{j=0}^{k} \binom{k}{j} (-1)^{k-j} j^n = \frac{1}{k!} \left[\Delta_x^k x^n \right]_{x=0}.$$

They have many amazing properties, including:

1. The exponential generating function [32, 26.8.12]

$$\sum_{n=0}^{\infty} \left\{ {n \atop k} \right\} \frac{t^n}{n!} = \frac{(e^t - 1)^k}{k!}.$$

Since $\left\{ {n \atop k} \right\} = 0$, for $k > n$, we can write

$$\sum_{n=0}^{\infty} \left\{ {n \atop k} \right\} \frac{t^n}{n!} = \sum_{n=k}^{\infty} \left\{ {n \atop k} \right\} \frac{t^n}{n!} = \sum_{n=0}^{\infty} \left\{ {n+k \atop k} \right\} \frac{t^{n+k}}{(n+k)!},$$

and therefore

$$\sum_{n=0}^{\infty} \left\{ {n+k \atop k} \right\} \frac{t^n}{(n+k)!} = \frac{1}{k!} \left(\frac{e^t - 1}{t} \right)^k.$$ (17)

2. The difference-differential transformation [32, 26.8.37]

$$\frac{1}{k!} \Delta_x^k = \sum_{n=0}^{\infty} \left\{ {n \atop k} \right\} \frac{1}{n!} \frac{d^n}{dx^n}.$$ (18)

Remark 2 In the next results, we will need some material from the theory of generating functions (see [39] for additional information).

1. Given a generating function

$$F(z) = \sum_{n=0}^{\infty} a_n z^n,$$ (19)

we define $[z^n] F(z)$ to be the coefficient of z^n in the Maclaurin series of $F(z)$, i.e.,

$$[z^n] F(z) = a_n. \tag{20}$$

2. The even part of the generating function (19) is given by

$$\frac{F(z) + F(-z)}{2} = \sum_{n=0}^{\infty} a_{2n} z^{2n}. \tag{21}$$

3. Given two sequences defined by their generating functions

$$F(z) = \sum_{n=0}^{\infty} a_n z^n, \quad G(z) = \sum_{n=0}^{\infty} b_n z^n,$$

the Cauchy product of the sequences is defined by $\left(a_j * b_j\right)_n = \sum_{j=0}^{n} a_j b_{n-j}$. The generating function of the Cauchy product of two sequences is the product of their generating functions,

$$\sum_{n=0}^{\infty} \left(a_j * b_j\right)_n z^n = F(z) G(z). \tag{22}$$

We have now all the elements to get new representations of the functions $q_n(k)$.

Proposition 3 *Let $q_n(k)$ be defined by (13). Then,*

$$q_n(k) = \sum_{j=0}^{2n} \left\{ \begin{matrix} j + 2k \\ 2k \end{matrix} \right\} \binom{2n + 2k}{2n - j} (-k)^{2n-j}. \tag{23}$$

Proof Using (18) in (14), we have

$$q_n(k) = \sum_{j=0}^{\infty} \left\{ \begin{matrix} j \\ 2k \end{matrix} \right\} \frac{1}{j!} \left[\frac{d^j}{dx^j} (x - k)^{2k+2n} \right]_{x=0}.$$

But

$$\frac{1}{j!}\left[\frac{d^j}{dx^j}(x-k)^{2k+2n}\right]_{x=0}=\left[x^j\right](x-k)^{2k+2n}$$

$$=\left[x^j\right]\sum_{j=0}^{2k+2n}\binom{2k+2n}{j}x^j(-k)^{2k+2n-j}=\binom{2k+2n}{j}(-k)^{2k+2n-j},$$

where $\left[x^j\right]$ was defined in (20).

Therefore,

$$q_n(k)=\sum_{j=0}^{\infty}\left\{\begin{matrix}j\\2k\end{matrix}\right\}\frac{1}{j!}\binom{2k+2n}{j}(-k)^{2k+2n-j}.$$

However, since

$$\left\{\begin{matrix}j\\2k\end{matrix}\right\}\binom{2k+2n}{j}=0,\quad j>2k+2n,$$

we have

$$q_n(k)=\sum_{j=2k}^{2k+2n}\left\{\begin{matrix}j\\2k\end{matrix}\right\}\frac{1}{j!}\binom{2k+2n}{j}(-k)^{2k+2n-j}=\sum_{j=0}^{2n}\left\{\begin{matrix}j+2k\\2k\end{matrix}\right\}\binom{2n+2k}{j+2k}(-k)^{2n-j},$$

and the result follows from the identity [32, 26.3.1]

$$\binom{n}{k}=\binom{n}{n-k}.$$

⊔

Corollary 1 *Let $q_n(k)$ be defined by (13). Then, $q_n(k)=(2k+1)_{2n}\,r_n(k)$, where $r_n(k)$ is defined by*

$$r_n(k)=\sum_{j=0}^{2n}\left\{\begin{matrix}2k+j\\2k\end{matrix}\right\}\frac{(2k)!}{(2k+j)!}\frac{(-k)^{2n-j}}{(2n-j)!}. \tag{24}$$

In particular, the first few $r_n(k)$ are given by,

$$r_0(k)=1,\quad r_1(k)=\frac{k}{12},\quad r_2(k)=\frac{k(5k-1)}{1440}. \tag{25}$$

Next, we find a generating function for the sequence $r_n(k)$.

Proposition 4 *Let $r_n(k)$ be defined by (24). Then, $r_n(k)$ has the ordinary generating function*

$$R_k(t) = \sum_{n=0}^{\infty} r_n(k) t^{2n} = \left[\frac{2}{t} \sinh \left(\frac{t}{2} \right) \right]^{2k}. \tag{26}$$

Proof From (24), we see that we can write $r_n(k)$ as a Cauchy product

$$r_n(k) = (2k)! \left(\begin{Bmatrix} 2k+j \\ 2k \end{Bmatrix} \frac{1}{(2k+j)!} * \frac{(-k)^j}{j!} \right)_{2n},$$

where $(x)_n$ denotes the Pochhammer symbol (also called shifted or rising factorial) [32, 5.2(iii)] defined by $(x)_0 = 1$ and

$$(x)_n = x(x+1) \cdots (x+n-1), \quad n \in \mathbb{N},$$

or as the ratio of two Gamma functions

$$(x)_n = \frac{\Gamma(x+n)}{\Gamma(x)}, \quad -(x+n) \notin \mathbb{N}_0.$$

Using (21), we get

$$\frac{1}{(2k)!} R_k(t) = \sum_{n=0}^{\infty} t^{2n} \left(\begin{Bmatrix} 2k+j \\ 2k \end{Bmatrix} \frac{1}{(2k+j)!} * \frac{(-k)^j}{j!} \right)_{2n} = \frac{G_k(t) + G_k(-t)}{2},$$

where

$$G_k(t) = \sum_{n=0}^{\infty} t^n \left(\begin{Bmatrix} 2k+j \\ 2k \end{Bmatrix} \frac{1}{(2k+j)!} * \frac{(-k)^j}{j!} \right)_n$$

$$= \left[\sum_{j=0}^{\infty} \begin{Bmatrix} 2k+j \\ 2k \end{Bmatrix} \frac{t^j}{(2k+j)!} \right] \left[\sum_{j=0}^{\infty} \frac{(-k)^j}{j!} t^j \right],$$

after using (22).

From (17), we have

$$\sum_{n=0}^{\infty} \begin{Bmatrix} n+2k \\ 2k \end{Bmatrix} \frac{t^n}{(n+2k)!} = \frac{1}{(2k)!} \left(\frac{e^t - 1}{t} \right)^{2k},$$

and clearly

$$\sum_{j=0}^{\infty} \frac{(-k)^j}{j!} t^j = e^{-kt}.$$

Thus,

$$(2k)! G_k(t) = \left(\frac{e^t - 1}{t} \right)^{2k} e^{-kt} = \left(\frac{e^t - 1}{t} \right)^{2k} e^{-2k \frac{t}{2}} = \left(\frac{e^{\frac{t}{2}} - e^{-\frac{t}{2}}}{t} \right)^{2k},$$

and we conclude that

$$G_k(t) = \frac{1}{(2k)!} \left[\frac{2}{t} \sinh \left(\frac{t}{2} \right) \right]^{2k}.$$

Since $\frac{2}{t} \sinh \left(\frac{t}{2} \right)$ is an even function, we get

$$R_k(t) = (2k)! \frac{G_k(t) + G_k(-t)}{2} = \left[\frac{2}{t} \sinh \left(\frac{t}{2} \right) \right]^{2k}.$$

$\square$

Corollary 2 *Let $r_n(k)$ be defined by (24). Then, $r_n \in \mathbb{Q}[k]$ and $\deg(r_n) = n$.*

Proof From (26), we have

$$\sum_{n=0}^{\infty} r_n(x+y) t^{2n} = \left[\frac{2}{t} \sinh \left(\frac{t}{2} \right) \right]^{2(x+y)} = \left[\frac{2}{t} \sinh \left(\frac{t}{2} \right) \right]^{2x} \left[\frac{2}{t} \sinh \left(\frac{t}{2} \right) \right]^{2y},$$

and using (22) we get

$$r_n(x+y) = \sum_{j=0}^{n} r_j(x) r_{n-j}(y). \tag{27}$$

In particular, setting $y = 1$

$$\Delta_x r_n(x) = r_n(x+1) - r_n(x) = \sum_{j=0}^{n-1} r_{n-j}(1) r_j(x),$$

where we have used (25). Using induction, the result follows. $\square$

To summarize, in this section, we have shown that

$$b_{n,k} = \frac{1}{2}\binom{2k}{k} q_n(k) + \frac{1}{2}\delta_{n+k,0}$$

and

$$q_n(k) = (2k+1)_{2n}\, r_n(k),$$

where $r_n \in \mathbb{Q}[k]$ and $\deg(r_n) = n$.

3 Main Result

In this section, we use our previous results to prove (10). We start with a few formulas that we will need in the sequel. All of them can be verified using Zeilberger's algorithm [33, 35]. We leave the proofs to the reader.

Lemma 2 *For all $j, k, n \in \mathbb{N}_0$, we have*

$$\binom{3n + \frac{1}{2}}{k - j}(-4)^{k-j}\binom{2j}{j}(2j+1)_{2n} \tag{28}$$

$$= \frac{4^{n+k}}{k!}\left(\frac{1}{2}\right)_{3n+1}\binom{k}{j}(-1)^{k-j}\left(j - k + 3n + \frac{3}{2}\right)_{k-2n-1}(j+1)_n.$$

Lemma 3 *For all $j \in \mathbb{N}_0$ we have*

$$\sum_{k=0}^{\infty}\binom{2k}{k}\binom{k}{j} z^k = \binom{2j}{j}\frac{z^j}{(1 - 4z)^{j+\frac{1}{2}}},\quad |z| < \frac{1}{4} \tag{29}$$

Corollary 3 *Let $u_m(k)$ be a polynomial in k of degree m. Then,*

$$\sum_{k=0}^{\infty}\binom{2k}{k} u_m(k)\, z^k = \frac{U_m(z)}{(1 - 4z)^{m+\frac{1}{2}}},\quad |z| < \frac{1}{4},$$

where $U_m(z)$ is a polynomial in z with $\deg(U_m) \le m$.

Proof Let's write $u_m(k)$ in the basis of binomial polynomials.

$$u_m(k) = \sum_{j=0}^{m} a_{m,j}\binom{k}{j}.$$

Using (29), we get

$$\sum_{k=0}^{\infty} \binom{2k}{k} u_m(k) \; z^k = \sum_{j=0}^{m} a_{m,j} \binom{2j}{j} \frac{z^j}{(1-4z)^{j+\frac{1}{2}}}$$

$$= \frac{1}{(1-4z)^{m+\frac{1}{2}}} \sum_{j=0}^{m} a_{m,j} \binom{2j}{j} z^j (1-4z)^{m-j},$$

and we conclude that

$$U_m(z) = \sum_{j=0}^{m} a_{m,j} \binom{2j}{j} z^j (1-4z)^{m-j}.$$

$\square$

We can now prove our main result.

Theorem 1 *Let $r_n(k)$ be a polynomial in k of degree n and $P_n(z)$ be defined by*

$$P_n(z) = (1-4z)^{3n+\frac{1}{2}} \sum_{k=0}^{\infty} \binom{2k}{k} (2k+1)_{2n} \; r_n(k) \; z^k.$$

Then, $P_n(z)$ is a polynomial in z of degree $2n$.

Proof We know from Corollary 3 that $P_n(z)$ is a polynomial with $\deg(P_n) \leq 3n$. Thus, we write

$$P_n(z) = \sum_{j=0}^{3n} c_{n,j} \; z^j.$$

Using the Cauchy product between power series, we have

$$c_{n,k} = \sum_{j=0}^{k} \binom{3n+\frac{1}{2}}{k-j} (-4)^{k-j} \binom{2j}{j} (2j+1)_{2n} \; r_n(j).$$

From (28), we get

$$c_{n,k} = \frac{4^{n+k}}{k!} \left(\frac{1}{2}\right)_{3n+1} \sum_{j=0}^{k} \binom{k}{j} (-1)^{k-j} \left(j-k+3n+\frac{3}{2}\right)_{k-2n-1} (j+1)_n \; r_n(j),$$

which we can write as the finite difference

$$c_{n,k} = \frac{4^{n+k}}{k!} \left(\frac{1}{2}\right)_{3n+1} \left[\Delta_x^k C_{n,k}(x)\right]_{x=0},$$

where

$$C_{n,k}(x) = \left(x - k + 3n + \frac{3}{2}\right)_{k-2n-1} (x+1)_n \, r_n(x).$$

$C_{n,k}(x)$ is a polynomial of degree $k-1$ for $k \geq 2n+1$ and therefore

$$\Delta_x^k C_{n,k}(x) = 0, \quad k \geq 2n+1.$$

We conclude that $c_{n,k} = 0$ for $k > 2n$, and the result is proved. $\square$

4 Symbolic Computation

In this section, we apply computer algebra methods to derive further results about the coefficient sequence $c_{n,k}$. Using algorithms for symbolic summation, it is possible to discover and prove a recurrence relation for fast computation of these coefficients. As a side result, we obtain a simple closed form for the leading coefficients that would otherwise not be easily discovered.

Holonomic functions form a class of functions for which a wide variety of algorithms is available to discover and prove non-trivial identities. In one variable, they are functions satisfying a linear difference or differential equation with polynomial coefficients. The classical (continuous) orthogonal polynomials are holonomic both in the degree n (satisfying a three term recurrence) and in the variable x (satisfying a second order ordinary differential equation with polynomial coefficients) and also holonomic as multivariate functions in n and x. For a non-expert introduction to holonomic functions in one and several variables as well as some algorithms for them, see [20].

Stirling numbers are an example of a sequence that is just outside the class of holonomic functions. They also satisfy recurrence relations, but of a different type. Methods like automated guessing of recurrence based on given data can certainly be applied to Stirling-type sequences, however tools for symbolic summation will not work the same way. There has been work on extending these algorithms [6] and these methods are also implemented in the Mathematica package Holonomic-Functions [21] by Christoph Koutschan. The sequence $b_{n,k}$ defined in (9) is of this Stirling-type and below we use automated guessing and a variation of Zeilberger's algorithm [40] to derive recurrence relations for it. The Mathematica notebook containing all computations carried out in this notebook can be found at https://www3.risc.jku.at/people/vpillwei/kapteyn/.

As a first step, we compute a recurrence relation for $b_{n,k}$ using HolonomicFunctions. There are different ways to write the sequence and it does make a difference for the algorithm. We use the definition (11),

$$b_{n,k} = \binom{2k}{k} \sum_{j=0}^{k} \frac{(-1)^{k-j}}{(k+j)!\,(k-j)!} j^{2k+2n}, \quad n, k \in \mathbb{N}_0,$$

instead of one involving Stirling numbers. Using the command

$$\text{Annihilator}[b[n, k], \{S[n], S[k]\}]$$

in HolonomicFunctions gives the recurrence,

$$(-k - 1)S_n S_k + 2(2k + 1)S_n + (k + 1)^3 S_k = 0.$$

The output is in operator form, where S_m denotes the forward shift in the variable m. The recurrence then reads as stated in the following lemma. To avoid the case distinction with the Kronecker delta for the case of both n and k being zero, in the following we always assume that $k \geq 1$. Note that $b_{n,0} = 0$ for $n \geq 1$. Hence, we may even consider $n, k \geq 1$.

Lemma 4 *Let the sequence $b_{n,k}$ be defined by (11). Then,*

$$(k + 1)b_{n+1,k+1} = 2(2k + 1)b_{n+1,k} + (k + 1)^3 b_{n,k+1}, \quad n \geq 0, \quad k \geq 1,$$

with initial values

$$b_{n,1} = 1, \quad b_{0,k} = \frac{1}{2}\binom{2k}{k}.$$

Proof The recurrence can be derived as shown above and the initial values $b_{n,1}$ are trivially verified for $k \geq 1$. It remains to show that $b_{0,k} = \frac{1}{2}\binom{2k}{k}$. For this first observe that for $n = 0$ and $k \geq 1$ we can rewrite

$$\sum_{j=0}^{k} \frac{(-1)^{k-j}}{(k+j)!\,(k-j)!} j^{2k} = \frac{1}{2}\frac{1}{(2k)!} \sum_{j=0}^{2k} \binom{2k}{j}(-1)^j (k - j)^{2k}.$$

Here we first reverse the order of summation and then using the fact that $k \geq 1$ extend the summation symmetrically to go up to $j = 2k$. Analogously as in the proof of Proposition 2. Note that for $j = k$ the summand vanishes if $k \geq 1$. Using

[15, (5.42)]

$$\sum_{j=0}^{m} \binom{m}{j} (-1)^j (a_0 + a_1 j + \cdots + a_m j^m) = (-1)^m m! a_m,$$

the result follows with $m = 2k$ and $a_{2k} = 1$.   □

The objects we are actually interested in are the polynomials $P_n(z)$ in the numerator of $g_n(z)$. Recall that they were defined as

$$P_n(z) = (1 - 4z)^{3n+1/2} \sum_{k \geq 1} b_{n,k} z^k = \sum_{k \geq 1} c_{n,k} z^k,$$

with

$$c_{n,k} = \sum_{j=1}^{k} \frac{(-3n - \frac{1}{2})_{k-j}}{(k-j)!} 4^{k-j} b_{n,j} = \sum_{j=1}^{k} a_{n,k-j} b_{n,j}. \qquad (30)$$

In order to derive a recurrence relation for the coefficient sequence $c_{n,k}$ we employ *creative telescoping* [41]. The basic principle is as follows: given the summand

$$f(n, k, j) = a_{n,k-j} b_{n,j},$$

an operator of the form

$$\mathcal{A} + (S_j - 1)\mathcal{D}$$

is determined that annihilates the input, i.e., when applied to the summand $f(n, k, j)$ gives zero. Moreover, $\mathcal{A}$ has coefficients depending only on n and k and not on the summation variable j and uses only shifts of f in n and k, i.e.,

$$\mathcal{A} = \sum_{a,b} \gamma_{a,b}(n, k) S_n^a S_k^b.$$

Note, that the summation runs over a finite index set only. Because of the nature of this operator and the factor $\Delta_j = S_j - 1$ in front of the second operator $\mathcal{D}$, one can sum over the equation and the delta-part can be evaluated using telescoping. In the ideal case, the summand has natural boundaries and the delta-part telescopes to zero. In this case the final recurrence for the sum is just $\mathcal{A} \cdot c_{n,k} = 0$. But even in a less lucky case, at least an inhomogeneous recurrence can be determined that possibly can be simplified further. Indeed, this is the case in our application.

The method of creative telescoping is implemented in the package Holonomic-Functions, even for the non-holonomic case. However, the size of the input for $c_{n,k}$ is too large and the computations are very expensive. Still, it is possible to guess a recurrence for $c_{n,k}$ first and use the support of the guessed recurrence as an input for creative telescoping. This speeds up the process considerably as the ansatz becomes much smaller. Of course the procedure is still rigorous—if there would not be an operator of this form, HolonomicFunctions will return the empty set.

For guessing we use the Mathematica implementation of Manuel Kauers [19] and find that

$$(k+3)c_{n+1,k+3} - (k+3)^3 c_{n,k+3} - 4(k-3n-1)c_{n+1,k+2}$$

$$+ 2\left(6k^3 - 18k^2 n + 33k^2 - 90kn + 57k - 114n + 29\right)c_{n,k+2}$$

$$- 4\left(12k^3 - 72k^2 n + 24k^2 + 108kn^2 - 144kn + 9k + 216n^2 - 24n + 2\right)c_{n,k+1}$$

$$+ 8(2k - 6n - 1)^3 c_{n,k} = 0, \qquad n, k \geq 1.$$

$$\tag{31}$$

From this we obtain an input for the support of the shifts in n and k in the method CreativeTelescoping of HolonomicFunctions. Once more note that a notebook with all these calculations can be downloaded and checked.

Given the summand as $a_{n,k-j}b_{n,k}$ in terms of their defining annihilators and the support

$$\{1, S_k, S_k^2, S_k^3, S_k^2 S_n, S_k^3 S_n\}$$

as an input, CreativeTelescoping returns the two following operators

$$\mathcal{A} = (k+3)S_k^3 S_n - (k+3)^3 S_k^3 - 4(k-3n-1)S_k^2 S_n$$

$$+ 2\left(6k^3 - 18k^2 n + 33k^2 - 90kn + 57k - 114n + 29\right)S_k^2$$

$$- 4\left(12k^3 - 72k^2 n + 24k^2 + 108kn^2 - 144kn + 9k + 216n^2 - 24n + 2\right)S_k$$

$$+ 8(2k - 6n - 1)^3,$$

which is the operator form of recurrence (30) above, and

$$\mathcal{D} = \frac{8j(2j - 2k + 6n + 3)(2j - 2k + 6n + 5)(2j - 2k + 6n + 7)}{(j - k - 3)(j - k - 2)(j - k - 1)}S_n$$

$$- \frac{24j^3(2n+1)(6n+5)(6n+7)}{(j - k - 3)(j - k - 2)(j - k - 1)}$$

for the delta part. In this case, we run into two difficulties. First, the summand $c_{n,k}$ does not have natural bounds for summation, i.e., it does not vanish outside the range of summation. On the other hand, we cannot sum j up to $k+3$ as we run into poles. Hence, we proceed by summing from $j = 1$ to $k - 1$ over the equation

$$\mathcal{A} \cdot (a_{n,k-j}b_{n,j}) + (S_j - 1)\mathcal{D} \cdot (a_{n,k-j}b_{n,j}) = 0.$$

In order to obtain a recurrence for $c_{n,k}$ we have to add and subtract the missing summands in the first part $\sum_{j=1}^{k-1} \mathcal{A} \cdot (a_{n,k-j}b_{n,j})$. As $b_{n,k}$ is given as a sum itself, it is easier to plug in only $a_{n,k-j}$ explicitly and use the recurrence satisfied by $b_{n,k}$ to simplify the equations. All this can be executed automatically in HolonomicFunctions. Moreover, all steps can also be easily verified using paper and pencil.

Theorem 2 *Let $c_{n,k}$ be defined by (30), then for $n, k \geq 1$, the sequence satisfies the recurrence (30) with initial values*

$$c_{n,1} = 1, \quad c_{n,2} = 2^{2n+2} - 3(4n + 1), \quad c_{n,2n} = 2^{4n-1}\frac{(\frac{1}{2})_n^3}{n!}, \quad c_{n,k} = 0, \quad k \geq 1.$$

Proof The recurrence can be computed as described above with computational details in the accompanying Mathematica notebook available at https://www3.risc. jku.at/people/vpillwei/kapteyn/.

The initial values $c_{n,1}$ and $c_{n,2}$ follow easily by plugging in the formula (30). In order to compute $c_{n,2n}$, first we plug in $k = 2n$ in the recurrence relation (30) and obtain

$$0 = -8\left(12n^3 + 12n^2 - 3n + 1\right)c_{n,2n+1} - 2\left(24n^3 + 48n^2 - 29\right)c_{n,2n+2}$$

$$- 8(2n + 1)^3 c_{n,2n} - (2n + 3)^3 c_{n,2n+3} + 4(n + 1)c_{n+1,2n+2} + (2n + 3)c_{n+1,2n+3}.$$

Next, observe that $c_{m,k} = 0$ for $k \geq 2m + 1$ by Theorem 1. Hence, $c_{n,2n+1}, c_{n,2n+2}, c_{n,2n+3}$, and $c_{n+1,2n+3}$ are all zero and the relation above simplifies to

$$4(n + 1)c_{n+1,2n+2} - 8(2n + 1)^3 c_{n,2n} = 0.$$

This recurrence can easily be solved and with $c_{1,2} = 1$ we obtain the result above.

□

Note that this recurrence with the given initial values can actually be used to compute the sequence $c_{n,k}$. In Fig. 1 the support of the recurrence is indicated by circles around the dots in the lattice, the dark gray area are the indices for which $c_{n,k} = 0$ and the light gray area depicts the non-zero initial values. The lattice is centered at $(1, 1)$. The first value to compute is $c_{2,3}$ and from there one always continues first along the $(n, 2n - 1)$-line and then downwards (n, i) for $2n - 2 \geq i \geq 3$. This way all values of the sequences can be computed recursively.

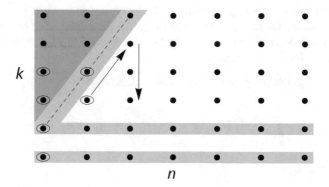

Fig. 1 Recurrence for $c_{n,k}$

Remark 3 It is worth remarking that the closed form of $c_{n,2n}$ is not easily proven without the recurrence relation and really gives the double sum evaluation

$$\sum_{j=1}^{2n} \frac{(-3n - \frac{1}{2})_{2n-j}}{(2n-j)!} 4^{2n-j} \binom{2j}{j} \sum_{i=0}^{j} \frac{(-1)^{j-i}}{(j+i)!\,(j-i)!} i^{2j+2n} = 2^{4n-1} \frac{(\frac{1}{2})_n^3}{n!}.$$

5 Conclusions

We have proved that the Kapteyn series of the second kind

$$g_n(z) = \sum_{k=0}^{\infty} k^{2n} J_k^2(2kz)$$

can be represented as

$$g_n(z) = \frac{P_n(z^2)}{\left(1 - 4z^2\right)^{3n+\frac{1}{2}}} + \frac{1}{2}\delta_{n,0}, \quad n \in \mathbb{N}_0,$$

where $P_n(x)$ is a polynomial of degree $2n$.
  Writing

$$P_n(z) = (1 - 4z)^{3n+\frac{1}{2}} \sum_{k=1}^{\infty} b_{n,k} z^k = \sum_{k=1}^{\infty} c_{n,k} z^k,$$

we have obtained several properties of the coefficients $b_{n,k}$, and a recurrence for the coefficients $c_{n,k}$.

Numerical evidence suggests that all coefficients $c_{n,k}$ should be nonnegative integers, but so far we haven't been able to prove this, except for particular cases such as the leading coefficients $c_{n,2n}$ based on the closed form representation. Thus, we propose the following conjecture.

Conjecture 1 Let the polynomials $P_n(z)$ be defined by

$$\sum_{k=0}^{\infty} k^{2n} \, \mathrm{J}_k^2 (2kz) = \frac{P_n\left(z^2\right)}{\left(1 - 4z^2\right)^{3n+\frac{1}{2}}}, \quad n \in \mathbb{N}.$$

Then, $P_n(x) \in \mathbb{N}_0[x]$.

Acknowledgements The first author was supported by the strategic program "Innovatives OÖ–2010 plus" from the Upper Austrian Government. The second author was partly funded by the Austrian Science Fund (FWF): W1214-N15 and under the grant SFB F50-07. We would like to thank the anonymous referees for valuable suggestions.

References

1. Armatte, M.: Robert Gibrat et la loi de l'effet proportionnel. Math. Inform. Sci. Humaines (129), 5–35 (1995)
2. Baricz, A., Jankov Maširević, D., Pogány, T.K.: Series of Bessel and Kummer-Type Functions. Lecture Notes in Mathematics, vol. 2207. Springer, Cham (2017)
3. Bessel, F.W.: Analytische Auflösung der Keplerschen Aufgabe. Abh. Preuß. Akad. Wiss. Berlin **XXV**, 49–55 (1819)
4. Carlini, F.: Ricerche sulla convergenza della serie che serve alla soluzione del problema di Keplero. Effem. Astron. **44**, 3–48 (1817)
5. Carlini, F.: Untersuchungen über die Convergenz der Reihe durch welche das Kepler'sche Problem gelöst wird. Schumacher Astron. Nachr. **30**(14), 197–212 (1850)
6. Chyzak, F., Kauers, M., Salvy, B.: A non-holonomic systems approach to special function identities. In: May, J. (ed.) Proceedings of ISSAC'09 (2009), pp. 111–118
7. Colwell, P.: Bessel functions and Kepler's equation. Am. Math. Month. **99**(1), 45–48 (1992)
8. Colwell, P.: Solving Kepler's Equation Over Three Centuries. Willmann-Bell, Richmond (1993)
9. Dominici, D.: A new Kapteyn series. Integral Transforms Spec. Funct. **18**(5–6), 409–418 (2007)
10. Dominici, D.: An application of Kapteyn series to a problem from queueing theory. PAMM **7**(1), 2050005–2050006 (2008)
11. Dominici, D.E.: On Taylor series and Kapteyn series of the first and second type. J. Comput. Appl. Math. **236**(1), 39–48 (2011)
12. Dutka, J.: On the early history of Bessel functions. Arch. Hist. Exact Sci. **49**(2), 105–134 (1995)
13. Eisinberg, A., Fedele, G., Ferrise, A., Frascino, D.: On an integral representation of a class of Kapteyn (Fourier-Bessel) series: Kepler's equation, radiation problems and Meissel's expansion. Appl. Math. Lett. **23**(11), 1331–1335 (2010)
14. Fröman, N., Fröman, P.O.: Physical Problems Solved by the Phase-Integral Method. Cambridge University Press, Cambridge (2002)

15. Graham, R.L., Knuth, D.E., Patashnik, O.: Concrete Mathematics, 2nd edn. Addison-Wesley, Reading (1994)
16. Jacobi, C.G.J.: Ueber die annähernde Bestimmung sehr entfernter Glieder in der Entwickelung der elliptischen Coordinaten nebst einer Ausdehnung der Laplaceschen Methode zur Bestimmung der Functionen gerader Zahlen. Schumacher Astron. Nachr. **28**(17), 257–272 (1849)
17. Kapteyn, W.: Recherches sur les fonctions de Fourier-Bessel. Ann. Sci. École Norm. Sup. **10**, 91–122 (1893)
18. Kapteyn, W.: On an expansion of an arbitrary function in a series of Bessel functions. Messenger Math. **35**, 122–125 (1906)
19. Kauers, M.: Guessing Handbook. RISC Report Series 09-07, Research Institute for Symbolic Computation (RISC), 4232. Johannes Kepler University Linz, Schloss Hagenberg, Hagenberg (2009)
20. Kauers, M.: The Holonomic Toolkit. In: Blümlein, J., Schneider, C. (eds.) Computer Algebra in Quantum Field Theory: Integration, Summation and Special Functions. Springer, Berlin (2013), pp. 119–144
21. Koutschan, C.: Advanced applications of the holonomic systems approach. PhD Thesis, RISC-Linz, Johannes Kepler University (2009)
22. Lagrange, J.L.: Nouvelle mèthode pour rèsoudre les èquations littèrales par le moyen des sèries. Mèm. de l'Acad. des Sci. **XXIV**, 251–326 (1768)
23. Lagrange, J.L.: Sur le Problème de Képler. Mèm. de l'Acad. des Sci. **XXV**, 204–233 (1771)
24. Lerche, I., Tautz, R.C.: A note on summation of Kapteyn series in astrophysical problems. Astrophys. J. **65**(2), 1288–1291 (2007)
25. Lerche, I., Tautz, R.C.: Kapteyn series arising in radiation problems. J. Phys. A **41**(3), 035202 (2008)
26. Lerche, I., Tautz, R.C.: Kapteyn series in high intensity Compton scattering. J. Phys. A **43**(11), 115207 (2010)
27. Lerche, I., Tautz, R.C., Citrin, D.S.: Terahertz-sideband spectra involving Kapteyn series. J. Phys. A **42**(36), 365206 (2009)
28. Lommel, E.C.J.V.: Studien über die Bessel'schen Funktionen. Leipzig (1868)
29. Meissel, E.: Neue Entwicklungen über die Bessel'schen Functionen. Astron. Nachr. **129**(3089), 281–284 (1892)
30. Meissel, E.: Weitere Entwicklungen über die Bessel'schen Functionen. Astron. Nachr. **130**(3116), 363–368 (1892)
31. Nielsen, N.: Handbuch der Theorie der Zylinderfunktionen. Druck und Verlag von B. G. Teubner, Leipzig (1904)
32. Olver, F.W.J., Lozier, D.W., Boisvert, R.F., Clark, C.W. (eds.): NIST handbook of mathematical functions. U.S. Department of Commerce, National Institute of Standards and Technology, Washington; Cambridge University Press, Cambridge (2010)
33. Paule, P., Schorn, M.: A Mathematica version of Zeilberger's algorithm for proving binomial coefficient identities. J. Symb. Comput. **20**(5–6), 673–698 (1995). Symbolic computation in combinatorics Δ_1 (Ithaca, NY, 1993)
34. Peetre, J.: Outline of a scientific biography of Ernst Meissel (1826–1895). Historia Math. **22**(2), 154–178 (1995)
35. Petkovšek, M., Wilf, H.S., Zeilberger, D.: $A = B$. A K Peters, Wellesley (1996)
36. Tautz, R.C., Dominici, D.: The analytical summation of a new class of Kapteyn series. Phys. Lett. A **374**(13–14), 1414–1419 (2010)
37. Tautz, R.C., Lerche, I., Dominici, D.: Methods for summing general Kapteyn series. J. Phys. A **44**(38), 385202 (2011)
38. Watson, G.N.: A treatise on the theory of Bessel functions. Cambridge Mathematical Library. Cambridge University Press, Cambridge (1995)
39. Wilf, H.S.: Generatingfunctionology, 3rd edn. A K Peters, Wellesley (2006)
40. Zeilberger, D.: A holonomic systems approach to special functions identities. J. Comput. Appl. Math. **32**(3), 321–368 (1990)
41. Zeilberger, D.: The method of creative telescoping. J. Symb. Comput. **11**, 195–204 (1991)

Comparative Analysis of Random Generators

Johannes vom Dorp, Joachim von zur Gathen, Daniel Loebenberger, Jan Lühr, and Simon Schneider

Dedicated to Peter Paule on his 60th birthday
Cogito ergo summo

1 Introduction

Randomness is an indispensable tool in computer algebra. Even for the basic and apparently simple task of factoring univariate polynomials over finite fields the only known efficient (= polynomial-time) algorithms are probabilistic, and finding a deterministic solution is the central theoretical problem in that area. For many, but not all, tasks of computational linear algebra the most efficient algorithms today use pre- and post-multiplication by random matrices, as introduced in Borodin et al. (1982) and refined in many ways since then; it is now a staple tool in that field.

Even greater is the importance in cryptography, say for generating all kinds of secret keys. Deterministic or predictable keys would allow an adversary to reproduce them and break the cryptosystem. Since the random keys are only known to the legitimate user, a brute-force attack would require an exhaustive search of a key space that is prohibitively large, thus preventing a feasible or practical search.

Now a fundamental problem is that we treat our computers as deterministic entities that, by their nature, cannot generate randomness. This is not literally true

J. vom Dorp
Fraunhofer FKIE, Bonn, Germany
e-mail: johannes.vom.dorp@fkie.fraunhofer.de

J. von zur Gathen (✉) · J. Lühr · S. Schneider
Bonn-Aachen International Center for Information Technology, Universität Bonn, Bonn, Germany
e-mail: gathen@bit.uni-bonn.de; luehr@cs.uni-bonn.de; schneider@cs.uni-bonn.de

D. Loebenberger
OTH Amberg-Weiden and Fraunhofer AISEC, Weiden, Germany
e-mail: d.loebenberger@oth-aw.de; daniel.loebenberger@aisec.fraunhofer.de

© Springer Nature Switzerland AG 2020
V. Pillwein, C. Schneider (eds.), *Algorithmic Combinatorics: Enumerative Combinatorics, Special Functions and Computer Algebra*, Texts & Monographs in Symbolic Computation, https://doi.org/10.1007/978-3-030-44559-1_10

because tiny random influences may come from effects like cosmic radiation, but these are easily controlled by error-correcting measures. Furthermore, quantum computers provide randomness naturally. Even more, they can factor integers in polynomial time and break most of the classical cryptosystems, say RSA, due to the famous algorithm by Shor (1999). But it is a matter of opinion whether or when scalable quantum computing will become a reality. Some central problems are described in Dyakonov (2018) and Clarke (2019), with a professionally optimistic view in the latter article.

How can we deal with this basic impossibility to generate randomness on our computers? After all, we do want secure internet connections and much more. A common solution works in two steps:

1. Produce values that are supposed to carry a reasonable amount of randomness, using an outside source, say measuring some physical process that looks chaotic to us.
2. Extend a small amount of true randomness to an arbitrarily large amount of *pseudorandomness*.

And what does that mean? True randomness refers to *uniform randomness*. A uniformly random source with values in a finite set produces each element of the set with the same probability. A pseudorandom source, usually called a *pseudorandom generator*, produces values that cannot be distinguished efficiently from uniformly random ones. That is, no efficient (polynomial-time) machine, deterministic or probabilistic, exists which can ask for an arbitrarily long stream of values, is given either a uniformly random stream or a stream generated by the pseudorandom generator, and then decides (with non-negligible probability of correctness) which of the two is the case.

Given a generator claimed to be pseudorandom, how do you prove that no such distinguishing machine exists? Unfortunately, we cannot, and there is no proof of any "provable security" in sight. The difficulty is embodied in the question $P \neq NP$ posed by Cook (1971) and, almost half a century later, is still an open one-million-dollar *millennium problem*. But computational complexity offers a solution: reductions. We take some algorithmic problem which is considered to be hard (not solvable in (random) polynomial time) and show that the existence of an efficient distinguisher implies a solution to the problem. A well-known such problem is the factorization of large integers. Many researchers have looked at it and no solution is known (except on the as yet hypothetical quantum computers). Such a reduction is currently the best way of establishing pseudorandomness.

Probability theory suggest a different approach: measure the entropy. It expresses the "amount" of randomness that a source produces. Unfortunately, entropy cannot be measured practically (Goldreich et al. 1999; von zur Gathen and Loebenberger 2018). As a way out, sometimes the *block entropy* is measured, see below. It will show large statistical abnormalities, if present, within the output stream, but cannot indicate their absence. In our context, this is rather useless, since even cryptographically weak generators may possess high block entropy.

An intermediate step before seeding a pseudorandom generator from a source is *randomness extraction*. Some of the methods in that area only require a lower bound on the source's *min-entropy*, a more intuitive measure for randomness. By their nature, physical random generators are not amenable to mathematically rigorous proofs of such bounds. Quite justifiably, reasonable engineering standards ignore such theoretical stipulations in practice, but we give some weight to them.

For physical hardware generators, applying a series of statistical tests like the above seems to be the only approach, and we also use it for lack of alternatives. For instance, lack of sufficient entropy caused severe weakening of Debian's OpenSSL implementation, see Schneier (2008). However, experts know the dangers of this approach quite well:

> The main part of a security evaluation considers the generic design and its implementation. The central goal is to quantify (at least a lower bound for) the entropy per random bit. Unfortunately, entropy cannot be measured as voltage or temperature. Instead, entropy is a property of random variables and not of observed realizations (here: random numbers). (Killmann and Schindler 2008)

This warning is often ignored in the literature.

Pseudorandom generators come in two flavors: based on a symmetric cryptosystem like the Advanced Encryption Standard (AES), or based on number-theoretic hard problems such as factoring integers. The general wisdom is that the latter are much slower than the former. The main goal of this paper is to examine this opinion which, to our surprise, turns out to be untenable.

We study one hardware generator; by its nature, it is out of scope for theoretical comparisons. Among the software pseudorandom generators, AES and some of the number-theoretic ones perform roughly equally well, provided they are run with *fair* implementations. We use corresponding home-brew code to run them, implemented with the same care. However, if the AES generator is run on specialized *AES-friendly* hardware, it outperforms the others by a large distance. This comes as no surprise.

Our comparative analysis covers some popular pseudorandom generators and two physical sources of randomness. Of course, the choice of possible generators is vast. We thus try to select examples of the respective classes to get a representative picture of the whole situation. Our measurements were reported in Burlakov et al. (2015a), so that their absolute values are somewhat outdated. But that is not the point here: we strive for a fair comparison of the generators, and that can be expected to carry through to later hardware versions, with a grain of salt.

An example of the insatiable thirst for randomness are TLS transactions, which consume at 43.000 new transactions per second (cipher suite ECDHE-ECDSA-AES256-GCM-SHA384) on a single Intel Xeon based system (cf. NGINX 2016 product information) 1376 KB/s of randomness to generate pre-master secrets of 256 bits each—ideally, using only negligible CPU resources.

In our setup we use as a source of random seeds one particular output of the hardware generator PRG310-4, which was analyzed in Schindler and Killmann (2003). On the software side we discuss several number-theoretic generators, namely the linear congruential generator, the RSA generator, and the Blum–Blum–Shub generator, all at carefully selected truncation rates of the output. The generators come with certain security reductions. For comparison we add to our analysis pseudorandom generators based on a well-studied block cipher, in our case AES in counter mode.

The article is structured as follows: We first present previous work on generator analysis in Sect. 2 before giving a detailed overview of the generators in Sect. 3. The main contribution is the evaluation regarding throughput and entropy consumption in Sect. 4. We conclude and elaborate on future work in Sect. 5.

All algorithms except the one employing AES-NI were implemented in a textbook manner using non-optimized C-code, thus providing a fair comparison. The source code of the algorithms is available at Burlakov et al. (2015b).

2 Related Work

Concerning physical generators, Killmann and Schindler (2008) analyze noisy diodes as a random source, providing a model for its entropy. One example of a noisy diodes based generator is the commercial generator PRG310-4, which is distributed by Bergmann (2019). Concerning non-physical true random generators, Linux' VirtIO generator as used in /dev/random is illustrated by Gutterman et al. (2006) and explained by Lacharme et al. (2012). Combined, they provide a clear picture of its inner workings. Additionally, there is the study by Müller (2019) in which the quality of /dev/random and /dev/urandom is studied with respect to the functionality classes for random generators as given by Killmann and Schindler (2011).

Referring to pseudorandom generators, the RSA based generator is explained in Shamir (1983), Fischlin and Schnorr (2000), and Steinfeld et al. (2006). Its cryptographic security is shown in Alexi et al. (1988) and extended in Steinfeld et al. (2006). Linear congruential generators were first proposed by Lehmer (1951). Attacks were discussed in Plumstead (1982) and Håstad and Shamir (1985). They all exploited its simple linear structure and come with a specific parameterization. Not all parameterizations—such as truncating its output to a single bit—have been attacked successfully as of today. Contini and Shparlinski (2005) analyze this in depth concluding that (for some cases)

> [...] we do not know if the truncated linear congruential generator can still be cryptanalyzed.

Blum et al. (1986) introduced the Blum–Blum–Shub generator. Alexi et al. (1988) and Fischlin and Schnorr (2000) show that the integer modulus can be factored, given a distinguisher for the generator.

A totally different approach for the construction of pseudorandom generators are the ones based on established cryptographic primitives. NIST (2015) specifies several standards for producing cryptographically secure random numbers. Besides hash-based techniques, there is also a standard employing a block cipher in counter mode, see also NIST (2001b) for this purpose.

RFC 4086, see Eastlake et al. (2005), compares different techniques and provides a de-facto standard focussing on internet engineering. There, several entropy pool techniques and randomness generation procedures are specified. However, RFC 4086 lacks recommendations for the ciphers to be used in OFB (output feedback) and CTR (counter mode) generation. We show here that such a general recommendation would also be ill-suited since the optimal choice depends heavily on the platform used.

We are not aware of any comprehensive fair benchmarking survey for all the generators mentioned above that integrates them into the Linux operating system.

3   The Generators

In the following, each generator which was implemented or applied for the comparative analysis is briefly presented. The output of a pseudorandom generator is, by definition, not efficiently distinguishable from uniform randomness, see for example Goldreich (2001). When assuming that certain problems in algorithmic number theory (such as factoring integers) are difficult to solve, the Blum–Blum–Shub, and RSA generators with suitable truncation have this property, but the linear congruential generator does not. Also the AES-based generator does not, but assuming AES to be a secure cipher, the AES-based generator is pseudorandom as well.

3.1   Linux /dev/random and /dev/urandom

The German Federal Office for Information Security[1] sets cryptographic standards in Germany and judges /dev/random to be a non-physical true random number generator (i.e., an NTG.1 generator in the terminology of Killmann and Schindler 2011) for most Linux kernel versions, see BSI (2019b).

/dev/urandom, however, does not fulfill the requirements for the class NTG.1, since property NTG.1.1 requires:

> The RNG shall test the external input data provided by a non-physical entropy source in order to estimate the entropy and to detect non-tolerable statistical defects [...], see Killmann and Schindler (2011).

Additionally, /dev/urandom violates NTG 1.6 which states

> The average Shannon entropy per internal random bit exceeds 0.997.

Both are clearly not met by /dev/urandom due to the fact that the device is non-blocking.

---

[1] Bundesamt für Sicherheit in der Informationstechnik (BSI).

However, `/dev/urandom` fulfills all other requirements of the class NTG.1, i.e. the conditions NTG 1.2 up to NTG 1.5. In particular, it is a DRG.3 generator if it is properly seeded.

As already mentioned, system events are used to gather entropy on Linux Systems. These events are post-processed and made available to the devices `/dev/random` and `/dev/urandom`. This includes estimating the entropy of the event and mixing.

However, `/dev/urandom` will still supply the user with "randomness" without checking whether the entropy-pool is still sufficiently filled. In fact, the user is instead supplied with pseudorandom data in favor of speed requirements.

In the OpenBSD operating system, none of the random devices is implemented in a blocking mode. The idea is that much potentially bad randomness is still better than the parsimonious use of high-quality randomness. This is in contrast to the opinion, as for example held by the BSI, that one should require all used randomness to be of guaranteed good quality. As of now, there is still no consensus on this issue.

Since the `/dev/urandom` device has undergone a major change introduced by Ts'o (2016) in kernel version 4.8, two kernel versions were benchmarked to test the differences. Namely the original Ubuntu 16.04 kernel 4.4.0 and the more recent version 4.10.0.

3.2   PRG310-4

The PRG310-4 gathers entropy from a system of two noisy diodes, see Bergmann (2019), and is connected to a computer via USB. Similar variants exist for different interface types. According to Bergmann (2019), its behavior follows the stochastic model in Killmann and Schindler (2008), who argue that

> [...] the true conditional entropies should be indeed very close to 1 [...], which gives an output of slightly more than 500 kBit internal random numbers per second.

Bergmann (2019) mentions that this device satisfies all requirements for class PTG.3, which are "hybrid physical random number generator with cryptographic post-processing" in the terminology of Killmann and Schindler (2011).

3.3   AES in Counter Mode

Due to the fact that since 2008 there is AES-NI,[2] realizing dedicated processor instructions on Intel and AMD processors for the block cipher AES as standardized by NIST (2001a), we add to our comparison the AES counter mode generator. This

---

[2]For a white paper of AES-NI, see Gueron (2010).

generator is also standardized by NIST (2015) and produces a sequence of 128 bit blocks. We aim at security level of 128 bits, thus employing AES-128 as the underlying block cipher.

The security of the AES generator directly reduces to the security of AES. Indeed, any distinguisher for the pseudorandom generator gives an equally good distinguisher for AES in counter mode. Assuming the latter to be secure, one concludes that also the pseudorandom generator is secure.

However, in contrast to the number-theoretic generators described below, we do not have any reductionist argument in our hands to actually prove that the generator is secure if some presumably hard mathematical problem is intractable. We need to trust that the cipher AES is secure—and the dedicated processor instructions on our CPU work as specified.

When one looks carefully at the definition of a DRG.3 generator in the sense of Killmann and Schindler (2011), AES in counter mode is *not* DRG.3. Specifically, it violates the condition DRG.3.3 of backward secrecy, since the NIST document allows in a single request multiple outputs before the transition function is applied, while the BSI requires that the state transition function of the generator is applied after each new random number.

3.4 Linear Congruential Generators

The linear congruential generator as presented in Lehmer (1951) produces for $i \geq 1$ a sequence of values in $x_i \in \mathbb{Z}_M$, generated by applying for $a, b \in \mathbb{Z}_M$ iteratively

$$x_i = a \cdot x_{i-1} + b \text{ in } \mathbb{Z}_M$$

to a secret random seed $x_0 \in \mathbb{Z}_M$ provided by an external source. The parameters a, b, and M are also kept secret and chosen from the external source.

While the bytes of linear congruential generator outputs are generally well-distributed, with byte entropy close to maximal, the generated sequences are predictable and therefore cryptographically insecure.

Plumstead (1982) describes how to recover the secrets a, b and M from the sequence of $(x_i)_{i \geq 0}$ alone. A possible mitigation against this attack is to output only some *least significant bits* of the x_i. Håstad and Shamir (1985) describe a lattice based attack on such truncated linear congruential generator where all parameters are public. Stern (1987) shows that also in the case when the parameters are kept secret. This attack can be used to predict linear congruential generators that output at least $\frac{1}{3}$ of the bits of the x_i. Contini and Shparlinski (2005) write that there is no cryptanalytic attack known when only approximately $k = \log_2 \log_2 M$ bits are output per round.

We are neither aware of a more powerful attack on the linear congruential generator nor of a more up-to-date security argument for truncated linear congruential generators.

In our evaluation we used a prime modulus M with 2048 bits. Per round we output $k = 11$ bits, which coincides with the value from Contini and Shparlinski (2005) mentioned above. For comparison, we also run the generator with modulus $M = 2^{2048}$ and full output, that is, no truncation, which is basically the fastest number-theoretic generator we can hope for.

The full linear congruential generator is not a pseudorandom generator in the terminology of Killmann and Schindler (2011), since it does not provide forward secrecy. If the sketched truncated version of the linear congruential generator can indeed not be cryptanalyzed then it belongs to the class DRG.3.

3.5   The Blum–Blum–Shub Generator

The Blum–Blum–Shub generator was introduced in 1982 to the cryptographic community and later published in Blum et al. (1986). The generator produces a pseudorandom bit sequence from a random seed by repeatedly squaring modulo a so called *Blum integer* $N = p \cdot q$, where p and q are distinct large random primes congruent to 3 mod 4. In its basic variant, in each round the least significant bit of the intermediate result is returned. Vazirani and Vazirani (1984) proved that the Blum–Blum–Shub generator is secure if $k = \log_2 \log_2 N$ least significant bits are output per round.

Alexi et al. (1988) proved that factoring the Blum integer N can be reduced to being able to guess the least significant bit of any intermediate square with non-negligible advantage. The output of this generator is thus cryptographically secure under the assumption that factoring Blum integers is a hard problem.

In our evaluation p and q are randomly selected 1024 bit primes with $p = q = 3$ mod 4, which corresponds—as above—to the security level of 128 bits following again the BSI (2019a) guideline TR-02102-1.

If factoring Blum integers is hard then the Blum–Blum–Shub generator—properly seeded—is a DRG.3 generator in the terminology of Killmann and Schindler (2011).

3.6   The RSA Generator

The RSA generator was first presented by Shamir (1983) and is one of the pseudorandom generators that are proven to be cryptographically secure under certain number-theoretical assumptions. Analogously to the RSA cryptosystem, the generator is initialized by choosing a modulus N as the product of two large random primes, and an exponent e with $1 < e < \varphi(N) - 1$ and $\gcd(e, \varphi(N)) = 1$. Here, φ denotes Euler's totient function. Starting from a seed x_0 provided by an external source, the generator iteratively computes

$$x_{i+1} = x_i^e \mod N,$$

extracts the least significant k bits of each intermediate result x_i and concatenates them as output.

Our implementation uses a random 2048-bit Blum integer (see Sect. 3.5) as modulus N and various choices for the parameters e and k.

In Alexi et al. (1988) it is shown that the RSA generator is pseudorandom for $k = \log_2 \log_2 N = 11$, under the assumption that the RSA inversion problem is hard. For our tests, we choose $e = 3$, as for small exponents the generator is expected to work fast and because it allows us to compare the results to the runtime of the Blum–Blum–Shub generator.

Under a stronger assumption called the SSRSA assumption, Steinfeld et al. (2006) prove the security of the generator for $k \leq n \cdot (\frac{1}{2} - \frac{1}{e} - \varepsilon - o(1))$ for any $\varepsilon > 0$, giving for $e = 3$ the parameter value $k = 238$. Additionally, we test the larger exponent $e = 2^{16} + 1$, which is widely used in practice, for it is a prime and its structure allows efficient exponentiation, with $k = 921$.

If the RSA inversion problem is hard then the RSA generator—properly seeded—is a DRG.3 generator in the terminology of Killmann and Schindler (2011).

4 Evaluation

To evaluate the efficiency of the generators considered, we developed a framework that runs the software generators based on seed data from the PRG310-4. To this end, we implemented the generators in C, using the GMP library, see Granlund (2014), to accomplish large integer arithmetic. The evaluation framework sequentially runs all generators, reading from one true random source file of 512 kB and producing 512 kB each, while measuring the runtime of each generator and the byte entropy of each output.

All algorithms were run on an Acer V Nitro notebook with a Intel Core i5-4210U CPU at 1.70 GHz with 8 GB RAM. We used Ubuntu 16.04 64-bit with kernel version 4.10.0-32, as well version 4.4.0-92 as reference for the kernel random devices.

This process was repeated 750 times specifically, so that the average runtime of the generators should not deviate considerably from its expectation.

To see this, let A be a randomized algorithm. Then the runtime $t(A)$ is a random variable. Without loss of generality let the runtime be bounded in the interval $I = [0..1]$. We write $t = \mathcal{E}t(A)$ for the expected runtime of A. Consider running the randomized algorithm k times. Then the average runtime of this experiment is

$$X_k = \frac{1}{k} \cdot (t(A)_1 + \ldots + t(A)_k).$$

For its expectation we have

$$\mathcal{E}X_k = \mathcal{E}t(A) = t \in I.$$

Thus, the expectation of the average runtime of k runs is equal to the expectation of a single run. If we observe after k runs an average runtime of X_k, then we can ask:

> How large should k be so that the probability that the observed value X_k significantly differs from its expectation $\mathcal{E}X_k$ is very small?

By Hoeffding's inequality we have

$$\text{prob}(|X_k - \mathcal{E}X_k| \geq \delta) \leq \epsilon$$

for a real number $\delta \in \mathbb{R}_{>0}$ and $\epsilon = 2\exp(-2k\delta^2)$. To be statistically significant, we set $\epsilon = \delta = 0.05$, as typically done in statistics. Then we require that $\text{prob}(|X_k - t| \geq 0.05) \leq 0.05 = 2\exp(-2k \cdot 0.05^2)$, i.e., $k > 737$.

Thus we need at least 737 runs of the algorithm so that the probability that the observed result deviates statistically significant from the actual expected runtime is smaller than 1/20. Thus, 750 runs will do the job.

In order to reduce the impact of other operating system components during our benchmarking, we decided to split up the initialization and generation processes and measure the time for the generation only after a certain amount of data was generated. This way, the throughput of the generators had time to stabilize and we thus omit possible noise that is produced when the generator is started up. To determine the appropriate amount of data to be generated before the measurement, we measured throughput for increasing amounts of data so that we could see at which point the throughput stabilizes.

Figure 1 shows the pseudorandom software generators along with the two versions of /dev/urandom as reference points. In the logarithmic scale on

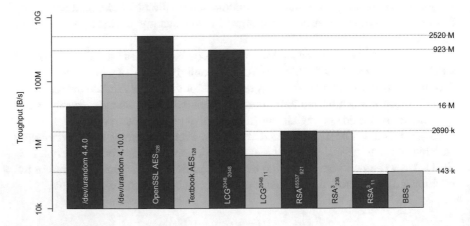

Fig. 1 Generator throughput after initialization for different output lengths

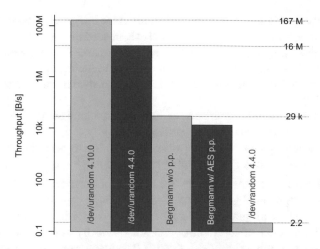

Fig. 2 Physical and non-physical true random generator throughput after initialization

the throughput axis, an AES implementation on AES-friendly hardware has a throughput of about 2.5 GB/s of pseudorandom data, while the RSA generator with 921 bit truncation and 65,537 as public exponent, i.e., the fastest number-theoretic generator assumed to be secure, provides about 2.7 MB/s of pseudorandom data. This makes the latter about 1000 times slower than the AES generator. The linear congruential generator can compete with the fast AES implementation, when not truncating the output, generating about 922 MB/s, but as explained in Sect. 3.4 without truncation the generator is not cryptographically secure. As a fairer comparison to the textbook implementations of the number theoretic generators, the textbook version of AES still generates 32.7 MB/s, beating the RSA implementation by a factor of ten.

A second benchmark was performed for the different physical generators considered, depicted in Fig. 2. Again a logarithmic scale is employed to allow having the /dev/urandom devices with up to 166.8 MB/s of output and the /dev/random device with 2.2 B/s of output in the same picture. The most surprising observation is the jump in performance regarding the /dev/urandom device introduced by the re-implementation described in Sect. 3.1. When only considering blocking physical devices, i.e., taking out /dev/urandom completely, the Bergmann generator outperforms the /dev/random device easily both with (13 kB/s) and without post-processing (29 kB/s).

The amount of randomness needed for seeding the software generators differs considerably. The least amount is needed by the AES based generators, which need 128 bits for the textbook and 256 bits for the OpenSSL implementation. The latter randomizes initial counter and key, whereas the former only randomizes the key. Both the RSA and the Blum–Blum–Shub generator need to generate two 1024 bit primes. The textbook method chooses uniformly random integers of the appropriate size and tests them successively for compositeness. This requires tests on expectedly $2 \cdot 1024 \cdot \ln 1024 \approx 14,200$ different integers by the prime number theorem, thus consuming approximately 1.8 MB of seed randomness. The primality tests themselves might consume additional randomness if a probabilistic variant is employed. There are cheaper methods, though, reducing the necessary amount of randomness to 2048 bits only. For details on this matter see Loebenberger and Nüsken (2014).

The linear congruential generator additionally randomizes the initial state and thus consumes further 2048 bits for seeding.

Taking the throughput of the physical generators into account, the amount of time needed between possible reseeding ranges from $\frac{128}{8} \cdot \frac{1}{29,000} = 0.00055$ s for the textbook AES generator seeded by the Bergmann generator to $\frac{6144}{8} \cdot \frac{1}{2.2} = 349$ s for the linear congruential generator seeded by /dev/random.

While the statistical quality of each generators output is not dependent on the reseeding, the amount of total entropy is not raised by any internal calculation, making regular reseeding sensible. Using the Bergmann generator for seeding, even the linear congruential generator can be reseeded every 0.026 s, which seems a reasonable time span especially in networking contexts.

5   Conclusion and Future Work

We implemented a number of software random generators and compared their performance to physical generators. A blocking /dev/random is way too slow to be of practical use as the only source of (pseudo-)randomness, except for seeding software generators. The generator PRG310-4 is roughly as fast as our Blum–Blum–Shub implementation. However, both are surpassed by the RSA generator when run with a fast parameter set, which offers the same level of security.

The most interesting result is the vast difference between blocking and non-blocking random devices. This illustrates in a nice way the still open question whether lots of potentially bad randomness surpass the parsimonious use of guaranteed high-quality randomness.

The results also suggest a profitable symbiosis of hardware-generated seeds and number-theoretic high throughput—rather the reverse of the situation in other cryptographic contexts, say, the Diffie–Hellman exchange of keys for fast AES encryption.

The speedup introduced by the AES-NI instruction-set allows to generate 151 MB/s on a laptop computer, surpassing the requirements of the NGINX cluster (1.3 MB/s) by far, implying a negligible CPU-load.

Practical use of our findings has not taken place yet. Depletion of /dev/random is a realistic issue—workarounds for implied problems even suggest using the non-blocking /dev/urandom as a physical generator, see Searle (2008). However, prohibiting the use of /dev/urandom for key generation is also under debate, see Bernstein (2014), and there seems to be no consensus in the near future.

As a next step, implementing and testing on kernel level using optimized implementations is recommended.

Implementing an AES based random generator in the Linux kernel appears to be reasonable, but other platforms (i.e. ARM) may favor other hardware-accelerated ciphers for better performance and less CPU load. Thus the cipher must be made configurable.

Acknowledgements This work arose from a course taught by Joachim von zur Gathen and Daniel Loebenberger at the University of Bonn. It was funded by the B-IT foundation and the state of North Rhine–Westphalia. Substantial improvements of the presentation of the article were conducted later at genua GmbH, Germany. We thank the anonymous referees for their fruitful comments.

References

WERNER ALEXI, BENNY CHOR, ODED GOLDREICH & CLAUS P. SCHNORR (1988). RSA and Rabin functions: Certain Parts are As Hard As the Whole. *SIAM Journal on Computing* **17**(2), 194–209. ISSN 0097-5397. https://doi.org/10.1137/0217013.

FRANK BERGMANN (2019). Professionelle Zufallsgeneratoren für kryptografisch sichere Zufallszahlen. http://www.ibbergmann.org/.

D. J. BERNSTEIN (2014). cr.yp.to: 2014.02.05: Entropy Attacks! http://blog.cr.yp.to/20140205-entropy.html. Last access 27 June 2015.

L. BLUM, M. BLUM & M. SHUB (1986). A simple unpredictable pseudo-random number generator. *SIAM Journal on Computing* **15**(2), 364–383. https://doi.org/10.1137/0215025.

A. BORODIN, J. VON ZUR GATHEN & J. E. HOPCROFT (1982). Fast parallel matrix and GCD Computations. *Information and Control* **52**, 241–256. http://dx.doi.org/10.1016/S0019-9958(82)90766-5. Extended Abstract in *Proceedings of the 23rd Annual IEEE Symposium on Foundations of Computer Science, Chicago IL* (1982).

BSI (2019a). Kryptographische Verfahren: Empfehlungen und Schlüssellängen. Technische Richtlinie BSI TR-02102-1, Bundesamt für Sicherheit in der Informationstechnik, Bonn, Germany. https://www.bsi.bund.de/SharedDocs/Downloads/DE/BSI/Publikationen/TechnischeRichtlinien/TR02102/BSI-TR-02102.pdf.

BSI (2019b). Overview of Linux kernels with NTG.1-compliant random number generator /dev/random. Technical report, Bundesamt für Sicherheit in der Informationstechnik. https://www.bsi.bund.de/SharedDocs/Downloads/DE/BSI/Publikationen/Studien/LinuxRNG/NTG1_Kerneltabelle_EN.pdf.

ALEKSEI BURLAKOV, JOHANNES VOM DORP, JOACHIM VON ZUR GATHEN, SARAH HILLMANN, MICHAEL LINK, DANIEL LOEBENBERGER, JAN LÜHR, SIMON SCHNEIDER & SVEN ZEMANEK (2015a). Comparative analysis of pseudorandom generators. In *Proceedings of the 22nd Crypto-Day, 9–10 July 2015, Munich*. Gesellschaft für Informatik. http://fg-krypto.gi.de/fileadmin/fg-krypto/Handout-22.pdf.

ALEKSEI BURLAKOV, JOHANNES VOM DORP, SARAH HILLMANN, MICHAEL LINK, JAN LÜHR, SIMON SCHNEIDER & SVEN ZEMANEK (2015b). Comparative analysis of pseudorandom generators: Source code. BitBucket. https://bitbucket.org/sirsimonrattle/15ss-taoc.

JIM CLARKE (2019). An Optimist's View of the 4 Challenges to Quantum Computing. *IEEE Spectrum* https://spectrum.ieee.org/tech-talk/computing/hardware/an-optimists-view-of-the-4-challenges-to-quantum-computing.

SCOTT CONTINI & IGOR E. SHPARLINSKI (2005). On Stern's Attack Against Secret Truncated Linear Congruential Generators. In *Information Security and Privacy—10th Australasian Conference, ACISP 2005, Brisbane, Australia, July 4–6, 2005. Proceedings*, COLIN BOYD & JUAN MANUEL GONZÁLEZ NIETO, editors, volume 3574 of *Lecture Notes in Computer Science*, 52–60. Springer-Verlag, Berlin, Heidelberg. ISBN 978-3-540-26547-4 (Print) 978-3-540-31684-8 (Online). ISSN 0302-9743. https://doi.org/10.1007/11506157_5.

STEPHEN A. COOK (1971). The Complexity of Theorem–Proving Procedures. In *Proceedings of the Third Annual ACM Symposium on Theory of Computing, Shaker Heights OH*, 151–158. ACM Press.

MIKHAIL DYAKONOV (2018). The Case Against Quantum Computing. *IEEE Spectrum*. https://spectrum.ieee.org/computing/hardware/the-case-against-quantum-computing.

D. EASTLAKE, J. SCHILLER & S. CROCKER (2005). Randomness Requirements for Security. BCP 106, RFC Editor. http://www.rfc-editor.org/rfc/rfc4086.txt.

R. FISCHLIN & C. P. SCHNORR (2000). Stronger Security Proofs for RSA and Rabin Bits. *Journal of Cryptology* **13**(2), 221–244. https://doi.org/10.1007/s001459910008. Communicated by Oded Goldreich.

JOACHIM VON ZUR GATHEN & DANIEL LOEBENBERGER (2018). Why one cannot estimate the entropy of English by sampling. *Journal of Quantitative Linguistics* **25**(1), 77–106. https://doi.org/10.1080/09296174.2017.1341724.

ODED GOLDREICH (2001). *Foundations of Cryptography*, volume I: Basic Tools. Cambridge University Press, Cambridge. ISBN 0-521-79172-3.

ODED GOLDREICH, AMIT SAHAI & SALIL VADHAN (1999). Can Statistical Zero Knowledge Be Made Non-interactive? Or On the Relationship of $\mathcal{SZK}$ and $\mathcal{NISZK}$. In *Advances in Cryptology: Proceedings of CRYPTO 1999, Santa Barbara, CA*, M. WIENER, editor, volume 1666 of *Lecture Notes in Computer Science*, 467–484. Springer-Verlag, Berlin, Heidelberg. ISBN 978-3-540-48405-9 (Online) 978-3-540-66347-8 (Print). ISSN 0302-9743. https://doi.org/10.1007/3-540-48405-1_30.

TORBJÖRN GRANLUND (2014). The GNU Multiple Precision Arithmetic Library. C library. https://gmplib.org/. Last access 25th June 2015.

SHAY GUERON (2010). Intel® Advanced Encryption Standard (AES) New Instructions Set: White paper. Technical report, Intel Corporation.

ZVI GUTTERMAN, BENNY PINKAS & TZACHY REINMAN (2006). Analysis of the Linux Random Number Generator. In *Proceedings of the 2006 IEEE Symposium on Security and Privacy*, 371–385. IEEE Computer Society. ISBN 0-7695-2574-1. https://doi.org/10.1109/SP.2006.5. Also available at http://eprint.iacr.org/2006/086.

JOHAN HÅSTAD & ADI SHAMIR (1985). The Cryptographic Security of Truncated Linearly Related Variables. In *Proceedings of the Seventeenth Annual ACM Symposium on Theory of Computing*, STOC '85, 356–362. ACM, New York, NY, USA. ISBN 0-89791-151-2. https://doi.org/10.1145/22145.22184.

WOLFGANG KILLMANN & WERNER SCHINDLER (2008). A Design for a Physical RNG with Robust Entropy Estimators. In *CHES 2008*, ELISABETH OSWALD & PANKAJ ROHATGI, editors, volume 5154 of *LNCS*, 146–163. ISBN 978-3-540-85052-6. https://doi.org/10.1007/978-3-540-85053-3_10.

WOLFGANG KILLMANN & WERNER SCHINDLER (2011). A proposal for: Functionality classes for random number generators. Anwendungshinweise und Interpretationen zum Schema AIS 20/AIS 31, Bundesamt für Sicherheit in der Informationstechnik, Bonn, Germany. https://www.bsi.bund.de/SharedDocs/Downloads/DE/BSI/Zertifizierung/Interpretationen/AIS_31_Functionality_classes_for_random_number_generators_e.pdf?__blob=publicationFile. Version 2.0.

PATRICK LACHARME, ANDREA RÖCK, VINCENT STRUBEL & MARION VIDEAU (2012). The Linux Pseudorandom Number Generator Revisited. https://eprint.iacr.org/2012/251.pdf. Last access 27 June 2015.

D. H. LEHMER (1951). Mathematical methods in large-scale computing units. In *Proceedings of a Second Symposium on Large-Scale Digital Calculating Machinery, 13–16 September*

1949, volume 26 of *Annals of the Computation Laboratory of Harvard University*, 141–146. Harvard University Press, Cambridge, Massachusetts. https://archive.org/details/proceedings_ of_a_second_symposium_on_large-scale_.

DANIEL LOEBENBERGER & MICHAEL NÜSKEN (2014). Notions for RSA integers. *International Journal of Applied Cryptography* **3**(2), 116–138. ISSN 1753-0571 (online), 1753-0563 (print). http://dx.doi.org/10.1504/IJACT.2014.062723.

STEPHAN MÜLLER (2019). Documentation and Analysis of the Linux Random Number Generator. Technical report, atsec information security GmbH for the Federal Office for Information Security. https://www.bsi.bund.de/SharedDocs/Downloads/EN/BSI/Publications/Studies/ LinuxRNG/LinuxRNG_EN.pdf.

NGINX (2016). Sizing Guide for Deploying NGINX Plus on Bare Metal Servers. Technical report, NGINX Inc. https://cdn-1.wp.nginx.com/wp-content/files/nginx-pdfs/Sizing-Guide-for-Deploying-NGINX-on-Bare-Metal-Servers.pdf. Last access 16 August 2017.

NIST (2001a). *Federal Information Processing Standards Publication 197—Announcing the ADVANCED ENCRYPTION STANDARD (AES)*. National Institute of Standards and Technology. http://csrc.nist.gov/publications/fips/fips197/fips-197.pdf. Federal Information Processings StandardsPublication 197.

NIST (2001b). NIST Special Publication 800-38A: Recommendation for Block Cipher Modes of Operation.

NIST (2015). NIST Special Publication 800-90A: Recommendation for Random Number Generation Using Deterministic Random Bit Generators. https://nvlpubs.nist.gov/nistpubs/ SpecialPublications/NIST.SP.800-90Ar1.pdf.

JOAN B. PLUMSTEAD (1982). Inferring a Sequence Generated by a Linear Congruence. *Proceedings of the 23rd Annual IEEE Symposium on Foundations of Computer Science, Chicago IL* 153–159. https://doi.org/10.1109/SFCS.1982.73. Published as Joan Boyar, *Inferring Sequences Produced by Pseudo-Random Number Generators*, Journal of the ACM **36** (1), 1989, pages 129–141, https://doi.org/10.1145/58562.59305.

WERNER SCHINDLER & WOLFGANG KILLMANN (2003). Evaluation Criteria for True (Physical) Random Number Generators Used in Cryptographic Applications. In *Cryptographic Hardware and Embedded Systems—CHES 2002*, JR. BURTON S. KALISKI, ÇETIN K. KOÇ & CHRISTOF PAAR, editors, volume 2523 of *LNCS*, 431–449. Springer-Verlag, Berlin, Heidelberg. ISSN 0302-9743 (Print), 1611-3349 (Online). https://doi.org/10.1007/3-540-36400-5_31.

BRUCE SCHNEIER (2008). Random Number Bug in Debian Linux. Weblog. https://www.schneier. com/blog/archives/2008/05/random_number_b.html.

CHRIS SEARLE (2008). Increase entropy on a 2.6 kernel Linux box. https://www.chrissearle.org/ 2008/10/13/Increase_entropy_on_a_2_6_kernel_linux_box/. Last access 27 June 2015.

ADI SHAMIR (1983). On the Generation of Cryptographically Strong Pseudorandom Sequences. *ACM Trans. Comput. Syst.* **1**(1), 38–44. ISSN 0734-2071. https://doi.org/10.1145/357353. 357357.

PETER W. SHOR (1999). Polynomial-Time Algorithms for Prime Factorization and Discrete Logarithms on a Quantum Computer. *SIAM Review* **41**(2), 303–332.

RON STEINFELD, JOSEF PIEPRZYK & HUAXIONG WANG (2006). On the Provable Security of an Efficient RSA-Based Pseudorandom Generator. In *Advances in Cryptology: Proceedings of ASIACRYPT 2006, Shanghai, China*, XUEJIA LAI & KEFEI CHEN, editors, volume 4284 of *Lecture Notes in Computer Science*, 194–209. Springer-Verlag, Berlin, Heidelberg. ISBN 978-3-540-49475-1. ISSN 0302-9743 (Print) 1611-3349 (Online). https://doi.org/10. 1007/11935230_13.

JACQUES STERN (1987). Secret linear congruential generators are not cryptographically secure. *Proceedings of the 28th Annual IEEE Symposium on Foundations of Computer Science, Los Angeles CA* 421–426.

THEODORE TS'O (2016). Commit of ChaCha based urandom to Linux kernel. https://git.kernel.org/pub/scm/linux/kernel/git/torvalds/linux.git/commit/\?id\= e192be9d9a30555aae2ca1dc3aad37cba484cd4a. Last access 29 August 2017.

UMESH V. VAZIRANI & VIJAY V. VAZIRANI (1984). Efficient and Secure Pseudo-Random Number Generation (Extended Abstract). In *Proceedings of the 25th Annual IEEE Symposium on Foundations of Computer Science, Singer Island FL*, 458–463. IEEE Computer Society Press. ISBN 0-8186-0591-X. ISSN 0272-5428.

Difference Equation Theory Meets Mathematical Finance

Stefan Gerhold and Arpad Pinter

Dedicated to Peter Paule on the occasion of his 60th birthday

1 Introduction

"Difference equations" and "mathematical finance" appearing in one sentence may evoke the association of numerical derivative pricing by discretizing PDEs. This is *not* what this paper is about. Rather, it deals with a nineteenth century result from complex analysis (Pringsheim's theorem) and two asymptotic methods (saddle point asymptotics; Hankel contour asymptotics) that have been applied to some problems from the theory of difference equations, and more recently in financial mathematics. Sections 2 and 3 of the present paper are surveys of articles that have appeared elsewhere, whereas most of Sect. 4 has not been published in a journal, but only in Arpad Pinter's PhD thesis [31]. The reader might be a bit surprised that the content of this paper is only peripherally related to Peter Paule's research interests. The reason is that my (Gerhold's) research during my PhD studies soon started to deviate from symbolic summation towards asymptotics and other problems, followed by a switch to mathematical finance. I am very grateful to Peter for tolerating this as my supervisor, for sparking my interest in combinatorics with his marvellous lectures and lively weekly seminar, and for many useful pieces of advice.

S. Gerhold (✉)
TU Wien, Vienna, Austria
e-mail: sgerhold@fam.tuwien.ac.at

Arpad Pinter
TU Wien Alumnus, Vienna, Austria
e-mail: apinter@fam.tuwien.ac.at

© Springer Nature Switzerland AG 2020
V. Pillwein, C. Schneider (eds.), *Algorithmic Combinatorics: Enumerative Combinatorics, Special Functions and Computer Algebra*, Texts & Monographs in Symbolic Computation, https://doi.org/10.1007/978-3-030-44559-1_11

2 Pringsheim's Theorem: Oscillations and the Rough Heston Model

A part of the PhD thesis [14] is devoted to proving inequalities by computer algebra. The proving method presented there and in [1, 17, 18] has received further attention, e.g. in [29, 30]. Here, we recall another problem on inequalities that was investigated in [14] and the subsequent paper [2]. With difference equations being a very common topic in Peter Paule's lectures and research seminar, it seemed to be a natural question to study the positivity of solutions of the simplest kind of linear difference equations: those with constant (real) coefficients, whose solutions are commonly referred to as *recurrence sequences*. In [2], the following result was established in this direction:

Theorem 1 (Bell, Gerhold 2007) *Let* $(f_n)_{n \in \mathbb{N}}$ *be a nonzero recurrence sequence with no positive dominating characteristic root. Then the sets* $\{n \in \mathbb{N}: f_n > 0\}$ *and* $\{n \in \mathbb{N}: f_n < 0\}$ *have positive density.*

The dominating characteristic roots are the roots of maximal modulus among the roots of the characteristic polynomial. They occur in the leading term of the well-known explicit representation of recurrence sequences. Applying the following theorem to the generating function $\sum_{n=1}^{\infty} f_n z^n$ immediately implies the weaker statement that these index sets are both infinite. This has been noted in Theorem 7.1.1 of [21].

Theorem 2 (Pringsheim's Theorem) *Suppose that the power series* $F(z) = \sum_{n=0}^{\infty} a_n z^n$ *has positive finite radius of convergence R, and that all the coefficients are non-negative real numbers. Then F has a singularity at R.*

Alfred Pringsheim (1850–1941), father in law of Thomas Mann, proved this result in 1894. For a proof, see Remmert [32, p. 235], or Flajolet and Sedgewick [8, p. 240]. I (Gerhold) must admit that I was not aware of Pringsheim's theorem when writing [2]. Shortly after the paper was published in 2007, Alan Sokal informed my coauthor Jason Bell and me that our proof of Theorem 1 can be shortened, since it is a corollary of Theorem 1 in [2] and a generalized version of Pringsheim's theorem (see p. 242 in [4]). This shortcut did not make our paper obsolete, since the *existence* of the densities in Theorem 1 is a non-trivial fact, and moreover there are further results in [2]. See also [15] and, for more recent results on the sign of recurrence sequences [28].

We now switch to an apparently completely unrelated topic. In mathematical finance, continuous time stochastic processes are used to model the unknown future behavior of assets such as stocks, FX rates, and others. In recent years, so-called *rough* models have received a lot of attention. We just mention that rough refers to the "low" Hölder continuity of the paths, and that these models feature excellent statistical properties when applied to real market data, while their numerical treatment poses some challenges. One particular such model is El Euch and Rosenbaum's rough Heston model [6], with parameters $\rho \in (-1, 1)$, and λ, ξ,

$\bar{v} > 0$, $\alpha \in (\frac{1}{2}, 1)$. It is defined by the SDE (stochastic differential equation)

$$dS_t = S_t \sqrt{V_t}\, dW_t, \quad S_0 > 0,$$

$$V_t = V_0 + \frac{1}{\Gamma(\alpha)} \int_0^t (t-s)^{\alpha-1} \lambda(\bar{v} - V_s)\, ds$$

$$+ \frac{1}{\Gamma(\alpha)} \int_0^t (t-s)^{\alpha-1} \xi \sqrt{V_s}\, dZ_s, \tag{1}$$

$$d\langle W, Z \rangle_t = \rho\, dt.$$

Here, W and Z are correlated Brownian motions, and $\langle \cdot, \cdot \rangle$ denotes the cross-variation (see Definition 5.5 in [23]). The process S models an asset price, and $\sqrt{V}$ its stochastic volatility. In [19], we investigated the *moment explosion time* of the rough Heston model. Briefly, this amounts to finding the domain of the map $(u, t) \mapsto \mathbb{E}[S_t^u]$. Knowing the time, depending on u, at which the moment $\mathbb{E}[S_t^u]$ ceases to exist is important when implementing option pricing, as we elucidate in [19]. In [6], it was shown that $\mathbb{E}[S_t^u]$ can be expressed by the solution of a fractional Riccati equation:

$$\mathbb{E}[S_t^u] = \exp\left(\bar{v}\lambda I_t^1 \psi(u, t) + v_0 I_t^{1-\alpha} \psi(u, t)\right),$$

where ψ satisfies

$$D_t^\alpha \psi(u, t) = R(u, \psi(u, t)) \tag{2}$$

with initial condition $I_t^{1-\alpha} \psi(u, 0) = 0$. Here, D and I denote the Riemann–Liouville fractional derivative resp. integral (see section 2.1 in [24]), and R is a certain polynomial whose coefficients depend on the model parameters. In [19] a fractional power series

$$\sum_{n=1}^{\infty} a_n(u) t^{\alpha n} \tag{3}$$

representing this solution was found. Thus, the problem of finding the moment explosion time is transferred to finding the explosion time of (3). The radius of convergence of the power series

$$\sum_{n=1}^{\infty} a_n(u) z^n \tag{4}$$

can be easily computed, because the fractional ODE (2) yields a recurrence that allows to compute the coefficients $a_n(u)$. However, a priori this need not yield the explosion time. To wit, the explosion time is related to the smallest singularity of (4) *on the positive real axis,* whereas there might be singularities closer to the origin that are negative or non-real, and therefore practically meaningless. This is where Pringsheim's theorem enters the stage. Under some restrictions on the parameters, we could show that $a_n(u) \geq 0$ holds, and so Theorem 2 guarantees that the explosion time can be computed from the radius of convergence of (4). Thus, we can determine the domain of finiteness of $\mathbb{E}[S_t^u]$, which is the basis for efficiently evaluating integrals needed to price options in the rough Heston model.

3 Saddle Point Asymptotics: Non-holonomic Sequences and the Heston Model

A holonomic sequence is a sequence of numbers that satisfies a linear difference equation with polynomial coefficients. This class of sequences, and their generating functions, has received a lot of attention, in particular from the viewpoint of automatic identity proving. Among a very large number of papers, we just cite [5, 25, 33]. When looking for problems to solve during my PhD thesis, I (Gerhold) started to think about the theoretical question of proving the *non*-holonomicity of certain sequences [3, 9, 13]. Asymptotic expansions are a very useful tool for this, because a holonomic generating function satisfies a linear ODE with polynomial coefficients, and it is well known that functions of this kind have a very restricted asymptotic behavior. This method was applied to a good deal of examples in [9]. In 2005, I sent an email to Philippe Flajolet, asking whether the approach could be used to prove that the sequence $e^{1/n}$ is not holonomic. I quote from his response:

This is interesting and here's a way we think it can be done. We didn't reflect too much about it however and didn't work out details. Take $f_n = \exp(1/n)$ and let $F(z) = \sum f_n(-z)^n$ be the corresponding OGF, taken for convenience with alternating signs. We want to prove, right in line with our joint paper, that there is some nonholonomic element in $F(z)$ as $z \to \infty$. Start from the Lindelöf integral

$$F(z) = \frac{1}{2i\pi} \int \exp(1/s) z^s \frac{\pi}{\sin(\pi s)} ds,$$

taken along $1/2 - i\infty$ to $1/2 + i\infty$. [Proved by residues upon closing by a large semicircle on the right, seems to work well here.] Then, move the integration line close to $\mathrm{Re}(s) = 0$ where the integrand blows up. There's a saddle point, a function of z, at

$$s_0 = 1/\sqrt{\log z}$$

roughly. Then, $F(z)$ should behave more or less like $\exp(2\sqrt{\log z})$ as $z \to +\infty$. This is nonholonomic. The full argument [to contradict the structure theorem] needs *making sure we can shake the argument of z a little, betwen some $[-\varepsilon, +\varepsilon]$, but usually the sin in the denominator of the integrand plays in your favour.*

This proof strategy worked, of course, although it took us some years of intermittent work to finish the corresponding paper [10], which contains several other asymptotic results (see also Sect. 4 of the present paper). Concerning the generating function of $e^{1/n}$, the above saddle point approach yields

$$\sum_{n\geq 1} e^{1/n}(-z)^n \sim -\frac{e^{2\sqrt{\log z}}}{2\sqrt{\pi}(\log z)^{1/4}}, \quad z \to \infty, \tag{5}$$

where the left hand side is to be understood in the sense of analytic continuation. This shows that $e^{1/n}$ is a non-holonomic sequence, since the right hand side cannot be asymptotically equivalent to any holonomic function. The latter statement follows from a well-known result on the asymptotic behavior of ODE solutions, summarized in Theorem 2 of [9]. Moreover, using (5), we evaluated the alternating sum

$$\sum_{k=1}^{n} \binom{n}{k}(-1)^k e^{1/k} \sim -\frac{e^{2\sqrt{\log n}}}{2\sqrt{\pi}(\log n)^{1/4}}$$

asymptotically for $n \to \infty$, which would be hard by elementary methods.

Again, we now jump to mathematical finance. Among the many asset price models that have been suggested and studied, sending $\alpha \to 1$ in the model from the previous section yields a particularly well-known one: The classical (non-rough) *Heston model*. Since its introduction in 1993 (see [22]), it has been used by many practitioners and studied by many researchers. The main advantages of this model are its explicit characteristic function, which allows for fast and easy option price computation, and its reasonable fit to market data. Its dynamics are as in (1), but with $\alpha = 1$, which removes the weakly singular kernel $(t-s)^{\alpha-1}$. This dramatically improves the regularity of the processes S, V and the numerical tractability of the model, at the price of a less satisfactory fit to financial market data.

My (Gerhold's) work on the Heston model began in 2009 at the ÖMG-DMV congress in Graz, when Peter Friz (TU Berlin) asked my colleague Friedrich Hubalek (TU Wien) about applying the saddle point method to option prices. Friedrich Hubalek directed Peter Friz to me, and we started to analyze option prices given by the Heston model asymptotically. A call option gives the option holder the right, but not the obligation, to buy a unit of the underlying asset at time T for the strike price K, where T and K are fixed. The payoff of this option at maturity T is $(S_T - K)^+$, because a share price $S_T > K$ yields a profit of $S_T - K$, whereas the option becomes worthless in the case $S_T \leq K$. At time zero, the price of the call option is

$$C(K, T) = \mathbb{E}[(S_T - K)^+].$$

We assumed zero interest rate here, and skipped the subtle but very important point that the expectation is to be taken under a special—so-called risk neutral—probability measure that does *not* coincide with the "real world" probability. The question we dealt with in [12] is the asymptotic behavior of $C(K, T)$ for large strike K, if S_T has the probability distribution given by the Heston model. The call price can be recovered by Fourier inversion from the moment generating function:

$$C(K, T) = \frac{K}{2\pi i} \int_{\beta - i\infty}^{\beta + i\infty} K^{-u} \frac{\mathbb{E}[S_T^u]}{u(u-1)} du. \tag{6}$$

Implementing this numerically requires knowing the domain of the characteristic function (equivalently, of the moment generating function), because this yields the possible values of β, the real part of the integration contour. This is the question we mentioned in Sect. 2 for the *rough* Heston model. For classical Heston, this domain is well-known, because the characteristic function has an explicit expression. It turns out that, at the border of this domain, it has a singularity of the form "exponential of a pole". Thus, a saddle point analysis with some similarities to the one above could be applied to the problem of approximating (6) (see also [16]). While the tail estimates are quite different, the local expansion, yielding the dominant term, is very similar. The formulas are somewhat tedious, and so we refer to [12] for details. We just mention that there are constants c_i, positive for $i = 1, 2, 3$, such that the Heston call price satisfies

$$C(K, T) \sim c_1 K^{-c_2} e^{c_3 \sqrt{\log K}} (\log K)^{c_4}, \quad K \to \infty,$$

for $T > 0$ fixed. The dominating factor K^{-c_2} was known before and follows quite easily from the explicit moment generating function. The sub-polynomial factor $e^{c_3 \sqrt{\log K}}$ was the main contribution of [12], improving numerical accuracy significantly. We recall here the role of the asymptotic factor $e^{2\sqrt{\log z}}$ in (5), which came from a very similar saddle point analysis, and proves the non-holonomicity of the function on the left hand side of (5).

4 Hankel Contour Asymptotics: Non-holonomic Sequences and the 3/2–Model

4.1 Setup

At the beginning of Sect. 3, we described an asymptotic evaluation of the generating function of $e^{1/n}$. In [10], we studied the natural extension e^{cn^θ} with parameters c

and θ, and also more general sequences and their generating functions.[1] We quote here the following result:

$$\sum_{n\geq 1} e^{\pm\sqrt{n}}(-z)^n = -1 \mp \frac{1}{\sqrt{\pi}\log z} + O\big((\log z)^{-3/2}\big). \quad z \to \infty. \tag{7}$$

As above, (7) not only proves non-holonomicity of $e^{\pm\sqrt{n}}$, but also approximations such as

$$\sum_{k=0}^{n}\binom{n}{k}(-1)^k e^{\pm\sqrt{k}} \sim -\frac{\pm 1}{\sqrt{\pi}\log n}, \quad n \to \infty.$$

The proof of (7) again starts with the Lindelöf representation

$$\sum_{n=1}^{\infty} e^{\pm\sqrt{n}}(-z)^n = -\frac{1}{2i\pi}\int_{1/2-i\infty}^{1/2+i\infty} e^{\pm\sqrt{s}} z^s \frac{\pi}{\sin\pi s}\,ds. \tag{8}$$

This time, a saddle point approach is not appropriate, because the singularity of $e^{\pm\sqrt{s}}$ at zero is too "tame", and a somewhat larger integration contour is needed to extract sufficient asymptotic information. The method of choice is to use a contour that goes around the branch cut of $e^{\pm\sqrt{s}}$, and part of which is transformed to a Hankel contour by a substitution. Recall that a well-known application of Hankel contour asymptotics is Flajolet and Odlyzko's singularity analysis of generating functions [7, 8]. We refer to [10] for details on the asymptotic analysis of (8), but use the same method in the present section on a different problem.

Maybe unsurprisingly at this point, the problem we consider comes from mathematical finance. The model we consider is again a stochastic volatility model, which goes under the name of 3/2–model. The logarithmic stock price process $X_t = \log S_t$ in this model solves the SDE (stochastic differential equation)

$$dX_t = -\tfrac{1}{2}V_t\,dt + \sqrt{V_t}\,dW_t, \qquad\qquad X_0 = x_0 \in \mathbb{R},$$

$$dV_t = \kappa V_t(\theta - V_t)\,dt + \xi V_t^{3/2}\,dZ_t, \qquad V_0 = v_0 > 0,$$

$$d\langle W, Z\rangle_t = \rho\,dt,$$

with correlated Brownian motions W and Z and parameters $\kappa > 0,\, \theta > 0,\, \xi > 0$ and $|\rho| < 1$. Define $\bar{\rho} := \sqrt{1-\rho^2}$ and $\bar{\kappa} := 2\kappa + \xi^2$. The moment-generating

---

[1]Needless to say, Philippe Flajolet and Bruno Salvy needed no help from a PhD student to set up the various asymptotic methods used in [10], but I (Gerhold) was of some use working out the technical estimates. Among many episodes worth remembering, I vividly recall Philippe's statement after having written the introduction of [10]: "We need brains (pointing at Bruno Salvy), we need strength (pointing at me), and we need blah-blah (pointing at himself)."

function (mgf) of X_T for $T > 0$ can be computed as

$$M(u, T) := \mathbb{E}[e^{uX_T}] = e^{ux_0} \frac{\Gamma(\mu_u - \alpha_u)}{\Gamma(\mu_u)} z_T^{\alpha_u} {}_1F_1(\alpha_u, \mu_u, -z_T), \tag{9}$$

at least for all $u \in \mathbb{C}$ in the vertical strip $a < \mathrm{Re}(u) < b$ with $a \leq 0$ and $b \geq 1$, and with the confluent hypergeometric function ${}_1F_1$ and the auxiliary functions

$$\alpha_u := \frac{1}{\xi^2}(\gamma_u - \chi_u), \qquad \gamma_u := \sqrt{\chi_u^2 - \xi^2 u(u - 1)},$$

$$\mu_u := \frac{1}{\xi^2}(\xi^2 + 2\gamma_u), \qquad \chi_u := \tfrac{1}{2}\bar{\kappa} - \rho\xi u, \tag{10}$$

$$z_T := \frac{2}{\xi^2 \beta_T}, \qquad \beta_T := \frac{v_0}{\kappa\theta}(e^{\kappa\theta T} - 1).$$

Without loss of generality, from now on, we assume $x_0 = 0$. Define the two real numbers

$$u_\pm := \frac{1}{2\xi\bar{\rho}^2} \left(\xi - \rho\bar{\kappa} \pm \sqrt{(\xi - \rho\bar{\kappa})^2 + \bar{\kappa}^2\bar{\rho}^2} \right), \tag{11}$$

which are the unique roots of the quadratic term under the square root of γ. After factorization of the polynomial, we have the following representation of γ

$$\gamma_u = \xi\bar{\rho}\sqrt{(u_+ - u)(u - u_-)}. \tag{12}$$

Throughout, we make the technical assumption

$$\mu_{u_+} - \alpha_{u_+} > 0$$

which is always satisfied if $\rho < 0$. Under this assumption, the right boundary b of the vertical strip, where Eq. (9) holds, can be extended until $b = u_+$. Note that the mgf has a branch cut along $[u_+, +\infty)$ due to the branch cut of (12). For further information on the 3/2–model, see e.g. Lewis [26].

4.2   Tail Asymptotics of the Density

We are interested in tail asymptotics of the density function $\varphi(k, T) := \varphi_{X_T}(k)$ of X_T for $T > 0$, i.e., the asymptotic behaviour as $k \to \infty$ for fixed $T > 0$. The density function φ can be expressed via Fourier-transform as

$$\varphi(k, T) = \frac{1}{2\pi i} \int_{a-i\infty}^{a+i\infty} e^{-ku} M(u, T)\, du, \quad k \in \mathbb{R}, \tag{13}$$

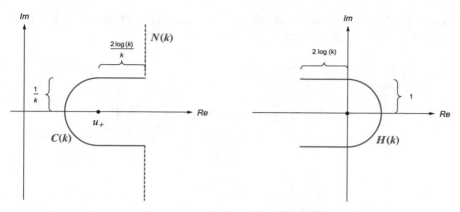

Fig. 1 In the left panel, the critical path $C(k)$ and the neglectable path $N(k)$ (dashed line) are illustrated in the complex plane, whereas the right panel displays the transformed path $\mathcal{H}(k)$ after the transformation $w \mapsto u_+ - \frac{w}{k}$

with $a \in (u_-, u_+)$. For the analysis, we adjust the integration path in (13) similarly to Friz and Gerhold [11] and split it into two parts, the critical path $C(k)$ and the neglectable path $N(k)$, depending on the strike parameter $k \geq 1$. The critical contour $C(k)$ embraces the critical moment u_+, see the left panel of Fig. 1.

The critical path $C(k)$ starts at $u_+ + 2\log(k)/k - i/k$, goes horizontally to $u_+ - i/k$, then clockwise along the half-circle with center u_+ and radius $1/k$ until it reaches $u_+ + i/k$, and again horizontally to the end point $u_+ + 2\log(k)/k + i/k$. The remaining part, denoted by $N(k)$, starts at the points $u_+ + 2\log(k)/k \pm i/k$ and goes straight to $u_+ + 2\log(k)/k + i\infty$ resp. $u_+ + 2\log(k)/k - i\infty$. This allows us to write the density function as

$$\varphi(k, T) = \frac{1}{2\pi i} \int_{C(k) \cup N(k)} e^{-ku} M(u, T) \, du, \quad k \in \mathbb{R}. \tag{14}$$

Theorem 3 (Tail Asymptotics) *Assume $\mu_{u_+} - \alpha_{u_+} > 0$. Then the first term in the tail expansion of the density function of X_T in the 3/2–model, with $T > 0$ fixed, is given by*

$$\varphi(k, T) \sim c \frac{e^{-ku_+}}{k^{3/2}}, \quad k \to \infty, \tag{15}$$

where $c = -m_1/(2\sqrt{\pi})$ with m_1 defined in (30).

Proof The integral over $N(k)$ in (14) is negligible; this will be proved in Lemma 1 below. Now consider the integral over the critical part $C(k)$ in (14). The change of variables $u = u_+ - w/k$ yields, as $k \to \infty$,

$$\frac{1}{2\pi i} \int_{C(k)} e^{-ku} M(u, T) \, du = \frac{e^{-ku_+}}{k} \frac{1}{2\pi i} \int_{\mathcal{H}(k)} e^w M\left(u_+ - \frac{w}{k}, T\right) dw, \tag{16}$$

where $\mathcal{H}(k)$ is the transformed path of $C(k)$, see the right panel of Fig. 1. In Lemma 2 below we give an expansion of M which yields

$$\frac{1}{2\pi i} \int_{\mathcal{H}(k)} e^w M\left(u_+ - \frac{w}{k}, T\right) dw$$

$$= M(u_+, T) \underbrace{\frac{1}{2\pi i} \int_{\mathcal{H}(k)} e^w dw}_{o\left(\frac{1}{k^2}\right)} + \frac{m_1}{\sqrt{k}} \underbrace{\frac{1}{2\pi i} \int_{\mathcal{H}(k)} e^w w^{1/2} dw}_{\to 1/\Gamma\left(-\frac{1}{2}\right) = -\frac{1}{2\sqrt{\pi}}}$$

$$+ \underbrace{\frac{1}{2\pi i} \int_{\mathcal{H}(k)} e^w O\left(\frac{w}{k}\right) dw}_{o\left(\frac{1}{k}\right)}.$$

The first integral is an easy computation. In the second and third integral, we used Hankel's integral representation for the gamma function, see [27]. Therefore,

$$\frac{1}{2\pi i} \int_{C(k)} e^{-ku} M(u, T) du \sim c \frac{e^{-ku_+}}{k^{3/2}}, \quad k \to \infty,$$

for $c = \frac{-m_1}{2\sqrt{\pi}}$.   □

While we established just first order asymptotics in Theorem 3, we note that the same method easily yields further terms in the asymptotic expansion, if desired.

Lemma 1 *The integral over $N(k)$ in (14) satisfies*

$$\frac{1}{2\pi i} \int_{N(k)} e^{-ku} M(u, T) du = o\left(e^{-ku_+} k^{-3/2}\right), \quad k \to \infty.$$

Proof By symmetry, it suffices to consider only the integral over the upper part of the contour $N(k)$. We define the path $u_k(t) := u_+ + 2 \log k / k + it$ with $t \in [1/k, \infty)$,

$$\frac{1}{2\pi i} \int_{u_k} e^{-ku} M(u, T) du = \frac{e^{-ku_+}}{k^2} \frac{1}{2\pi} \int_{1/k}^{\infty} e^{-itk} M(u_k(t), T) dt.$$

By showing the boundedness of the latter integral, the proof is finished. We use the triangular inequality for integrals and split the integral into two parts,

$$\left| \int_{1/k}^{\infty} e^{-itk} M(u_k(t), T) dt \right| \leq \int_{1/k}^{t_1} |M(u_k(t), T)| dt + \int_{t_1}^{\infty} |M(u_k(t), T)| dt,$$

$$(17)$$

where $t_1 \geq 1$ will be determined later. For the first integral in (17), note that $2 \log k / k \in [0, 1]$ for any $k \geq 1$. Recall that $M(\cdot, T)$ has a branch cut along $[u_+, \infty)$, but a continuous extension $\tilde{M}$ of M exists on the half-plane $\Im(s) \geq 0$. Hence $|M(\cdot, T)|$ attains a maximum value on $[u_+, u_+ + 1] + i(0, t_1]$,

$$\int_{1/k}^{t_1} |M(u_k(t), T)| \, dt \leq t_1 \max_{u \in [u_+, u_+ + 1] + i(0, t_1]} |M(u, T)| < \infty$$

In order to show the boundedness of the second integral in (17) and to determine $t_1 \geq 1$, we have to take a closer look at the mgf and the auxiliary functions defined in (10). The fact that $2 \log k / k \in [0, 1]$ for $k > 1$ ensures $u_k(t) = it + O(1)$ for $t \to \infty$ uniformly for all $k \geq 0$. Thus, the following asymptotic expansions of the auxiliary functions χ and γ in (10) hold

$$\chi(u_k(t)) = -i\xi\rho t + O(1),$$

$$\gamma(u_k(t)) = \sqrt{-\xi^2 \rho^2 t^2 + \xi^2 t^2 + O(t)} = \xi \bar{\rho} t + O(1),$$

and simple computations then yield

$$\alpha(u_k(t)) = \tfrac{1}{\xi}(\bar{\rho} + i\rho)t + O(1), \tag{18}$$

$$\mu(u_k(t)) = \tfrac{2}{\xi}\bar{\rho} t + O(1),$$

$$\mu(u_k(t)) - \alpha(u_k(t)) = \tfrac{1}{\xi}(\bar{\rho} - i\rho)t + O(1), \tag{19}$$

for $t \to \infty$ uniformly for all $k \geq 1$. Due to (18), (19) and $\bar{\rho} > 0$, there exists $t_0 \geq 1$, such that $\mathrm{Re}(\mu(u_k(t)) - \alpha(u_k(t))) > 1$ and $\mathrm{Re}(\alpha(u_k(t))) > 1$ for all $k \geq 1$ and $t \geq t_0$. In particular, in this region we have

$$\mathrm{Re}(\mu(u_k(t))) > \mathrm{Re}(\alpha(u_k(t))) > 0,$$

and so we can use the representation (34) of the confluent hypergeometric function, which reduces the mgf to

$$M(u_k(t), T) = \frac{z_T^{\alpha(u_k(t))}}{\Gamma(\alpha(u_k(t)))} \int_0^1 e^{-z_T y} y^{\alpha(u_k(t)) - 1}(1 - y)^{\mu(u_k(t)) - \alpha(u_k(t)) - 1} \, dy. \tag{20}$$

Note that the absolute value of the integral is bounded by 1. Furthermore, we have uniformly for all $k \geq 1$

$$|z_T^{\alpha(u_k(t))}| = \exp\left(\tfrac{1}{\xi}\bar{\rho}\log(z_T)t(1 + o(1))\right), \quad t \to \infty. \tag{21}$$

Our choice $\text{Re}(\alpha(u_k(t))) > 1$ guarantees $|\arg(\alpha(u_k(t)))| < \frac{\pi}{2}$ and Stirling's formula (35) is applicable to $\Gamma(\alpha(u_k(t)))$ for all $t \geq t_0$ and all $k \geq 1$. Combining with (18) we have, uniformly for all $k \geq 1$,

$$|\Gamma(\alpha(u_k(t)))| \sim \sqrt{2\pi} |e^{-z} z^z z^{-1/2}|_{z=\frac{1}{\xi}(\bar{\rho}+i\rho)t}$$

$$= \sqrt{2\pi\xi}\, x^{-1/2}\, \exp\left(\tfrac{1}{\xi}\bar{\rho}t\log(\tfrac{t}{\xi}) - \tfrac{1}{\xi}\rho\arg(\bar{\rho}+i\rho)t - \tfrac{1}{\xi}\bar{\rho}t\right)$$

$$= \exp\left(\tfrac{1}{\xi}\bar{\rho}t\log t\big(1+o(1)\big)\right), \quad t \to \infty. \tag{22}$$

Putting (21) and (22) back into formula (20), we can find a sufficiently large $t_1 \geq t_0$ such that

$$|M(u_k(t), T)| \leq \exp\left(-(1+\varepsilon)\tfrac{1}{\xi}\bar{\rho}t\log t\right) \tag{23}$$

for all $t \geq t_1$ and all $k \geq 1$, with a constant $\varepsilon > 0$. The integrability of the right-hand side of (23) proves that the third integral in (17) is bounded. $\qquad\square$

Lemma 2 *Assume $\mu_{u_+} - \alpha_{u_+} > 0$. Near the critical moment u_+, the following expansion of the mgf holds uniformly for all $w \in \mathcal{H}(k)$,*

$$M\left(u_+ - \frac{w}{k}, T\right) = M(u_+, T) + m_1\sqrt{\frac{w}{k}} + O\left(\frac{w}{k}\right), \quad k \to \infty,$$

where m_1 is defined in (30).

Proof First, we expand the functions χ and γ in a neighborhood of u_+. Using the representation (12) of γ we only have to expand $\sqrt{u - u_-} = \sqrt{(u_+ - u_-) - (u_+ - u)}$ near u_+. Thus, as $u \to t_+$,

$$\gamma_u = \xi\bar{\rho}\sqrt{u_+ - u_-}(u_+ - u)^{1/2} + O\left((u_+ - u)^{3/2}\right), \tag{24}$$

$$\chi_u = \chi_{u_+} + \rho\xi(u_+ - u). \tag{25}$$

With these results, expansions for α and μ near u_+ can easily be computed,

$$\alpha_u = \alpha_{u_+} + \frac{\bar{\rho}}{\xi}\sqrt{u_+ - u_-}(u_+ - u)^{1/2} + O(u_+ - u), \quad u \to u_+ \tag{26}$$

$$\mu_u = \mu_{u_+} + 2\frac{\bar{\rho}}{\xi}\sqrt{u_+ - u_-}(u_+ - u)^{1/2} + O\left((u_+ - u)^{3/2}\right), \quad u \to u_+. \tag{27}$$

Define $u_k(w) := u_+ - \frac{w}{k}$, $w \in \mathcal{H}(k)$, for $k \geq 1$. From the uniform convergence $\sup_{w \in \mathcal{H}(k)} |u_k(w) - u_+| \to 0$ for $k \to \infty$, we have

$$\Delta\alpha := \alpha(u_k(w)) - \alpha_{u_+} = \frac{\bar{\rho}}{\xi}\sqrt{u_+ - u_-}\left(\frac{w}{k}\right)^{1/2} + O\left(\frac{w}{k}\right), \quad k \to \infty \quad (28)$$

$$\Delta\mu := \mu(u_k(w)) - \mu_{u_+} = 2\frac{\bar{\rho}}{\xi}\sqrt{u_+ - u_-}\left(\frac{w}{k}\right)^{1/2} + O\left(\left(\frac{w}{k}\right)^{3/2}\right), \quad k \to \infty,$$
$$(29)$$

uniformly for all $w \in \mathcal{H}(k)$. Define the function

$$\tilde{M}(\alpha, \mu) := \frac{\Gamma(\mu - \alpha)}{\Gamma(\mu)}(z_T)^\alpha {}_1F_1(\alpha, \mu, -z_T),$$

for all $(\alpha, \mu) \in \mathbb{C}^2$ where $\mu - \alpha, \mu \notin \mathbb{Z}_0^-$. In this region $\tilde{M}$ is jointly analytic in both variables. Note the relation $M(u, T) = \tilde{M}(\alpha_u, \mu_u)$. Since $\mu_{u_+} = 1$ and $\mu_{u_+} - \alpha_{u_+} > 0$, we can make a Taylor expansion of $\tilde{M}$ at the point $(\alpha_{u_+}, \mu_{u_+})$. Combining this with (28) and (29) gives us, uniformly for all $w \in \mathcal{H}(k)$,

$$M(u_k(w), T) = \tilde{M}(\alpha(u_k(w)), \mu(u_k(w)))$$

$$= \tilde{M}(\alpha_{u_+}, \mu_{u_+}) + \Delta\alpha\frac{\partial}{\partial\alpha}\tilde{M}(\alpha_{u_+}, \mu_{u_+}) + \Delta\mu\frac{\partial}{\partial\mu}\tilde{M}(\alpha_{u_+}, \mu_{u_+})$$

$$+ O\left((\Delta\alpha)^2\right) + O\left((\Delta\mu)^2\right)$$

$$= M(u_+, T) + \underbrace{\left(\frac{\partial\tilde{M}}{\partial\alpha} + 2\frac{\partial\tilde{M}}{\partial\mu}\right)(\alpha_{u_+}, \mu_{u_+})\frac{\bar{\rho}}{\xi}\sqrt{u_+ - u_-}\left(\frac{w}{k}\right)^{1/2}}_{=:m_1} + O\left(\frac{w}{k}\right), \quad k \to \infty.$$
$$(30)$$

$\square$

4.3   Large-Strike Asymptotics for the Implied Volatility

From tail asymptotics for the density function, it is possible to obtain large strike asymptotics for the implied volatility, see Gulisashvili [20] and Friz, Gerhold, Gulisashvili and Sturm [12]. Recall that the implied volatility is the volatility parameter that has to be used in the Black–Scholes model to recover given option prices. The statement is, that if the density function φ satisfies, for fixed $T > 0$,

$$c_1 k^{-\xi}h(k) \leq \varphi(k) \leq c_2 k^{-\xi}h(k),$$

for all sufficiently large k, with $\xi > 2$, h slowly varying and constants $c_1, c_2 > 0$, then the implied volatility $\sigma_{\mathrm{imp}}(K, T)$ satisfies

$$\sigma_{\mathrm{imp}}(K, T)\frac{\sqrt{T}}{\sqrt{2}} = \sqrt{\log K + \log \frac{1}{K^{2-\xi}h(K)} + \tfrac{1}{2}\log\log \frac{1}{K^{2-\xi}h(K)}} \tag{31}$$

$$- \sqrt{\log \frac{1}{K^{2-\xi}h(K)} + \tfrac{1}{2}\log\log \frac{1}{K^{2-\xi}h(K)}}$$

$$+ O\big((\log K)^{-1}\big),$$

as $K \to \infty$. In Theorem 3, we have established tail asymptotics for the density φ_{X_T} of the log-price $X_T = \log(S_T)$ in the 3/2–model. Because the density φ_{S_T} of S_T is given by

$$\varphi_{S_T}(K) = \frac{\varphi_{X_T}(\log K)}{K}, \quad K > 0,$$

we clearly have the tail asymptotics for φ_{S_T}

$$\varphi_{S_T}(K) \sim cK^{-(u_+ + 1)}h(K), \quad K \to \infty, \tag{32}$$

with the slowly varying function $h(K) = (\log K)^{-3/2}$. Note that the critical moment always satisfies $u_+ \geq 1$, and $u_+ = 1$ if and only if $2\xi\rho = \bar{\kappa}$. Hence, the previous statement is applicable.

Theorem 4 *Assume $\mu_{u_+} - \alpha_{u_+} > 0$ and $2\xi\rho \neq \bar{\kappa}$. The large-strike expansion of the implied volatility function in the 3/2–model, with $T > 0$ fixed, is given by, as $K \to \infty$,*

$$\sigma_{\mathrm{imp}}(K, T)\frac{\sqrt{T}}{\sqrt{2}} = (\sqrt{u_+} - \sqrt{u_+ - 1})\sqrt{\log K} \tag{33}$$

$$+ \frac{1}{2}\left(\frac{1}{\sqrt{u_+}} - \frac{1}{\sqrt{u_+ - 1}}\right)\frac{\log\log K}{\sqrt{\log K}} + O\left(\frac{\log\log\log K}{\sqrt{\log K}}\right).$$

Proof A straightforward calculation, using (31) and (32), shows

$$\sigma_{\mathrm{imp}}(K, T)\frac{\sqrt{T}}{\sqrt{2}} = \sqrt{\log K - \log(K^{1-u_+}h(K)) - \tfrac{1}{2}\log\big(-\log(K^{1-u_+}h(K))\big)}$$

$$- \sqrt{-\log(K^{1-u_+}h(K)) - \tfrac{1}{2}\log\big(-\log(K^{1-u_+}h(K))\big)}$$

$$+ O\big((\log K)^{-1}\big)$$

$$= \sqrt{u_+\log K + \log\log K + O(\log\log\log K)}$$

$$- \sqrt{(u_+ - 1) \log K + \log \log K + O(\log \log \log K)}$$
$$+ O((\log K)^{-1}),$$

$$= \sqrt{u_+} \sqrt{\log K} \left(1 + \frac{\log \log K}{u_+ \log K} + O\left(\frac{\log \log \log K}{\log K}\right)\right)$$
$$- \sqrt{(u_+ - 1)} \sqrt{\log K} \left(1 + \frac{\log \log K}{(u_+ - 1) \log K} + O\left(\frac{\log \log \log K}{\log K}\right)\right)$$
$$+ O((\log K)^{-1}),$$

as $K \to \infty$. This easily yields the statement. $\qquad\qquad\qquad\qquad\square$

We state the following lemmas which are used in the proof of the tail asymptotics of the density function. The first lemma describes a representation of the confluent hypergeometric function $_1F_1$, whereas the second lemma is the well-known Stirling formula for the Gamma function. For further details, see e.g. [27].

Lemma 3 *If* $\mathrm{Re}(\mu) > \mathrm{Re}(\alpha) > 0$, *then the confluent hypergeometric function* $_1F_1$ *has the integral representation*

$$_1F_1(\alpha, \mu, z) = \frac{\Gamma(\mu)}{\Gamma(\alpha)\Gamma(\mu - \alpha)} \int_0^1 e^{zy} y^{\alpha - 1} (1 - y)^{\mu - \alpha - 1} \, dy. \qquad (34)$$

Lemma 4 (Stirling) *The Gamma function satisfies*

$$\Gamma(z) = \sqrt{2\pi} e^{-z} z^z z^{-1/2} (1 + o(1)), \quad z \to \infty \quad with \ |\arg(z)| < \pi - \varepsilon, \qquad (35)$$

where $\varepsilon > 0$ *is arbitrary.*

Acknowledgements We gratefully acknowledge financial support from the Austrian Science Fund (FWF) under grant P 24880.

References

1. Alzer, H., Gerhold, S., Kauers, M., Lupas, A.: On Turán's inequality for Legendre polynomials. Expo. Math. **25**, 181–186 (2007)
2. Bell, J.P., Gerhold, S.: On the positivity set of a linear recurrence sequence. Isr. J. Math. **157**, 333–345 (2007)
3. Bell, J.P., Gerhold, S., Klazar, M., Luca, F.: Non-holonomicity of sequences defined via elementary functions. Ann. Combin. **12**, 1–16 (2008)
4. Boas, R.P. Jr.: Entire Functions. Academic, New York (1954)
5. Chyzak, F.: An extension of Zeilberger's fast algorithm to general holonomic functions. Discrete Math. **217**, 115–134 (2000). Formal power series and algebraic combinatorics (Vienna, 1997)
6. El Euch, O., Rosenbaum, M.: The characteristic function of rough Heston models. Math. Finance **29**(1), 3–38 (2019)

7. Flajolet, P., Odlyzko, A.: Singularity analysis of generating functions. SIAM J. Discrete Math. **3**, 216–240 (1990)
8. Flajolet, P., Sedgewick, R.: Analytic Combinatorics. Cambridge University Press, Cambridge (2009)
9. Flajolet, P., Gerhold, S., Salvy, B.: On the non-holonomic character of logarithms, powers and the nth prime function. Electron. J. Combin. **11**, 1–16 (2005)
10. Flajolet, P., Gerhold, S., Salvy, B.: Lindelöf representations and (non-)holonomic sequences. Electron. J. Combin. **17**, 1–28 (2010)
11. Friz, P., Gerhold, S.: Extrapolation analytics for Dupire's local volatility. In: Large Deviations and Asymptotic Methods in Finance. Springer Proceedings in Mathematics and Statistics, vol. 110. Springer, Cham (2015), pp. 273–286
12. Friz, P., Gerhold, S., Gulisashvili, A., Sturm, S.: On refined volatility smile expansion in the Heston model. Quant. Finance **11**, 1151–1164 (2011)
13. Gerhold, S.: On some non-holonomic sequences. Electron. J. Combin. **11**, 1–8 (2004)
14. Gerhold, S.: Combinatorial sequences: non-holonomicity and inequalities. PhD Thesis, J. Kepler University Linz (2005)
15. Gerhold, S.: Point lattices and oscillating recurrence sequences. J. Differ. Equ. Appl. **11**, 515–533 (2005)
16. Gerhold, S.: Counting finite languages by total word length. Integers **11**, 863–872 (2011)
17. Gerhold, S., Kauers, M.: A procedure for proving special function inequalities involving a discrete parameter. In: Kauers, M. (ed.) Proceedings of ISSAC'05. ACM Press, New York (2005), pp. 156–162
18. Gerhold, S., Kauers, M.: A computer proof of Turán's inequality. J. Inequalities Pure Appl. Math. **7**, 1–4 (2006)
19. Gerhold, S., Gerstenecker, C., Pinter, A.: Moment explosions in the rough Heston model. Decis. Econ. Finan. **42**(2), 575–608 (2019)
20. Gulisashvili, A.: Asymptotic formulas with error estimates for call pricing functions and the implied volatility at extreme strikes. SIAM J. Financial Math. **1**, 609–641 (2010)
21. Győri, I., Ladas, G.: Oscillation Theory of Delay Differential Equations. Oxford Mathematical Monographs. The Clarendon Press/Oxford University Press, New York (1991). With applications, Oxford Science Publications
22. Heston, S.: A closed-form solution for options with stochastic volatility with applications to bond and currency options. Rev. Financial Stud. **6**, 327–343 (1993)
23. Karatzas, I., Shreve, S.E.: Brownian Motion and Stochastic Calculus. Graduate Texts in Mathematics, 2nd edn., vol. 113. Springer, New York (1991)
24. Kilbas, A.A., Srivastava, H.M., Trujillo, J.J.: Theory and applications of fractional differential equations. North-Holland Mathematics Studies, vol. 204. Elsevier, Amsterdam (2006)
25. Koutschan, C.: Advanced Applications of the Holonomic Systems Approach. PhD Thesis, J. Kepler University Linz (2009)
26. Lewis, A.L.: Option Valuation Under Stochastic Volatility. Finance Press, Newport Beach (2000)
27. Olver, F.W.J., Lozier, D.W., Boisvert, R.F., Clark, C.W. (eds.): NIST Handbook of Mathematical Functions. U.S. Department of Commerce National Institute of Standards and Technology, Washington (2010)
28. Ouaknine, J., Worrell, J.: Positivity problems for low-order linear recurrence sequences. In: Proceedings of the Twenty-Fifth Annual ACM-SIAM Symposium on Discrete Algorithms. ACM, New York (2014), pp. 366–379
29. Pillwein, V.: Termination conditions for positivity proving procedures. In: ISSAC 2013— Proceedings of the 38th International Symposium on Symbolic and Algebraic Computation. ACM, New York (2013), pp. 315–321
30. Pillwein, V., Schussler, M.: An efficient procedure deciding positivity for a class of holonomic functions. ACM Commun. Comput. Algebra **49**, 90–93 (2015)

31. Pinter, A.: Small-time asymptotics, moment explosion and the moderate deviations regime. PhD Thesis, TU Wien (2017)
32. Remmert, R.: Theory of Complex Functions. Graduate Texts in Mathematics, vol. 122. Springer, New York (1991)
33. Zeilberger, D.: A holonomic systems approach to special functions identities. J. Comput. Appl. Math. **32**, 321–368 (1990)

Evaluations as L-Subsets

Adalbert Kerber

Dedicated to Professor Peter Paule on the occasion of his 60th birthday

1 Introduction

In "Pure Mathematics" a statement like $x \in X$ is either true or false, in formal terms: it has a *truth value*, "yes" or "no", in numerical terms: $tv(x \in X) \in L = \{0, 1\}$. But in "Applied Mathematics" multivalued parameters are used. For example, if a given refrigerant is ecologically worthwhile may be answered by its values of ODP (ozone depletion potential), GWP (general warming potential) and ALT (atmospheric lifetime), i.e. by a triple of real numbers. Let us call such answers *evaluations*. They are elements of a lattice L, in this particular case of $L = [0, 1]^3$, if the parameters are normalized. And we can consider such evaluations as L-subsets (see below) of the cartesian product $O \times A$ of the set O of objects and the set A of attributes. The crucial point is that this allows a *choice* of a suitable set theory together with the corresponding logic, in a *problem-orientable way*, since we can use a more or a less strict argumentation.

2 The Usual Model of Evaluation

Assume, together with a *lattice L*, a set of *objects* $o \in O$ and a set of *attributes* $a \in A$. By an *evaluation $\mathcal{E}$*, of the o w.r.t. the a and over L, we mean a mapping

$$\mathcal{E} \colon O \times A \to L \colon (o, a) \mapsto \mathcal{E}((o, a)) = tv(o \text{ has } a),$$

A. Kerber (✉)
Mathematics Department, University of Bayreuth, Bayreuth, Germany
e-mail: kerber@uni-bayreuth.de

© Springer Nature Switzerland AG 2020
V. Pillwein, C. Schneider (eds.), *Algorithmic Combinatorics: Enumerative Combinatorics, Special Functions and Computer Algebra*, Texts & Monographs in Symbolic Computation, https://doi.org/10.1007/978-3-030-44559-1_12

i.e., an *L-subset* of $O \times A$, containing (o, a) with the *truth value* $tv(o$ has $a) = \mathcal{E}((o, a))$. Here is an example: A fictive evaluation of reading and writing devices a, b, c, d for CDs and DVDs,

$$L = \{\ominus\ominus / \ominus\ominus, \ldots, \oplus\oplus / \oplus\oplus\} \equiv \{-2, -1, 0, 1, 2\}^2,$$

in the usual notation of the computer journal c't, where the attributes are reading and writing abilities of the devices, both on CDs and DVDs. A test result may look like that:

	wDVD	wCD	rDVD	rCD
a	$\oplus\oplus/\oplus$	$\ominus/\ominus$	$\ominus\ominus/\bigcirc$	$\oplus/\ominus$
b	$\oplus/\ominus$	$\oplus/\ominus$	$\ominus/\ominus$	$\bigcirc/\ominus$
c	$\bigcirc/\bigcirc$	$\oplus\oplus/\oplus\oplus$	$\bigcirc/\oplus$	$\oplus/\bigcirc$
d	$\oplus/\oplus$	$\bigcirc/\ominus$	$\bigcirc/\bigcirc$	$\oplus/\bigcirc$

and L looks as follows:

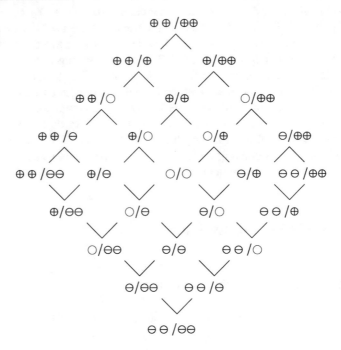

$\mathcal{E}$ can be considered as an *L*-subset of $O \times A$, where $\mathcal{E}((o, a)) = tv((o, a) \in \mathcal{E})$, the truth value of (o, a) being an element of $\mathcal{E}$. This Ansatz has advantages over the standard situation, where the values are contained in $\{0, 1\}$ as we shall see.

3 Evaluation as an L-Subset

Instead of $Y^X = \{0, 1\}^{O \times A}$, the set of all mappings from $O \times A$ to $\{0, 1\}$, or the set of *(classical) subsets* of $O \times A$ (if we identify a mapping with its inverse image of 1), we consider $L^{O \times A}$, the set of all *L-subsets* of $O \times A$, $\mathcal{E} \in L^{O \times A}$, for a given lattice L, where $\mathcal{E}$ is a formal description of the set of pairs (o, a) where $\mathcal{E}((o, a)) \neq 0$. The advantage of this Ansatz is that *we can choose a set theory (with its logic) over L*, in order to allow a problem-orientation, and we can *explore the evaluation* in order to deduce "all the knowledge" contained in it.

We recall that we can identify the set of classical subsets of a set X with the following set of mappings:

$$\{0, 1\}^X = \{S \mid S : X \to \{0, 1\}\},$$

associating with S the subset $\{x \in X \mid S(x) \neq 0\} \subseteq X$. Correspondingly, the set of *L-subsets* of X can be identified with the set

$$L^X = \{S \mid S : X \to L\},$$

when we associate with S the subset $\{x \in X \mid S(x) \neq 0\} \subseteq_L X$. The *$L$-inclusion* of two such L-subsets is defined in terms of the partial order $\leq$ on L in the following way:

$$S \leq S' \iff \forall x \in X : S(x) \leq S'(x).$$

The crucial point is that we can define *various* set theories on L^X since the intersections of two such sets can be introduced using different *t-norms* $\tau : L \times L \to L$. These are the mappings which are *symmetric, monotone* (in both the coordinates), *associative* and fulfill the *side condition* $\tau(\lambda, 1_L) = \lambda$. Each one of these mappings defines an *L-intersection* S of M, $N \in L^X$, where

$$S(x) = (M \cap_\tau N)(x) = \tau(M(x), N(x)).$$

Here are the most important t-norms:

– The *standard norm* is defined by

$$s(\lambda, \mu) = \lambda \wedge \mu.$$

- The *drastic norm* is

$$d(\lambda, \mu) = \begin{cases} \lambda & \mu = 1_L, \\ \mu & \lambda = 1_L, \\ 0_L & \text{otherwise.} \end{cases}$$

- And if $L = [0, 1]$, we have the *algebraic product* and *the bounded difference*, also called the *Lukasiewicz-norm*

$$a(\lambda, \mu) = \lambda \cdot \mu, \quad b(\lambda, \mu) = \text{Max}\{0, \lambda + \mu - 1\}.$$

- In particular the following is true:

$$d(\lambda, \mu) \le \tau(\lambda, \mu) \le s(\lambda, \mu).$$

Specific *t*-norms lead to a corresponding logic:

- $\tilde{\tau} : L \times L \to L$ is a *residuum* of τ, iff

$$\tau(\lambda, \mu) \le \nu \iff \lambda \le \tilde{\tau}(\mu, \nu).$$

- If $\tau(\alpha, \bigvee M) = \bigvee_{\beta \in M} \tau(\alpha, \beta)$ holds, then

$$\tilde{\tau}(\alpha, \beta) = \bigvee \{\gamma \mid \tau(\alpha, \gamma) \le \beta\}.$$

In this case τ is called a *residual t*-norm.
- *This yields a logic $\tilde{\tau}$, corresponding to L and τ.*

Example Models $\mathcal{SA}$ of strong and $\mathcal{WA}$ of weak acid

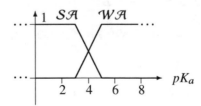

If we choose the *t*-norm $\tau = s$, an acid with pK_a-value 4 is both strong and weak, while, if $\tau = d$, $(\mathcal{SA} \cap_d \mathcal{WA})(r) = 0$ (although $\mathcal{SA}(4) = \mathcal{WA}(4) = 0.5$). We are in fact using kind of *semantic notion of truth*, based on τ!

Residua, for $L = [0, 1]$

$$\tilde{s}(\alpha, \beta) = \begin{cases} 1 & \text{if } \alpha \leq \beta, \\ \beta & \text{otherwise}, \end{cases}$$

$$\tilde{d}(\alpha, \beta) = \begin{cases} \beta & \text{if } \alpha = 1, \\ 1 & \text{otherwise}, \end{cases}$$

$$\tilde{a}(\alpha, \beta) = \begin{cases} \beta/\alpha & \text{if } \alpha \neq 0, \\ 1 & \text{otherwise}, \end{cases}$$

$$\tilde{b}(\alpha, \beta) = \text{Min}\{1, 1 - \alpha + \beta\}.$$

Hence, we have several choices and this opens a way to

Problem-Orientation Choose a suitable lattice L as set of values, pick a suitable residual *t*-norm τ obtaining a set theory, its residuum $\tilde{\tau}$ gives the corresponding logic. Apply that to $\mathcal{E} \in L^{O \times A}$, the evaluation considered, and get a basis of the implications (see below).

4 Mathematical Tools for the Exploration

We say that object $o \in O$ *has* the attribute $a \in A$ if and only if $\mathcal{E}((o, a)) > 0$. Moreover, we introduce, for $\mathcal{A} \in L^A$, an $\mathcal{A}' \in L^O$ by putting

$$\mathcal{A}'(o) = \tilde{\tau}(\mathcal{A} \Rightarrow \mathcal{E}) = \bigwedge_{a \in A} \tilde{\tau}(\mathcal{A}(a), \mathcal{E}(o, a)).$$

And we evaluate "$\mathcal{A} \in L^A$ implies $\mathcal{B} \in L^A$ in $\mathcal{E}$" by:

$$\tilde{\tau}(\mathcal{A} \Rightarrow \mathcal{B}) = \bigwedge_{o \in O} \tilde{\tau}(\mathcal{A}'(o), \mathcal{B}'(o)).$$

$\mathcal{A} \Rightarrow \mathcal{B}$ *holds* in $\mathcal{E}$ if and only if $\tilde{\tau}(\mathcal{A} \Rightarrow \mathcal{B}) = 1$, i.e., if and only if $\mathcal{A}' \leq \mathcal{B}'$. A basis for the implications is obtained as follows: We define *pseudo-contents* $\mathcal{P}$ by

$$\mathcal{P} \neq \mathcal{P}'' \text{ and for each pseudo-content} \mathcal{Q} \subset_L \mathcal{P}: \mathcal{Q}'' \leq \mathcal{P}.$$

The *Duquenne/Guigues-basis*,

$$\mathbb{P} = \{\mathcal{P} \Rightarrow (\mathcal{P}'' \setminus \mathcal{P}) \mid \mathcal{P} \text{ pseudo-content}\},$$

implies every attribute implication following from $\mathcal{E}$.

Example An exploration of an evaluation. We consider an extended evaluation of the refrigerants, adding molecular substructures, Cl-, F-, Br-, I-atoms, and using simplified binary parameters, so that the interested reader can check the basis online. The simplified parameters are denoted $nODP^*$, $nGWP^*$, $nALT^*$, obtaining the following evaluation:

C	nODP*	nGWP*	nALT*	nC	Cl	F	Br	I	ether	CO$_2$	NH$_3$
1	1	0	0	0	1	1	0	0	0	0	0
2	0	1	0	0	1	1	0	0	0	0	0
6	0	0	0	1	1	1	0	0	0	0	0
7	0	0	0	1	1	1	0	0	0	0	0
8	0	1	1	0	0	1	0	0	0	0	0
16	0	0	0	1	0	0	0	0	0	0	0
21	0	0	0	0	0	0	0	0	0	1	0
22	1	0	0	0	1	1	1	0	0	0	0
23	0	1	1	1	0	1	0	0	0	0	0
29	0	1	1	1	0	1	0	0	1	0	0
32	0	0	0	0	1	0	0	0	0	0	0
33	1	0	0	1	1	1	0	0	0	0	0
35	1	0	1	1	1	1	0	0	0	0	0
36	0	0	0	0	0	1	0	1	0	0	0
37	0	0	0	1	0	0	0	0	1	0	0
38	0	0	0	0	0	0	0	0	0	0	1
39	0	0	0	1	0	1	0	0	1	0	0
40	0	0	0	1	0	1	0	0	1	0	0

As it is binary, the reader can obtain the Duquenne/Guigues basis of it *online*, using **CONEXP-1.3** by Yevtushenko [1], getting:

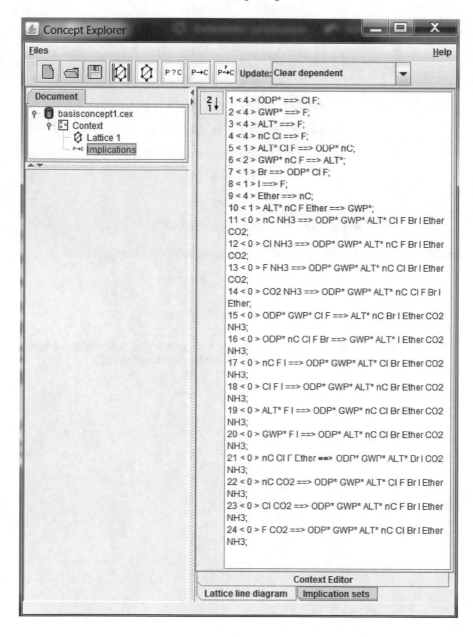

The *reduced* Duquenne/Guigues basis is

$$\{nODP^*\} \implies \{Cl, F\}$$
$$\{nGWP^*\} \implies \{F\}$$
$$\{nALT^*\} \implies \{F\}$$
$$\{nC, Cl\} \implies \{F\}$$
$$\{nALT^*, Cl, F\} \implies \{nODP^*, nC\}$$
$$\{nGWP^*, nC, F\} \implies \{nALT^*\}$$
$$\{Br\} \implies \{nODP^*, Cl, F\}$$
$$\{I\} \implies \{F\}$$
$$\{ether\} \implies \{nC\}$$
$$\{nALT^*, nC, F, ether\} \implies \{nGWP^*\}$$

and it can be considered as *a set of hypotheses on refrigerants in general*!

Reference

1. Yevtushenko, S.A.: System of data analysis "concept explorer". In: Proceedings of the 7th National Conference on Artificial Intelligence, K II-2000, pp. 127–134 (2000)

Exact Lower Bounds for Monochromatic Schur Triples and Generalizations

Christoph Koutschan and Elaine Wong

*Dedicated to Peter Paule, our academic father and grandfather.
Peter, we wish you many more happy, healthy, and productive
years.*

1 Introduction and Historical Background

Let $\mathbb{N}$ denote the set of positive integers. A triple $(x, y, z) \in \mathbb{N}^3$ is called a Schur triple if its entries satisfy the equation $x + y = z$. The set $\{1, \ldots, n\}$ of all positive integers up to n will be denoted by $[n]$. A coloring of $[n]$ is a map $\chi : [n] \to C$ for some finite set C of colors. For example, a map $\chi : [n] \to \{\text{red, blue}\}$ is a 2-coloring. We say that a Schur triple is monochromatic (with respect to a given coloring) if all of its entries have been assigned the same color; we will abbreviate "monochromatic Schur triple" by MST.

With these notations, one can ask questions like: given $n \in \mathbb{N}$ and a coloring χ of $[n]$, how many MSTs are there in $[n]^3$? Let us denote this number as follows:

$$\mathcal{M}(n, \chi) := \left| \left\{ (x, y, z) \in [n]^3 : z = x + y \wedge \chi(x) = \chi(y) = \chi(z) \right\} \right|. \tag{1}$$

For our purposes, two Schur triples $(x, y, x+y)$ and $(y, x, x+y)$ are considered distinct if $x \neq y$. We emphasize this convention since sometimes in the literature these two triples are counted only once, which is equivalent to imposing the extra condition $x \leq y$. For example, there are exactly four monochromatic Schur triples on $[6] = \{1, \ldots, 6\}$ when 2 and 4 are colored red and $1, 3, 5, 6$ are colored blue, namely $(1, 5, 6)$, $(2, 2, 4)$, $(3, 3, 6)$, and $(5, 1, 6)$. We will use a short-hand notation for 2-colorings, namely as words on the alphabet $\{R, B\}$: the i-th letter is R if the

C. Koutschan (✉) · E. Wong
Johann Radon Institute for Computational and Applied Mathematics (RICAM), Austrian
Academy of Sciences, Linz, Austria
e-mail: christoph.koutschan@ricam.oeaw.ac.at

© Springer Nature Switzerland AG 2020
V. Pillwein, C. Schneider (eds.), *Algorithmic Combinatorics: Enumerative
Combinatorics, Special Functions and Computer Algebra*, Texts & Monographs
in Symbolic Computation, https://doi.org/10.1007/978-3-030-44559-1_13

223

integer i is colored red and B if it is blue. So the above 2-coloring would be denoted by $BRBRBB$. We will also make use of the power notation for words, e.g., $R^2B^3 = RRBBB$.

The namesake of the triples in this work refers to Issai Schur [11], who in 1917 studied a modular version of Fermat's last theorem (first formulated and proved by Leonard Dickson). In order to give a simpler proof of the theorem, Schur introduced a *Hilfssatz* confirming the existence of a least positive integer $n = n(m)$ such that for any m-coloring of $[n]$ an MST exists (this is nowadays known as Schur's theorem). In 1927, Van der Waerden [15] generalized this result to monochromatic arithmetic progressions of any length k. Then in 1928, Ramsey proved his eponymous theorem, showing the existence of a least positive integer n such that every edge-coloring of a complete graph on n vertices, with the colors red and blue, admits either a complete red subgraph or a complete blue subgraph. However, a real increase in the popularity of these kinds of Ramsey-theoretic problems came with the rediscovery of Ramsey's theorem in a 1935 paper of Erdős and Szekeres [4], which ultimately led to a simpler proof of Schur's theorem, indicating their close connections. For the curious reader, this rich history is beautifully depicted in a book by Landman and Robertson [8].

We now arrive at a point of more than just questions of existence. In 1959, Alan Goodman [5] studied the *minimum* number of monochromatic triangles under a 2-edge coloring of a complete graph on n vertices. Then in 1996, Graham, Rödl, and Ruciński [6] found it natural to extend the problem of "determining the minimum number under any 2-coloring" to Schur triples. In fact, Graham offered a prize of 100 USD for an answer to such a question; it has subsequently been successfully answered many times over, in an asymptotic sense. In order to give some more context to this problem, we first introduce some additional notation.

We start by wondering about what we can say about the number of MSTs on $[n]$ if we do not prescribe a particular coloring. It is not difficult to calculate that there are exactly $\sum_{i=1}^{n-1} i = \frac{1}{2}n(n-1) = \binom{n}{2}$ Schur triples on $[n]$. Trivially, this yields an upper bound for the number of MSTs, which can be achieved by coloring all numbers with the same color. This is the reason why it is more natural (and more interesting!) to ask for a lower bound for $\mathcal{M}(n, \chi)$, that is: for given $n \in \mathbb{N}$, what is the "best" lower bound for the number of MSTs regardless of the choice of coloring? Of course, 0 is a trivial such lower bound, but we are aiming for something sharp, in the sense that for each n there exists a coloring for which this bound is actually attained. Differently stated, we are looking for the minimal number of monochromatic Schur triples among all possible colorings of $[n]$:

$$\mathcal{M}(n) := \min_{\chi : [n] \to \{R, B\}} \mathcal{M}(n, \chi). \qquad (2)$$

For example, for $n = 6$, one cannot avoid the occurrence of monochromatic Schur triples, but there exists a 2-coloring for which only a single such triple occurs, namely the triple $(1, 1, 2)$ for the coloring $RRBBBR$. Therefore, we have $\mathcal{M}(6) = \mathcal{M}(6, RRBBBR) = 1$.

As mentioned before, this problem was only studied from an asymptotic point of view: Robertson and Zeilberger [9] was first to give the lower bound $\frac{1}{22}n^2 + O(n)$ as $n \to \infty$ (and consequently won Graham's cash prize), where it has to be noted that they count only Schur triples $(x, y, x + y)$ with the condition $x \leq y$ imposed. This lower bound was independently confirmed by Datskovsky [3], Schoen [10], and Thanatipanonda [13]. Schoen also provided a proof of an "optimal" coloring of $[n]$ that would give such a minimum number, and such a coloring is what we assume later in this paper. The asymptotic lower bounds for the generalized Schur triples case $(x, y, x + ay)$ for $a \geq 2$ is $\frac{1}{2a(a^2+2a+3)}n^2 + O(n)$ as $n \to \infty$, without the requirement of $x \leq y$. This was conjectured by Thanatipanonda [13] and Butler et al. [1], and subsequently proven in 2017 by Thanatipanonda and Wong [14].

In this paper, we take a slightly different approach by using known computer algebra techniques and creative simplifications to develop exact formulas for the minimum number of such triples (in both the Schur triples case and the generalized Schur triples case) and give an analysis of the transitional behavior between the cases. Thus, in order to keep some consistency for comparison, we will remove the assumption of $x \leq y$ when counting MSTs. In this way, we can explain why the behavior of the minimum number of triples jumps when moving from the case $a = 1$ to the case $a \geq 2$ (note that the above asymptotic formula does not specialize to the expected prefactor $\frac{1}{11}$ when $a = 1$ is substituted).

The overall plan is to systematically exploit the full force of symbolic computation and perform a complete analysis of determining the minimum number of monochromatic triples $(x, y, x + ay)$ in both the discrete context $(a \in \mathbb{N})$ and the continuous context $(a \in \mathbb{R}^+)$. This requires three courses of a mathematical meal. We serve an appetizer in Sect. 2, showing how to derive an exact formula for the minimum in the classic Schur triple case (corresponding to $a = 1$ in the general equation). This sets us up for the main course in Sect. 3, where we perform a full analysis for $a > 0$, illustrating that a global minimum can always be found. Interesting transitional behaviors occur at many locations for $a \in (0, 1)$ and one key transition occurs at $a \approx 1.17$. Admittedly, this course may be a bit difficult to swallow, and we hope that the reader will not suffer from indigestion. For dessert, we follow the procedure described in Sect. 2, and illustrate how it can systematically produce (ostensibly, an infinite number) of exact formulas for the minimum number of generalized Schur triples. Accordingly, in Sect. 4, we leave the reader with exact formulas for the minimum number of generalized Schur triples for $a = 2, 3, 4$, and $a = \frac{1}{2}$, with the hope that s/he will leave satisfied.

For the reader's convenience, all computations and diagrams are in the Mathematica notebook [7] that accompanies this paper, freely available at the first author's website. This material may also be of independent interest, since we believe that also other problems can be attacked in a similar fashion, see for example the recent study on the peaceable queens problem [16].

2 Exact Lower Bound for Monochromatic Schur Triples

It has been shown previously [9, 10] that for fixed n the number $\mathcal{M}(n, \chi)$ is minimized when χ consists of three blocks of numbers with the same color ("runs"), i.e., when χ is of the form $R^s B^{t-s} R^{n-t}$, where s and t are approximately $\frac{4}{11}n$ and $\frac{10}{11}n$, respectively. In this section, we derive exact expressions for the optimal choice of s and t, as well as for the corresponding minimum $\mathcal{M}(n)$.

Lemma 1 *Let $n, s, t \in \mathbb{N}$ be such that $1 \leq s \leq t \leq n$. Moreover, assume that the inequalities $t \geq 2s$ and $s \geq n - t$ hold. Then the number of monochromatic Schur triples on $[n]$ under the coloring $R^s B^{t-s} R^{n-t}$, denoted by $\mathcal{M}(n, s, t)$, is exactly*

$$\mathcal{M}(n, s, t) = \frac{s(s-1)}{2} + \frac{(t-2s)(t-2s-1)}{2} + (n-t)(n-t-1). \tag{3}$$

Proof In Fig. 1 the situation is depicted for $n = 33$, $s = 12$, and $t = 30$. One sees that the dots representing the MSTs are arranged in four regions of right triangular shape. The triangles arise as follows:

1. The dots in the lower left corner correspond to red MSTs, whose components are taken from the first block of red numbers; hence there are $s - 1$ dots in the first row of this triangle.
2. The central triangle contains all blue MSTs, whose first two components (x, y) satisfy the inequalities $x > s$, $y > s$, and $x + y \leq t$. Note that such MSTs only exist if $t \geq 2s + 2$ (for $t = 2s + 1$ and $t = 2s$ the second term in (3) vanishes and the formula is still correct). The number of dots on each side is therefore $t - 2s - 1$.

Fig. 1 All $\mathcal{M}(33) = 87$ monochromatic Schur triples for $s = 12$ and $t = 30$ with corresponding coloring $R^{12} B^{18} R^3$; each triple $(x, y, x + y)$ is represented by a dot at position (x, y). The vertical lines are given by $x = s$, $x = t$, and $x = n$, the horizonal ones by $y = s$, $y = t$, and $y = n$. The three diagonal lines visualize the equations $x + y = s$, $x + y = t$, and $x + y = n$

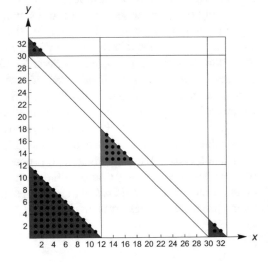

3. The two triangles in the upper left and lower right corners correspond to red MSTs, whose first two entries belong to different blocks of red numbers. By symmetry they have the same shape and they have $n - t - 1$ dots on their sides. Here we use the condition $s \geq n - t$, because otherwise these two regions would no longer be triangles and we would be counting different things beyond the scope of our assumptions.

Adding up the contributions from these three cases, one obtains the claimed formula. $\square$

The optimal values for s and t are easily derived using the techniques of multivariable calculus, once the form $R^s B^{t-s} R^{n-t}$ is assumed: by letting n go to infinity and by scaling the square $[0, n]^2 \subset \mathbb{R}^2$ to the unit square $[0, 1]^2$, we see that the portion of pairs $(x, y) \in [n]^2$ for which $(x, y, x + y)$ is an MST among all pairs in $[n]^2$ equals the area of a certain region in the unit square; for example, see the shaded regions in Fig. 1. In this limit process, the integers s and t turn into real numbers satisfying $0 \leq s \leq t \leq 1$. According to (3) the area of the shaded region in Fig. 1 is given by the formula

$$A(s, t) = \frac{s^2}{2} + \frac{(t - 2s)^2}{2} + 2 \cdot \frac{(1 - t)^2}{2} = \frac{5s^2}{2} + \frac{3t^2}{2} - 2st - 2t + 1.$$

Equating the gradient

$$\left(\frac{\partial A}{\partial s}, \frac{\partial A}{\partial t} \right) = (5s - 2t, 3t - 2s - 2)$$

to zero, one immediately gets the location of the minimum $(s, t) = \left(\frac{4}{11}, \frac{10}{11} \right)$.

Lemma 2 *For fixed $n \in \mathbb{N}$, the integers s_0 and t_0 that minimize the function $M(n, s, t)$ are given by*

$$s_0 = \left\lfloor \frac{4n + 2}{11} \right\rfloor \quad and \quad t_0 = \left\lfloor \frac{10n}{11} \right\rfloor.$$

Proof Strictly speaking, we prove the minimality of the function $M(n, s, t)$ under the additional assumption $t \geq 2s \wedge s \geq n - t$ from Lemma 1. The fact that this is also the global minimum for all $1 \leq s \leq t \leq n$ follows as a special case from the more general discussion as described in the proof of Lemma 4.

The statement is proven by case distinction into 11 cases, according to the remainder n modulo 11. Here we show details for the case $n = 11k + 5$, and the remaining cases can be similarly verified with a computer; for these cases we refer the reader to the accompanying electronic material [7].

By setting $n = 11k + 5$ we can eliminate the floors from the definitions of s_0 and t_0; we obtain $s_0 = \lfloor \frac{1}{11}(4n + 2) \rfloor = 4k + 2$ and $t_0 = \lfloor \frac{10}{11}n \rfloor = 10k + 4$. Our goal is to show that among all integers $i, j \in \mathbb{Z}$ the expression $M(n, s_0 + i, t_0 + j)$ is minimal for $i = j = 0$. Using (3) one gets

$$M(11k+5, 4k+2+i, 10k+4+j) = \frac{1}{2}\left(2+5i+5i^2-3j-4ij+3j^2+12k+22k^2\right).$$

The stated goal is equivalent to showing that the polynomial

$$p(i, j) = 5i + 5i^2 - 3j - 4ij + 3j^2$$

is nonnegative for all $(i, j) \in \mathbb{Z}^2$. Such a task can, in principle, be routinely executed by cylindrical algebraic decomposition (CAD) [2]. In this method, the variables i and j are treated as real variables, which causes some problems in the present application. The reason is that $p(i, j) \geq 0$ does not hold for all $i, j \in \mathbb{R}$. The situation is depicted in Fig. 2, where the ellipse represents the zero set of $p(i, j)$ and its inside consists of values (i, j) for which the polynomial $p(i, j)$ is negative. To our relief, we see that no integer lattice points lie inside the ellipse, since such points would be counterexamples to our claim.

Our strategy now is the following: we prove that $p(i, j) \geq 0$ for all integer points that are close to $(0, 0)$, e.g., for all (i, j) with $-2 \leq i \leq 2$ and $-2 \leq j \leq 2$. These points are shown in Fig. 2, with the respective value of $p(i, j)$ attached to them. In particular, we see that the minimum $p(i, j) = 0$ is attained several times, namely on the three points that lie exactly on the boundary of the ellipse.

Fig. 2 Zero set of the polynomial $p(i, j)$ from Lemma 2 and its values at integer lattice points $(i, j) \in \mathbb{Z}^2$

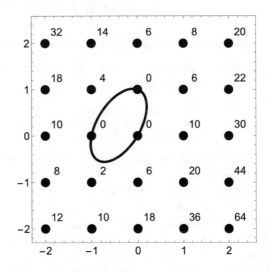

Then we invoke cylindrical algebraic decomposition on the formula

$$\forall i, j \in \mathbb{R}: (-2 \leq i \leq 2 \wedge -2 \leq j \leq 2) \vee p(i, j) \geq 0, \tag{4}$$

which states that if the point (i, j) lies outside the square that we have already considered, then $p(i, j) \geq 0$ holds. Calling the Mathematica command `CylindricalDecomposition` with input (4), we immediately get `True`. □

We are ready to state the main theorem of this section, which is an exact formula for the minimal number of MSTs for any 2-coloring of $[n]$. Apart from the asymptotic results mentioned in Sect. 1, there is only one paper [10] where a similar result is stated, but only for the case $n = 22k$ and for Schur triples $(x, y, x+y)$ with $x \leq y$. In contrast, we consider all $x, y \in [n]$ and our formula holds for all $n \in \mathbb{N}$.

Theorem 1 *The minimal number of monochromatic Schur triples that can be attained under any 2-coloring of $[n]$ is*

$$M(n) = \left\lfloor \frac{n^2 - 4n + 6}{11} \right\rfloor.$$

Proof As in Lemma 2, we argue by case distinction $n = 11k + \ell, 0 \leq \ell \leq 10$. Using $s_0 = \lfloor \frac{1}{11}(4n + 2) \rfloor$ and $t_0 = \lfloor \frac{10}{11}n \rfloor$ from the lemma, we obtain the following values for $M(n, s_0, t_0)$:

$\ell = 0: M(11k, 4k, 10k)$ $= 11k^2 - 4k$ $= \frac{1}{11}(n^2 - 4n)$

$\ell = 1: M(11k + 1, 4k, 10k)$ $= 11k^2 - 2k$ $= \frac{1}{11}(n^2 - 4n + 3)$

$\ell = 2: M(11k + 2, 4k, 10k + 1)$ $= 11k^2$ $= \frac{1}{11}(n^2 - 4n + 4)$

$\ell = 3: M(11k + 3, 4k + 1, 10k + 2) = 11k^2 + 2k$ $= \frac{1}{11}(n^2 - 4n + 3)$

$\ell = 4: M(11k + 4, 4k + 1, 10k + 3) = 11k^2 + 4k$ $= \frac{1}{11}(n^2 - 4n)$

$\ell = 5: M(11k + 5, 4k + 2, 10k + 4) = 11k^2 + 6k + 1 = \frac{1}{11}(n^2 - 4n + 6)$

$\ell = 6: M(11k + 6, 4k + 2, 10k + 5) = 11k^2 + 8k + 1 = \frac{1}{11}(n^2 - 4n - 1)$

$\ell = 7: M(11k + 7, 4k + 2, 10k + 6) = 11k^2 + 10k + 2 = \frac{1}{11}(n^2 - 4n + 1)$

$\ell = 8: M(11k + 8, 4k + 3, 10k + 7) = 11k^2 + 12k + 3 = \frac{1}{11}(n^2 - 4n + 1)$

$\ell = 9: M(11k + 9, 4k + 3, 10k + 8) = 11k^2 + 14k + 4 = \frac{1}{11}(n^2 - 4n - 1)$

$\ell = 10: M(11k + 10, 4k + 3, 10k + 9) = 11k^2 + 16k + 6 = \frac{1}{11}(n^2 - 4n + 6)$

One easily observes that in each case, the result is of the form $\frac{1}{11}(n^2 - 4n) + \delta_\ell$, where $-\frac{1}{11} \leq \delta_\ell \leq \frac{6}{11}$ holds for all ℓ. Hence the claimed formula follows.   □

The first 25 terms of the sequence $\left(\mathcal{M}(n)\right)_{n \geq 1}$ are

$$0, 0, 0, 0, 1, 1, 2, 3, 4, 6, 7, 9, 11, 13, 15, 18, 20, 23, 26, 29, 33, 36, 40, 44, 48, \ldots$$

We have added this sequence to the Online Encyclopedia of Integer Sequences [12] under the number A321195.

3 Asymptotic Lower Bound for Generalized Schur Triples

We now turn to generalized Schur triples, i.e., triples (x, y, z) subject to $z = x + ay$ for some parameter $a \in \mathbb{N}$, as studied by Thanatipanonda and Wong [14]. Here, we allow a to be even more general, i.e., $a \in \mathbb{R}^+$. Consequently, we have to adapt the definition of generalized Schur triples: we use the condition $z = x + \lfloor ay \rfloor$. The case $a < 0$ does not add new aspects to the analysis, as it can be transformed to the $a > 0$ case by exchanging the roles of x and z and by changing the floor function to a ceiling.

Again, we choose to use the assumption that the minimal number of monochromatic generalized Schur triples (MGSTs) occurs at a coloring in the form of three blocks $R^s B^{t-s} R^{n-t}$. We justify using this assumption with the experimental evidence of Butler, Costello, and Graham [1] (who argued for the generalized Schur triple case $a > 1$) and adapting the intuition in the argument of Schoen [10] (who only argued for the Schur triple case $a = 1$).

We would like to know for which choice of s and t (depending on n and a) the minimum occurs. Similar to the previous section, we let n go to infinity and correlate the number of MGSTs with the area of polygonal regions in the unit square. We then define a function $A(s, t, a)$ that determines this area, and minimize it. Hence, throughout this section, s and t are real numbers with $0 \leq s \leq t \leq 1$.

Figure 3 shows two situations for different choices of a, s, t. In contrast to the previous section, we do a very careful case analysis and do not impose extra conditions on s and t as in Lemma 1, at the cost of introducing a "few" more case distinctions. The full case analysis for normal Schur triples then follows by specializing to $a = 1$ in the resulting formulas.

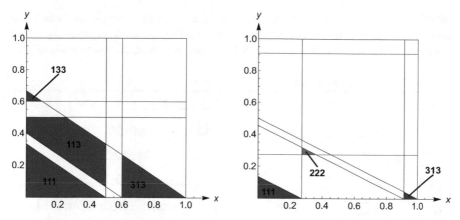

Fig. 3 Regions (in red and blue) corresponding to monochromatic generalized Schur triples for $a = \frac{3}{2}$, $s = \frac{1}{2}$, $t = \frac{3}{5}$ (left) and $a = 2$, $s = \frac{3}{11}$, $t = \frac{10}{11}$ (right); their area being measured by $A(s, t, a)$ from Lemma 3

In the process of analyzing the different cases, we encounter several conditions on a, s, t. For our referencing convenience, we distinguish these conditions here using the following abbreviations:

$$
\begin{array}{ll}
C_1 \equiv 1 - as \geq 0, & C_2 \equiv 1 - as - s \geq 0, \\
C_3 \equiv 1 - as - t \geq 0, & C_4 \equiv t - as \geq 0, \\
C_5 \equiv t - as - s \geq 0, & C_6 \equiv 1 - at \geq 0, \\
C_7 \equiv 1 - at - s \geq 0, & C_8 \equiv 1 - at - t \geq 0, \\
C_9 \equiv 1 - a \geq 0, & C_{10} \equiv 1 - a - s \geq 0, \\
C_{11} \equiv s - a \geq 0, & C_{12} \equiv 1 - a - t \geq 0, \\
C_{13} \equiv t - a \geq 0, & C_{14} \equiv t - a - s \geq 0, \\
C_{15} \equiv s - at \geq 0, & C_{16} \equiv t - at - s \geq 0.
\end{array}
\tag{5}
$$

In Figs. 5 and 6, the lines that represent some of these conditions are depicted. They split the triangle $0 \leq s \leq t \leq 1$ into several regions, depending on the value of a.

Lemma 3 *Let $a, s, t \in \mathbb{R}$ with $a > 0$ and $0 \leq s \leq t \leq 1$. Then the area $A(s, t, a)$ of the region*

$$
\{(x, y) \in \mathbb{R}^2 : (x, y, x + ay) \in ([0, s] \cup (t, 1])^3 \vee (x, y, x + ay) \in (s, t]^3\}
$$

is given by a piecewise defined function, where 70 case distinctions have to be made. For the sake of brevity, only the first 17 cases are listed below, since they will be the *most important ones in the subsequent analysis; in fact they are sufficient to describe $A(s, t, a)$ for $a \geq 1$. We label the region corresponding to the i-th case as (R_i). They are expressed in terms of the conditions (5) (where overlines denote negations):*

	conditions on a, s, t	$A(s, t, a)$
(R_1)	$\overline{C_1}$	$\frac{s^2 - 2ts + 2s + t^2 - 2t + 1}{2a}$
(R_2)	$C_3 \wedge C_4 \wedge \overline{C_6}$	$\frac{2as^2 + 2s^2 + 2as - 4ats - 2ts + t^2}{2a}$
(R_3)	$C_3 \wedge \overline{C_4} \wedge \overline{C_6}$	$\frac{-a^2s^2 + 2as^2 + 2s^2 + 2as - 2ats - 2ts}{2a}$
(R_4)	$\overline{C_2} \wedge C_4 \wedge \overline{C_6}$	$\frac{s^2 + 2as - 2ats - 2ts + 2s + 2t^2 - 2t}{2a}$
(R_5)	$\overline{C_2} \wedge \overline{C_4} \wedge C_6$	$\frac{-a^2s^2 + s^2 + 2as - 2ts + 2s + a^2t^2 + t^2 - 2at - 2t + 1}{2a}$
(R_6)	$C_1 \wedge \overline{C_2} \wedge \overline{C_4} \wedge \overline{C_6}$	$\frac{-a^2s^2 + s^2 + 2as - 2ts + 2s + t^2 - 2t}{2a}$
(R_7)	$C_2 \wedge \overline{C_3} \wedge C_4 \wedge \overline{C_6}$	$\frac{a^2s^2 + 2as^2 + 2s^2 - 2ats - 2ts + 2t^2 - 2t + 1}{2a}$
(R_8)	$C_2 \wedge \overline{C_3} \wedge \overline{C_4} \wedge C_6$	$\frac{2as^2 + 2s^2 - 2ts + a^2t^2 + t^2 - 2at - 2t + 2}{2a}$
(R_9)	$C_2 \wedge \overline{C_3} \wedge \overline{C_4} \wedge \overline{C_6}$	$\frac{2as^2 + 2s^2 - 2ts + t^2 - 2t + 1}{2a}$
(R_{10})	$C_3 \wedge C_4 \wedge C_6 \wedge \overline{C_7}$	$\frac{2as^2 + 2s^2 + 2as - 4ats - 2ts + a^2t^2 + t^2 - 2at + 1}{2a}$
(R_{11})	$C_3 \wedge \overline{C_4} \wedge C_6 \wedge \overline{C_7}$	$\frac{-a^2s^2 + 2as^2 + 2s^2 + 2as - 2ats - 2ts + a^2t^2 - 2at + 1}{2a}$
(R_{12})	$\overline{C_4} \wedge C_8$	$\frac{(1 + 2a - a^2)s^2 + 2s(1 - 2at + a - t) + (at + t - 1)^2}{2a}$
(R_{13})	$\overline{C_4} \wedge C_7 \wedge \overline{C_8}$	$\frac{-a^2s^2 + 2as^2 + s^2 + 2as - 4ats - 2ts + 2s}{2a}$
(R_{14})	$C_4 \wedge C_8 \wedge \overline{C_9}$	$\frac{(a^2 + 2a + 2)t^2 - 2t(3as + a + s + 1) + (s + 1)(2as + s + 1)}{2a}$
(R_{15})	$C_4 \wedge C_7 \wedge \overline{C_8} \wedge C_9$	$\frac{2as^2 + s^2 + 2as - 6ats - 2ts + 2s + t^2}{2a}$
(R_{16})	$\overline{C_2} \wedge C_4 \wedge C_6 \wedge \overline{C_9}$	$\frac{s^2 + 2as - 2ats - 2ts + 2s + a^2t^2 + 2t^2 - 2at - 2t + 1}{2a}$
(R_{17})	$C_2 \wedge \overline{C_3} \wedge C_4 \wedge C_6 \wedge \overline{C_9}$	$\frac{a^2s^2 + 2as^2 + 2s^2 - 2ats - 2ts + a^2t^2 + 2t^2 - 2at - 2t + 2}{2a}$

Proof As can be seen in Fig. 3, the region whose area we would like to determine is the union of several polygons. Let $I_1 = [0, s]$, $I_2 = (s, t]$, and $I_3 = (t, 1]$ denote the intervals that correspond to the different blocks of the coloring (I_1 and I_3 being red and I_2 being blue). Then $x, y \in I_1 \wedge x + ay \in I_3$ is allowed while $x, y \in I_1 \wedge x + ay \in I_2$ is not. From this point on, we will refer to the case $(x, y, x + ay) \in I_i \times I_j \times I_k$ by ijk. It is easy to see that we have to consider only seven cases: 111, 222, 113, 131, 133, 313, 333. The cases 311 and 331 are clearly impossible since $x \geq t$ contradicts $x + ay \leq s$. All other combinations of 1, 2, 3 violate the monochromatic coloring condition.

In both parts of Fig. 3, case 111 corresponds to the triangle that touches the origin. The coordinates of its other two vertices are $(s, 0)$ and $(0, \frac{s}{a})$, hence its area is $\frac{1}{2} \cdot s \cdot \frac{s}{a}$. However, this is valid only for $a \geq 1$. If $a < 1$, then the point $(0, \frac{s}{a})$ is above the line $y = s$ and so the top of the triangle is cut off. As a result, one obtains a quadrilateral with vertices $(0, 0)$, $(s, 0)$, $(s - as, s)$, $(0, s)$, whose area is given by $\frac{1}{2} \cdot s \cdot (2s - as)$.

The case 222 is similar, with the difference being that the corresponding polygon disappears if $\frac{t-s}{a} < s$; in the right part of Fig. 3 the polygon 222 is present while in the left part it is not. The polygons 313, 333, and 131 are characterized by comparably simple case distinctions, while 133 and 113 require a much more involved analysis. In Fig. 4, we present such an analysis for 133, and refer to the accompanying electronic material [7] for 113.

What we have achieved so far is a representation of $A(s, t, a)$ as a sum of seven piecewise functions. However, what is required is a representation of $A(s, t, a)$ as a single piecewise function, since that will be needed for determining the location of the minimum.

The conditions that are used to characterize the different pieces in Fig. 4 (and in the remaining cases that have not been discussed explicitly), are listed in (5). In order to combine the seven piecewise functions, we need a common refinement of the regions on which they are defined. We start with the finest possible refinement, which is obtained by considering all $2^{16} = 65536$ logical combinations of C_i and $\overline{C_i}$ for $1 \leq i \leq 16$. Using Mathematica's simplification procedures, we remove those cases that contain contradictory combinations of conditions, such as $C_1 \wedge \overline{C_2}$ for example. After this purging, we are left with a subdivision of the set

$$\{(s, t, a) : 0 \leq s \leq t \leq 1 \wedge a \geq 0\} \subset \mathbb{R}^3, \tag{6}$$

which is an infinite triangular prism, into 114 polyhedral regions. Finally, we merge regions on which $A(s, t, a)$ is defined by the same expression into a single region, yielding a representation of $A(s, t, a)$ as a piecewise function defined by 70 different expressions. Each of them is of the form $\frac{1}{a} p(s, t, a)$ where p is a polynomial in s, t, a of degree at most 2 in each of the variables. For more details, and to see the definition of $A(s, t, a)$ in its full glory, see the accompanying electronic material [7].

□

We have seen that the different domains of definition for $A(s, t, a)$ are polyhedra in $\mathbb{R}^3$ (some of which are unbounded). In Figs. 5 and 6 two 2-dimensional slices of the set (6) for particular choices of a are shown. Note that in Fig. 5 condition C_5 is not shown since it was eliminated in the process of merging regions on which A is defined by the same expression. Moreover, $C_9 \equiv a \leq 1$ is not visible since its plane $a = 1$ is parallel to the depicted cross section $a = 1.4$.

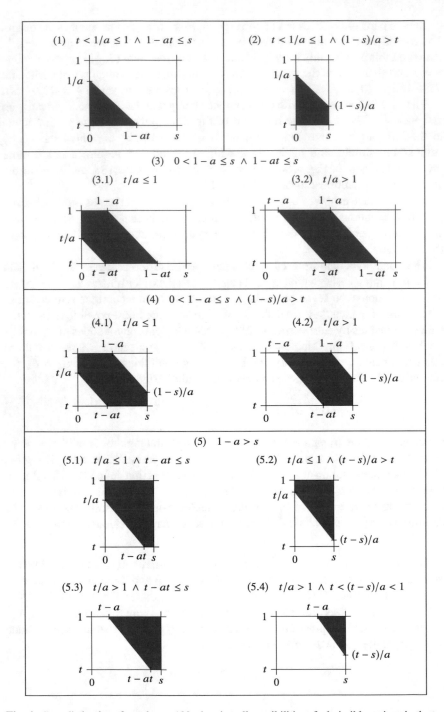

Fig. 4 Case distinctions for polygon 133, showing all possibilities of admissible regions in the top left corner (depending on conditions for a, s, t). The empty cases (not shown) correspond to the conditions $1/a \leq t$ or $t - a \geq s$

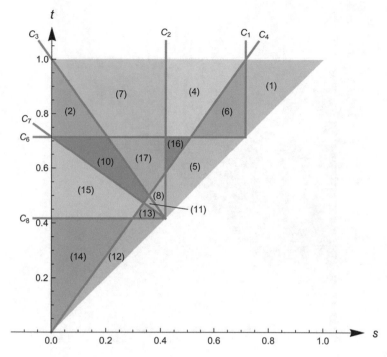

Fig. 5 Domains of definition of $A(s, t, a)$ for $a = 1.4$, according to Lemma 3. Note that not all 17 cases listed in the lemma are present for this particular choice of a

Lemma 4 *For $a > 0$, the minimum of the function $A(s, t, a)$ (defined in Lemma 3) on the triangle $0 \le s \le t \le 1$,*

$$m(a) := \min_{0 \le s \le t \le 1} A(s, t, a)$$

is given by a piecewise rational function, depending on a, according to the following case distinctions (where we also give the location (s_0, t_0) of the minimum):

	s_0	t_0	$m(a)$
$0 \le a \le \alpha_1$	$\dfrac{(a-4)a}{a^3-a-4}$	$\dfrac{-2a^2+4a+2}{-a^3+a+4}$	$\dfrac{-a^4+2a^3-2a^2+6a-4}{2(a^3-a-4)}$
$\alpha_1 \le a \le \alpha_2$	$\dfrac{a(a^2-3)}{a^4-8a-1}$	$\dfrac{a^3+a^2-5a-1}{a^4-8a-1}$	$\dfrac{a^3-2a^2+a-2}{2(a^4-8a-1)}$
$\alpha_2 \le a \le \alpha_3$	$\dfrac{-2a^3+2a+1}{-a^4+8a+3}$	$\dfrac{2a^3+a^2-6a-2}{a^4-8a-3}$	$\dfrac{a^6+a^4-12a^3+4a^2-1}{2a(a^4-8a-3)}$
$\alpha_3 \le a \le \alpha_4$	$\dfrac{-2a^2+a+1}{-4a^3+5a^2+6a+1}$	$\dfrac{-2a^3+a^2+4a+1}{-4a^3+5a^2+6a+1}$	$\dfrac{4a^4-9a^3+2a^2+a-2}{2(4a^3-5a^2-6a-1)}$

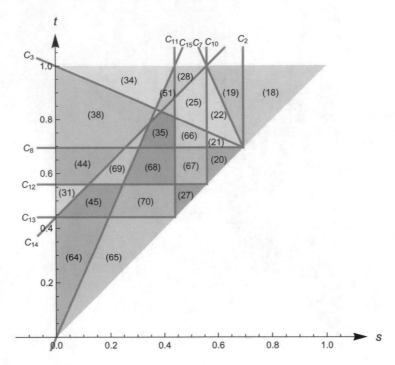

Fig. 6 Domains of definition of the area function $A(s, t, a)$ for $a = 0.44$

$\alpha_4 \leq a \leq \alpha_5$	$\dfrac{a^3+a+1}{-4a^3+3a^2+6a+1}$	$\dfrac{2a^2+4a+1}{-4a^3+3a^2+6a+1}$	$\dfrac{4a^4-4a^3+a-2}{2(4a^3-3a^2-6a-1)}$
$\alpha_5 \leq a \leq \alpha_6$	$-\dfrac{3a^2+a-1}{4a^3-4a^2-4a+1}$	$\dfrac{-4a^2-2a+1}{4a^3-4a^2-4a+1}$	$\dfrac{8a^3-4a^2-5a+2}{2(4a^3-4a^2-4a+1)}$
$\alpha_6 \leq a \leq \alpha_7$	$\dfrac{2a+1}{7a+1}$	$\dfrac{8a^2+6a+1}{7a^2+8a+1}$	$\dfrac{-2a^2+3a+2}{2(a+1)(7a+1)}$
$\alpha_7 \leq a \leq 1$	$\dfrac{(a+1)^2}{a(7a+4)}$	$\dfrac{(a+1)(4a+1)}{a(7a+4)}$	$\dfrac{-7a^4+6a^3+6a^2-2a-1}{2a^2(7a+4)}$
$1 \leq a \leq \alpha_8$	$\dfrac{(a+1)^2}{a^4+2a^3+3a^2+2a+3}$	$\dfrac{(a+1)(a^2+2a+2)}{a^4+2a^3+3a^2+2a+3}$	$\dfrac{a^4-a^2-2a+4}{2a(a^4+2a^3+3a^2+2a+3)}$
$\alpha_8 \leq a$	$\dfrac{a+1}{a^2+2a+3}$	$\dfrac{a^2+2a+2}{a^2+2a+3}$	$\dfrac{1}{2a(a^2+2a+3)}$

Here, the quantities $\alpha_1, \ldots, \alpha_8$ stand for the following algebraic numbers, where Root(p, I) *denotes the unique real root of the polynomial p in the interval I:*

$$\alpha_1 = 0.295597\ldots = \text{Root}(a^3 + a^2 + 3a - 1, [0, 1]),$$

$$\alpha_2 = 0.395065\ldots = \text{Root}(a^5 - 9a^2 + a + 1, [0, 1]),$$

$$\alpha_3 = 0.405669\ldots = \text{Root}(2a^4 - a^3 - 6a^2 + 1, [0, 1]),$$

$$\alpha_4 = 0.553409\ldots = \text{Root}(12a^4 - 15a^3 - 24a^2 + 5a + 6, [0, 1]),$$

$$\alpha_5 = 0.622179\ldots = \mathrm{Root}\!\left(4a^3 - 8a^2 - 3a + 4, [0, 1]\right),$$

$$\alpha_6 = 0.647363\ldots = \mathrm{Root}\!\left(8a^2 + a - 4, [0, 1]\right) = \tfrac{1}{16}\!\left(\sqrt{129} - 1\right),$$

$$\alpha_7 = 0.931478\ldots = \mathrm{Root}\!\left(7a^3 - 5a - 1, [0, 1]\right),$$

$$\alpha_8 = 1.174559\ldots = \mathrm{Root}\!\left(a^3 + a^2 - 3, [1, 2]\right).$$

Proof We locate the minimum in a similar fashion as in Sect. 2, by identifying points (s, t) where the gradient of the area function A vanishes. What complicates our task is the additional parameter a. Since A is defined in pieces, it may not be differentiable at the boundaries between different regions, and therefore, we should be aware that such locations could contain the minimum. For each region (R_i), $1 \le i \le 70$, on which $A(s, t, a)$ is defined, we perform the following steps:

- compute the gradient $\left(\frac{\partial A}{\partial s}, \frac{\partial A}{\partial t}\right)$,
- find all points (s, t) where the gradient is zero, and
- for each such point determine for which values of a it actually lies in (R_i).

On the region (R_1) from Lemma 3, the gradient of A is $\frac{1}{a}(s - t + 1, t - s - 1)$, which vanishes on all points $(s, s + 1)$; however, since the region (R_1) is characterized by $\overline{C_1} \equiv s > \frac{1}{a}$ (and the general condition $s \le t \le 1$), one sees that none of these points lie in it. Continuing in this manner, we find that in each of the regions $(R_2) - (R_{70})$ there is exactly one point (s, t) for which the gradient of A vanishes, but in most cases this point lies outside the region for all a. For example, on (R_2) the gradient is $\frac{1}{a}(2as - 2at + 2s - t + a, t - 2as - s)$, which equals zero for

$$(s, t) = \left(\frac{a}{4a^2 + 2a - 1}, \frac{a(2a + 1)}{4a^2 + 2a - 1}\right). \tag{7}$$

In order to find the values of a that give us that $(s, t) \in (R_2)$, the conditions defining (R_2) (plus the global assumptions) need to be satisfied, namely:

$$as + t \le 1 \ \wedge\ t \ge as \ \wedge\ at > 1 \ \wedge\ 0 < s < t < 1.$$

After substituting s and t with the right hand side of (7) and clearing denominators, one gets a system of polynomial inequalities, involving only the variable a. Cylindrical algebraic decomposition simplifies it to

$$a \ge \mathrm{Root}\!\left(2a^3 - 3a^2 - 2a + 1, [1, 2]\right) = 1.889228559\ldots$$

Hence, for each a satisfying this condition we have a local minimum at the point given in (7).

We proceed in similar fashion and identify 17 local minima, each occurring only for a in a certain interval. Some of these intervals partly overlap, which means that we have to study a subdivision of the positive real line that is a refinement of all 17

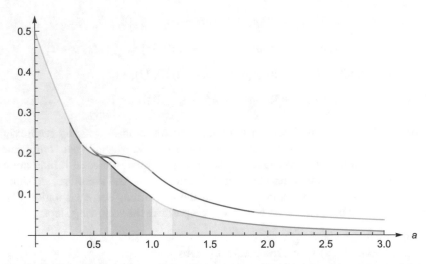

Fig. 7 Plot of $A(s, t, a)$ on the 17 different intervals of a identified from the 17 local minima in the proof of Lemma 4 for $0 \le a \le 3$; the shading under the graph indicates the main 10 intervals that are needed to describe the global minimum function $m(a)$

intervals. When two functions intersect in the interior of an interval, it is split into two subintervals. CAD is once again employed to find the smallest among the local minima; this is done individually for each of the refined intervals. As a result, we obtain the piecewise description of the function $m(a)$ given above; see Fig. 7 and the accompanying electronic material [7] for details.

It is clear from construction that $A(s, t, a)$ must be a continuous function, since the admissible polygons (shaded regions in Fig. 3) cannot jump or disappear if the parameters a, s, t are changed infinitesimally, i.e., if the lines in Fig. 3 are shifted or slanted by a little bit. In contrast, it is not obvious why it should be differentiable. Therefore, there is a possibility that the minimum can occur where the derivative does not exist. Hence, it is necessary to study the values of $A(s, t, a)$ along the boundaries of the different domains of definition. To accomplish this task, we view A as a bivariate function in s and t, with a parameter a. For each inequality in the list of conditions (5), the corresponding equation defines a line in $\mathbb{R}^2$. For each such line, we proceed to determine the range of a for which the line intersects the triangle $0 \le s \le t \le 1$. On the resulting line segment, the pieces of $A(s, t, a)$ are given by univariate polynomials, still involving the parameter a. Equating their derivatives to zero, we find all of the local minima on this line segment, which could give rise to local minima of $A(s, t, a)$. After looking at all 16 lines, each of which splits into at most 70 segments, we find 225 candidates for minima. CAD confirms that none of them are actually smaller than the one given by $m(a)$. This fact also becomes apparent by plotting these candidates against the function $m(a)$, as shown in Fig. 8 (top part).

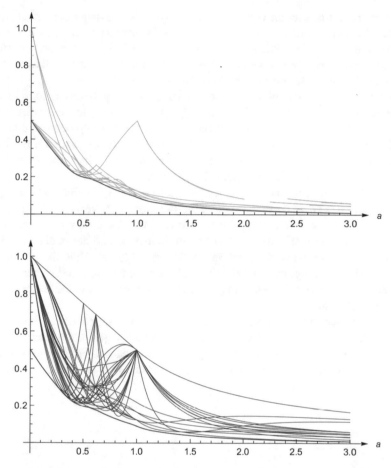

Fig. 8 Global minimum of $A(s, t, a)$ (red curve) compared to potential minima along lines (green curves, top part) and potential minima on intersection points (blue curves, bottom part)

Finally, we should also check all points where any two lines defined by (5) intersect. We find 54 points that lie inside the triangle $0 \leq s \leq t \leq 1$, at least for certain choices of a. The value of $A(s, t, a)$ at a particular point is given by a piecewise function depending on a. Assembling all pieces for all points, we obtain 348 cases. For each of them, CAD confirms (rigorously!) that the value of $A(s, t, a)$ does not go below $m(a)$. A "non-rigorous proof" of this fact is shown in Fig. 8 (bottom part).

Summarizing, we have shown that, for each particular choice of $a > 0$, the minimum of the function $A(s, t, a)$ on the triangle $0 \leq s \leq t \leq 1$ is given by $m(a)$, and we have determined the location (s_0, t_0) where this minimum is attained. This immediately establishes an asymptotic lower bound for MGSTs on$[n]$, as n goes to infinity. □

We wrap up this section with some remarks on the consequences of Lemma 4 and on what appears to be erratic (jumpy) behavior for some values of a in Fig. 8. We assure the reader that it is not due to the amount of alcohol that was consumed throughout this meal, but rather an indication of the appearance and disappearance of certain admissible regions for the MGSTs as a changes.

First, we would like to note that Lemma 4 explains why the asymptotic formula for MGSTs for integral $a \geq 2$ given in [1, 13, 14] does not specialize to the previously known case $a = 1$: this phenomenon is due to the piecewise definition of $m(a)$, with a transition at $1 < \alpha_8 < 2$. Geometrically speaking, α_8 marks the point where the polygon 133 (see Fig. 3) disappears, when a increases from 1 to 2, and $s = s_0(a)$ and $t = t_0(a)$ are updated constantly.

A second interesting finding that follows from Lemma 4 is that there is a jump of $(s_0(a), t_0(a))$ at $a = \alpha_4 = 0.5534\ldots$; the function $m(a)$ however is continuous. In Fig. 7 one sees that at $a = \alpha_4$ the functions of two local minima intersect, and therefore this point marks the jump from one branch to another one. In Fig. 9 the situation is shown for two different values of a close to α_4: while the shaded area in both parts of the figure is almost the same, the values of s and t change quite dramatically. We invite the reader to play with such transitions in the accompanying electronic material [7].

In the next section, we bring up the fact that the coloring pattern of three blocks that we generously assumed for $a > 0$ does not actually give the global minimum on $0 < a < 1$ over any 2-coloring of $[n]$ and we take care to emphasize this in the statement of the theorems. This will therefore explain the erratic behavior at $a = 1$ in both graphs of Fig. 8.

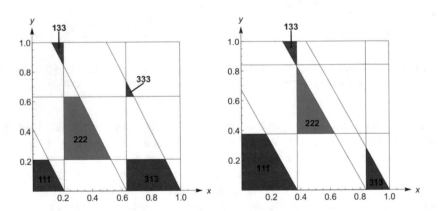

Fig. 9 The red and blue polygons correspond to monochromatic generalized Schur triples for $a = \frac{1}{2}, s = \frac{4}{19}, t = \frac{12}{19}$ (left) and $a = 0.56, s = 0.377, t = 0.841$ (right)

4 Exact Bounds for Generalized Schur Triples

In this section we apply the results from the last section, i.e., from the continuous setting, to the discrete enumeration problem of monochromatic generalized Schur triples (MGSTs). Hence, s and t are now integers with $1 \leq s \leq t \leq n$ that describe the coloring $R^s B^{t-s} R^{n-t}$ of $[n]$. Throughout this section we use the convention that a sum whose lower bound is greater than its upper bound is zero, i.e.,

$$\sum_{x=i}^{j} f(x) = \begin{cases} f(i) + \cdots + f(j), & \text{if } i \leq j, \\ 0, & \text{if } i > j. \end{cases}$$

Analogous to Sect. 2 we use the notation $\mathcal{M}^{(a)}$ to count MGSTs. More precisely, we define $\mathcal{M}^{(a)}(n, s, t)$ and $\mathcal{M}^{(a)}(n)$, as follows:

$$\mathcal{M}^{(a)}(n, s, t) := \left| \{ T = (x, y, x + \lfloor ay \rfloor) \in [n]^3 : \right.$$

$$\left. T \in ([s] \cup \{t + 1, \dots, n\})^3 \vee T \in \{s + 1, \dots, t\}^3 \} \right|,$$

$$\mathcal{M}^{(a)}(n) := \min_{1 \leq s \leq t \leq n} \mathcal{M}^{(a)}(n, s, t).$$

In contrast to the previous section, we will now mostly look at special cases for a, since we cannot hope to get an exact formula for the minimal number of MGSTs for general $a \in \mathbb{R}^+$.

Lemma 5 *Let $a \in \mathbb{R}$ with $a \geq 1$ and let $n, s, t \in \mathbb{N}$ with $1 \leq s \leq t \leq n$. Furthermore, assume that the inequalities $as + t \geq n$, $t \geq as$, and $s + as \leq t$ hold. Then the number $\mathcal{M}^{(a)}(n, s, t)$ of monochromatic generalized Schur triples of $[n]$ under the coloring $R^s B^{t-s} R^{n-t}$ is given by*

$$\sum_{y=1}^{\lfloor s/a \rfloor} \sum_{x=1}^{s - \lfloor ay \rfloor} 1 + \sum_{y=s+1}^{\lfloor (t-s)/a \rfloor} \sum_{x=s+1}^{t - \lfloor ay \rfloor} 1 + \sum_{y=1}^{\lfloor (n-t)/a \rfloor} \sum_{x=t+1}^{n - \lfloor ay \rfloor} 1 + \sum_{y=t+1}^{\lfloor n/a \rfloor} \sum_{x=1}^{n - \lfloor ay \rfloor} 1.$$

Moreover, the explicit list of these MGSTs $(x, y, x + \lfloor ay \rfloor)$ can be directly read off from the above formula.

Proof Under the given assumptions, we have to consider monochromatic triples of types 111, 222, 313, and 133, see, e.g., Fig. 3. Obviously, the four sums correspond exactly to these four cases. Note that if $at > n$, then the case 133 is not present, which is reflected by the fact that the corresponding sum is zero in this case. □

The assumed inequalities in Lemma 5 tell us that we are either in (R_7) (when $at > n$) or in (R_{17}) (when $at \leq n$); these regions were introduced in Lemma 3. Recall $\alpha_8 = 1.174559\dots$ from Lemma 4, and also that the global minimum of the area function $A(s, t, a)$ is located in (R_7) (when $a \geq \alpha_8$) or in (R_{17}) (when $1 \leq a \leq \alpha_8$).

Theorem 2 *The minimal number of monochromatic generalized Schur triples of the form* $(x, y, x + 2y)$ *that can be attained under any 2-coloring of* $[n]$ *of the form* $R^s B^{t-s} R^{n-t}$ *is*

$$M^{(2)}(n) = \left\lfloor \frac{n^2 - 10n + 33}{44} \right\rfloor.$$

Proof For $a = 2$ we clearly have $\alpha_8 \leq a$, and by Lemma 4 it follows that the optimal choice for s and t is expected around the point

$$n \cdot \left(\frac{a+1}{a^2 + 2a + 3}, \frac{a^2 + 2a + 2}{a^2 + 2a + 3} \right) = \left(\frac{3n}{11}, \frac{10n}{11} \right).$$

The three conditions $2s + t \geq n, t \geq 2s, 3s \leq t$ are satisfied (at least for large n), and therefore we can use Lemma 5 to compute the exact number of MGSTs:

$$M^{(2)}(n, s, t) = \sum_{y=1}^{\lfloor s/2 \rfloor} \sum_{x=1}^{s-2y} 1 + \sum_{y=s+1}^{\lfloor (t-s-1)/2 \rfloor} \sum_{x=s+1}^{t-2y} 1 + \sum_{y=1}^{\lfloor (n-t)/2 \rfloor} \sum_{x=t+1}^{n-2y} 1 =$$

$$= \left\lfloor \frac{s}{2} \right\rfloor \left\lfloor \frac{s-1}{2} \right\rfloor + \left\lfloor \frac{n-t}{2} \right\rfloor \left\lfloor \frac{n-t-1}{2} \right\rfloor + \left\lfloor \frac{t-s}{2} \right\rfloor \left\lfloor \frac{t-s-1}{2} \right\rfloor + 2s^2 - st + s.$$

From now on, we proceed in an analogous fashion as in the proofs of Lemma 2 and Theorem 1. Empirically, we find that for each $n \in \mathbb{N}$, the minimum of $M^{(2)}(n, s, t)$ is attained at

$$s_0 = \left\lfloor \frac{3n+1}{11} \right\rfloor, \qquad t_0 = \left\lfloor \frac{10n}{11} \right\rfloor + \begin{cases} -1, & \text{if } n = 22k + 10, \\ 0, & \text{otherwise.} \end{cases}$$

When we plug in $s_0 + i$ and $t_0 + j$ into the above formula for $M^{(2)}(n, s, t)$, we need to make a case distinction $n = 22k + \ell$ for $0 \leq \ell \leq 21$ in order to get rid of the floors. Moreover, we need to distinguish even and odd i (resp. j). Evaluating and simplifying

$$M^{(2)}(22k + \ell, s_0 + 2i_1 + i_2, t_0 + 2j_1 + j_2), \quad 0 \leq \ell \leq 21, \ i_2, j_2 \in \{0, 1\},$$

we obtain 88 polynomials in i_1, j_1, k. Applying CAD individually to each of these polynomials and by checking a few values explicitly (not unlike what we did in the proof of Lemma 5), one proves that the minimum is indeed attained at (s_0, t_0). Finally, one evaluates $M^{(2)}(22k + \ell, s_0, t_0)$ for all $\ell = 0, \ldots, 21$ and finds that it is always of the form $\frac{1}{44}(n^2 - 10n) + \delta_\ell$, where the values $\delta_0, \ldots, \delta_{21}$ are

$$0, \frac{9}{44}, \frac{4}{11}, \frac{21}{44}, \frac{6}{11}, \frac{25}{44}, \frac{6}{11}, \frac{21}{44}, \frac{4}{11}, \frac{9}{44}, 0, \frac{3}{4}, \frac{5}{11}, \frac{5}{44}, \frac{8}{11}, \frac{13}{44}, -\frac{2}{11}, \frac{13}{44}, \frac{8}{11}, \frac{5}{44}, \frac{5}{11}, \frac{3}{4}.$$

Since the largest value is $\frac{3}{4}$ and since the smallest value is greater than $-\frac{1}{4}$ (i.e., all values δ_ℓ lie inside an interval of length 1), the claimed formula follows.

One last detail: we still have to examine for which n the conditions $2s + t \geq n$, $t \geq 2s$, $3s \leq t$ are satisfied, as it could happen that for small n the point (s_0, t_0) lies not inside the correct region (R_{17}), due to the rounding errors. With the (somewhat generous) assumptions $\frac{3n+1}{11} - 1 \leq s \leq \frac{3n+1}{11}$ and $\frac{10n}{11} - 2 \leq t \leq \frac{10n}{11}$ we find that the above conditions are satisfied for all $n \geq 25$. For the remaining values $n < 25$, the claimed formula can be verified by an explicit computation. $\qquad\square$

Theorem 3 *The minimal number of monochromatic generalized Schur triples of the form $(x, y, x + 3y)$ that can be attained under any 2-coloring of $[n]$ of the form $R^s B^{t-s} R^{n-t}$ is*

$$M^{(3)}(n) = \left\lfloor \frac{n^2 - 18n + 101}{108} \right\rfloor + \begin{cases} 1, & \text{if } n = 54k + 36, \\ -1, & \text{if } n = 54k + 30 \text{ or } n = 54k + 42, \\ 0, & \text{otherwise.} \end{cases}$$

Proof For $a = 3$, it follows by Lemma 4 that the optimal choice for s and t is expected around the point

$$n \cdot \left(\frac{a+1}{a^2 + 2a + 3}, \frac{a^2 + 2a + 2}{a^2 + 2a + 3} \right) = \left(\frac{4n}{18}, \frac{17n}{18} \right).$$

This means that the proof will require $18 \cdot a = 54$ case distinctions $n = 54k + \ell$ for $0 \leq \ell \leq 53$. Empirically, we find that for each $n \in \mathbb{N}$, the minimum of $M^{(3)}(n, s, t)$ is attained at

$$s_0 = \left\lfloor \frac{4n}{18} \right\rfloor - \begin{cases} 1, & \text{if } n = 54k + 18, \\ 0, & \text{otherwise,} \end{cases}$$

$$t_0 = \left\lfloor \frac{17n}{18} \right\rfloor - \begin{cases} 1, & \text{if } n = 9k + i \text{ for } i \in \{3, 4, 7, 8\}, \\ 2, & \text{if } n = 54k + 18, \\ 0, & \text{otherwise.} \end{cases}$$

Applying CAD to the 486 polynomials

$$M^{(3)}(54k + \ell, s_0 + 3i_1 + i_2, t_0 + 3j_1 + j_2), \quad 0 \leq \ell \leq 53, \ i_2, j_2 \in \{0, 1, 2\},$$

proves that our choice of (s_0, t_0) locates the minimum. Evaluating $M^{(3)}(n, s_0, t_0)$ for $n = 54k + \ell$, one obtains $\frac{1}{108}(n^2 - 18n) + \delta_\ell$, where $\delta_{36} = 1$, $\delta_{30} = \delta_{42} = -\frac{1}{3}$, and all remaining δ_ℓ range from $-\frac{1}{27}$ to $\frac{101}{108}$. Hence, the claimed formula follows. $\qquad\square$

Theorem 4 *The minimal number of monochromatic generalized Schur triples of the form* $(x, y, x + 4y)$ *that can be attained under any 2-coloring of* $[n]$ *of the form* $R^s B^{t-s} R^{n-t}$ *is*

$$
\mathcal{M}^{(4)}(n) = \left\lfloor \frac{n^2 - 28n + 245}{216} \right\rfloor - \begin{cases} 1, & \text{if } n = 108k + i \text{ for } i \in I, \\ 0, & \text{otherwise}, \end{cases}
$$

where $I = \{0, 1, 27, 28, 43, 47, 48, 53, 58, 63, 67, 68, 69, 73, 78, 83, 88, 89, 93\}$.

Proof For $a = 4$, it follows by Lemma 4 that the optimal choice for s and t is expected around the point

$$
n \cdot \left(\frac{a+1}{a^2 + 2a + 3}, \frac{a^2 + 2a + 2}{a^2 + 2a + 3} \right) = \left(\frac{5n}{27}, \frac{26n}{27} \right).
$$

This means that the proof will require $27 \cdot a = 108$ case distinctions of the form $n = 108k + \ell$ for $0 \le \ell \le 107$. Empirically, we find that for each $n \in \mathbb{N}$, the minimum of $\mathcal{M}^{(4)}(n, s, t)$ is attained at

$$
s_0 = \left\lfloor \frac{5n - 4}{27} \right\rfloor + \begin{cases} -1, & \text{if } n = 108k + 28, \\ 1, & \text{if } n = 108k + i \text{ for } i \in \{0, 87, 103\}, \\ 0, & \text{otherwise}. \end{cases}
$$

$$
t_0 = \left\lfloor \frac{26n - 34}{27} \right\rfloor + \begin{cases} -1, & \text{if } n = 108k + i \text{ for } i \in \{28, 33, 38, 43\}, \\ 1, & \text{if } n = 108k + i \\ & \quad \text{for } i \in \{1, 77, 78, 82, 83, 88, 93, 98, 104\}, \\ 2, & \text{if } n = 108k + i \text{ for } i \in \{0, 87, 103\}, \\ 0, & \text{otherwise}. \end{cases}
$$

Applying CAD to the 1728 polynomials

$$
\mathcal{M}^{(4)}(108k + \ell, s_0 + 4i_1 + i_2, t_0 + 4j_1 + j_2), \quad 0 \le \ell \le 107, \ i_2, j_2 \in \{0, 1, 2, 3\},
$$

proves that our choice of (s_0, t_0) locates the minimum. Evaluating $\mathcal{M}^{(4)}(n, s_0, t_0)$ for $n = 108k + \ell$, $0 \le \ell \le 107$, one obtains 108 polynomials of the form $\frac{1}{216}(n^2 - 28n) + \delta_\ell$. At this point, the analysis deviates a bit from the previous two theorems, because we observe that the range of the computed δ_ℓ's is much larger than 1. Therefore, we would like to choose an appropriate interval to contain the largest number of δ_ℓ such that we minimize the number of exceptional cases (i.e., the necessary corrections resulting from applying the floor function to numbers that are out of range).

To accomplish this, we find that shifting all of the values down by $\frac{29}{216}$ gives the minimum number (19, to be precise) of δ_ℓ that are not within range (i.e., not in $[0, 1)$). We now realize that these are the values that give us our desired count, so we add 1 to make sure it is recognized by the floor function. Hence, the optimal delta is $\frac{29}{216} + 1 = \frac{245}{216}$. Finally, for each of the 19 δ_ℓ's that are out of bounds (in this case, less than 0), we remove 1 and this gives us our claimed formula. $\qquad\square$

Theorem 5 *The minimal number of monochromatic generalized Schur triples of the form* $\left(x, y, x + \lfloor \frac{1}{2}y \rfloor\right)$ *that can be attained under any 2-coloring of* $[n]$ *of the form* $R^s B^{t-s} R^{n-t}$ *is given by*

$$
M^{(1/2)}(n) = \left\lfloor \frac{15n^2 + 72}{76} \right\rfloor + \begin{cases} 1, & \text{if } n = 38k + 18 \text{ or } n = 38k + 20, \\ -1, & \text{if } n = 38k + 19, \\ 0, & \text{otherwise.} \end{cases}
$$

Proof For $a = \frac{1}{2}$, it follows by Lemma 4 that the optimal choice for s and t is expected around the point

$$
n \cdot \left(\frac{-2a^2 + a + 1}{-4a^3 + 5a^2 + 6a + 1}, \frac{-2a^3 + a^2 + 4a + 1}{-4a^3 + 5a^2 + 6a + 1} \right) = \left(\frac{4n}{19}, \frac{12n}{19} \right).
$$

For this choice of parameters we end up in region (R_{69}) (see Fig. 6). Under the conditions that characterize this region, more precisely

$$
\frac{n}{2} \le t \le \frac{2n}{3} \ \wedge \ t - s \le \frac{n}{2} \ \wedge \ 2s \le t,
$$

the number of MGSTs is given by

$$
M^{(1/2)}(n, s, t) = \sum_{y=1}^{s} \sum_{x=1}^{s-\lfloor y/2 \rfloor} 1 + \sum_{y=s+1}^{t} \sum_{x=s+1}^{t-\lfloor y/2 \rfloor} 1 + \sum_{y=1}^{s} \sum_{x=t+1}^{n-\lfloor y/2 \rfloor} 1 +
$$

$$
+ \sum_{y=2t-2s+1}^{n} \sum_{x=t+1-\lfloor y/2 \rfloor}^{s} 1 + \sum_{y=t+1}^{2n-2t-1} \sum_{x=t+1}^{n-\lfloor y/2 \rfloor} 1.
$$

The five double sums correspond to the cases 111, 222, 313, 133, 333, respectively, and the summation ranges are chosen such that they actually agree with the first two coordinates of the monochromatic triples in question, see Fig. 9.

In order to eliminate all floor functions, a case distinction $n = 38k + \ell$ is made. It is conjectured that the minimum is attained at $(s, t) = (s_0, t_0)$ with

$$
s_0 = \left\lfloor \frac{4n + 7}{19} \right\rfloor + \begin{cases} 1, & \text{if } n = 19k + 17, \\ 0, & \text{otherwise}, \end{cases}
$$

$$
t_0 = \left\lfloor \frac{12n + 6}{19} \right\rfloor + \begin{cases} 1, & \text{if } n = 19k + 4, \\ 0, & \text{otherwise}. \end{cases}
$$

This conjecture is proven by case distinction and CAD, as in Theorem 2. As a final result, one obtains the claimed formula, see [7] for the details.   □

It has to be noted that all results presented so far in this section (Theorems 2–5) are based on the assumption of the optimal coloring being of the form $R^s B^{t-s} R^{n-t}$. While we have strong evidence that this assumption is valid for $a > 1$ (and in fact we know it to be true [10] for $a = 1$), it seems to be inappropriate for $0 < a < 1$. More concretely, we can construct explicit examples where we get fewer MGSTs for $a = \frac{1}{2}$ than predicted in Theorem 5: the first instance is $n = 4$, where Theorem 5 yields four MGSTs for the coloring $RBBR$, namely $(1, 1, 1)$, $(4, 1, 4)$, $(2, 2, 3)$, $(2, 3, 3)$, but where the better coloring $RBRB$ exists, that allows only three MGSTs, namely $(1, 1, 1)$, $(3, 1, 3)$, and $(2, 4, 4)$. Note, however, that this is not a counterexample to the theorem because the coloring $RBRB$ is not of the form $R^s B^{t-s} R^{n-t}$.

We close this section by stating a conjecture about what we believe is the true minimum for $a = \frac{1}{2}$.

Conjecture 1 For $n \geq 12$, the minimal number of monochromatic generalized Schur triples of the form $\left(x, y, x + \lfloor \frac{1}{2}y \rfloor\right)$ that can be attained under any 2-coloring of $[n]$ is given by

$$
\left\lfloor \frac{n^2 + 5}{6} \right\rfloor,
$$

and it occurs at the coloring $R^s B^{t-s} R^{u-t} B^{n-u}$ for

$$
s = \left\lfloor \frac{n + 3}{6} \right\rfloor, \qquad t = \left\lfloor \frac{n + 1}{2} \right\rfloor, \qquad u = \left\lfloor \frac{5n + 3}{6} \right\rfloor.
$$

Curiously, the conjectured formula is not valid for $n = 11$, where it would give a minimum number of 21 MGSTs with a four-block coloring. The true minimum is 20 and it is attained at the coloring $RBRBBRRBRBB$.

5   Conclusions and Outlook

In this paper we have presented, for the first time, exact formulas for the minimum number of monochromatic (generalized) Schur triples. We give such formulas explicitly only for the few cases $a = 1, 2, 3, 4$, but we want to point out that we could do many more special cases, say $a = 5, 6, 7, \ldots$ or $a = \frac{3}{2}, \frac{5}{4}, \ldots$, based on the general analysis carried out in Sect. 3. In fact, the proofs would be done in completely analogous fashion, requiring only little human interaction, but an increasing amount of computation time. In this sense, our paper contains a hidden treasure, which is an infinite set of theorems that just have to be unveiled.

For future research, we propose to look more closely at the cases of generalized Schur triples $(x, y, x + \lfloor ay \rfloor)$ with $0 < a < 1$. Our analysis is based on the assumption that the optimal coloring that produces the least number of monochromatic triples consists of three blocks. Computational experiments suggest that this assumption is not valid for $0 < a < 1$. For example, we believe that four blocks are necessary to capture the minimum in the case $a = \frac{1}{2}$, as conjectured in the previous section. For some less nice rational numbers $a < 1$ we were even not able to detect a block pattern in the optimal coloring, but that may be an artifact due to the limited size of n for which we can do exhaustive searches (note that there are 2^n possible colorings).

Our results are heavily based on symbolic computation techniques, such as cylindrical algebraic decomposition and symbolic summation. Often our proofs require case distinctions into several dozens or even several hundred cases, and it would be too tedious to check all of them by hand. The reader should be convinced by now that symbolic computation can be very useful and that it could be adapted to solve problems in other areas of mathematics. We provide all details of our calculations in the supplementary electronic material [7], which we hope is instructive for readers who would like to become more acquainted with the techniques that we used here.

Acknowledgements   We would like to thank our colleagues Thotsaporn Thanatipanonda, Thibaut Verron, Herwig Hauser, and Carsten Schneider for inspiring discussions, comments, and encouragement. The first author was supported by the Austrian Science Fund (FWF): P29467-N32. The second author was supported by the Austrian Science Fund (FWF): F5011-N15.

References

1. Butler, S., Costello, K.P., Graham, R.:   Finding patterns avoiding many monochromatic constellations. Exp. Math. **19**(4), 399–411 (2010)
2. Collins, G.E.:   Quantifier elimination for the elementary theory of real closed fields by cylindrical algebraic decomposition. In: Lecture Notes in Computer Science, vol. 33, pp. 134–183 (1975)
3. Datskovsky, B.A.: On the number of monochromatic Schur triples. Adv. Appl. Math. **31**(1), 193–198 (2003)

4. Erdős, P., Szekeres, G.: A combinatorial problem in geometry. Compos. Math. **2**, 463–470 (1935)
5. Goodman, A.: On sets of acquaintances and strangers at any party. Am. Math. Mon. **66**(9), 778–783 (1959)
6. Graham, R., Rödl, V., Ruciński, A.: On Schur properties of random subsets of integers. J. Number Theory **61**, 388–408 (1996)
7. Koutschan, C., Wong, E.: Mathematica notebook SchurTriples.nb (2019). http://www.koutschan.de/data/schur/
8. Landman, B., Robertson, A.: Ramsey Theory on the Integers. AMS, Providence (2004)
9. Robertson, A., Zeilberger, D.: A 2-coloring of $[1, N]$ can have $(1/22)N^2 + O(N)$ monochromatic Schur triples, but not less! Electron. J. Comb. **5**, 1–4 (1998)
10. Schoen, T.: The number of monochromatic Schur triples. Eur. J. Comb. **20**(8), 855–866 (1999)
11. Schur, I.: Über die Kongruenz $x^m + y^m \equiv z^m$ (mod. p). Jahresber. Deutsch. Math.-Verein. **25**, 114–116 (1917)
12. Sloane, N.J.A.: The on-line encyclopedia of integer sequences. OEIS Foundation Inc. http://oeis.org
13. Thanatipanonda, T.: On the monochromatic Schur triples type problem. Electron. J. Comb. **16**, article #R14 (2009)
14. Thanatipanonda, T., Wong, E.: On the minimum number of monochromatic generalized Schur triples. Electron. J. Comb. **24**(2), article #P2.20 (2017)
15. van der Waerden, B.L.: Beweis einer Baudetschen Vermutung. Nieuw. Arch. Wisk. **15**, 212–216 (1927)
16. Yao, Y., Zeilberger, D.: Numerical and symbolic studies of the peaceable queens problem. Exp. Math. (2019). DOI: http://10.1080/10586458.2019.1616338

Evaluation of Binomial Double Sums Involving Absolute Values

Christian Krattenthaler and Carsten Schneider

Dedicated to Peter Paule on the occasion of his 60th birthday

1 Introduction

Motivated by work in [3] concerning the Hadamard maximal determinant problem [10], Brent and Osborn [2] proved the double sum evaluation

$$\sum_{i,j=-n}^{n} |i^2 - j^2| \binom{2n}{n+i} \binom{2n}{n+j} = 2n^2 \binom{2n}{n}^2. \tag{1}$$

It should be noted that the difficulty in evaluating this sum lies in the appearance of the absolute value. Without the absolute value, the summand would become antisymmetric in i and j so that the sum would trivially vanish. Together with Ohtsuka and Prodinger, they went on in [6] (see [5] for the published version) to

Research partially supported by the Austrian Science Foundation (FWF) grant SFB F50 (F5005-N15 and F5009-N15) in the framework of the Special Research Program "Algorithmic and Enumerative Combinatorics".

C. Krattenthaler
Fakultät für Mathematik, Universität Wien, Vienna, Austria
e-mail: Christian.Krattenthaler@univie.ac.at

C. Schneider (✉)
Johannes Kepler University Linz, Research Institute for Symbolic Computation (RISC), Linz, Austria
e-mail: Carsten.Schneider@risc.jku.at

© Springer Nature Switzerland AG 2020
V. Pillwein, C. Schneider (eds.), *Algorithmic Combinatorics: Enumerative Combinatorics, Special Functions and Computer Algebra*, Texts & Monographs in Symbolic Computation, https://doi.org/10.1007/978-3-030-44559-1_14

consider more general double sums of the form

$$\sum_{i,j=-n}^{n} |i^s j^t (i^k - j^k)^\beta| \binom{2n}{n+i}\binom{2n}{n+j}, \tag{2}$$

mostly for small positive integers s, t, k, β. Again, without the absolute value, the summation would not pose any particular problem since it could be carried out separately in i and j by means of a relatively straightforward application of the binomial theorem. In several cases, they found explicit evaluations of such sums—sometimes with proof, sometimes conjecturally.

The purpose of the current paper is to provide a complete treatment of double sums of the form (2) and of the more general form

$$\sum_{i,j} |i^s j^t (i^k - j^k)^\beta| \binom{2n}{n+i}\binom{2m}{m+j}, \tag{3}$$

with an independent parameter m. More precisely, using the computer algebra package Sigma [15], we were led to the conjecture that these double sums of the form (2) can always be expressed in terms of a linear combination of just four functions, namely $\binom{4n}{2n}$, $\binom{2n}{n}^2$, $4^n\binom{2n}{n}$, and 16^n, with coefficients that are rational in n, while in many instances double sums of the form (3) can be expressed in terms of a linear combination of the four functions $\binom{2n+2m}{n+m}$, $\binom{2n}{n}\binom{2m}{m}$, $4^n\binom{2m}{m}$, and $4^m\binom{2n}{n}$, with coefficients that are rational in n and m. We demonstrate this observation in Theorems 1–4, in a much more precise form.

It is not difficult to see that the problem of evaluation of double sums of the form (2) and (3) can be reduced to the evaluation of sums of the form

$$\sum_{0 \le i \le j} i^s j^t \binom{2n}{n+i}\binom{2m}{m+j} \tag{4}$$

(and a few simpler *single* sums). See the proofs of Theorems 1–4 in Sect. 7 and Remark 3(1). We furthermore show (see the proofs of Propositions 1 and 2 in Sect. 5, which may be considered as the actual main result of the present paper) that for the evaluation of double sums of the form (4) it suffices to evaluate four *fundamental* double sums, given in Lemmas 1–4 in Sect. 2. While Lemmas 2–4 are relatively easy to prove by telescoping arguments (see the proofs in Sect. 2), the proof of Lemma 1 is more challenging. We provide two different proofs, one using computer algebra, and one using complex contour integrals. We believe that both proofs are of intrinsic interest. The algorithmic proof is described in Sect. 3. There, we explain that the computer algebra package Sigma can be used in a completely automatic fashion to evaluate double sums of the form (4). In particular, the reader can see how we empirically discovered our main results in Sects. 5 and 7. The second proof, based on the power of complex integration, is explained in Sect. 4.

We close our paper by proving another conjecture from [6, Conj. 3.1], namely the inequality (see Theorem 5 in Sect. 8)

$$\sum_{i,j} \left| j^2 - i^2 \right| \binom{2n}{n+i} \binom{2m}{m+j} \geq 2nm \binom{2n}{n} \binom{2m}{m}.$$

We show moreover that equality holds if and only if $m = n$, in which case the evaluation (1) applies. Although Lemmas 1–4 would provide a good starting point for a proof of the inequality, we prefer to use a more direct approach, involving an application of Gosper's algorithm [7] at a crucial point.

We wish to point out that Bostan et al. [1] have developed an algorithmic approach—based on contour integrals—that is capable of automatically finding a recurrence for the double sum (2) for any particular choice of s, t, k, β, and, thus, is able to establish an evaluation of such a sum (such as (1), for example) once the right-hand side is found.

Our final remark is that some of the double sums (2) and (3) can be embedded into infinite families of multidimensional sums that still allow for closed form evaluations, see [4].

2   The Fundamental Lemmas

In this section, we state the summation identities which form the basis of the evaluation of double sums of the form (4) (and, thus, of double sums of the form (2) and (3)). As it turns out, Lemmas 2–4 are very easy to prove since at least one summation of the double sum can be put in telescoping form, see the proofs below. Lemma 1 is much more subtle. We provide two different proofs, the first being algorithmic—see Sect. 3, the second making use of complex integration—see Sect. 4.

Lemma 1   *For all non-negative integers n and m, we have*

$$\sum_{0 \leq i \leq j} \binom{2n}{n+i} \binom{2m}{m+j} = 2^{2n+2m-3} + \frac{1}{4} \binom{2n+2m}{n+m} + \frac{1}{2} \binom{2n}{n} \binom{2m}{m}$$

$$+ 2^{2m-2} \binom{2n}{n} - \frac{1}{8} \sum_{\ell=0}^{n-m} \binom{2n-2\ell}{n-\ell} \binom{2m+2\ell}{m+\ell}, \qquad (5)$$

where the sum on the right-hand side has to be interpreted as explained in Lemma 7.

Lemma 2 *For all non-negative integers n and m, we have*

$$\sum_{0 \leq i \leq j} i \binom{2n}{n+i}\binom{2m}{m+j} = -\frac{n}{4}\binom{2n+2m}{n+m} + n\, 2^{2m-2}\binom{2n}{n} + \frac{nm}{4(n+m)}\binom{2n}{n}\binom{2m}{m}.$$

(6)

Lemma 3 *For all non-negative integers n and m, we have*

$$\sum_{0 \leq i \leq j} j \binom{2n}{n+i}\binom{2m}{m+j} = \frac{m}{4}\binom{2n+2m}{n+m} + \frac{m(m+2n)}{4(n+m)}\binom{2n}{n}\binom{2m}{m}.$$

(7)

Lemma 4 *For all non-negative integers n and m, we have*

$$\sum_{0 \leq i \leq j} i\, j \binom{2n}{n+i}\binom{2m}{m+j} = \frac{mn}{2(n+m)}\binom{2n+2m-2}{n+m-1} + \frac{nm^2}{4(n+m)}\binom{2n}{n}\binom{2m}{m}.$$

(8)

Proof of Lemma 2 We have[1]

$$i\binom{2n}{n+i} = \frac{n+i}{2}\binom{2n}{n+i} - \frac{n+i+1}{2}\binom{2n}{n+i+1}.$$

Thus, we obtain

$$\sum_{0 \leq i \leq j} i \binom{2n}{n+i}\binom{2m}{m+j} = \frac{1}{2}\sum_{j \geq 0}\left(n\binom{2n}{n} - (n+j+1)\binom{2n}{n+j+1}\right)\binom{2m}{m+j}$$

$$= \frac{n}{2}\binom{2n}{n}\sum_{j \geq 0}\binom{2m}{m+j} - \frac{1}{2}\sum_{j \geq 0}(n-j)\binom{2n}{n+j}\binom{2m}{m+j}.$$

The first sum is, essentially, one half of a binomial theorem,

$$\sum_{j \geq 0}\binom{2m}{m+j} = \frac{1}{2}\binom{2m}{m} + 2^{2m-1}.$$

---

[1]The informed reader will have guessed that the telescoping form of the summand was discovered by using Gosper's algorithm [7] (see also [14]). The particular implementation that we applied is the one due to Paule and Schorn [13].

In order to evaluate the second sum, we observe that[2]

$$\sum_{j\geq 0}(n-j)\binom{2n}{n+j}\binom{2m}{m+j} = n\sum_{j\geq 0}\binom{2n}{n+j}\binom{2m}{m+j} - \sum_{j\geq 0}j\binom{2n}{n+j}\binom{2m}{m+j}$$

$$= \frac{n}{2}\sum_{j=-\infty}^{\infty}\binom{2n}{n+j}\binom{2m}{m+j} + \frac{n}{2}\binom{2n}{n}\binom{2m}{m}$$

$$- \sum_{j\geq 0}\left(\frac{(n+j)(m+j)}{2(m+n)}\binom{2n}{n+j}\binom{2m}{m+j}\right.$$

$$\left. - \frac{(n+j+1)(m+j+1)}{2(m+n)}\binom{2n}{n+j+1}\binom{2m}{m+j+1}\right)$$

$$= \frac{n}{2}\sum_{j=-\infty}^{\infty}\binom{2n}{n+j}\binom{2m}{m-j} + \frac{n}{2}\binom{2n}{n}\binom{2m}{m} - \frac{nm}{2(m+n)}\binom{2n}{n}\binom{2m}{m}.$$

The sum in the last line can be evaluated by means of the Chu–Vandermonde summation formula (cf. [9, Sec. 5.1, (5.27)]). Substitution of these findings and little simplification then leads to the right-hand side of (6). □

Proof of Lemma 3 We have

$$j\binom{2m}{m+j} = \frac{m+j}{2}\binom{2m}{m+j} - \frac{m+j+1}{2}\binom{2m}{m+j+1}. \tag{9}$$

Thus, we obtain

$$\sum_{0\leq i\leq j}j\binom{2n}{n+i}\binom{2m}{m+j} = \frac{1}{2}\sum_{i\geq 0}\binom{2n}{n+i}(m+i)\binom{2m}{m+i}$$

$$= \frac{m}{2}\sum_{i\geq 0}\binom{2n}{n+i}\binom{2m}{m+i} + \frac{1}{2}\sum_{i\geq 0}i\binom{2n}{n+i}\binom{2m}{m+i}.$$

We have evaluated the same sums in the previous proof. We leave it to the reader to fill in the details in order to arrive at the right-hand side of (7). □

---

[2]For the finding of the telescoping form of the sum over $j \geq 0$ below see footnote 1.

Proof of Lemma 4 Using (9), we have

$$\sum_{0\leq i\leq j} i\,j \binom{2n}{n+i}\binom{2m}{m+j} = \frac{1}{2}\sum_{i\geq 0} i \binom{2n}{n+i}(m+i)\binom{2m}{m+i}$$

$$= \frac{1}{2}\sum_{i\geq 0}(n+i)\binom{2n}{n+i}(m+i)\binom{2m}{m+i}$$

$$- \frac{n}{2}\sum_{i\geq 0}\binom{2n}{n+i}(m+i)\binom{2m}{m+i}$$

$$= 2nm\sum_{i\geq 0}\binom{2n-1}{n+i-1}\binom{2m-1}{m+i-1} - \frac{n}{2}\sum_{i\geq 0}\binom{2n}{n+i}(m+i)\binom{2m}{m+i}.$$

We have evaluated the second sum in the previous proof. In order to evaluate the first sum, we do the substitution $i \to -i + 1$ and obtain

$$\sum_{i\geq 0}\binom{2n-1}{n+i-1}\binom{2m-1}{m+i-1} = \frac{1}{2}\sum_{i\geq 0}\binom{2n-1}{n+i-1}\binom{2m-1}{m+i-1}$$

$$+ \frac{1}{2}\sum_{i\leq 1}\binom{2n-1}{n-i}\binom{2m-1}{m-i}$$

$$= \frac{1}{2}\sum_{i=-\infty}^{\infty}\binom{2n-1}{n+i-1}\binom{2m-1}{m+i-1}$$

$$+ \frac{1}{2}\binom{2n-1}{n-1}\binom{2m-1}{m-1} + \frac{1}{2}\binom{2n-1}{n}\binom{2m-1}{m}$$

$$= \frac{1}{2}\sum_{i=-\infty}^{\infty}\binom{2n-1}{n+i-1}\binom{2m-1}{m-i} + \binom{2n-1}{n}\binom{2m-1}{m}.$$

Again, the sum can be evaluated by means of the Chu–Vandermonde summation formula, and then substitution of these findings and little simplification leads to the right-hand side of (8).   □

3 Proof of Lemma 1 Using the Computer Algebra Package **Sigma**

Here we show how Lemma 1 can be established by using the algorithmic tools provided by the summation package Sigma [15] of the second author. Algorithmic proofs of Lemmas 2–4 are much simpler and could be obtained completely analogously.

We seek an alternative representation of the double sum

$$S(n, m) = \sum_{0 \le i \le j} \binom{2n}{n+i}\binom{2m}{m+j} \tag{10}$$

for all non-negative integers m, n with the following property: if one specialises m (respectively n) to a non-negative integer or if one knows the distance between n and m, then the evaluation of the double sum should be performed in a direct and simple fashion. In order to accomplish this task, we utilise the summation package Sigma [15].

The sum (10) can be rewritten in the form

$$S(n, m) = \sum_{j=0}^{m} f(n, m, j) \tag{11}$$

with

$$f(n, m, j) = \binom{2m}{j+m} \sum_{i=0}^{j} \binom{2n}{i+n}. \tag{12}$$

Given this sum representation we will exploit the following summation spiral that is built into Sigma:

1. Calculate a linear recurrence in m of order d (for an appropriate positive integer d) for the sum $S(n, m)$ by the creative telescoping paradigm;
2. solve the recurrence in terms of (indefinite) nested sums over hypergeometric products with respect to m (the corresponding sequences are also called *d'Alembertian solutions*, see [14]);
3. combine the solutions into an expression $\mathrm{RHS}(n, m)$ such that $S(n, l) = \mathrm{RHS}(n, l)$ holds for all n and $l = 0, 1, \ldots, d - 1$.

Then this implies that $S(n, m) = \mathrm{RHS}(n, m)$ holds for all non-negative integers m, n.

Remark 1 This summation engine can be considered as a generalisation of [14] that works not only for hypergeometric products but for expressions in terms of nested sums over such hypergeometric products. It is based on a constructive summation theory of difference rings and fields [17, 18] that enhances Karr's summation approach [11] in various directions.

In the following paragraphs, we assume that $m \le n$. We activate Sigma's summation spiral.

Step 1.   Observe that our sum (11) with summand given in (12) is already in the
right input form for $\texttt{Sigma}$: the summation objects of (12) are given in terms
of nested sums over hypergeometric products. More precisely, let S_j denote the
shift operator with respect to j, that is, $S_j F(j) := F(j+1)$. Then, if one applies
this shift operator to the arising objects of $f(n, m, j)$, one can rewrite them again
in their non-shifted versions:

$$S_j \binom{2m}{j+m} = \frac{m-j}{1+j+m} \binom{2m}{j+m},$$

$$S_j \sum_{i=0}^{j} \binom{2n}{i+n} = \sum_{i=0}^{j} \binom{2n}{i+n} + \frac{n-j}{1+j+n} \binom{2n}{j+n}. \tag{13}$$

With the help of these identities, we can look straightforwardly for a linear
recurrence in the free integer parameter m as follows. First, we load $\texttt{Sigma}$ into
the computer algebra system *Mathematica*,

In[1]:= << **Sigma.m**

> Sigma - A summation package by Carsten Schneider © RISC-Linz

and enter our definite sum $S(n, m)$:

In[2]:= **mySum = SigmaSum[Binomial[2m, j + m]SigmaSum[Binomial[2n, i + n], {i, 0, j}], {j, 0, m}]**

Out[2]= $\displaystyle \sum_{j=0}^{m} \binom{2m}{j+m} \sum_{i=0}^{j} \binom{2n}{i+n}$

Then we compute a recurrence in m by executing the function call

In[3]:= **rec = GenerateRecurrence[mySum, m][[1]]**

Out[3]= $\text{SUM}[m+1] - 4\,\text{SUM}[m] == -\dfrac{1}{1+m+n} \displaystyle\sum_{i=0}^{m} \binom{2m}{i+m}\binom{2n}{i+n} + \dfrac{mn}{(m+1)(1+m+n)}\binom{2m}{m}\binom{2n}{n}$

This means that $\text{SUM}[m] = S(n, m)(= \texttt{mySum})$ is a solution of the output recurrence. But what is going on behind the scenes? Roughly speaking, Zeilberger's
creative telescoping paradigm [14] is carried out in the setting of difference rings.
More precisely, one tries to compute a recurrence for the summand $f(n, m, j)$
of the form

$$c_0(n, m) f(n, m, j) + c_1(n, m) f(n, m+1, j) + \cdots + c_d(n, m) f(n, m+d, j)$$

$$\overset{.}{=} g(n, m, j+1) - g(n, m, j), \tag{14}$$

for $d = 0, 1, 2, \ldots$. In our particular instance, Sigma is successful for $d = 1$ and delivers the solution $c_0(n, m) = -4$, $c_1(n, m) = 1$, and

$$
g(n, m, j) = \frac{(2j - 1)}{-1 + j - m} \binom{2m}{j + m} \sum_{i=0}^{j} \binom{2n}{i + n}
$$

$$
+ \frac{j - n}{1 + m + n} \binom{2m}{j + m} \binom{2n}{j + n} + \frac{1}{-1 - m - n} \sum_{i=0}^{j} \binom{2m}{i + m} \binom{2n}{i + n},
$$

$$(15)$$

which holds for all non-negative integers j, m, n with $0 \leq j \leq m \leq n$. The correctness can be verified by substituting the right-hand side of (12) into (14), rewriting the summation objects in terms of $\binom{2m}{j+m}$ and $\sum_{i=0}^{j} \binom{2n}{i+n}$ using the relations given in (13) and $S_m\binom{2m}{j+m} = \frac{2(m+1)(2m+1)}{(m-j+1)(1+j+m)} \binom{2m}{j+m}$, and applying simple rational function arithmetic. We recall that we assumed $m \leq n$, and this restriction is indeed essential for being allowed to use Sigma in the described setup. However, the above check reveals that the result is in fact correct without any restriction on the relative sizes of m and n.

Finally, by summing (14) over j from 0 to m, we obtain the linear recurrence

$$
\sum_{j=0}^{m} f(n, m + 1, j) - 4 \sum_{j=0}^{m} f(n, m, j) = - \sum_{j=0}^{m+1} \binom{2n}{i + n} + \frac{1}{-1 - m - n}
$$

$$
\sum_{i=0}^{m} \binom{2m}{i + m} \binom{2n}{i + n} + \frac{mn}{(m + 1)(1 + m + n)} \binom{2m}{m} \binom{2n}{n}.
$$

which, by the above remark, holds for all non-negative integers m, n. As is straightforward to see, this is indeed equivalent to Out[3].

Step 2.   We now apply our summation toolbox to the definite sum $\sum_{i=0}^{m} \binom{2m}{i+m} \binom{2n}{i+n}$ and obtain

$$
\sum_{i=0}^{m} \binom{2m}{m + i} \binom{2n}{n + i} = \frac{1}{2} \binom{2m}{m} \binom{2n}{n} + \frac{1}{2} \binom{2m + 2n}{m + n}.
$$

$$(16)$$

Note that the calculations can be verified rigorously and as a consequence we obtain a proof that the identity holds for all non-negative integers m, n. Since we remain in this particular case purely in the hypergeometric world, one could also use the classical toolbox described in [14]. Yet another (classical) proof consists in observing that the sum on the left-hand side of (16) can be rewritten as

$$\frac{1}{2}\left(\sum_{i=0}^{m}\binom{2m}{m+i}\binom{2n}{n-i}+\sum_{i=0}^{m}\binom{2m}{m-i}\binom{2n}{n+i}\right)=\frac{1}{2}\left(\sum_{i=0}^{2m}\binom{2m}{i}\binom{2n}{n+m-i}\right.$$

$$\left.+\binom{2m}{m}\binom{2n}{n}\right),$$

and then evaluating the sum on the right-hand side by means of the Chu–Vandermonde summation formula.

As a consequence, we arrive at the linear recurrence

$$\textsf{In[4]:= rec = rec/.}\sum_{i=0}^{m}\binom{2m}{i+m}\binom{2n}{i+n}\rightarrow\frac{1}{2}\binom{2m}{m}\binom{2n}{n}+\frac{1}{2}\binom{2m+2n}{m+n}$$

$$\textsf{Out[4]= SUM}[m+1]-4\textsf{SUM}[m]==-\frac{\binom{2m+2n}{m+n}}{1+m+n}\frac{1}{2}+\frac{(-1-m+2mn)\binom{2m}{m}\binom{2n}{n}}{2(m+1)(1+m+n)}$$

Now we can activate Sigma's recurrence solver with the function call

$$\textsf{In[5]:= recSol = SolveRecurrence[rec, SUM[m]]}$$

$$\textsf{Out[5]= }\left\{\left\{0,2^{2m}\right\},\left\{1,\frac{1}{4}\binom{2m}{m}\binom{2n}{n}+\frac{1}{4}\binom{2m+2n}{m+n}+2^{2m}\binom{2n}{n}\left(-\frac{1}{4}+\frac{1}{4}n\sum_{i=0}^{m}\frac{2^{-2i}\binom{2i}{i}}{i+n}\right)\right\}\right\}$$

This means that the first entry of the output is the solution of the homogeneous version of the recurrence, and the second entry is a solution of the recurrence itself. Hence, the general solution is

$$c\,2^{2m}+\frac{1}{4}\binom{2m}{m}\binom{2n}{n}+\frac{1}{4}\binom{2m+2n}{m+n}+2^{2m}\binom{2n}{n}\left(-\frac{1}{4}+\frac{1}{4}n\sum_{i=0}^{m}\frac{2^{-2i}\binom{2i}{i}}{i+n}\right),$$

(17)

where the constant c (free of m) can be freely chosen. We note that this solution can be easily verified by substituting it into **rec** computed in Out[4] and using the relations

$$S_m\binom{2m}{m}=\frac{2(2m+1)}{m+1}\binom{2m}{m},$$

$$S_m\binom{2m+2n}{m+n}=\frac{2(2m+2n+1)}{m+n+1}\binom{2m+2n}{m+n},$$

$$S_m\sum_{i=0}^{m}\frac{2^{-2i}\binom{2i}{i}}{i+n}=\sum_{i=0}^{m}\frac{2^{-2i}\binom{2i}{i}}{i+n}+\frac{2^{-2m}(2m+1)}{2(m+1)(1+m+n)}\binom{2m}{m}.$$

Step 3.   Looking at the initial value $S(n,0)=\binom{2n}{n}$, we conclude that the specialisation $c=\frac{1}{2}\binom{2n}{n}$ in (17) equals $S(n,m)$ for all $n\geq 0$ and $m=0$.

Summarising, we have found (together with a proof) the representation

$$S(n, m) = 2^{2m-2}\binom{2n}{n}n\sum_{i=0}^{m}\frac{2^{-2i}\binom{2i}{i}}{i+n} + 2^{2m-2}\binom{2n}{n} + \frac{1}{4}\binom{2m}{m}\binom{2n}{n} + \frac{1}{4}\binom{2m+2n}{m+n},$$

(18)

which holds for all non-negative integers m, n. This last calculation step can be also carried out within $\mathtt{Sigma}$, by making use of the function call

In[6]:= **FindLinearCombination[recSol, {0, {($\binom{2n}{n}$)}}, m, 1]**

Out[6]= $2^{2m-2}\binom{2n}{n}n\sum_{i=0}^{m}\frac{2^{-2i}\binom{2i}{i}}{i+n} + 2^{2m-2}\binom{2n}{n} + \frac{1}{4}\binom{2m}{m}\binom{2n}{n} + \frac{1}{4}\binom{2m+2n}{m+n}$

Strictly speaking, the above derivations contained one "human" (= non-automatic) step, namely at the point where we checked (15) and observed that this relation actually holds without the restriction $m \le n$. For the algorithmic "purist" we point out that it is also possible to set up the problem appropriately under the restriction $m > n$ (by splitting the double sum $S(n, m)$ into two parts) so that $\mathtt{Sigma}$ is applicable. Not surprisingly, $\mathtt{Sigma}$ finds (18) again.

In this article, we are particularly interested in the evaluation of $S(n, m)$ if one fixes the distance $r = n - m \ge 0$ (or $r = m - n \ge 0$). In order to find such a representation for the case $m \le n$, we manipulate the obtained sum

$$\sum_{i=0}^{m}\frac{2^{-2i}\binom{2i}{i}}{i+n} = \sum_{i=0}^{m}\frac{2^{-2i}\binom{2i}{i}}{i+r+m} := T(m, r) \tag{19}$$

in (18) further by applying once more $\mathtt{Sigma}$'s summation spiral (where r takes over the role of m).

Step 1.   Using $\mathtt{Sigma}$ (alternatively one could use the Paule and Schorn implementation [13] of Zeilberger's algorithm), we obtain the recurrence

$$2(m + r)T(m, r) + (-1 - 2m - 2r)T(m, r + 1) = \frac{2^{-2m}(2m + 1)\binom{2m}{m}}{2m + r + 1}.$$

Step 2.   Using $\mathtt{Sigma}$'s recurrence solver we obtain the general solution

$$d \frac{2^{2r}m\binom{2m}{m}}{\binom{2m+2r}{m+r}(m+r)} + \frac{2^{-2m}\binom{2m}{m}}{m+r} - \frac{2^{-2m+2r}(4m+1)\binom{2m}{m}^2}{2\binom{2m+2r}{m+r}(m+r)}$$

$$- \frac{2^{2r-2m}m\binom{2m}{m}}{\binom{2m+2r}{m+r}(m+r)}\sum_{i=0}^{r}\frac{2^{-2i}\binom{2m+2i}{m+i}}{2m+i},$$

where the constant d (free of r) can be freely chosen.

Step 3.   Looking at the initial value

$$T(m, 0) = \sum_{i=0}^{m} \frac{2^{-2i} \binom{2i}{i}}{i + m} = \frac{2^{2m-1}}{m \binom{2m}{m}} + \frac{2^{-2m-1} \binom{2m}{m}}{m},$$

which we simplified by another round of Sigma's summation spiral, we conclude that we have to specialise d to

$$d = \frac{2^{2m-1}}{m \binom{2m}{m}} + \frac{2^{-2m-1}(4m + 1) \binom{2m}{m}}{m}.$$

With this choice, we end up at the identity

$$T(m, r) = -\frac{2^{2r-2m} m \binom{2m}{m}}{\binom{2m+2r}{m+r}(m + r)} \sum_{i=0}^{r} \frac{2^{-2i} \binom{2i+2m}{i+m}}{i + 2m} + \frac{2^{-2m} \binom{2m}{m}}{m + r} + \frac{2^{2m+2r-1}}{\binom{2m+2r}{m+r}(m + r)},$$

being valid for all non-negative integers r, m. Finally, performing the substitution $r \to n - m$, we find the identity

$$T(m, n-m) = -\frac{2^{2n-4m} \binom{2m}{m}}{n \binom{2n}{n}} m \sum_{i=0}^{n-m} \frac{2^{-2i} \binom{2i+2m}{i+m}}{i + 2m} + \frac{2^{2n-1}}{n \binom{2n}{n}} + \frac{2^{-2m} \binom{2m}{m}}{n}, \qquad (20)$$

which holds for all non-negative integers n, m with $n \geq m$. By substituting this result into (18), we see that we have discovered *and* proven that

$$S(n, m) = -2^{-2m+2n-2} \binom{2m}{m} m \sum_{i=0}^{n-m} \frac{2^{-2i} \binom{2i+2m}{i+m}}{i + 2m}$$

$$+ 2^{2m-2} \binom{2n}{n} + \frac{1}{2} \binom{2m}{m} \binom{2n}{n} + \frac{1}{4} \binom{2m + 2n}{m + n} + 2^{2m+2n-3}, \qquad (21)$$

which is valid for all non-negative integers n, m with $n \geq m$. In a similar fashion, if $m \geq n$, we obtain

$$S(n, m) = 2^{2m-2n-2} \binom{2n}{n} n \sum_{i=0}^{m-n} \frac{2^{-2i} \binom{2i+2n}{i+n}}{i + 2n}$$

$$+ 2^{2m-2} \binom{2n}{n} + \frac{1}{4} \binom{2m}{m} \binom{2n}{n} + \frac{1}{4} \binom{2m + 2n}{m + n} + 2^{2m+2n-3}. \qquad (22)$$

We note that the interaction of the summation steps 1–3 is carried out at various places in a recursive manner. In order to free the user from all these mechanical but rather subtle calculation steps, the additional package `EvaluateMultiSums` [16] has been developed recently. It coordinates all these calculation steps cleverly and discovers identities as above completely automatically whenever such a simplification in terms of nested sums over hypergeometric products is possible. For instance, after loading the package

In[7]:= **<< EvaluateMultiSum.m**

> EvaluateMultiSums by Carsten Schneider © RISC-Linz

we can transform the sum (10) into the desired form by executing the function call

In[8]:= **res = EvaluateMultiSum[** $\binom{2n}{n+i}\binom{2m}{m+j}$, {{i, 0, j}, {j, 0, m}}, {m, n}, {0, 0}, {n, ∞}]

Out[8]= $\dfrac{(2n+1)2^{2m-3}(2n)!}{n^2((n-1)!)^2} \displaystyle\sum_{i=1}^{m} \dfrac{2^{-2i}\binom{2i}{i}}{1+i+n}$

$+ \dfrac{(4n+3)2^{2m-3}(2n)!}{n^2(n+1)((n-1)!)^2} + \dfrac{(3+4m+2n)\binom{2m}{m}(2n)!}{8n^2(1+m+n)((n-1)!)^2} + \dfrac{(2m+2n)!}{4n^2((n-1)!)^2((n+1)_m)^2}$

Here, `Sigma` uses the *Pochhammer symbol* $(\alpha)_m$ defined by

$$(\alpha)_m = \begin{cases} \alpha(\alpha+1)(\alpha+2)\cdots(\alpha+m-1), & \text{for } m > 0, \\ 1, & \text{for } m = 0, \qquad\qquad (23) \\ 1/(\alpha-1)(\alpha-2)(\alpha-3)\cdots(\alpha+m), & \text{for } m < 0, \end{cases}$$

which we shall also use later. The parameters m, n in the calculation above are bounded from below by $0, 0$ and from above by n, ∞, respectively. If one prefers a representation purely in terms of binomial coefficients, one may execute the following function calls:

In[9]:= **res = SigmaReduce[res, m, Tower →** {$\binom{2m}{m}$, $\binom{2n+2m}{n+m}$}];

In[10]:= **res = SigmaReduce[res, n, Tower →** {$\binom{2n}{n}$}];

Out[10]= $2^{2m-3}(2n+1)\binom{2n}{n}\displaystyle\sum_{i=1}^{m}\dfrac{2^{-2i}\binom{2i}{i}}{1+i+n} + \dfrac{(4n+3)2^{2m-3}\binom{2n}{n}}{n+1} + \dfrac{(3+4m+2n)\binom{2m}{m}\binom{2n}{n}}{8(1+m+n)} +$

$\dfrac{1}{4}\binom{2m+2n}{m+n}$

If one rewrites the arising sum manually by means of the function call below, one finally ends up exactly at the result given in (18):

In[11]:= **res = SigmaReduce[res, m, Tower →** {$\displaystyle\sum_{i=1}^{m}\dfrac{2^{-2i}\binom{2i}{i}}{i+n}$}]

Out[11]= $2^{2m-2}\binom{2n}{n}n\displaystyle\sum_{i=1}^{m}\dfrac{2^{-2i}\binom{2i}{i}}{i+n} + 2^{2m-1}\binom{2n}{n} + \dfrac{1}{4}\binom{2m}{m}\binom{2n}{n} + \dfrac{1}{4}\binom{2m+2n}{m+n}$

Analogously one can carry out these calculation steps to calculate the simplification given in (20) automatically.

Comparison with Lemma 1 reveals that (21) or (22) do not quite agree with the right-hand side of (5). For example, in order to prove that (21) is equivalent with (5), we would have to establish the identity

$$\frac{1}{8} \sum_{l=0}^{n-m} \binom{2m+2l}{m+l}\binom{2n-2l}{n-l} = 2^{-2m+2n-2}\binom{2m}{m} m \sum_{i=0}^{n-m} \frac{2^{-2i}\binom{2i+2m}{i+m}}{i+2m}.$$

This can, of course, be routinely achieved by using the Paule and Schorn implementation [13] of Zeilberger's algorithm. Alternatively, we may use our Sigma summation technology again. Let

$$T'(n,m) := \sum_{l=0}^{n-m} \binom{2m+2l}{m+l}\binom{2n-2l}{n-l}.$$

The above described summation spiral leads to

$$T'(n,m) = -2^{2m+1} n \binom{2n}{n} \sum_{i=0}^{m} \frac{2^{-2i}\binom{2i}{i}}{i+n} + 2\binom{2m}{m}\binom{2n}{n} + 2^{2m+2n}.$$

If this relation is substituted in (18), then we arrive exactly at the assertion of Lemma 1.

Clearly, the case where $m \geq n$ can be treated in a similar fashion. This finishes the algorithmic proof of Lemma 1.   □

4   Proof of Lemma 1 Using Complex Contour Integrals

In this section, we show how to prove Lemma 1 by making use of complex contour integrals. Before we can embark on the proof of the lemma, we need to establish several auxiliary evaluations of specific contour integrals.

Remark In order to avoid a confusion of the summation index i with the usual short notation for $\sqrt{-1}$, throughout this section we write $\mathbf{i}$ for $\sqrt{-1}$.

Lemma 5 *For all non-negative integers n, we have*

$$\frac{1}{2\pi\mathbf{i}} \int_C \frac{dz}{z^{n+1}(1-z)^{n+1}} \frac{1}{(1-2z)} = 2^{2n}, \tag{24}$$

where C is a contour close to 0, *which encircles* 0 *once in the positive direction.*

Proof Let I_1 denote the expression on the left-hand side of (24). We blow up the contour C so that it is sent to infinity. While doing this, we must pass over the poles $z = 1/2$ and $z = 1$ of the integrand. This must be compensated by taking the residues at these points into account. Since the integrand is of the order $O(z^{-2})$ as $|z| \to \infty$, the integral along the contour near infinity vanishes. Thus, we obtain

$$I_1 = - \mathrm{Res}_{z=1/2} \frac{1}{z^{n+1}(1-z)^{n+1}} \frac{1}{(1-2z)} - \mathrm{Res}_{z=1} \frac{1}{z^{n+1}(1-z)^{n+1}} \frac{1}{(1-2z)}$$

$$= 2^{2n+1} - \frac{1}{2\pi i} \int_C \frac{1}{(1+z)^{n+1}(1-(1+z))^{n+1}} \frac{1}{(1-2(1+z))} \, dz.$$

As the substitution $z \to -z$ shows, the last integral is identical with I_1. Thus, we have obtained an equation for I_1, from which we easily get the claimed result. $\qquad\square$

Lemma 6 *For all non-negative integers n and m, we have*

$$\frac{1}{(2\pi i)^2} \int_{C_1} \int_{C_2} \frac{1}{(u-t)} \frac{du}{u^{n+1}(1-u)^{n+1}} \frac{dt}{t^m(1-t)^m} = -\frac{1}{2}\binom{2n+2m}{n+m}, \qquad (25)$$

where C_1 and C_2 are contours close to 0, which encircle 0 once in the positive direction, and C_2 is entirely in the interior of C_1.

Proof We treat here the case where $n \geq m$. The other case can be disposed of completely analogously.

Let I_2 denote the expression on the left-hand side of (25). Clearly, interchange of u and t in the integrand does not change I_2. In that case however, we must also interchange the corresponding contours. Hence, I_2 is also equal to one half of the sum of the original expression and the one where u and t are exchanged, that is,

$$I_2 = \frac{1}{2(2\pi i)^2} \int_{C_1} \int_{C_2} \frac{1}{(u-t)} \frac{du}{u^{n+1}(1-u)^{n+1}} \frac{dt}{t^m(1-t)^m}$$

$$- \frac{1}{2(2\pi i)^2} \int_{C_2} \int_{C_1} \frac{1}{(u-t)} \frac{dt}{t^{n+1}(1-t)^{n+1}} \frac{du}{u^m(1-u)^m}.$$

We would like to put both expressions under one integral. In order to do so, we must blow up the contour C_2 in the second integral (the contour for t) so that it passes across C_1. When doing so, the term $u-t$ in the denominator will vanish, and so we shall collect a residue at $t = u$. This yields

$$I_2 = \frac{1}{2(2\pi i)^2} \int_{C_1} \int_{C_2} \frac{du\, dt}{(u-t)\big(u(1-u)\, t(1-t)\big)^{n+1}}$$

$$\cdot \left(\big(t(1-t)\big)^{n-m+1} - \big(u(1-u)\big)^{n-m+1} \right)$$

$$+ \frac{1}{2(2\pi i)} \int_{C_1} \mathrm{Res}_{t=u} \frac{1}{(u-t)} \frac{dt}{t^{n+1}(1-t)^{n+1}} \frac{du}{u^m(1-u)^m}$$

$$= \frac{1}{2\,(2\pi i)^2} \int_{C_1} \int_{C_2} \frac{du\,dt\,(u+t-1)}{\left(u(1-u)\,t(1-t)\right)^{n+1}} \sum_{\ell=0}^{n-m} \left(t(1-t)\right)^{\ell} \left(u(1-u)\right)^{n-m-\ell}$$

$$- \frac{1}{2\,(2\pi i)} \int_{C_1} \frac{du}{u^{n+m+1}(1-u)^{n+m+1}}$$

$$= \sum_{\ell=0}^{n-m} \frac{1}{2\,(2\pi i)^2} \int_{C_1} \int_{C_2} \frac{du\,dt}{u^{m+\ell}(1-u)^{m+\ell+1}\left(t(1-t)\right)^{n-\ell+1}}$$

$$- \sum_{\ell=0}^{n-m} \frac{1}{2\,(2\pi i)^2} \int_{C_1} \int_{C_2} \frac{du\,dt}{\left(u(1-u)\right)^{m+\ell+1} t^{n-\ell+1}(1-t)^{n-\ell}} - \frac{1}{2}\binom{2n+2m}{n+m}$$

$$= \frac{1}{2} \sum_{\ell=0}^{n-m} \binom{2n-2\ell}{n-\ell}\binom{2m+2\ell-1}{m+\ell} - \frac{1}{2} \sum_{\ell=0}^{n-m} \binom{2n-2\ell-1}{n-\ell-1}\binom{2m+2\ell}{m+\ell}$$

$$- \frac{1}{2}\binom{2n+2m}{n+m} = -\frac{1}{2}\binom{2n+2m}{n+m},$$

the last equality following from $\binom{2k}{k} = 2\binom{2k-1}{k}$. $\square$

Lemma 7 *For all non-negative integers n and m with $n \geq m$, we have*

$$\frac{1}{(2\pi i)^2} \int_{C_1} \int_{C_2} \frac{1}{(u-t)(1-2t)} \frac{du}{u^{n+1}(1-u)^{n+1}} \frac{dt}{t^m(1-t)^m}$$

$$= -\frac{1}{4} \sum_{\ell=0}^{n-m} \binom{2n-2\ell}{n-\ell}\binom{2m+2\ell}{m+\ell} - 3 \cdot 2^{2n+2m-2}, \qquad (26)$$

where C_1 and C_2 are contours close to 0, which encircle 0 once in the positive direction, and C_2 is entirely in the interior of C_1. The sum on the right-hand side must be interpreted according to

$$\sum_{k=M}^{N-1} Expr(k) = \begin{cases} \sum_{k=M}^{N-1} Expr(k), & N > M, \\ 0, & N = M, \\ -\sum_{k=N}^{M-1} Expr(k), & N < M. \end{cases} \qquad (27)$$

Proof Again, here we treat the case where $n \geq m$. The other case can be disposed of completely analogously.

Let I_3 denote the expression on the left-hand side of (26). We apply the same trick as in the proof of Lemma 6 and observe that I_3 is equal to one half of the sum of the original expression and the one where u and t are exchanged, plus the residue of the latter at $t = u$. To be precise,

$$I_3 = \frac{1}{2\,(2\pi\mathrm{i})^2} \int_{C_1} \int_{C_2} \frac{du\,dt}{(u-t)\,(1-2u)\,(1-2t)\,\big(u(1-u)\,t(1-t)\big)^{n+1}}$$

$$\cdot \Big((1-2u)\,\big(t(1-t)\big)^{n-m+1} - (1-2t)\,\big(u(1-u)\big)^{n-m+1}\Big)$$

$$+ \frac{1}{2\,(2\pi\mathrm{i})^2} \int_{C_1} \mathrm{Res}_{t=u} \frac{1}{(u-t)(1-2u)}\frac{1}{t^{n+1}(1-t)^{n+1}}\frac{du}{u^m(1-u)^m}$$

$$= \frac{1}{2\,(2\pi\mathrm{i})^2} \int_{C_1} \int_{C_2} \frac{du\,dt}{(u-t)\,(1-2t)\,\big(u(1-u)\,t(1-t)\big)^{n+1}}$$

$$\cdot \Big(\big(t(1-t)\big)^{n-m+1} - \big(u(1-u)\big)^{n-m+1}\Big)$$

$$- \frac{1}{(2\pi\mathrm{i})^2} \int_{C_1} \int_{C_2} \frac{du\,dt}{(1-2u)\,(1-2t)\,\big(u(1-u)\big)^m \big(t(1-t)\big)^{n+1}}$$

$$- \frac{1}{2\,(2\pi\mathrm{i})^2} \int_{C_1} \frac{1}{(1-2u)}\frac{du}{u^{n+m+1}(1-u)^{n+m+1}}$$

$$= \frac{1}{2\,(2\pi\mathrm{i})^2} \int_{C_1} \int_{C_2} \frac{du\,dt\,(u+t-1)}{(1-2t)\,\big(u(1-u)\,t(1-t)\big)^{n+1}} \sum_{\ell=0}^{n-m} \big(t(1-t)\big)^{\ell}\big(u(1-u)\big)^{n-m-\ell}$$

$$- 2^{2m-2+2n} - 2^{2n+2m-1}$$

$$= \sum_{\ell=0}^{n-m} \frac{1}{2\,(2\pi\mathrm{i})^2} \int_{C_1'} \int_{C_2'} \frac{du\,dt}{(1-2t)\,u^{m+\ell}(1-u)^{m+\ell+1}\big(t(1-t)\big)^{n-\ell+1}}$$

$$- \sum_{\ell=0}^{n-m} \frac{1}{2\,(2\pi\mathrm{i})^2} \int_{C_1'} \int_{C_2'} \frac{du\,dt}{(1-2t)\,\big(u(1-u)\big)^{m+\ell+1}t^{n-\ell+1}(1-t)^{n-\ell}}$$

$$- 3 \cdot 2^{2m+2n-2}$$

$$= \frac{1}{2} \sum_{\ell=0}^{n-m} \binom{2m+2\ell-1}{m+\ell} 2^{2n-2\ell}$$

$$- \frac{1}{2} \sum_{\ell=0}^{n-m} \binom{2m+2\ell}{m+\ell}\left(2^{2n-2\ell-1} + \frac{1}{2}\binom{2n-2\ell}{n-\ell}\right) - 3 \cdot 2^{2n+2m-2}$$

$$= -\frac{1}{4} \sum_{\ell=0}^{n-m} \binom{2n-2\ell}{n-\ell}\binom{2m+2\ell}{m+\ell} - 3 \cdot 2^{2n+2m-2},$$

which is again seen by observing $\binom{2k}{k} = 2\binom{2k-1}{k}$.   □

We are now in the position to prove Lemma 1 from Sect. 2.

Proof of Lemma 1 Using complex contour integrals, we may write

$$
\sum_{0 \le i \le j} \binom{2n}{n+i}\binom{2m}{m+j} = \sum_{0 \le i \le j} \binom{2n}{n-i}\binom{2m}{m-j}
$$

$$
= \sum_{0 \le i \le j} \frac{1}{(2\pi i)^2} \int_{C_1}\int_{C_2} \frac{(1+x)^{2n}}{x^{n-i+1}} \frac{(1+y)^{2m}}{y^{m-j+1}}\, dx\, dy
$$

$$
= \frac{1}{(2\pi i)^2} \int_{C_1}\int_{C_2} \frac{(1+x)^{2n}}{x^{n+1}} \frac{(1+y)^{2m}}{y^{m+1}} \frac{dx\, dy}{(1-xy)(1-y)},
$$

where C_1 and C_2 are contours close to 0, which encircle 0 once in the positive direction.

Now we do the substitutions $x = u/(1-u)$ and $y = t/(1-t)$, implying $dx = du/(1-u)^2$ and $dy = dt/(1-t)^2$. This leads to

$$
\sum_{0 \le i \le j} \binom{2n}{n+i}\binom{2m}{m+j}
$$

$$
= \frac{1}{(2\pi i)^2} \int_{C_1'}\int_{C_2'} \frac{du}{u^{n+1}(1-u)^{n+1}} \frac{dt}{t^{m+1}(1-t)^{m+1}} \frac{(1-u)(1-t)^2}{(1-u-t)(1-2t)}
$$

$$
= \frac{1}{2\,(2\pi i)^2} \int_{C_1'}\int_{C_2'} \frac{du}{u^{n+1}(1-u)^{n+1}} \frac{dt}{t^{m+1}(1-t)^{m+1}}
$$

$$
- \frac{1}{(2\pi i)^2} \int_{C_1'}\int_{C_2'} \frac{du}{u^{n+1}(1-u)^{n+1}} \frac{dt}{t^m(1-t)^m} \frac{1}{(1-2t)}
$$

$$
+ \frac{1}{2\,(2\pi i)^2} \int_{C_1'}\int_{C_2'} \frac{du}{u^{n+1}(1-u)^{n+1}} \frac{dt}{t^{m+1}(1-t)^{m+1}} \frac{1}{(1-2t)}
$$

$$
+ \frac{1}{2\,(2\pi i)^2} \int_{C_1'}\int_{C_2'} \frac{du}{u^{n+1}(1-u)^{n+1}} \frac{dt}{t^m(1-t)^m} \frac{1}{(1-u-t)}
$$

$$
+ \frac{1}{2\,(2\pi i)^2} \int_{C_1'}\int_{C_2'} \frac{du}{u^{n+1}(1-u)^{n+1}} \frac{dt}{t^m(1-t)^m} \frac{1}{(1-u-t)(1-2t)}. \tag{28}
$$

We now discuss the evaluation of the five integrals on the right-hand side one by one. First of all, we have

$$
\frac{1}{2\,(2\pi i)^2} \int_{C_1'}\int_{C_2'} \frac{du}{u^{n+1}(1-u)^{n+1}} \frac{dt}{t^{m+1}(1-t)^{m+1}}
$$

$$
= \frac{1}{2} \langle u^n \rangle (1-u)^{-n-1} \langle t^m \rangle (1-t)^{-m-1}
$$

$$
= \frac{1}{2}\binom{2n}{n}\binom{2m}{m}. \tag{29}
$$

Next, by Lemma 5, we have

$$\frac{1}{(2\pi i)^2} \int_{C_1'} \int_{C_2'} \frac{du}{u^{n+1}(1-u)^{n+1}} \frac{dt}{t^m(1-t)^m} \frac{1}{(1-2t)} = 2^{2m-2}\binom{2n}{n} \tag{30}$$

and

$$\frac{1}{2(2\pi i)^2} \int_{C_1'} \int_{C_2'} \frac{du}{u^{n+1}(1-u)^{n+1}} \frac{dt}{t^{m+1}(1-t)^{m+1}} \frac{1}{(1-2t)} = 2^{2m-1}\binom{2n}{n}. \tag{31}$$

In order to evaluate

$$I_4 := \frac{1}{2(2\pi i)^2} \int_{C_1'} \int_{C_2'} \frac{du}{u^{n+1}(1-u)^{n+1}} \frac{dt}{t^m(1-t)^m} \frac{1}{(1-u-t)},$$

we blow up the contour C_1' (the contour for u) so that it is sent to infinity. While doing this, we pass over the poles $u = 1 - t$ and $u = 1$ of the integrand. This must be compensated by taking the residues at these points into account. Since the integrand is of the order $O(u^{-2})$ as $|u| \to \infty$, the integral along the contour near infinity vanishes. Thus, we obtain

$$
\begin{aligned}
I_4 &= -\frac{1}{2(2\pi i)} \int_{C_2'} \operatorname{Res}_{u=1-t} \frac{1}{u^{n+1}(1-u)^{n+1}} \frac{dt}{t^m(1-t)^m} \frac{1}{(1-u-t)} \\
&\quad - \frac{1}{2(2\pi i)} \int_{C_2'} \operatorname{Res}_{u=1} \frac{1}{u^{n+1}(1-u)^{n+1}} \frac{dt}{t^m(1-t)^m} \frac{1}{(1-u-t)} \\
&= \frac{1}{2(2\pi i)} \int_{C_2'} \frac{dt}{t^{n+m+1}(1-t)^{n+m+1}} \\
&\quad - \frac{1}{2(2\pi i)^2} \int_{C_1'} \int_{C_2'} \frac{du}{(1+u)^{n+1}(1-(1+u))^{n+1}} \frac{dt}{t^m(1-t)^m} \frac{1}{(1-(1+u)-t)} \\
&= \frac{1}{2}\binom{2n+2m}{n+m} - \frac{1}{4}\binom{2n+2m}{n+m} = \frac{1}{4}\binom{2n+2m}{n+m}, \tag{32}
\end{aligned}
$$

which is seen by performing the substitution $u \to -u$ in the second expression in the next-to-last line and applying Lemma 6.

Finally, in order to evaluate

$$I_5 := \frac{1}{2(2\pi i)^2} \int_{C_1'} \int_{C_2'} \frac{du}{u^{n+1}(1-u)^{n+1}} \frac{dt}{t^m(1-t)^m} \frac{1}{(1-u-t)(1-2t)} \tag{33}$$

we again blow up the contour C_1 so that it is sent to infinity. While doing this, we pass over the poles $u = 1-t$ and $u = 1$ of the integrand. This must be compensated by taking the residues at these points into account. Since the integrand is of the order

$O(u^{-2})$ as $|u| \to \infty$, the integral along the contour near infinity vanishes. Thus, we obtain

$$
\begin{aligned}
I_5 &= -\frac{1}{2\,(2\pi i)} \int_{C_2'} \operatorname{Res}_{u=1-t} \frac{1}{u^{n+1}(1-u)^{n+1}} \frac{dt}{t^m(1-t)^m} \frac{1}{(1-u-t)(1-2t)} \\
&\quad - \frac{1}{2\,(2\pi i)} \int_{C_2'} \operatorname{Res}_{u=1} \frac{1}{u^{n+1}(1-u)^{n+1}} \frac{dt}{t^m(1-t)^m} \frac{1}{(1-u-t)(1-2t)} \\
&= \frac{1}{2\,(2\pi i)} \int_{C_2'} \frac{dt}{t^{n+m+1}(1-t)^{n+m+1}} \frac{1}{(1-2t)} \\
&\quad - \frac{1}{2(2\pi i)^2} \int_{C_1'} \int_{C_2'} \frac{du}{(1+u)^{n+1}(1-(1+u))^{n+1}} \frac{dt}{t^m(1-t)^m} \frac{1}{(1-(1+u)-t)(1-2t)} \\
&= 2^{2n+2m-1} - \frac{1}{8} \sum_{\ell=0}^{n-m} \binom{2n-2\ell}{n-\ell} \binom{2m+2\ell}{m+\ell} - 3 \cdot 2^{2n+2m-3}, \quad\quad (34)
\end{aligned}
$$

which is seen by applying Lemma 5 to the first expression in the next-to-last line, performing the substitution $u \to -u$ in the second expression, and applying Lemma 7. By combining (28)–(34) and simplifying, we obtain the right-hand side of (5). $\qquad\square$

5 Main Results

This section contains our main results concerning double sums of the form

$$
\sum_{0 \le i \le j} i^s j^t \binom{2n}{n+i} \binom{2m}{m+j}.
$$

If both s and t are even, then we are only able to provide a result in the special case where $m = n$. (It would also be possible to provide a similar result for the case where the difference $n - m$ is some fixed integer.) The reason is that the identity in Lemma 1, on which an evaluation of the above sum will have to be based, contains the sum over ℓ that cannot be simplified if n and m are generic. Proposition 1 restricts attention to this special case. On the other hand, if s and t are not both even, then it is possible to provide a general result for the above double sum without any restriction on n and m. The evaluations are then based on Lemmas 2–4, and the corresponding results are presented in Proposition 2. It should be noted that, for the three cases of parity of s and t that are treated in both propositions, it is not true that Proposition 1 is a direct consequence of Proposition 2 as the assertions in Proposition 1 are more refined.

Proposition 1 *For all non-negative integers* s, t, k *and* n, *we have*

$$
\sum_{0 \le i \le j \le n} i^s j^t \binom{2n}{n+i}\binom{2n}{n+j} = \frac{P_{s,t}^{(1)}(n)}{(4n-1)(4n-3)\cdots(4n-2S-2T+1)}\binom{4n}{2n}
$$

$$
+ \frac{P_{s,t}^{(2)}(n)}{(2n-1)(2n-3)\cdots(2n-2\lfloor(S+T)/2\rfloor+1)}\binom{2n}{n}^2 + P_{s,t}^{(3)}(n)\cdot 4^n\binom{2n}{n} + P_{s,t}^{(4)}(n)\cdot 16^n,
\tag{35}
$$

where the $P_{s,t}^{(i)}(n)$, $i = 1, 2, 3, 4$, *are polynomials in* n, $S = \lfloor s/2 \rfloor$ *and* $T = \lfloor t/2 \rfloor$.
More specifically,

1. *if* s *and* t *are even, then, as polynomials in* n, $P_{s,t}^{(1)}(n)$ *is of degree at most* $3S+3T$,
   $P_{s,t}^{(2)}(n)$ *is of degree at most* $2S+2T+\lfloor(S+T)/2\rfloor$, $P_{s,t}^{(3)}(n)$ *is identically zero if*
   $s \ne 0$, $P_{0,t}^{(3)}(n)$ *is of degree at most* $2T$, *and* $P_{s,t}^{(4)}(n)$ *is of degree at most* $2S+2T$;
2. *if* s *is odd and* t *is even, then, as polynomials in* n, $P_{s,t}^{(1)}(n)$ *is of degree at most*
   $3S + 3T + 1$, $P_{s,t}^{(2)}(n)$ *is of degree at most* $2S + 2T + 1 + \lfloor(S+T)/2\rfloor$, $P_{s,t}^{(3)}(n)$
   is of degree at most $2S + 2T + 1$, *and* $P_{s,t}^{(4)}(n)$ *is identically zero;*
3. *if* s *is even and* t *is odd, then, as polynomials in* n, $P_{s,t}^{(1)}(n)$ *is of degree at most*
   $3S + 3T + 1$, $P_{s,t}^{(2)}(n)$ *is of degree at most* $2S + 2T + 1 + \lfloor(S+T)/2\rfloor$, *and*
   $P_{s,t}^{(3)}(n)$ *and* $P_{s,t}^{(4)}(n)$ *are identically zero;*
4. *if* s *and* t *are odd, then, as polynomials in* n, $P_{s,t}^{(1)}(n)$ *is of degree at most* $3S +$
   $3T + 2$, $P_{s,t}^{(2)}(n)$ *is of degree at most* $2S + 2T + 2 + \lfloor(S+T)/2\rfloor$, *and* $P_{s,t}^{(3)}(n)$
   and $P_{s,t}^{(4)}(n)$ *are identically zero.*

Remark 2 As the proof below shows, explicit formulae for the polynomials $P_{s,t}^{(i)}(n)$,
$i = 1, 2, 3, 4$, can be given that involve the coefficients $c_{a,S}(n)$ and $c_{b,T}(n)$ in (36)
and (37), for which an explicit formula exists as well, see Lemma 10. Admittedly,
these explicit formulae are somewhat cumbersome, and therefore we refrain from
presenting them in full here.

Proof of Proposition 1 We start with the case in which both s and t are even. With
the notation of the proposition, we have $s = 2S$ and $t = 2T$. We write

$$
i^{2S} = \sum_{a=0}^{S} c_{a,S}(n)\left(n^2 - i^2\right)\left((n-1)^2 - i^2\right)\cdots\left((n-a+1)^2 - i^2\right),
\tag{36}
$$

where $c_{a,S}(n)$ is a polynomial in n of degree $2S - 2a$, $a = 0, 1, \ldots, S$, and

$$
j^{2T} = \sum_{b=0}^{T} c_{b,T}(n)\left(n^2 - j^2\right)\left((n-1)^2 - j^2\right)\cdots\left((n-b+1)^2 - j^2\right),
\tag{37}
$$

where $c_{b,T}(n)$ is a polynomial in n of degree $2T - 2b$, $b = 0, 1, \ldots, T$. It should be noted that $c_{S,S}(n) = (-1)^S$ and $c_{T,T}(m) = (-1)^T$. For an explicit formula for the coefficients $c_{a,S}(n)$ see Lemma 10.

If we use the expansions (36) and (37) on the left-hand side of (35), then we obtain the expression

$$\sum_{a=0}^{S}\sum_{b=0}^{T} c_{a,S}(n)\, c_{b,T}(n)\left((2n-2a+1)_{2a}\,(2n-2b+1)_{2b} \sum_{0 \leq i \leq j} \binom{2n-2a}{n+i-a}\binom{2n-2b}{n+j-b}\right)$$

$$= \sum_{a=0}^{S}\sum_{b=0}^{T} c_{a,S}(n)\, c_{b,T}(n)\left((2n-2a+1)_{2a}\,(2n-2b+1)_{2b}\right.$$

$$\cdot\left(2^{4n-2a-2b-3} + \frac{1}{4}\binom{4n-2a-2b}{2n-a-b} + \frac{1}{2}\binom{2n-2a}{n-a}\binom{2n-2b}{n-b}\right.$$

$$\left.\left.+2^{2n-2b-2}\binom{2n-2a}{n-a} - \frac{1}{8}\sum_{\ell=0}^{b-a}\binom{2n-2a-2\ell}{n-a-\ell}\binom{2n-2b+2\ell}{n-b+\ell}\right)\right),$$

due to Lemma 1 with n replaced by $n - a$ and $m = n - b$. This expression can be further simplified by noting that

$$\sum_{a=0}^{S} c_{a,S}(n)\,(2n-2a+1)_{2a}\binom{2n-2a}{n-a} = 0^{2S}\binom{2n}{n}, \tag{38}$$

which is equivalent to the expansion (36) for $i = 0$. Thus, we obtain

$$\frac{1}{2}0^{2S+2T}\binom{2n}{n}^2 + 0^{2S}\binom{2n}{n}\sum_{b=0}^{T} c_{b,T}(n)\,2^{2n-2b-2}\,(2n-2b+1)_{2b}$$

$$+\sum_{a=0}^{S}\sum_{b=0}^{T} c_{a,S}(n)\, c_{b,T}(n)\left((2n-2a+1)_{2a}\,(2n-2b+1)_{2b}\right.$$

$$\left.\cdot\left(2^{4n-2a-2b-3} + \frac{1}{4}\binom{4n-2a-2b}{2n-a-b} - \frac{1}{8}\sum_{\ell=0}^{b-a}\binom{2n-2a-2\ell}{n-a-\ell}\binom{2n-2b+2\ell}{n-b+\ell}\right)\right).$$

Taking into account the properties of $c_{a,S}(n)$ and $c_{b,T}(n)$, from this expression it is clear that $P_{s,t}^{(4)}(n)$, the coefficient of $2^{4n} = 16^n$, has degree at most $2S + 2T$ as a polynomial in n. It is furthermore obvious that, due to the term $0^{2S} = 0^s$, the polynomial $P_{s,t}^{(3)}(n)$, the coefficient of $2^{2n}\binom{2n}{n} = 4^n\binom{2n}{n}$, vanishes for $s \neq 0$, while its degree is at most $2T$ if $s = 0$.

In order to verify the claim about $P_{s,t}^{(1)}(n)$, the coefficient of $\binom{4n}{2n}$, we write

$$c_{a,S}(n)\, c_{b,T}(n)\, (2n - 2a + 1)_{2a}\, (2n - 2b + 1)_{2b} \binom{4n - 2a - 2b}{2n - a - b}$$

$$= c_{a,S}(n)\, c_{b,T}(n)\, \frac{(2n - 2a + 1)_{2a}\, (2n - 2b + 1)_{2b}\, (2n - a - b + 1)_{a+b}^2}{(4n - 2a - 2b + 1)_{2a+2b}} \binom{4n}{2n}.$$

It is easy to see that $(2n - a - b + 1)_{a+b}$ divides numerator and denominator. After this division, the denominator becomes

$$2^{a+b}(4n - 1)(4n - 3) \cdots (4n - 2a - 2b + 1),$$

that is, part of the denominator below $P^{(1)}(n)$ in (35). The terms which are missing are

$$(4n - 2a - 2b - 1)(4n - 2a - 2b - 3) \cdots (4n - 2S - 2T + 1).$$

Thus, if we put everything on the denominator

$$(4n - 1)(4n - 3) \cdots (4n - 2S - 2T + 1),$$

then we see that the numerator of the coefficient of $\binom{4n}{2n}$ has degree at most

$$(2S - 2a) + (2T - 2b) + 2a + 2b + 2(a + b) + (S + T - a - b) - (a + b) = 3S + 3T,$$

as desired.

Finally, we turn our attention to $P_{s,t}^{(2)}(n)$, the coefficient of $\binom{2n}{n}^2$. We have

$$c_{a,S}(n)\, c_{b,T}(n)\, (2n - 2a + 1)_{2a}\, (2n - 2b + 1)_{2b} \binom{2n - 2a - 2\ell}{n - a - \ell} \binom{2n - 2b + 2\ell}{n - b + \ell}$$

$$= c_{a,S}(n)\, c_{b,T}(n)\, \frac{(n - a - \ell + 1)_{a+\ell}^2\, (n - b + \ell + 1)_{b-\ell}^2\, (2n - 2b + 1)_{2\ell}}{(2n - 2a - 2\ell + 1)_{2\ell}} \binom{2n}{n}^2 \tag{39a}$$

$$= c_{a,S}(n)\, c_{b,T}(n)$$

$$\times \frac{(n - a - \ell + 1)_{a+\ell}^2\, (n - b + \ell + 1)_{b-\ell}^2\, (2n - 2b + 1)_{2b-2a-2\ell}}{(2n - 2b + 2\ell + 1)_{2b-2a-2\ell}} \binom{2n}{n}^2. \tag{39b}$$

Let us assume $a \le b$, in which case we need to consider non-negative indices ℓ. (If $a > b$, then, according to the convention (27), we have to consider negative ℓ. Using the definition (23) of the Pochhammer symbol for negative indices, the arguments would be completely analogous.) We make the further assumption that $\ell \leq \frac{1}{2}(b-a)$ and use expression (39a). (If $\ell > \frac{1}{2}(b-a)$, then analogous arguments work starting from expression (39b).)

It is easy to see that $(n - a - \ell + 1)_\ell$ divides numerator and denominator (as polynomials in n) of the prefactor in (39a). Second, the (remaining) factor $2^{2\ell}(n - a - \ell + \frac{1}{2})_\ell$ in the denominator and the factor $(2n - 2b + 1)_{2\ell}$ in the numerator do not have common factors for $\ell \leq \frac{1}{2}(b-a)$. The denominator is a factor of the denominator below $P_{s,t}^{(2)}(n)$ in (35). If in (39a) we extend denominator and numerator by the "missing" factor

$$(n - \lfloor (S+T)/2 \rfloor + \frac{1}{2})_{\lfloor (T+S)/2 \rfloor - \lfloor b+a \rfloor/2} (n - a + \frac{1}{2})_a,$$

then, due to the properties of $c_{a,S}(n)$ and $c_{b,T}(n)$, the numerator polynomial is of degree at most

$$(2S - 2a) + (2T - 2b) + 2(a + \ell) + 2(b - \ell) + 2\ell - \ell$$
$$+ \lfloor (T+S)/2 \rfloor - \lfloor (b+a)/2 \rfloor + a$$
$$= 2S + 2T + \ell + \lfloor (T+S)/2 \rfloor - \lfloor (b+a)/2 \rfloor + a$$
$$\leq 2S + 2T + \lfloor (b-a)/2 \rfloor + \lfloor (T+S)/2 \rfloor - \lfloor (b+a)/2 \rfloor + a$$
$$\leq 2S + 2T + \lfloor (S+T)/2 \rfloor,$$

as desired.

For the other cases, namely (s, t) being (odd, even), (even, odd), respectively (odd, odd), we proceed in the same way. That is, we apply the expansions (36) and (37) on the left-hand side of (35). Then, however, instead of Lemma 1, we apply Lemma 2, Lemma 3, and Lemma 4, respectively. The remaining arguments are completely analogous to those from the case of (s, t) being (even,even) (and, in fact, much simpler since the right-hand sides of the identities in Lemmas 2–4 are simpler than the one in Lemma 1). $\qquad\square$

Proposition 2 *Let s, t and n, m be non-negative integers.*
If s and t are not both even or both odd, then

$$\sum_{0 \leq i \leq j} i^s j^t \binom{2n}{n+i} \binom{2m}{m+j}$$

$$= \frac{Q_{s,t}^{(1)}(n, m)}{(2n + 2m - 1)(2n + 2m - 3) \cdots (2n + 2m - 2S - 2T + 1)} \binom{2n + 2m}{n + m}$$

$$+ \frac{Q_{s,t}^{(2)}(n, m)}{(n + m)(n + m - 1)(n + m - 2) \cdots (n + m - S - T)} \binom{2n}{n} \binom{2m}{m}$$

$$+ Q_{s,t}^{(3)}(n, m) \cdot 4^m \binom{2n}{n}, \qquad (40)$$

where the $Q_{s,t}^{(i)}(n, m)$, $i = 1, 2, 3$, are polynomials in n and m, $S = \lfloor s/2 \rfloor$ and $T = \lfloor t/2 \rfloor$. More specifically,

1. *if s is odd and t is even, then, as polynomials in n and m, $Q_{s,t}^{(1)}(n, m)$ is of degree at most $3S+3T+1$, $Q_{s,t}^{(2)}(n, m)$ is of degree at most $3S+3T+2$, and $Q_{s,t}^{(3)}(n, m)$ is of degree at most $2S + 2T + 1$;*
2. *if s is even and t is odd, then, as polynomials in n and m, $Q_{s,t}^{(1)}(n, m)$ is of degree at most $3S+3T+1$, $Q_{s,t}^{(2)}(n, m)$ is of degree at most $3S+3T+2$, and $Q_{s,t}^{(3)}(n, m)$ is identically zero.*

If s and t are odd, then

$$
\sum_{0 \leq i \leq j} i^s j^t \binom{2n}{n+i}\binom{2m}{m+j}
$$

$$
= \frac{Q_{s,t}^{(1)}(n, m)}{(2n + 2m - 1)(2n + 2m - 3)\cdots(2n + 2m - 2S - 2T - 1)}\binom{2n + 2m}{n + m}
$$

$$
+ \frac{Q_{s,t}^{(2)}(n, m)}{(n + m)(n + m - 1)(n + m - 2)\cdots(n + m - S - T)}\binom{2n}{n}\binom{2m}{m},
$$
(41)

where $S = \lfloor s/2 \rfloor$ and $T = \lfloor t/2 \rfloor$, and, as polynomials in n and m, $Q_{s,t}^{(1)}(n, m)$ and $Q_{s,t}^{(2)}(n, m)$ are of degree at most $3S + 3T + 3$.

The proof of this proposition is completely analogous to the proof of Proposition 1 and is therefore left to the reader. Also here (cf. Remark 2), explicit formulae for the polynomials $Q_{s,t}^{(i)}(n, m)$, $i = 1, 2, 3$, can be given that involve coefficients $c_{a,S}(n)$ and $c_{b,T}(m)$ for which an explicit formula exists (see Lemma 10).

6 Some More Auxiliary Results

In this section we derive some single sum evaluations that we shall need in the proofs in Sect. 7.

Lemma 8 *For all non-negative integers n and k, we have*

$$
\sum_{j=1}^{n} j^{2k}\binom{2n}{n+j} = -\frac{0^{2k}}{2}\binom{2n}{n} + 4^n \sum_{b=0}^{k} c_{b,k}(n)\,(2n - 2b + 1)_{2b}\,2^{-2b-1}, \quad (42)
$$

and

$$\sum_{j=1}^{n} j^{2k+1}\binom{2n}{n+j} = \frac{1}{2}\binom{2n}{n}\sum_{b=0}^{k} c_{b,k}(n)\,(n-b)_{b+1}\,(n-b+1)_b\,, \tag{43}$$

where the coefficients $c_{b,k}(n)$ are defined in (37) (with explicit formula provided in Lemma 10).

Proof We use the expansion (37) with $T = k$ on the left-hand side of (42). This gives

$$\sum_{j=1}^{n} j^{2k}\binom{2n}{n+j} = \sum_{j=1}^{n}\sum_{b=0}^{k} c_{b,k}(n)\,(2n-2b+1)_{2b}\binom{2n-2b}{n+j-b}$$

$$= \sum_{b=0}^{k} c_{b,k}(n)\,(2n-2b+1)_{2b}\left(2^{2n-2b-1} - \frac{1}{2}\binom{2n-2b}{n-b}\right)$$

$$= -\frac{0^{2k}}{2}\binom{2n}{n} + \sum_{b=0}^{k} c_{b,k}(n)\,(2n-2b+1)_{2b}\,2^{2n-2b-1},$$

where we used (37) with $T = k$ and $j = 0$ in the last line. This is exactly the right-hand side of (42).

Now we do the same on the left-hand side of (43). This leads to

$$\sum_{j=1}^{n} j^{2k+1}\binom{2n}{n+j} = \sum_{j=1}^{n} j \cdot \sum_{b=0}^{k} c_{b,k}(n)\,(2n-2b+1)_{2b}\binom{2n-2b}{n+j-b}$$

$$= \sum_{b=0}^{k} c_{b,k}(n)\,(2n-2b)_{2b+1}\sum_{j=1}^{n}\left(\binom{2n-2b-1}{n+j-b-1} - \frac{1}{2}\binom{2n-2b}{n+j-b}\right)$$

$$= \sum_{b=0}^{k} c_{b,k}(n)\,(2n-2b)_{2b+1}\left(2^{2n-2b-2} - \frac{1}{2}2^{2n-2b-1} + \frac{1}{4}\binom{2n-2b}{n-b}\right)$$

$$= \frac{1}{2}\sum_{b=0}^{k} c_{b,k}(n)\,(n-b)_{b+1}\,(n-b+1)_b\binom{2n}{n}.$$

This is exactly the right-hand side of (43).   □

Lemma 9 *For all non-negative integers n and h, k, we have*

$$\sum_{j\geq 1} j^{2k} \binom{2n}{n+j}\binom{2m}{m+j}$$

$$= -\frac{0^{2k}}{2}\binom{2n}{n}\binom{2m}{m} + \frac{1}{2}\sum_{b=0}^{k} c_{b,k}(n)\,(2n-2b+1)_{2b}\binom{2n+2m-2b}{n+m-b} \tag{44}$$

and

$$\sum_{j\geq 1} j^{2h+2k+1}\binom{2n}{n+j}\binom{2m}{m+j}$$

$$= \sum_{a=0}^{h}\sum_{b=0}^{k} c_{a,h}(n)\,c_{b,k}(m)\,(n-a+1)_a^2\,(m-b+1)_b^2\,\frac{(n-a)(m-b)}{2(n+m-a-b)}\binom{2n}{n}\binom{2m}{m}, \tag{45}$$

where the coefficients $c_{a,h}(n)$ and $c_{b,k}(m)$ are defined in (36) (with explicit formula provided in Lemma 10).

Proof We start by using the expansion (37) with $T = k$ on the left-hand side of (44). This gives

$$\sum_{j\geq 1} j^{2k}\binom{2n}{n+j}\binom{2m}{m+j} = \sum_{j\geq 1}\sum_{b=0}^{k} c_{b,k}(n)\,(2n-2b+1)_{2b}\binom{2n-2b}{n+j-b}\binom{2m}{m+j}. \tag{46}$$

We have

$$\sum_{j\geq 1}\binom{2n-2b}{n+j-b}\binom{2m}{m+j} = \sum_{j\leq -1}\binom{2n-2b}{n+j-b}\binom{2m}{m+j}$$

and hence

$$\sum_{j\geq 1}\binom{2n-2b}{n+j-b}\binom{2m}{m+j} = -\frac{1}{2}\binom{2n-2b}{n-b}\binom{2m}{m} + \frac{1}{2}\sum_{j}\binom{2n-2b}{n+j-b}\binom{2m}{m+j}$$

$$= -\frac{1}{2}\binom{2n-2b}{n-b}\binom{2m}{m} + \frac{1}{2}\binom{2n+2m-2b}{n+m-b},$$

due to the Chu–Vandermonde summation. We substitute this back into (46) and obtain

$$
\sum_{j\geq 1} j^{2k}\binom{2n}{n+j}\binom{2m}{m+j} = \sum_{b=0}^{k} c_{b,k}(n)\,(2n-2b+1)_{2b}
$$

$$
\cdot\left(-\frac{1}{2}\binom{2n-2b}{n-b}\binom{2m}{m}+\frac{1}{2}\binom{2n+2m-2b}{n+m-b}\right)
$$

$$
= -\frac{0^{2k}}{2}\binom{2n}{n}\binom{2m}{m}+\frac{1}{2}\sum_{b=0}^{k} c_{b,k}(n)\,(2n-2b+1)_{2b}\binom{2n+2m-2b}{n+m-b},
$$

where we used (37) with $T=k$ and $j=0$ in the last line.

In order to establish (45), we write $j^{2h+2k+1}=j\cdot j^{2h}\cdot j^{2k}$ and use (37) with $T=h$ and with $T=k$. This leads to

$$
\sum_{j=1}^{n} j^{2h+2k+1}\binom{2n}{n+j}\binom{2m}{m+j}
$$

$$
= \sum_{j\geq 1} j\cdot\sum_{a=0}^{h}\sum_{b=0}^{k} c_{a,h}(n)\,c_{b,k}(m)\,(2n-2a+1)_{2a}\,(2m-2b+1)_{2b}\binom{2n-2a}{n+j-a}\binom{2m-2b}{m+j-b}.
$$

$$
(47)
$$

Using the standard hypergeometric notation

$$
{}_pF_q\!\left[\begin{matrix} a_1,\ldots,a_p \\ b_1,\ldots,b_q \end{matrix};z\right]=\sum_{m=0}^{\infty}\frac{(a_1)_m\cdots(a_p)_m}{m!\,(b_1)_m\cdots(b_q)_m}z^m\,,
$$

where the Pochhammer symbol $(\alpha)_m$ is defined in (23), we have

$$
\sum_{j\geq 1} j\binom{2n-2a}{n+j-a}\binom{2m-2b}{m+j-b}
$$

$$
= \binom{2n-2a}{n-a+1}\binom{2m-2b}{m-b+1}{}_3F_2\!\left[\begin{matrix} 2,-n+a+1,-m+b+1 \\ n-a+2,m-b+2 \end{matrix};1\right].
$$

This $_3F_2$-series can be evaluated by means of (the terminating version of) Dixon's summation (see [19, Appendix (III.9)])

$$
{}_3F_2\!\left[\begin{matrix} A,B,-N \\ 1+A-B,1+A+N \end{matrix};1\right]=\frac{(1+A)_N\,(1+\frac{A}{2}-B)_N}{(1+\frac{A}{2})_N\,(1+A-B)_N},
$$

where N is a non-negative integer. Indeed, if we choose $A = 2$, $B = -n + a + 1$, and $N = m - b - 1$ in this summation formula, then we obtain

$$\sum_{j=1}^{n} j \binom{2n - 2a}{n + j - a} \binom{2m - 2b}{m + j - b} = \frac{(n - a + 1)(m - b + 1)}{2(n + m - a - b)} \binom{2n - 2a}{n - a + 1} \binom{2m - 2b}{m - b + 1}.$$

If this is substituted back in (47), then we obtain the right-hand side of (45) after little manipulation. □

For the proof of our theorems it is not necessary to have an explicit formula for the coefficients $c_{a,S}(n)$ in the expansion (36)—the coefficients that appeared in the proof of Proposition 1, and in Lemmas 8 and 9—at our disposal. However, it is still of intrinsic interest to provide such an explicit formula.

Lemma 10 *The coefficient $c_{a,S}(n)$ in the expansion (36) is given by*

$$c_{a,S}(n) = \sum_{r=0}^{a} \frac{2(-1)^{a+r}(n - r)^{2S+1}}{r!\,(a - r)!\,(2n - a - r)_{a+1}}.$$ (48)

Proof Substituting $n - b$ for i in (36), $b = 0, 1, \ldots, S$, we obtain the triangular system of linear equations for the $c_{a,S}$'s

$$(n - b)^{2S} = \sum_{a=0}^{b} c_{a,S}(n)\,(2n - a - b + 1)_a\,(b - a + 1)_a, \quad b = 0, 1, \ldots, S. \quad (49)$$

Now, using the inversion formula of Gould and Hsu [8] (in the statement [12, Eq. (1.1)] of the formula put $n = b$, $k = a$, $l = r$, $a_j = 2n - j$, $b_j = -1$, in this order), we see that the matrix

$$\big((2n - a - b + 1)_a\,(b - a + 1)_a\big)_{b,a \geq 0}$$

is inverse to the matrix

$$\left(\frac{(-1)^{a+r}(2n - 2r)}{r!\,(a - r)!\,(2n - a - r)_{a+1}}\right)_{a,r \geq 0}.$$

Hence, if the system (49) is inverted, that is, the coefficients $c_{a,S}(n)$, $a = 0, 1, \ldots, S$, are expressed in terms of the $(n - b)^{2S}$, $b = 0, 1, \ldots, S$, then the connection coefficients are given by the latter matrix. This proves (48). □

7 Summation Formulae for Binomial Double Sums Involving Absolute Values

In this section we present the implications of Propositions 1 and 2 on sums of the form (2) and (3) with $\beta = 1$. As we point out in Remark 3(1) below, it would also be possible to derive similar theorems for arbitrary β. (An example of an evaluation with $\beta = 3$ is given in (66).)

We start with results for double sums of the form (3) with even k (and $\beta = 1$). First, we also let $m = n$. The corresponding evaluations are given in Theorem 1 below. In Theorem 2 we address these same double sums for generic n and m. Similarly to Proposition 2, for that case we have results only if s and t are not both even.

Theorem 1 *Let s, t, k and n be non-negative integers.*
If s and t are even, then

$$\sum_{-n \le i, j \le n} \left| i^s j^t (j^{2k} - i^{2k}) \right| \binom{2n}{n+i} \binom{2n}{n+j}$$

$$= \frac{U^{(2)}_{s,t,k}(n)}{(2n-1)(2n-3) \cdots (2n - 2\lfloor (S+T+k)/2 \rfloor + 1)} \binom{2n}{n}^2, \quad (50)$$

where $U^{(2)}_{s,t,k}(n)$ is of degree at most $2S + 2T + 2k + \lfloor (S+T+k)/2 \rfloor$.
If s and t are both odd, then

$$\sum_{-n \le i, j \le n} \left| i^s j^t (j^{2k} - i^{2k}) \right| \binom{2n}{n+i} \binom{2n}{n+j}$$

$$= \frac{U^{(2)}_{s,t,k}(n)}{(2n-1)(2n-3) \cdots (2n - 2\lceil (S+T+k)/2 \rceil + 1)} \binom{2n}{n}^2, \quad (51)$$

where $U^{(2)}_{s,t,k}(n)$ is of degree at most $2S + 2T + 2k + \lceil (S+T+k)/2 \rceil$.
If s and t have different parity, then

$$\sum_{-n \le i, j \le n} \left| i^s j^t (j^{2k} - i^{2k}) \right| \binom{2n}{n+i} \binom{2n}{n+j}$$

$$= \frac{U^{(1)}_{s,t,k}(n)}{(4n-1)(4n-3) \cdots (4n - 2S - 2T - 2k + 1)} \binom{4n}{2n} + U^{(3)}_{s,t,k}(n) \cdot 4^n \binom{2n}{n}, \quad (52)$$

where $U^{(1)}_{s,t,k}(n)$ and $U^{(3)}_{s,t,k}(n)$ are polynomials in n, $S = \lfloor s/2 \rfloor$ and $T = \lfloor t/2 \rfloor$.

More specifically,

1. *if s is odd and t is even, then, as polynomials in n, $U^{(1)}_{s,t,k}(n)$ is of degree at most $3S + 3T + 3k + 1$, and $U^{(3)}_{s,t,k}(n)$ is of degree at most $2S + 2T + 2k + 1$;*
2. *if s is even and t is odd, then, as polynomials in n, $U^{(1)}_{s,t,k}(n)$ is of degree at most $3S + 3T + 3k + 1$, and $U^{(3)}_{s,t,k}(n)$ is of degree at most $2S + 2T + 2k + 1$.*

Remark As the proof below shows, also here (cf. Remark 2) explicit formulae for the polynomials $U^{(i)}_{s,t,k}(n)$, $i = 1, 2, 3$, can be given that involve coefficients $c_{a,A}(n)$, for various specific choices of A. As we pointed out at several places already, Lemma 10 provides an explicit formula for these coefficients.

Proof of Theorem 1 The claim is trivially true for $k = 0$. Therefore we may assume from now on that $k > 0$.

Using the operations $(i, j) \rightarrow (-i, j)$, $(i, j) \rightarrow (i, -j)$, and $(i, j) \rightarrow (j, i)$, which do not change the summand, we see that

$$\sum_{-n \le i, j \le n} \left| i^s j^t (j^{2k} - i^{2k}) \right| \binom{2n}{n+i} \binom{2n}{n+j}$$

$$= 4 \sum_{0 \le i \le j \le n} \alpha(i = 0) \alpha(j = 0) \left(i^s j^t + i^t j^s \right) \left(j^{2k} - i^{2k} \right) \binom{2n}{n+i} \binom{2n}{n+j}$$

$$= 4 \sum_{0 \le i \le j \le n} \left(i^s j^t + i^t j^s \right) \left(j^{2k} - i^{2k} \right) \binom{2n}{n+i} \binom{2n}{n+j}$$

$$- 2 \binom{2n}{n} \sum_{j=1}^{n} \left(0^s j^t + 0^t j^s \right) j^{2k} \binom{2n}{n+j}, \tag{53}$$

where $\alpha(A) = \frac{1}{2}$ if A is true and $\alpha(A) = 1$ otherwise. Now one splits the sums into several sums of the form

$$\sum_{0 \le i \le j \le n} i^A j^B \binom{2n}{n+i} \binom{2n}{n+j}, \quad \text{respectively} \quad \sum_{j=1}^{n} j^B \binom{2n}{n+j}.$$

To sums of the second form, we apply Lemma 8. In order to evaluate the sums of the first form, we proceed as in the proof of Proposition 1. That is, we apply the expansions (36) and (37), and subsequently we use Lemmas 1–4 to evaluate the sums over i and j. Inspection of the result makes all assertions of the theorem obvious, except for the implicit claims in (50) and (51) that the term $4^n \binom{2n}{n}$ does not appear.

In order to verify these claims, we have to figure out what the coefficients of $4^n \binom{2n}{n}$ of the various sums in (53) are precisely. For the case of even s and t, from Lemma 1 we obtain that the coefficient of $4^n \binom{2n}{n}$ in the expression (53) equals

$$4 \sum_{a=0}^{S} \sum_{b=0}^{T+k} c_{a,S}(n)\, c_{b,T+k}(n)\, (2n-2a+1)_{2a}\, (2n-2b+1)_{2b}\, 2^{-2b-2} \binom{2n-2a}{n-a} \binom{2n}{n}^{-1}$$

$$+4 \sum_{a=0}^{T} \sum_{b=0}^{S+k} c_{a,T}(n)\, c_{b,S+k}(n)\, (2n-2a+1)_{2a}\, (2n-2b+1)_{2b}\, 2^{-2b-2} \binom{2n-2a}{n-a} \binom{2n}{n}^{-1}$$

$$-4 \sum_{a=0}^{S+k} \sum_{b=0}^{T} c_{a,S+k}(n)\, c_{b,T}(n)\, (2n-2a+1)_{2a}\, (2n-2b+1)_{2b}\, 2^{-2b-2} \binom{2n-2a}{n-a} \binom{2n}{n}^{-1}$$

$$-4 \sum_{a=0}^{T+k} \sum_{b=0}^{S} c_{a,T+k}(n)\, c_{b,S}(n)\, (2n-2a+1)_{2a}\, (2n-2b+1)_{2b}\, 2^{-2b-2} \binom{2n-2a}{n-a} \binom{2n}{n}^{-1}$$

$$-2 \cdot 0^{2S} \sum_{b=0}^{T+k} c_{b,T+k}(n)\, (2n-2b+1)_{2b}\, 2^{-2b-1}$$

$$-2 \cdot 0^{2T} \sum_{b=0}^{S+k} c_{b,S+k}(n)\, (2n-2b+1)_{2b}\, 2^{-2b-1}.$$

We may use (38) to simplify the double sums. In this manner, we arrive at the expression

$$0^{2S} \sum_{b=0}^{T+k} c_{b,T+k}(n)\, (2n-2b+1)_{2b}\, 2^{-2b} + 0^{2T} \sum_{b=0}^{S+k} c_{b,S+k}(n)\, (2n-2b+1)_{2b}\, 2^{-2b}$$

$$-0^{2S+2k} \sum_{b=0}^{T} c_{b,T}(n)\, (2n-2b+1)_{2b}\, 2^{-2b} - 0^{2T+2k} \sum_{b=0}^{S} c_{b,S}(n)\, (2n-2b+1)_{2b}\, 2^{-2b}$$

$$-0^{2S} \sum_{b=0}^{T+k} c_{b,T+k}(n)\, (2n-2b+1)_{2b}\, 2^{-2b} - 0^{2T} \sum_{b=0}^{S+k} c_{b,S+k}(n)\, (2n-2b+1)_{2b}\, 2^{-2b},$$

which visibly vanishes due to our assumption that $k > 0$.

The proof for the analogous claim in the case of odd s and t proceeds along the same lines. The only difference is that, instead of Lemma 1, here we need Lemma 4, and instead of (42) we need (43). $\square$

Theorem 2 *Let s, t, k and n, m be non-negative integers. If s and t are not both even, then*

$$\sum_{i,j} \left| i^s j^t (j^{2k} - i^{2k}) \right| \binom{2n}{n+i}\binom{2m}{m+j}$$

$$= \frac{V_{s,t,k}^{(1)}(n, m)}{(2n + 2m - 1)(2n + 2m - 3) \cdots (2n + 2m - 2S - 2T - 2k + 1)} \binom{2n + 2m}{n + m}$$

$$+ \frac{V_{s,t,k}^{(2)}(n, m)}{(n + m - 1)(n + m - 2) \cdots (n + m - S - T - k)} \binom{2n}{n}\binom{2m}{m}$$

$$+ V_{s,t,k}^{(3)}(n, m) \cdot 4^m \binom{2n}{n} + V_{s,t,k}^{(4)}(n, m) \cdot 4^n \binom{2m}{m}, \qquad (54)$$

where the $V_{s,t,k}^{(i)}(n, m)$, $i = 1, 2, 3, 4$, are polynomials in n and m, $S = \lfloor s/2 \rfloor$ and $T = \lfloor t/2 \rfloor$.

More specifically,

1. *if s is odd and t is even, then, as polynomials in n and m, $V_{s,t,k}^{(1)}(n, m)$ is of degree at most $3S + 3T + 3k + 1$, $V_{s,t,k}^{(3)}(n, m)$ is of degree at most $2S + 2T + 2k + 1$, and $V_{s,t,k}^{(2)}(n, m)$ and $V_{s,t,k}^{(4)}(n, m)$ are identically zero,*
2. *if s is even and t is odd, then, as polynomials in n and m, $V_{s,t,k}^{(1)}(n, m)$ is of degree at most $3S + 3T + 3k + 1$, $V_{s,t,k}^{(4)}(n, m)$ is of degree at most $2S + 2T + 2k + 1$, and $V_{s,t,k}^{(2)}(n, m)$ and $V_{s,t,k}^{(3)}(n, m)$ are identically zero,*
3. *if s and t are odd, then, as polynomials in n and m, $V_{s,t,k}^{(2)}(n, m)$ is of degree at most $3S + 3T + 3k + 2$, and $V_{s,t,k}^{(1)}(n, m)$, $V_{s,t,k}^{(3)}(n, m)$, and $V_{s,t,k}^{(4)}(n, m)$ are identically zero.*

Remark Again (cf. Remark 2), from the proof below it is obvious that explicit formulae for the polynomials $V_{s,t,k}^{(i)}(n, m)$, $i = 1, 2, 3, 4$, are available in terms of coefficients $c_{a,A}(n)$ and $c_{b,B}(m)$, for various specific choices of A and B, with Lemma 10 providing an explicit formula for these coefficients.

Proof of Theorem 2 Again, the claim is trivially true for $k = 0$. Therefore we may assume from now on that $k > 0$.

We follow the same idea as in the proof of Theorem 1, that is, we observe that the operations $(i, j) \to (-i, j)$ and $(i, j) \to (i, -j)$ leave the summand invariant. However, a notable difference here is that the interchange of summation indices

$(i, j) \to (j, i)$ does not leave the summand invariant. Consequently, here we see that

$$\sum_{i,j} \left| i^s j^t (j^{2k} - i^{2k}) \right| \binom{2n}{n+i} \binom{2m}{m+j}$$

$$= 4 \sum_{0 \le i \le j} \alpha(i = 0) \alpha(j = 0) i^s j^t \left(j^{2k} - i^{2k} \right) \binom{2n}{n+i} \binom{2m}{m+j}$$

$$+ 4 \sum_{0 \le i \le j} \alpha(i = 0) \alpha(j = 0) i^t j^s \left(j^{2k} - i^{2k} \right) \binom{2n}{n+j} \binom{2m}{m+i}$$

$$= 4 \sum_{0 \le i \le j} i^s j^t \left(j^{2k} - i^{2k} \right) \binom{2n}{n+i} \binom{2m}{m+j}$$

$$+ 4 \sum_{0 \le i \le j} i^t j^s \left(j^{2k} - i^{2k} \right) \binom{2n}{n+j} \binom{2m}{m+i}$$

$$- 2 \binom{2n}{n} \sum_{j=1}^{m} 0^s j^{t+2k} \binom{2m}{m+j} - 2 \binom{2m}{m} \sum_{j=1}^{n} 0^t j^{s+2k} \binom{2n}{n+j},$$

$$(55)$$

where $\alpha(\mathcal{A})$ has the same meaning as in the proof of Theorem 1. Now one splits the sums into several sums of the form

$$\sum_{0 \le i \le j} i^A j^B \binom{2n}{n+i} \binom{2m}{m+j} \quad \text{and} \quad \sum_{0 \le i \le j} i^A j^B \binom{2m}{m+i} \binom{2n}{n+j},$$

$$\text{respectively} \quad \sum_{j=1}^{n} j^B \binom{2n}{n+j} \quad \text{and} \quad \sum_{j=1}^{m} j^B \binom{2m}{m+j}.$$

To sums of the second form, we apply Lemma 8. In order to evaluate the sums of the first form, we proceed as in the proof of Proposition 1. That is, we apply the expansions (36) and (37) (with n replaced by m if appropriate), and subsequently we use Lemmas 2–4 to evaluate the sums over i and j. Inspection of the result makes all assertions of the theorem obvious, except for the claims in Items (1) and (2) that the polynomial $V^{(2)}_{s,t,k}(n, m)$, the coefficient of $\binom{2n}{n}\binom{2m}{m}$ in (54), vanishes.

Below we treat Item (1), that is, the case where s is odd and t is even. Item (2) can be handled completely analogously.

After having done the above described manipulations, we see that, for odd s and even t, the coefficient of $\binom{2n}{n}\binom{2m}{m}$ in the expression (55) equals

$$\sum_{a=0}^{S}\sum_{b=0}^{T+k} c_{a,S}(n)\, c_{b,T+k}(m)\,(2n-2a+1)_{2a}\,(2m-2b+1)_{2b}\frac{(n-a)(m-b)}{n+m-a-b}$$

$$\times \binom{2n-2a}{n-a}\binom{2m-2b}{m-b}\binom{2n}{n}^{-1}\binom{2m}{m}^{-1}$$

$$-\sum_{a=0}^{S+k}\sum_{b=0}^{T} c_{a,S+k}(n)\, c_{b,T}(m)\,(2n-2a+1)_{2a}\,(2m-2b+1)_{2b}\frac{(n-a)(m-b)}{n+m-a-b}$$

$$\times \binom{2n-2a}{n-a}\binom{2m-2b}{m-b}\binom{2n}{n}^{-1}\binom{2m}{m}^{-1}$$

$$+\sum_{a=0}^{T}\sum_{b=0}^{S+k} c_{a,T}(m)\, c_{b,S+k}(n)\,(2m-2a+1)_{2a}\,(2n-2b+1)_{2b}$$

$$\times \frac{(n-b)(n-b+2(m-a))}{n+m-a-b}\binom{2n-2b}{n-b}\binom{2m-2a}{m-a}\binom{2n}{n}^{-1}\binom{2m}{m}^{-1}$$

$$-\sum_{a=0}^{T+k}\sum_{b=0}^{S} c_{a,T+k}(m)\, c_{b,S}(n)\,(2m-2a+1)_{2a}\,(2n-2b+1)_{2b}$$

$$\times \frac{(n-b)(n-b+2(m-a))}{n+m-a-b}\binom{2n-2b}{n-b}\binom{2m-2a}{m-a}\binom{2n}{n}^{-1}\binom{2m}{m}^{-1}$$

$$-0^t\sum_{b=0}^{S+k} c_{b,S+k}(n)\,(n-b)_{b+1}\,(n-b+1)_b.$$

In the last two double sums above, we interchange the summation indices a and b. Then the first and fourth double sum can be combined into one double sum, as well as the second and third double sum. Thus, the above expression simplifies to

$$-\sum_{a=0}^{S}\sum_{b=0}^{T+k} c_{a,S}(n)\, c_{b,T+k}(m)\,(2n-2a+1)_{2a}\,(2m-2b+1)_{2b}(n-a)$$

$$\times \binom{2n-2a}{n-a}\binom{2m-2b}{m-b}\binom{2n}{n}^{-1}\binom{2m}{m}^{-1}$$

$$+\sum_{a=0}^{S+k}\sum_{b=0}^{T} c_{a,S+k}(n)\, c_{b,T}(m)\,(2n-2a+1)_{2a}\,(2m-2b+1)_{2b}(n-a)$$

$$\times \binom{2n-2a}{n-a}\binom{2m-2b}{m-b}\binom{2n}{n}^{-1}\binom{2m}{m}^{-1}$$

$$-0^t\sum_{b=0}^{S+k} c_{b,S+k}(n)\,(n-b)_{b+1}\,(n-b+1)_b.$$

In both double sums, the sum over b can be evaluated by means of (38). This leads us to the expression

$$-0^{2T+2k} \sum_{a=0}^{S} c_{a,S}(n)\,(n-a)_{a+1}\,(n-a+1)_a + 0^{2T} \sum_{a=0}^{S+k} c_{a,S+k}(n)\,(n-a)_{a+1}\,(n-a+1)_a$$

$$-0^t \sum_{b=0}^{S+k} c_{b,S+k}(n)\,(n-b)_{b+1}\,(n-b+1)_b,$$

which visibly vanishes due to our assumptions that $k > 0$ and that t is even. □

We now turn to our results for double sums of the form (3) with odd k (and $\beta = 1$). We first state our results for $m = n$ and immediately thereafter the one we obtain for generic n and m in the case where s and t are both odd. We then indicate the proofs of both theorems.

Theorem 3 *Let s, t, k and n be non-negative integers.*
If s and t are not both odd, then

$$\sum_{-n \le i,j \le n} \left| i^s j^t (j^{2k+1} - i^{2k+1}) \right| \binom{2n}{n+i}\binom{2n}{n+j}$$

$$= \frac{X_{s,t,k}^{(1)}(n)}{(4n-1)(4n-3)\cdots(4n-2S-2T-2k+1)}\binom{4n}{2n}$$

$$+ \frac{X_{s,t,k}^{(2)}(n)}{(2n-1)(2n-3)\cdots(2n-2\lceil(S+T+k)/2\rceil+1)}\binom{2n}{n}^2$$

$$+ X_{s,t,k}^{(3)}(n) \cdot 4^n \binom{2n}{n} + X_{s,t,k}^{(4)}(n) \cdot 16^n, \qquad (56)$$

where the $X_{s,t,k}^{(i)}(n)$, $i = 1, 2, 3, 4$, are polynomials in n, $S = \lfloor s/2 \rfloor$ and $T = \lfloor t/2 \rfloor$.
More specifically,

1. *if s and t are even, then, as polynomials in n, $X_{s,t,k}^{(1)}(n)$ is of degree at most $3S + 3T + 3k$, and $X_{s,t,k}^{(2)}(n)$, $X_{s,t,k}^{(3)}(n)$, and $X_{s,t,k}^{(4)}(n)$ are identically zero;*
2. *if s is odd and t is even, then, as polynomials in n, $X_{s,t,k}^{(2)}(n)$ is of degree at most $2S + 2T + 2k + 1 + \lceil(S+T+k)/2\rceil$, $X_{s,t,k}^{(4)}(n)$ is of degree at most $2S + 2T + 2k + 1$, and $X_{s,t,k}^{(1)}(n)$ and $X_{s,t,k}^{(3)}(n)$ are identically zero;*
3. *if s is even and t is odd, then, as polynomials in n, $X_{s,t,k}^{(2)}(n)$ is of degree at most $2S + 2T + 2k + 1 + \lceil(S+T+k)/2\rceil$, $X_{s,t,k}^{(4)}(n)$ is of degree at most $2S + 2T + 2k + 1$, and $X_{s,t,k}^{(1)}(n)$ and $X_{s,t,k}^{(3)}(n)$ are identically zero.*

If s and t are odd, then

$$\sum_{-n \le i,j \le n} \left| i^s j^t (j^{2k+1} - i^{2k+1}) \right| \binom{2n}{n+i}\binom{2n}{n+j}$$

$$= \frac{X_{s,t,k}^{(1)}(n)}{(4n-1)(4n-3)\cdots(4n-2S-2T-2k-1)}\binom{4n}{2n} + X_{s,t,k}^{(3)}(n) \cdot 4^n \binom{2n}{n},$$

$$(57)$$

where $S = \lfloor s/2 \rfloor$ and $T = \lfloor t/2 \rfloor$, and, as polynomials in n, $X_{s,t,k}^{(1)}(n)$ is of degree at most $3S + 3T + 3k + 2$, and $X_{s,t,k}^{(3)}(n)$ is of degree at most $2S + 2T + 2k + 2$.

Theorem 4 *Let s, t, k and n, m be non-negative integers. If s and t are both odd, then*

$$\sum_{i,j} \left| i^s j^t (j^{2k+1} - i^{2k+1}) \right| \binom{2n}{n+i}\binom{2m}{m+j}$$

$$= \frac{Y_{s,t,k}^{(1)}(n,m)}{(2n+2m-1)(2n+2m-3)\cdots(2n+2m-2S-2T-2k-1)}\binom{2n+2m}{n+m}$$

$$+ Y_{s,t,k}^{(3)}(n,m) \cdot 4^m \binom{2n}{n} + Y_{s,t,k}^{(4)}(n,m) \cdot 4^n \binom{2m}{m}, \qquad (58)$$

where $S = \lfloor s/2 \rfloor$ and $T = \lfloor t/2 \rfloor$, and, as polynomials in n and m, $Y_{s,t,k}^{(1)}(n,m)$ is of degree at most $3S + 3T + 3k + 3$, and $Y_{s,t,k}^{(3)}(n,m)$ and $Y_{s,t,k}^{(4)}(n,m)$ are of degree at most $2S + 2T + 2k + 1$.

Remark From the proof below it is obvious that also here (cf. Remark 2) explicit formulae for the polynomials $X_{s,t,k}^{(i)}(n)$ and $Y_{s,t,k}^{(i)}(n,m)$, $i = 1, 2, 3, 4$, exist in terms of coefficients $c_{a,A}(n)$ and $c_{b,B}(m)$, for various specific choices of A and B, with Lemma 10 providing an explicit formula for these coefficients.

Proof of Theorems 3 and 4 We use the operations $(i, j) \to (-i, j)$ and $(i, j) \to (i, -j)$ (but not $(i, j) \to (j, i)$). What we get is (for the proof of Theorem 3 we have to assume that $m = n$)

$$\sum_{i,j} \left| i^s j^t (j^{2k+1} - i^{2k+1}) \right| \binom{2n}{n+i}\binom{2m}{m+j}$$

$$= \frac{1}{2}\sum_{i,j} \left(\left| i^s j^t (j^{2k+1} - i^{2k+1}) \right| + \left| i^s j^t (j^{2k+1} + i^{2k+1}) \right| \right)\binom{2n}{n+i}\binom{2m}{m+j}$$

$$= 2 \sum_{0 \le i, j} \alpha(i = 0) \, \alpha(j = 0) \left(\left| i^s j^t (j^{2k+1} - i^{2k+1}) \right| \right.$$

$$\left. + \left| i^s j^t (j^{2k+1} + i^{2k+1}) \right| \right) \binom{2n}{n+i} \binom{2m}{m+j}$$

$$= 2 \sum_{0 \le i \le j} \left(\left| i^s j^t (j^{2k+1} - i^{2k+1}) \right| + \left| i^s j^t (j^{2k+1} + i^{2k+1}) \right| \right) \binom{2n}{n+i} \binom{2m}{m+j}$$

$$+ 2 \sum_{0 \le i < j} \left(\left| i^t j^s (j^{2k+1} - i^{2k+1}) \right| + \left| i^t j^s (j^{2k+1} + i^{2k+1}) \right| \right) \binom{2n}{n+j} \binom{2m}{m+i}$$

$$- 2 \binom{2n}{n} 0^s \sum_{0 \le j} j^{t+2k+1} \binom{2m}{m+j} - 2 \binom{2m}{m} 0^t \sum_{0 \le i} i^{s+2k+1} \binom{2n}{n+i}$$

$$= 4 \sum_{0 \le i \le j} i^s j^{t+2k+1} \binom{2n}{n+i} \binom{2m}{m+j} + 4 \sum_{0 \le i \le j} i^t j^{s+2k+1} \binom{2n}{n+j} \binom{2m}{m+i}$$

$$- 2 \binom{2n}{n} 0^s \sum_{0 \le j} j^{t+2k+1} \binom{2m}{m+j} - 2 \binom{2m}{m} 0^t \sum_{0 \le i} i^{s+2k+1} \binom{2n}{n+i}$$

$$- 4 \sum_{j \ge 1} j^{s+t+2k+1} \binom{2n}{n+j} \binom{2m}{m+j}, \tag{59}$$

where $\alpha(\mathcal{A})$ has the same meaning as in the proof of Theorem 1. To the single sums over i and over j, we apply Lemmas 8 and 9. In order to evaluate the sums over $0 \le i \le j$, we proceed as in the proof of Proposition 1. That is, we apply the expansions (36) and (37) (with n replaced by m if appropriate), and subsequently we use Lemmas 1–4 to evaluate the sums over $0 \le i \le j$. Inspection of the result makes all assertions of the theorem obvious, except for the claims of the vanishing of the polynomial $X^{(2)}_{s,t,k}(n)$ in Theorem 3, Item (1), of the vanishing of the polynomial $X^{(1)}_{s,t,k}(n)$ in Theorem 3, Items (2) and (3), and of the claim that the coefficient of $\binom{2n}{n}^2$ in Theorem 3, right-hand side of (57), vanishes, as well as the coefficient of $\binom{2n}{n}\binom{2m}{m}$ in Theorem 4, right-hand side of (58).

Below we treat the last case, that is, the case of generic n and m where s and t are both odd. The other claims can be handled completely analogously.

Following the above described procedure, using (45) with $h = S + T + k + 1$ and $k = 0$ for the evaluation of the sum over j in the last line of (59), we obtain from

Lemma 2 that the coefficient of $\binom{2n}{n}\binom{2m}{m}$ in the expression (59) equals

$$4\sum_{a=0}^{S}\sum_{b=0}^{T+k+1} c_{a,S}(n)\, c_{b,T+k+1}(m)\, (n-a+1)_a^2\, (m-b+1)_b^2\, \frac{(n-a)(m-b)}{4(n+m-a-b)}$$

$$+4\sum_{b=0}^{T}\sum_{a=0}^{S+k+1} c_{b,T}(m)\, c_{a,S+k+1}(n)\, (n-a+1)_a^2\, (m-b+1)_b^2\, \frac{(n-a)(m-b)}{4(n+m-a-b)}$$

$$-4\sum_{b=0}^{S+T+k+1} c_{b,S+T+k+1}(n)\, (n-b)_{b+1}\, (n-b+1)_b\, \frac{m}{2(n+m-b)}.$$

If we now use (45) with $(S, T+k+1)$, $(S+k+1, T)$, and $(S+T+k+1, 0)$ in place of (h, k), we see that the above expression vanishes. This establishes the assertion about the "non-appearance" of the term $\binom{2n}{n}\binom{2m}{m}$ in Theorem 4, and thus also the assertion about the "non-appearance" of $\binom{2n}{n}^2$ in Eq. (57) of Theorem 3.   □

Remark 3

(1) It is obvious from the proofs of Theorems 1–4 that we could deduce analogous theorems for the more general sums (2) and (3). We omit this here for the sake of brevity, but provide an example of such an evaluation in (66) below.

(2) Theorems 1–4 imply an obvious algorithm to evaluate a sum of the form (2) or (3) for any given s, t, k and $\beta = 1$. (Again, an extension to arbitrary β would be possible.) Namely, addressing the case of odd k and $m = n$, one makes an indeterminate Ansatz for the polynomials $X_{s,t}^{(1)}(n)$, $X_{s,t}^{(2)}(n)$, $X_{s,t}^{(3)}(n)$, $X_{s,t}^{(4)}(n)$ in Theorem 3, one evaluates the sum on the left-hand side of (56) for $n = S + T + k, \ldots, N + S + T + k$, where N is the number of indeterminates involved in the Ansatz, giving rise to a system of $N+1$ linear equations for the N indeterminates. One solves the system and substitutes the solutions on the right-hand side of (56).

In this manner, we can establish any of the proved or conjectured double sum evaluations in [6]. For example, we obtain

$$\sum_{-n\le i,j\le n} \left|j^3 - i^3\right|\binom{2n}{n+i}\binom{2n}{n+j} = \frac{4n^2(5n-2)}{4n-1}\binom{4n-1}{2n-1}, \tag{60}$$

$$\sum_{-n\le i,j\le n} \left|j^5 - i^5\right|\binom{2n}{n+i}\binom{2n}{n+j} = \frac{8n^2(43n^3 - 70n^2 + 36n - 6)}{(4n-2)(4n-3)}\binom{4n-2}{2n-2}, \tag{61}$$

$$\sum_{i,j} \left|ij(j^2 - i^2)\right|\binom{2n}{n+i}\binom{2m}{m+j} = \frac{mn(n^2 - n + m^2 - m)}{n+m-1}\binom{2n}{n}\binom{2m}{m}, \tag{62}$$

$$\sum_{i,j} \left| i^3 j^3 (j^2 - i^2) \right| \binom{2n}{n+i} \binom{2m}{m+j} = \frac{2n^2 m^2 P_1(n,m)}{(n+m-1)(n+m-2)(n+m-3)}$$

$$\times \binom{2n}{n} \binom{2m}{m},$$

(63)

$$\sum_{-n \leq i,j \leq n} \left| j^7 - i^7 \right| \binom{2n}{n+i} \binom{2n}{n+j} = \frac{16n^2 P_2(n)}{(4n-3)(4n-4)(4n-5)} \binom{4n-3}{2n-3},$$

(64)

where

$$P_1(n,m) = n^4 + 2n^3 m - 6n^3 - 6n^2 m + 11n^2 + 2nm^3 - 6nm^2 + 12nm$$

$$- 10n + m^4 - 6m^3 + 11m^2 - 10m + 4$$

and

$$P_2(n) = 531n^5 - 1960n^4 + 2800n^3 - 1952n^2 + 668n - 90.$$

These identities (with $m = n$ for (62) and (63)) establish the conjectured identities (5.7)–(5.9), (5.12), (5.14) from [6]. However, our machinery also yields

$$\sum_{-n \leq i,j \leq n} \left| i^4 j^3 (j^5 - i^5) \right| \binom{2n}{n+i} \binom{2n}{n+j}$$

$$= \frac{n^4 \left(414n^6 - 2968n^5 + 8332n^4 - 11853n^3 + 9105n^2 - 3592n + 565 \right)}{2(2n-5)(2n-3)(2n-1)} \binom{2n}{n}^2$$

$$+ \frac{1}{128} n^2 (3n-1) \left(105n^3 - 210n^2 + 147n - 34 \right) 16^n \qquad (65)$$

or

$$\sum_{-n \leq i,j \leq n} \left| ij(j^3 - i^3)^3 \right| \binom{2n}{n+i} \binom{2n}{n+j}$$

$$= \frac{1}{16} n^2 \left(1377n^4 - 3870n^3 + 4503n^2 - 2442n + 496 \right) 4^n \binom{2n}{n}$$

$$- \frac{4n^3 P_3(n)}{(4n-7)(4n-5)(4n-3)(4n-1)} \binom{4n}{2n}, \qquad (66)$$

where

$$P_3(n) = 1917n^7 - 11160n^6 + 26439n^5 - 33189n^4 + 23945n^3 - 9951n^2 + 2206n - 201,$$

for example. Obviously, one could also use the summation tools described in Sect. 3 to simplify the left-hand sides to their right-hand sides.

(3) In case the reader wonders what would happen if, instead of double sums of the form (3), we would consider double sums of the form

$$\sum_{i,j} |i^s j^t (i^k - j^k)^\beta| \binom{2n+1}{n+i} \binom{2m+1}{m+j} \tag{67}$$

or mixed sums

$$\sum_{i,j} |i^s j^t (i^k - j^k)^\beta| \binom{2n+1}{n+i} \binom{2m}{m+j}, \tag{68}$$

we point out that

$$\binom{2n+1}{n+i} = \frac{n+i+1}{2n+2} \binom{2(n+1)}{n+1+i} = \frac{1}{2} \binom{2(n+1)}{n+1+i} + \frac{i}{2n+2} \binom{2(n+1)}{n+1+i},$$

and thus double sums of the form (67) or (68) can be written as a linear combination of our familiar double sums (3).

8 An Inequality for a Binomial Double Sum

In this final section, we establish Conjecture 3.1 from [6], which provides a lower bound on sums of the form (3) with $s = t = 0$, $k = 2$, $\beta = 1$.

Theorem 5 *For all non-negative integers m and n, we have*

$$\sum_{i,j} |j^2 - i^2| \binom{2n}{n+i} \binom{2m}{m+j} \geq 2nm \binom{2n}{n} \binom{2m}{m}, \tag{69}$$

and equality holds if and only if m = n.

Proof Without loss of generality, we assume $m \geq n$.

Using the operations $(i, j) \to (-i, j)$ and $(i, j) \to (i, -j)$, which do not change the summand, we see that (69) is equivalent to

$$\sum_{0 \leq i,j} \alpha(i = 0) \alpha(j = 0) |j^2 - i^2| \binom{2n}{n+i} \binom{2m}{m+j} \geq \frac{nm}{2} \binom{2n}{n} \binom{2m}{m}, \tag{70}$$

where $\alpha(i = 0)$ has the same meaning as in the proof of Proposition 1. By Lemma 11, we see that the claim would be established if we were able to show that

$$\sum_{0 \le i < j} \alpha(i = 0) \left(\binom{2n}{n+i}\binom{2m-2}{m+j-1} - \binom{2n-2}{n+j-1}\binom{2m}{m+i} \right) \tag{71}$$

is non-negative, with equality holding only if $m = n$. Indeed, Lemma 13 says that these two last assertions hold even for each summand in (71) individually. (It is at this point that our assumption $m \ge n$ comes into play.) This completes the proof of the theorem. $\qquad \square$

Lemma 11 *For all non-negative integers m and n, we have*

$$\sum_{0 \le i,j} \alpha(i = 0)\alpha(j = 0) \left| j^2 - i^2 \right| \binom{2n}{n+i}\binom{2m}{m+j}$$

$$= \frac{nm}{2} \binom{2n}{n}\binom{2m}{m}$$

$$+ 2(m-n) \sum_{0 \le i < j} \alpha(i = 0) \left(\binom{2n}{n+i}\binom{2m-2}{m+j-1} - \binom{2n-2}{n+j-1}\binom{2m}{m+i} \right). \tag{72}$$

Proof We write

$$j^2 - i^2 = (n^2 - i^2) - (m^2 - j^2) + (m^2 - n^2)$$

and decompose the sum on the left-hand side of (72) into two parts according to whether $i < j$ or $i > j$. Thereby, the sum on the left-hand side of (72) becomes

$$(2n-1)_2 \sum_{0 \le i < j} \alpha(i = 0) \binom{2n-2}{n+i-1}\binom{2m}{m+j}$$

$$- (2m-1)_2 \sum_{0 \le i < j} \alpha(i = 0) \binom{2n}{n+i}\binom{2m-2}{m+j-1}$$

$$- (2n-1)_2 \sum_{0 \le j < i} \alpha(j = 0) \binom{2n-2}{n+i-1}\binom{2m}{m+j}$$

$$+ (2m-1)_2 \sum_{0 \le j < i} \alpha(j = 0) \binom{2n}{n+i}\binom{2m-2}{m+j-1}$$

$$+ (m^2 - n^2) \sum_{0 \le i < j} \alpha(i = 0) \left(\binom{2n}{n+i}\binom{2m}{m+j} - \binom{2n}{n+j}\binom{2m}{m+i} \right). \tag{73}$$

We next show how to evaluate the first two (double) sums in (73). In the first line of (73), we use the decomposition

$$
\binom{2m}{m+j} = \binom{2m-2}{m+j} + 2\binom{2m-2}{m+j-1} + \binom{2m-2}{m+j-2}, \tag{74}
$$

while in the second line we use the same decomposition with m replaced by n and j by i. This leads to

$$
(2n-1)_2 \sum_{0 \le i < j} \alpha(i=0) \binom{2n-2}{n+i-1}\binom{2m}{m+j}
$$

$$
- (2m-1)_2 \sum_{0 \le i < j} \alpha(i=0) \binom{2n}{n+i}\binom{2m-2}{m+j-1}
$$

$$
= (2n-1)_2 \sum_{0 \le i < j} \alpha(i=0) \binom{2n-2}{n+i-1}\binom{2m-2}{m+j}
$$

$$
+ (2n-1)_2 \sum_{0 \le i < j} \alpha(i=0) \binom{2n-2}{n+i-1}\binom{2m-2}{m+j-2}
$$

$$
- (2n-1)_2 \sum_{0 \le i < j} \alpha(i=0) \binom{2n-2}{n+i}\binom{2m-2}{m+j-1}
$$

$$
- (2n-1)_2 \sum_{0 \le i < j} \alpha(i=0) \binom{2n-2}{n+i-2}\binom{2m-2}{m+j-1}
$$

$$
+ \big((2n-1)_2 - (2m-1)_2\big) \sum_{0 \le i < j} \alpha(i=0) \binom{2n}{n+i}\binom{2m-2}{m+j-1}.
$$

By a simultaneous shift of i and j by one, one sees that the first and fourth sum on the right-hand side cancel each other largely, and the same is true for the second and the third sum. Thus, we have

$$
(2n-1)_2 \sum_{0 \le i < j} \alpha(i=0) \binom{2n-2}{n+i-1}\binom{2m}{m+j}
$$

$$
- (2m-1)_2 \sum_{0 \le i < j} \alpha(i=0) \binom{2n}{n+i}\binom{2m-2}{m+j-1}
$$

$$
= -\frac{1}{2}(2n-1)_2 \sum_{0 < j} \binom{2n-2}{n-1}\binom{2m-2}{m+j}
$$

$$-\frac{1}{2}(2n-1)_2 \sum_{0<j} \binom{2n-2}{n-2}\binom{2m-2}{m+j-1}$$

$$+\frac{1}{2}(2n-1)_2 \sum_{0<j} \binom{2n-2}{n-1}\binom{2m-2}{m+j-2}$$

$$+\frac{1}{2}(2n-1)_2 \sum_{0<j} \binom{2n-2}{n}\binom{2m-2}{m+j-1}$$

$$+\big((2n-1)_2-(2m-1)_2\big) \sum_{0\le i<j} \alpha(i=0)\binom{2n}{n+i}\binom{2m-2}{m+j-1}.$$

Here, there is more cancellation: the second and fourth sum on the right-hand side cancel each other, while the first and third cancel each other in large parts, with only two terms remaining. As a result, we obtain

$$(2n-1)_2 \sum_{0\le i<j} \alpha(i=0)\binom{2n-2}{n+i-1}\binom{2m}{m+j}$$

$$-(2m-1)_2 \sum_{0\le i<j} \alpha(i=0)\binom{2n}{n+i}\binom{2m-2}{m+j-1}$$

$$=\frac{1}{2}(2n-1)_2\binom{2n-2}{n-1}\binom{2m-1}{m}$$

$$+\big((2n-1)_2-(2m-1)_2\big) \sum_{0\le i<j} \alpha(i=0)\binom{2n}{n+i}\binom{2m-2}{m+j-1}$$

$$=\frac{n^2}{4}\binom{2n}{n}\binom{2m}{m}$$

$$+\big((2n-1)_2-(2m-1)_2\big) \sum_{0\le i<j} \alpha(i=0)\binom{2n}{n+i}\binom{2m-2}{m+j-1}.$$

The same calculation, with n and m interchanged, yields

$$-(2n-1)_2 \sum_{0\le j<i} \alpha(j=0)\binom{2n-2}{n+i-1}\binom{2m}{m+j}$$

$$+(2m-1)_2 \sum_{0\le j<i} \alpha(j=0)\binom{2n}{n+i}\binom{2m-2}{m+j-1}$$

$$= \frac{m^2}{4} \binom{2n}{n} \binom{2m}{m}$$

$$+ ((2m-1)_2 - (2n-1)_2) \sum_{0 \le i < j} \alpha(i=0) \binom{2m}{m+i} \binom{2n-2}{n+j-1}.$$

If we put everything together, then we have shown that the sum on the left-hand side of (72) equals

$$\frac{n^2 + m^2}{4} \binom{2n}{n} \binom{2m}{m}$$

$$+ \left(4(m^2 - n^2) - 2(m - n)\right)$$

$$\times \sum_{0 \le i < j} \alpha(i=0) \left(\binom{2n-2}{n+j-1} \binom{2m}{m+i} - \binom{2n}{n+i} \binom{2m-2}{m+j-1} \right)$$

$$+ (m^2 - n^2) \sum_{0 \le i < j} \alpha(i=0) \left(\binom{2n}{n+i} \binom{2m}{m+j} - \binom{2n}{n+j} \binom{2m}{m+i} \right).$$

If we finally use Lemma 12 in this expression, then the result is the right-hand side of (72). $\qquad\square$

Lemma 12 *For all non-negative integers m and n, we have*

$$4 \sum_{0 \le i < j} \alpha(i=0) \left(\binom{2n-2}{n+j-1} \binom{2m}{m+i} - \binom{2n}{n+i} \binom{2m-2}{m+j-1} \right)$$

$$+ \sum_{0 \le i < j} \alpha(i=0) \left(\binom{2n}{n+i} \binom{2m}{m+j} - \binom{2n}{n+j} \binom{2m}{m+i} \right)$$

$$= - \frac{m-n}{4(m+n)} \binom{2n}{n} \binom{2m}{m}. \qquad (75)$$

Proof Using the decomposition (74) in the second line of (75), we compute

$$4 \sum_{0 \le i < j} \alpha(i=0) \left(\binom{2n-2}{n+j-1} \binom{2m}{m+i} - \binom{2n}{n+i} \binom{2m-2}{m+j-1} \right)$$

$$+ \sum_{0 \le i < j} \alpha(i=0) \left(\binom{2n}{n+i} \binom{2m}{m+j} - \binom{2n}{n+j} \binom{2m}{m+i} \right)$$

$$= \sum_{0 \le i < j} \alpha(i=0) \left(2 \binom{2n-2}{n+j-1} \binom{2m}{m+i} \right)$$

$$-\binom{2n-2}{n+j}\binom{2m}{m+i}-\binom{2n-2}{n+j-2}\binom{2m}{m+i}$$

$$+\binom{2n}{n+i}\binom{2m-2}{m+j}+\binom{2n}{n+i}\binom{2m-2}{m+j-2}-2\binom{2n}{n+i}\binom{2m-2}{m+j-1}\biggr)$$

$$=\sum_{0\le i}\alpha(i=0)\left(\binom{2n-2}{n+i}\binom{2m}{m+i}-\binom{2n-2}{n+i-1}\binom{2m}{m+i}\right.$$

$$\left.+\binom{2n}{n+i}\binom{2m-2}{m+i-1}-\binom{2n}{n+i}\binom{2m-2}{m+i}\right)$$

$$=\frac{m-n}{m+n}\sum_{0\le i}\alpha(i=0)\left(\frac{(2n-2)!\,(2m-2)!\,(4nm-4(i+1)n-4(i+1)m+1)}{(n+i)!\,(n-i-1)!\,(m+i)!\,(m-i-1)!}\right.$$

$$\left.-\frac{(2n-2)!\,(2m-2)!\,(4nm-4in-4im+1)}{(n+i-1)!\,(n-i)!\,(m+i-1)!\,(m-i)!}\right)$$

$$=\frac{m-n}{m+n}\left(-\frac{1}{2}\frac{(2n-2)!\,(2m-2)!\,(4nm-4n-4m+1)}{n!\,(n-1)!\,m!\,(m-1)!}\right.$$

$$\left.-\frac{1}{2}\frac{(2n-2)!\,(2m-2)!\,(4nm+1)}{(n-1)!\,n!\,(m-1)!\,m!}\right)$$

$$=-\frac{m-n}{4(m+n)}\binom{2n}{n}\binom{2m}{m},$$

which is the desired result.[3]   □

Lemma 13 *For all non-negative integers m, n, i, j with $m \ge n$ and $i < j$, we have*

$$\binom{2n}{n+i}\binom{2m-2}{m+j-1}\ge\binom{2n-2}{n+j-1}\binom{2m}{m+i},$$

with equality if and only if $m = n$.

Proof We have

$$\frac{\binom{2n}{n+i}\binom{2m-2}{m+j-1}}{\binom{2n-2}{n+j-1}\binom{2m}{m+i}}=\frac{2n(2n-1)}{2m(2m-1)}\frac{(m-j+1)(m-j)}{(n-j+1)(n-j)}\prod_{k=i+1}^{j-1}\frac{(n+k)(m-k+1)}{(n-k+1)(m+k)}$$

$$=\frac{\left(2+\frac{2j-2}{n-j+1}\right)\left(2+\frac{2j-1}{n-j}\right)}{\left(2+\frac{2j-2}{m-j+1}\right)\left(2+\frac{2j-1}{m-j}\right)}\prod_{k=i+1}^{j-1}\frac{nm+km-(k-1)n-k(k-1)}{nm-(k-1)m+kn-k(k-1)}\ge 1,$$

and visibly equality holds if and only if $m = n$.   □

---

[3] For the finding of the telescoping form of the sum over i see footnote 1.

Acknowledgements The authors thank an anonymous referee for an extremely careful reading of the original manuscript and for the many suggestions leading to an improved presentation.

References

1. Bostan, A., Lairez, P., Salvy, B.: Multiple binomial sums. J. Symb. Comput. **80**, 351–386 (2017)
2. Brent, R.P., Osborn, J.H.: Note on a double binomial sum relevant to the Hadamard maximal determinant problem (2013). arxiv:1309.2795. http://arxiv.org/abs/1309.2795
3. Brent, R.P., Osborn, J.H., Smith, W.D.: Lower bounds on maximal determinants of binary matrices via the probabilistic method (2014). arxiv:1402.6817. http://arxiv.org/abs/1402.6817
4. Brent, R.P., Krattenthaler, C., Warnaar, S.O.: Discrete analogues of Macdonald–Mehta integrals. J. Combin. Theory A **144**, 80–138 (2016)
5. Brent, R.P., Ohtsuka, H., Osborn, J.H., Prodinger, H.: Some binomial sums involving absolute values. J. Integer Seq. **19**, 14 pp., Art. 16.3.7 (2016)
6. Brent, R.P., Ohtsuka, H., Osborn, J.H., Prodinger, H.: Some binomial sums involving absolute values. Unpublished manuscript. arxiv:1411.1477v1. http://arxiv.org/abs/1411.1477v1
7. Gosper, R.W.: Decision procedure for indefinite hypergeometric summation. Proc. Natl. Acad. Sci. U. S. A. **75**, 40–42 (1978)
8. Gould, H.W., Hsu, L.C.: Some new inverse series relations. Duke Math. J. **40**, 885–891 (1973)
9. Graham, R.L., Knuth, D.E., Patashnik, O.: Concrete Mathematics. Addison-Wesley, Reading (1989)
10. Hadamard, J.: Résolution d'une question relative aux déterminants. Bull. Sci. Math. **17**, 240–246 (1893)
11. Karr, M.: Summation in finite terms. J. Assoc. Comput. Mach. **28**, 305–350 (1981)
12. Krattenthaler, C.: A new matrix inverse. Proc. Am. Math. Soc. **124**, 47–59 (1996)
13. Paule, P., Schorn, M.: A Mathematica version of Zeilberger's algorithm for proving binomial coefficient identities. J. Symbol. Comput. **20**, 673–698 (1995)
14. Petkovšek, M., Wilf, H., Zeilberger, D.: A=B. A. K. Peters, Wellesley (1996)
15. Schneider, C.: Symbolic summation assists combinatorics. Sém. Lothar. Combin. **56**, 36 pp., Article B56b (2007)
16. Schneider, C.: Simplifying multiple sums in difference fields. In: Schneider C., Blümlein J. (eds.) Computer Algebra in Quantum Field Theory: Integration, Summation and Special Functions. Texts and Monographs in Symbolic Computation. Springer, Wien, pp. 325–360 (2013). arxiv.1304.4134. http://arxiv.org/abs/1304.4134
17. Schneider, C.: A difference ring theory for symbolic summation. J. Symb. Comput. **72**, 82–127 (2016)
18. Schneider, C.: Summation theory II: characterizations of $R\Pi\Sigma^*$-extensions and algorithmic aspects. J. Symb. Comput. **80**, 616–664 (2017)
19. Slater, L.J.: Generalized Hypergeometric Functions. Cambridge University Press, Cambridge (1966)

On Two Subclasses of Motzkin Paths and Their Relation to Ternary Trees

Helmut Prodinger, Sarah J. Selkirk, and Stephan Wagner

To Peter Paule, a vibrant mathematician and a true innovator, on the occasion of his 60th birthday.

1 Introduction

A Motzkin path is a non-negative lattice path with steps from the step set $\{\_, \diagup, \diagdown\}$ such that the path starts and ends on the x-axis. By placing further restrictions on Motzkin paths we obtain an interesting subclass.

Definition 1 An *S-Motzkin path* is a Motzkin path of length $3n$ with n of each type of step such that the following conditions hold

1. The initial step must be $\_$, and
2. $\_$ and $\diagup$ steps alternate.

This definition was inspired by a question at the recent International Mathematics Competition [10] involving restricted three-dimensional walks which can be translated into the two-dimensional S-Motzkin paths. These paths are enumerated by the generalized Catalan number, $\frac{1}{2n+1}\binom{3n}{n}$, and thus are bijective to ternary trees and non-crossing trees, as well as many other combinatorial objects [2, 5, 8, 11, 13]. We define another subclass of Motzkin paths which is related to both S-Motzkin paths and ternary trees.

Definition 2 A *T-Motzkin path* is a Motzkin path of length $3n$ with n of each type of step such that

1. The initial step is $\diagup$, and
2. $\diagup$ and $\_$ steps alternate.

H. Prodinger (✉) · S. J. Selkirk · S. Wagner
Stellenbosch University, Department of Mathematical Sciences, Stellenbosch, South Africa
e-mail: hproding@sun.ac.za; sjselkirk@sun.ac.za; swagner@sun.ac.za

© Springer Nature Switzerland AG 2020
V. Pillwein, C. Schneider (eds.), *Algorithmic Combinatorics: Enumerative Combinatorics, Special Functions and Computer Algebra*, Texts & Monographs in Symbolic Computation, https://doi.org/10.1007/978-3-030-44559-1_15

Note that although similar in definition, the class of T-Motzkin paths is larger than the class of S-Motzkin paths. Interchanging the $\_$ and $/$ steps in an arbitrary S-Motzkin path provides a T-Motzkin path, but the converse is not true. T-Motzkin paths of length $3n$ are enumerated by $\frac{1}{n+1}\binom{3n+1}{n}$ and thus bijective to the class of ordered pairs of ternary trees introduced by Knuth [6]. There are several other equinumerous objects which can be found on the Online Encyclopedia of Integer Sequences A006013 [13].

Introducing another type of path is necessary for finding generating function equations for S-Motzkin and T-Motzkin paths, and thus we define a *U-path* to be an S-Motzkin path without the initial $\_$ step. Symbolic equations for T-Motzkin paths and U-paths can be obtained in terms of each other by making use of a decomposition based on the first return of the path. Since S-Motzkin paths and U-paths are 'almost' the same, the generating function for S-Motzkin paths can be easily obtained from that of U-paths.

Various parameters associated with different types of lattice paths have been studied [1, 9, 12] and we provide analysis of the number of returns, peaks, valleys, and valleys on the x-axis in both S-Motzkin and T-Motzkin paths. This analysis is done using the symbolic equations and generating functions that are derived, as well as methods from the seminal book *Analytic Combinatorics* by Flajolet and Sedgewick [4]. During this analysis some interesting identities were found and are discussed briefly in Sect. 5.

The study of these paths as well as parameters related to them has resulted in some generalizations and developments which will be reported in further publications.

2 Bijections

2.1 S-Motzkin Paths and Ternary Trees

A bijection between S-Motzkin paths of length $3n$ and ternary trees with n nodes is provided.

2.1.1 S-Motzkin Paths to Ternary Trees

We define $\varnothing$ to be the empty path. For an arbitrary S-Motzkin path $\mathcal{M}$, the canonical decomposition is given by

$$\Phi(\mathcal{M}) = (\mathcal{A}, \mathcal{B}, C),$$

where $\mathcal{A}$, $\mathcal{B}$, and C represent the S-Motzkin paths associated with the left, middle, and right subtrees respectively. Furthermore,

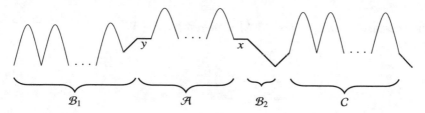

Fig. 1 Canonical decomposition of an arbitrary S-Motzkin path

- C is the path from the penultimate to the final return of M, with the initial and final step removed,
- $\mathcal{A}$ is the path from y to x (not including x), where x is the first $\_$ to the left of C, y is a $\_$ step, and the path from y to x is a Motzkin path of maximal length, and
- $\mathcal{B}$ is what remains of M after removing the path from the penultimate to the final return of M, as well as the path from y to x (including x).

This process is performed recursively and terminates at an empty path. Note that each application of Φ adds one node and removes one of each type of step. This proves inductively that an S-Motzkin path of length $3n$ maps to a ternary tree with n (internal) nodes (Fig. 1).

2.1.2 Ternary Trees to S-Motzkin Paths

The inverse mapping is performed recursively on the end nodes as follows. Each node of a ternary tree has three (possibly empty) subtrees. Call the paths associated with the left, middle, and right subtrees $\mathcal{A}$, $\mathcal{B}$, and C respectively.

Starting at the end nodes, replace each node with $\mathcal{B}_1\,\mathcal{A}\,\_\,\mathcal{B}_2\,/\,C\,\diagdown$, where $\mathcal{B}_1$ is the path from the start of $\mathcal{B}$ to the final $/$ step of $\mathcal{B}$. The path $\mathcal{B}_2$ is what remains of $\mathcal{B}$ after removing $\mathcal{B}_1$. This process is continued recursively on each set of end nodes and terminates at the root to produce an S-Motzkin path. Note that for each node that is removed one of each type of step is added, and thus a ternary tree with n nodes produces an S-Motzkin path of length $3n$.

2.1.3 Example

As an example, we map the following S-Motzkin path into a ternary tree. Since the steps are reversible, the inverse mapping can be seen by reading the example in reverse. Let M be the S-Motzkin path

The canonical decomposition of M is then $\Phi(M) = \left(\text{—}\wedge, \text{—}\wedge\wedge, \text{—}\wedge\wedge \right)$.
Hence

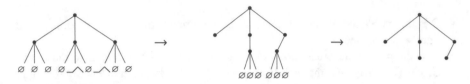

Continuing recursively:

$$\underset{\varnothing\ \varnothing\ \varnothing\ \varnothing\ \wedge\!\varnothing\ \wedge\!\varnothing\ \varnothing\ \varnothing}{\text{tree}} \rightarrow \underset{\varnothing\varnothing\varnothing\ \varnothing\varnothing\varnothing}{\text{tree}} \rightarrow \text{tree}$$

2.2 T-Motzkin Paths and Pairs of Ternary Trees

2.2.1 T-Motzkin Paths to Pairs of Ternary Trees

Since a bijection between S-Motzkin paths and ternary trees is already provided, we show that every T-Motzkin path can be decomposed uniquely into an ordered pair of S-Motzkin paths (possibly including an empty path) (Table 1).

Given an arbitrary T-Motzkin path N, we perform a canonical decomposition $\Omega(N) = (\mathcal{A}, \mathcal{B})$ where

- $\mathcal{B}$ is the path from y to x (not including x) where x is the rightmost ___ step of N, y is a ___ step, and the path from y to x is a Motzkin path of maximal length, and
- $\mathcal{A}$ is what remains of N after removing the path from y to x (including x), with an additional ___ step at the start of the path. In Fig. 2 this is the path ___ $\mathcal{A}_1\mathcal{A}_2$.

Note that both $\mathcal{A}$ and $\mathcal{B}$ are S-Motzkin paths.

2.2.2 Pairs of Ternary Trees to T-Motzkin Paths

Given an arbitrary pair of ternary trees, we can use the bijection given in Sect. 2.1 to obtain an ordered pair of S-Motzkin paths, $(\mathcal{A}, \mathcal{B})$. All S-Motzkin paths start with a ___ step and end in an $\diagup$ step followed by a series of $\diagdown$ steps. To obtain a T-Motzkin path from $(\mathcal{A}, \mathcal{B})$ we

- remove the initial ___ step from $\mathcal{A}$, and
- insert the path $\mathcal{B}$ ___ immediately after the final $\diagup$ step of $\mathcal{A}$.

Table 1 Bijection for $n = 3$

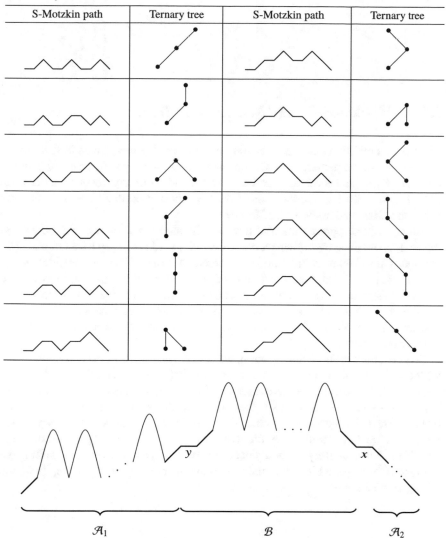

Fig. 2 Canonical decomposition of an arbitrary T-Motzkin path

2.2.3 Example

We provide an example of the mapping from T-Motzkin paths to ternary trees. The inverse mapping can be seen by reading this example in reverse. Let $\mathcal{N}$ be

.

Then $\Omega(\mathcal{N})$ is given by

$$\left(\underline{\quad} \diagup\diagup\diagdown\diagdown \ , \ \diagup\diagup\diagdown \right) \quad \longrightarrow \quad \left(\diagdown \ , \ \diagdown \right).$$

2.3 S-Motzkin Paths and Non-crossing Trees

We use the definition and representation of non-crossing trees given in [8]. For the convenience of the reader, these are repeated here. A *non-crossing tree* with n nodes is a tree whose nodes are arranged on a circle and numbered (counter-clockwise) from 1 to n, with 1 being the root of the tree. Furthermore, all edges lie entirely inside the circle and no two edges intersect.

An equivalent representation of this, and the representation that will be used in this text, is obtained by drawing a plane tree with markers to separate left and right children. Consider an arbitrary node numbered i and one of its children numbered j. If a child is a left child, then $j < i$, and a right child is a child such that $j > i$. Note that we do not distinguish between left and right children at the root. Below is the usual representation of a non-crossing tree as well as the equivalent representation that we will use. Numbering the nodes in the second tree is not necessary, but done in this case for clarity (Fig. 3).

To assist in describing the bijection, we define a *piece* to be a maximal subpath of a Motzkin path consisting of (in order) one up step, a series of down steps (possibly empty), one horizontal step, and a series of down steps (possibly empty) (Fig. 4). Note that an arbitrary S-Motzkin path of length $3n$ consists of an initial $\underline{\quad}$ step followed by $n - 1$ pieces, and a final $\diagup$ step followed by a series of $\diagdown$ steps. Each piece is uniquely determined by the number of $\diagdown$ steps and the position of the $\underline{\quad}$ step. The *characteristic pair* of a piece is the ordered pair (t, i) with t denoting the number of $\diagdown$ steps in the piece, and i denoting the position of the $\underline{\quad}$ step (with the $\diagup$ step in position 0).

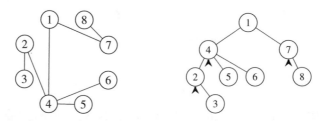

Fig. 3 A non-crossing tree represented in two different ways

Fig. 4 The four pieces in the given S-Motzkin path of length 15

2.3.1 S-Motzkin Paths to Non-crossing Trees

Given an arbitrary S-Motzkin path of length $3n$, we let c denote the number of steps in the final series of $\searrow$ steps of the path. In the resulting non-crossing tree, draw a root with c children. Considering the pieces of the path from right to left, we find the characteristic pair (t, i) and draw $i - 1$ left subtrees and $t - i + 1$ right subtrees on the rightmost available node.

2.3.2 Non-crossing Trees to S-Motzkin Paths

Given an arbitrary non-crossing tree, let c denote the number of children of the root. Associate with each non-root node an ordered pair $(u, j + 1)$ where u equals the number of children of the node and j equals the number of left subtrees of the node. Then remove the leftmost leaf. Draw an initial $\_$ step and repeat the following until only the root remains: consider the leftmost leaf's ordered pair $(u, j + 1)$ and add an $\nearrow$ step, $j \searrow$ steps, a $\_$ step and $u - j \searrow$ steps to the path, and then remove the leftmost leaf from the tree. Finally, add an $\nearrow$ step, and $c \searrow$ steps to the path (Table 2).

3 Generating Functions and Related Paths

Let $\mathcal{T}$ be the class of T-Motzkin paths, $\mathcal{U}$ be the class of U-paths, and

$$T(z) = \sum_{n \geq 0} t_n z^n \quad \text{and} \quad U(z) = \sum_{n \geq 0} u_n z^n$$

be their respective generating functions, where t_n and u_n represent the number of paths of length n in the given class.

We derive symbolic equations for the two types of paths based on a first return decomposition. Note that the only U-path of length less than five is given by $\wedge$. Taking into account the first return of a U-path, it is clear that a U-path of length five or more can be decomposed as either

$$(a) \quad \diagup^{\;x} \diagdown_y \quad \text{or} \quad (b) \quad \wedge\_z\; .$$

Table 2 Bijection for $n = 3$

S-Motzkin path	Non-crossing tree	S-Motzkin path	Non-crossing tree

In (a), $\mathcal{X}$ can either be a U-path or a T-Motzkin path. If $\mathcal{X}$ is a U-path, then $\mathcal{Y}$ is either empty or $\mathcal{Y}$ is a — step followed by a U-path. If $\mathcal{X}$ is a T-Motzkin path, then $\mathcal{Y}$ is a U-path. In (b), $\mathcal{Z}$ has to be a U-path. This then results in the symbolic equation

$$\mathcal{U} \;=\; \wedge \;+\; {\overset{\mathcal{T}}{\diagup}\!\diagdown}\,\mathcal{U} \;+\; \wedge\mathcal{U} \;+\; \diagup^{\,\mathcal{U}}\!\diagdown \;+\; \diagup^{\,\mathcal{U}}\!\diagdown\,\mathcal{U},$$

from which we obtain the equation

$$U(z) = z^2 + z^3 T(z)U(z) + 2z^3 U(z) + z^4 U(z)^2. \tag{1}$$

Again, any T-Motzkin path of length $3n$ is either empty or, considering the first return of the path, of the form (a) or (b) as given in the U-path case.

In (a), $\mathcal{X}$ can either be a U-path or a T-Motzkin path. If $\mathcal{X}$ is a U-path, then an 'extra' — step needs to appear in $\mathcal{Y}$, and thus $\mathcal{Y}$ is given by a — step followed by a T-Motzkin path. If $\mathcal{X}$ is a T-Motzkin path, then $\mathcal{Y}$ is also a T-Motzkin path. With analogous reasoning we can see that for (b) the only possibility for $\mathcal{Z}$ is a T-Motzkin path. Using this we obtain the symbolic equation

$$\mathcal{T} \;=\; \varepsilon \;+\; {\overset{\mathcal{T}}{\diagup}\!\diagdown}\,\mathcal{T} \;+\; \wedge\mathcal{T} \;+\; \diagup^{\,\mathcal{U}}\!\diagdown\,\mathcal{T}$$

which results in the equation

$$T(z) = 1 + z^3 T(z)^2 + z^3 T(z) + z^4 U(z) T(z). \tag{2}$$

Solving the system of equations given by (1) and (2) yields

$$T(z) = 1 + 2z^3 T(z)^2 - z^6 T(z)^3,$$

$$U(z) = z^2 + 3z^3 U(z) + 3z^4 U(z)^2 + z^5 U(z)^3$$

which, with substitutions, is amenable to application of the Lagrange inversion formula [4, Theorem A.2]. To demonstrate this, consider the equation

$$T(z) = 1 + 2z^3 T(z)^2 - z^6 T(z)^3.$$

This can be factorised as $T(z)(1 - z^3 T(z))^2 = 1$, and with substitutions $R = z^3 T(z)$ and $x = z^3$ we find that $x = R(1 - R)^2$. Therefore

$$[z^{3n}] T(z) = [x^{n+1}] R = \frac{1}{n+1}[w^n]\frac{1}{(1-w)^{2n+2}} = \frac{1}{n+1}\binom{3n+1}{n},$$

which results in

$$T(z) = \sum_{n \geq 0} \frac{1}{n+1}\binom{3n+1}{n} z^{3n} \quad \text{and similarly} \quad U(z) = \sum_{n \geq 1} \frac{1}{2n+1}\binom{3n}{n} z^{3n-1}.$$

Since U-paths are S-Motzkin paths without the initial horizontal step, the generating function for S-Motzkin paths is given by $S(z) = \sum_{n \geq 1} \frac{1}{2n+1}\binom{3n}{n} z^{3n}$. Note that $(1 + S(z))^2 = T(z)$, which was pointed out by Knuth in his 2014 Christmas lecture [6]. We have proved this by means of the bijection provided in Sect. 2.2.

4 Analysis of Various Parameters

In this section the analysis of the number of returns is done in detail, and results for the number of peaks, the number of valleys, and the number of valleys on the x-axis are done similarly. The study of these parameters in Dyck paths can be found in [1, 7].

4.1   The Number of Returns

From the generating functions for U-paths and T-Motzkin paths along with the substitutions $x = z^3$ and $x = t(1 - t)^2$, we obtain

$$T(z) = \frac{1}{(1 - t)^2} \quad \text{and} \quad S(z) = \frac{t}{1 - t}.$$

We introduce the variable u to count the number of returns, and also count the right end of a ⌐step on the x-axis as a return in this context. From the symbolic equations for U-paths and T-Motzkin paths we obtain the bivariate generating functions:

$$S(z, u) = u^2 z^3 + u z^3 S(z, u) T(z, 1) + u^2 z^3 S(z, u) + u^2 z^3 S(z, 1) + u^2 z^3 S(z, 1) S(z, u),$$

$$T(z, u) = 1 + u z^3 T(z, 1) T(z, u) + u^2 z^3 T(z, u) + u^2 z^3 S(z, 1) T(z, u).$$

Solving this system of equations we find that

$$S(z, u) = \frac{(1 - t) t u^2}{1 - tu - tu^2 + t^2 u^2} \quad \text{and} \quad T(z, u) = \frac{1}{1 - tu - tu^2 + t^2 u^2}.$$

4.1.1   Mean and Variance

For a bivariate generating function $K(z, u)$ with u representing the parameter of interest, we obtain the mean and variance as follows. The mean is given by

$$K_{\text{ave}} = [z^n] \frac{\partial}{\partial u} K(z, u) \Big|_{u=1} \Big/ [z^n] K(z, 1),$$

and the variance is

$$K_{\text{var}} = [z^n] \frac{\partial^2}{(\partial u)^2} K(z, u) \Big|_{u=1} \Big/ [z^n] K(z, 1) + K_{\text{ave}} - \left(K_{\text{ave}} \right)^2.$$

In the sections that follow some simplifications occur when calculating variances. These are discussed in more detail in Sect. 5.

To determine the average number of returns we calculate the derivative of $S(z, u)$ and $T(z, u)$ with respect to u,

$$\frac{\partial}{\partial u} S(z, u) \Big|_{u=1} = \frac{(2 - t) t}{(1 - t)^3} \quad \text{and} \quad \frac{\partial}{\partial u} T(z, u) \Big|_{u=1} = \frac{t(3 - 2t)}{(1 - t)^4}.$$

The total number of returns in all paths of length $3n$ is then obtained by extracting the coefficients of these expressions by means of Cauchy's integral formula. For S-Motzkin paths this results in

$$[x^n]\frac{(2-t)t}{(1-t)^3} = \frac{1}{2\pi i}\oint \frac{1}{(t(1-t)^2)^{n+1}} \cdot \frac{t(2-t)}{(1-t)^3} \cdot (1-t)(1-3t)\,dt$$

$$= \frac{1}{2\pi i}\oint \frac{1}{t^n} \cdot \frac{2-7t+3t^2}{(1-t)^{2n+4}}\,dt = [t^{n-1}]\frac{2-7t+3t^2}{(1-t)^{2n+4}}$$

$$= 2\binom{3n+2}{n-1} - 7\binom{3n+1}{n-2} + 3\binom{3n}{n-3},$$

and for T-Motzkin paths we obtain

$$[x^n]\frac{t(3-2t)}{(1-t)^4} = 3\binom{3n+3}{n-1} - 11\binom{3n+2}{n-2} + 6\binom{3n+1}{n-3}.$$

Therefore in S-Motzkin paths the average number of returns for paths of length $3n$ is

$$\frac{2\binom{3n+2}{n-1} - 7\binom{3n+1}{n-2} + 3\binom{3n}{n-3}}{\frac{1}{2n+1}\binom{3n}{n}} = \frac{n(23n+17)}{2(2n+3)(n+1)} = \frac{23}{4} - \frac{81}{8n} + O\left(\frac{1}{n^2}\right)$$

and for T-Motzkin paths the average number of returns is

$$\frac{3\binom{3n+3}{n-1} - 11\binom{3n+2}{n-2} + 6\binom{3n+1}{n-3}}{\frac{1}{n+1}\binom{3n+1}{n}} = \frac{(19n+26)n}{2(2n+3)(n+2)} = \frac{19}{4} - \frac{81}{8n} + O\left(\frac{1}{n^2}\right).$$

To calculate the variance in the number of returns for paths of length $3n$, we find the second derivatives of $S(z,u)$ and $T(z,u)$ with respect to u:

$$\frac{\partial^2}{(\partial u)^2}S(z,u)\bigg|_{u=1} = \frac{2t(1+3t-4t^2+t^3)}{(1-t)^5}$$

and

$$\frac{\partial^2}{(\partial u)^2}T(z,u)\bigg|_{u=1} = \frac{2t(1+6t-9t^2+3t^3)}{(1-t)^6}.$$

We again determine the coefficients using Cauchy's integral formula,

$$[x^n]\frac{2t(1+3t-4t^2+t^3)}{(1-t)^5} = 2\left[\binom{3n+4}{n-1} - 13\binom{3n+2}{n-3} + 13\binom{3n+1}{n-4} - 3\binom{3n}{n-5}\right]$$

and

$$[x^n]\frac{2t(1+6t-9t^2+3t^3)}{(1-t)^6} = 2\left[\binom{3n+5}{n-1} + 3\binom{3n+4}{n-2} - 27\binom{3n+3}{n-3}\right.$$
$$\left. + 30\binom{3n+2}{n-4} - 9\binom{3n+1}{n-5}\right],$$

with which we find that the variance for the number of returns for S-Motzkin paths of length $3n$ is

$$\frac{2(313n^3 + 652n^2 + 53n - 178)n}{(2n+5)(2n+4)(2n+3)(2n+2)} + \frac{n(23n+17)}{2(2n+3)(n+1)} - \left(\frac{n(23n+17)}{2(2n+3)(n+1)}\right)^2$$
$$= \frac{3(14n^2 + 31n + 8)(3n+2)(3n+1)(n-1)n}{4(2n+5)(2n+3)^2(n+2)(n+1)^2}.$$

Similarly, the variance for the number of returns for T-Motzkin paths of length $3n$ is given by

$$\frac{3(79n^3 + 252n^2 + 91n - 142)n}{2(2n+5)(2n+3)(n+3)(n+2)} + \frac{(19n+26)n}{2(2n+3)(n+2)} - \left(\frac{(19n+26)n}{2(2n+3)(n+2)}\right)^2$$
$$= \frac{3(14n^3 + 45n^2 + 19n - 18)(3n+4)(3n+2)n}{4(2n+5)(2n+3)^2(n+3)(n+2)^2}.$$

4.1.2 Limiting Distributions

We have defined t implicitly by $t(1-t)^2 = x$. It is well known that this type of implicit equation leads to a square root singularity [4, Section VII.4]. In this particular case, the singularity occurs at $x = \frac{4}{27}$, $t = \frac{1}{3}$, where $\frac{d}{dt}t(1-t^2) = (1-t)(1-3t) = 0$. At this point, the singular expansion of t with respect to x is

$$t = \frac{1}{3} - \frac{2}{3\sqrt{3}}\left(1 - \frac{27x}{4}\right)^{1/2} + O\left(1 - \frac{27x}{4}\right).$$

The generating function for the number of returns in S-Motzkin paths is given by

$$S(z,u) = \frac{(1-t)tu^2}{1 - tu - tu^2 + t^2u^2}.$$

Note that for $|x| \le \frac{4}{27}$ and $|u| \le 1$, we have $|t| \le \frac{1}{3}$ and thus

$$|1 - tu - tu^2 + t^2u^2| \ge 1 - |t||u| - |t||u|^2 - |t|^2|u|^2 \ge 1 - \frac{1}{3} - \frac{1}{3} - \frac{1}{9} = \frac{2}{9} > 0,$$

so the denominator is nonzero and the singularity of t remains the dominant singularity. This generating function has the Taylor expansion (with substitution of the singular expansion of t):

$$S(z, u) = \frac{2u^2}{9 - 3u - 2u^2} + \frac{9u^2}{27 - 27u + 4u^3}\left(t - \frac{1}{3}\right) + O\left(\left(t - \frac{1}{3}\right)^2\right)$$

$$= \frac{2u^2}{9 - 3u - 2u^2} - \frac{2\sqrt{3}\,u^2}{27 - 27u + 4u^3}\left(1 - \frac{27x}{4}\right)^{\frac{1}{2}} + O\left(1 - \frac{27x}{4}\right).$$

Applying singularity analysis [4, Section VI], we obtain

$$[x^n]S(z, u) \sim \frac{2\sqrt{3}\,u^2}{27 - 27u + 4u^3} \cdot \frac{1}{2\sqrt{\pi}} \cdot n^{-3/2}\left(\frac{27}{4}\right)^n.$$

Therefore, the probability generating function for the number of returns in S-Motzkin paths of length $3n$, which is given by $[x^n]S(x, u)/[x^n]S(x, 1)$, converges to

$$\frac{4u^2}{27 - 27u + 4u^3} = \frac{4u^2}{(2u - 3)^2(u + 3)} = \frac{4}{27}u^2 + \frac{4}{27}u^3 + \frac{4}{27}u^4 + \frac{92}{729}u^5 + \cdots.$$

By [4, Theorem IX.1], the distribution of the number of returns in S-Motzkin paths converges to the discrete distribution given by this probability generating function. The probability that the number of returns is precisely k converges to

$$[u^k]\frac{4u^2}{(2u - 3)^2(u + 3)} = \frac{4}{3^{k+3}}(3k \cdot 2^{k-1} - 2^k + (-1)^k).$$

In a similar manner, we find that the limiting probability generating function for the number of returns in T-Motzkin paths of length $3n$ is given by

$$\frac{4u}{(2u - 3)^2(u + 3)} = \frac{4}{27}u + \frac{4}{27}u^2 + \frac{4}{27}u^3 + \frac{92}{729}u^4 + \frac{76}{729}u^5 + \cdots,$$

and the probability that the number of returns is precisely k converges to

$$[u^k]\frac{4u}{(2u - 3)^2(u + 3)} = \frac{4}{3^{k+4}}(3k \cdot 2^k + 2^k - (-1)^k).$$

The convergence in both cases is demonstrated in the figures below.

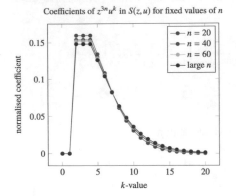

Coefficients of $z^{3n}u^k$ in $S(z, u)$ for fixed values of n

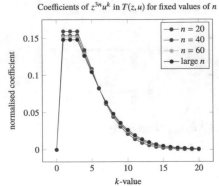

Coefficients of $z^{3n}u^k$ in $T(z, u)$ for fixed values of n

4.2 The Number of Peaks

There are two possible types of peaks:

$$(1) \quad \wedge \quad \text{and} \quad (2) \quad \text{\Large\frown}$$

We first consider peaks of type (1) and again use the variable u to count them. Then from the symbolic equations

$$S(z, u) = uz^3 + z^3 T(z, u)S(z, u) + uz^3 S(z, u) + z^3 S(z, u) + z^3 S(z, u)^2,$$

$$T(z, u) = 1 + z^3 T(z, u)^2 + uz^3 T(z, u) + z^3 S(z, u)T(z, u)$$

we obtain the results in Table 3.

Table 3 Results for peaks of type (1)

$K(z, u)$	$S(z, u)$	$T(z, u)$
$\left. \frac{\partial}{\partial u} K(z, u) \right|_{u=1}$	$\frac{t(1-2t)}{(1-3t)(1-t)}$	$\frac{t}{(1-3t)(1-t)}$
$\left. [x^n] \frac{\partial}{\partial u} K(z, u) \right|_{u=1}$	$\binom{3n}{n-1} - 2\binom{3n-1}{n-2}$	$\binom{3n}{n-1}$
Mean	$\frac{n}{3} + \frac{2}{3}$	$\frac{n(n+1)}{3n+1}$
$\left. \frac{\partial^2}{(\partial u)^2} K(z, u) \right|_{u=1}$	$\frac{2t^2(1-5t+8t^2-3t^3)}{(1-3t)^3(1-t)}$	$\frac{2t^2(1-2t)}{(1-3t)^3(1-t)}$
$\left. [x^n] \frac{\partial^2}{(\partial u)^2} K(z, u) \right|_{u=1}$	$\binom{3n-2}{n-3} \frac{n(n+3)}{(n-2)}$	$\binom{3n-1}{n-2} n$
Variance	$\frac{2(2n+1)(n-1)}{9(3n-1)} \sim \frac{4}{27}n$	$\frac{2(2n+1)(n+1)n}{3(3n+1)^2} \sim \frac{4}{27}n$

The system of equations for $S(z, u)$ and $T(z, u)$ satisfies the technical conditions of [3], where it is shown that we have convergence to a normal law in a rather general setting. By the main result of [3], the number of peaks (of both types) asymptotically follows a Gaussian distribution.

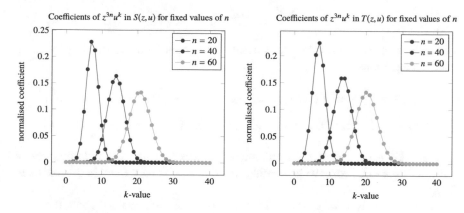

We now consider peaks of type (2), and again use the variable u to count them. From the symbolic equations we obtain

$$S(z, u) = z^3 + z^3 T(z, u) S(z, u) + u z^3 S(z, u) + z^3 S(z, u) + z^3 S(z, u)^2,$$

$$T(z, u) = 1 + z^3 T(z, u)^2 + u z^3 T(z, u) + z^3 S(z, u) T(z, u).$$

Note here that $\mathcal{T}$ contains an empty path. As a result, for paths of the form

$$\overset{\mathcal{T}}{\diagup}\!\!\diagdown_{\mathcal{U}} \qquad \text{and} \qquad \overset{\mathcal{T}}{\diagup}\!\!\diagdown_{\mathcal{T}},$$

we obtain $u z^3 S(z, u)$ and $u z^3 T(z, u)$ respectively. If the path is not empty we obtain $z^3 (T(z, u) - 1) S(z, u)$ and $z^3 (T(z, u) - 1) T(z, u)$. Using the generating function equations we obtain the results in Table 4.

Table 4 Results for peaks of type (2)

$K(z, u)$	$S(z, u)$	$T(z, u)$
$\left.\frac{\partial}{\partial u} K(z, u)\right\|_{u=1}$	$\frac{t^2}{1-3t}$	$\frac{t}{1-3t}$
$\left.[x^n]\frac{\partial}{\partial u} K(z, u)\right\|_{u=1}$	$\binom{3n-2}{n-2}$	$\binom{3n-1}{n-1}$
Mean	$\frac{(2n+1)(n-1)}{3(3n-1)} \sim \frac{2}{9} n$	$\frac{(2n+1)(n+1)}{3(3n+1)} \sim \frac{2}{9} n$
$\left.\frac{\partial^2}{(\partial u)^2} K(z, u)\right\|_{u=1}$	$\frac{2(1-2t)(1-t)t^3}{(1-3t)^3}$	$\frac{2(1-3t+3t^2)(1-t)t^2}{(1-3t)^3}$
$\left.[x^n]\frac{\partial^2}{(\partial u)^2} K(z, u)\right\|_{u=1}$	$\binom{3n-3}{n-3}\frac{2n}{3}$	$\binom{3n-3}{n-2}n$
Variance	$\frac{2(10n^2-11n+2)(2n+1)(n-1)}{9(3n-1)^2(3n-2)}$ $\sim \frac{40}{243} n$	$\frac{2(30n^3-23n^2-3n+2)(2n+1)(n+1)}{9(3n+1)^2(3n-1)(3n-2)}$ $\sim \frac{40}{243} n$

4.3 Valleys

There are two possible types of valleys:

$$(1) \quad \vee \quad \text{and} \quad (2) \quad \diagdown\!\diagup$$

For valleys of type (1), using the variable u to count them we obtain generating function equations

$$S(z, u) = z^3 + uz^3 T(z, u) S(z, u) + z^3 S(z, u) + z^3 S(z, u) + z^3 S(z, u)^2,$$

$$T(z, u) = 1 + uz^3 T(z, u)(T(z, u) - 1) + 2z^3 T(z, u) + z^3 S(z, u) T(z, u).$$

This is taking into account that the empty T-Motzkin path does not contribute a valley of type (1). From the generating function equations we obtain the results in Table 5.

As in our analysis of peaks in the previous subsection, we can apply the main result of [3] to prove that the number of valleys (of both types) asymptotically follows a Gaussian distribution.

The generating function equations for valleys of type (2), again using u to count the number of valleys, are given by

$$S(z, u) = z^3 + z^3 T(z, u) S(z, u) + uz^3 S(z, u) + z^3 S(z, u) + uz^3 S(z, u)^2,$$

$$T(z, u) = 1 + z^3 T(z, u)^2 + uz^3 (T(z, u) - 1) + z^3 + uz^3 S(z, u)(T(z, u) - 1)$$
$$+ z^3 S(z, u).$$

Again, we take into account the absence of a valley in the case of an empty T-Motzkin path. The equations yield the results in Table 6.

Table 5 Results for valleys of type (1)

$K(z, u)$	$S(z, u)$	$T(z, u)$
$\left. \frac{\partial}{\partial u} K(z, u) \right|_{u=1}$	$\frac{t^2}{(1-3t)(1-t)}$	$\frac{2t^2}{(1-3t)(1-t)^2}$
$\left. [x^n] \frac{\partial}{\partial u} K(z, u) \right|_{u=1}$	$\binom{3n-1}{n-2}$	$2\binom{3n}{n-2}$
Mean	$\frac{n}{3} - \frac{1}{3}$	$\frac{n(n-1)}{3n+1}$
$\left. \frac{\partial^2}{(\partial u)^2} K(z, u) \right|_{u=1}$	$\frac{2(1-t-3t^2)t^3}{(1-3t)^3(1-t)}$	$\frac{2(2-t-9t^2)t^3}{(1-3t)^3(1-t)^2}$
$\left. [x^n] \frac{\partial^2}{(\partial u)^2} K(z, u) \right|_{u=1}$	$(n-1)\binom{3n-2}{n-3}$	$\frac{(n-1)(n-2)}{(n+1)}\binom{3n-1}{n-2}$
Variance	$\frac{2(2n+1)(n-1)}{9(3n-1)} \sim \frac{4}{27}n$	$\frac{2(2n+1)(n+1)(n-1)}{3(3n+1)^2} \sim \frac{4}{27}n$

Table 6 Results for valleys of type (2)

$K(z, u)$	$S(z, u)$	$T(z, u)$
$\frac{\partial}{\partial u} K(z, u)\Big|_{u=1}$	$\frac{t^2}{(1-3t)}$	$\frac{2t^2}{(1-3t)(1-t)}$
$[x^n]\frac{\partial}{\partial u} K(z, u)\Big|_{u=1}$	$\binom{3n-2}{n-2}$	$2\binom{3n-1}{n-2}$
Mean	$\frac{(n-1)(2n+1)}{3(3n-1)} \sim \frac{2}{9}n$	$\frac{2(n+1)(n-1)}{3(3n+1)} \sim \frac{2}{9}n$
$\frac{\partial^2}{(\partial u)^2} K(z, u)\Big|_{u=1}$	$\frac{2(1-2t)(1-t)t^3}{(1-3t)^3}$	$\frac{2(2-3t-3t^2)t^3}{(1-3t)^3}$
$[x^n]\frac{\partial^2}{(\partial u)^2} K(z, u)\Big|_{u=1}$	$\frac{2n}{3}\binom{3n-3}{n-3}$	$2(n-1)\binom{3n-3}{n-3}$
Variance	$\frac{2(10n^2-11n+2)(2n+1)(n-1)}{9(3n-1)^2(3n-2)} \sim \frac{40}{243}n$	$\frac{4(15n^2-19n+8)(2n+1)(n+1)(n-1)}{9(3n+1)^2(3n-1)(3n-2)} \sim \frac{40}{243}n$

Table 7 Results for valleys on the x-axis of type (1)

$K(z, u)$	$S(z, u)$	$T(z, u)$
$\frac{\partial}{\partial u} K(z, u)\Big|_{u=1}$	$\frac{t^2}{(1-t)^3}$	$\frac{(2-t)t^2}{(1-t)^4}$
$[x^n]\frac{\partial}{\partial u} K(z, u)\Big|_{u=1}$	$\binom{3n+1}{n-2} - 3\binom{3n}{n-3}$	$2\binom{3n+2}{n-2} - 7\binom{3n+1}{n-3} + 3\binom{3n}{n-4}$
Mean	$\frac{7(n-1)n}{2(2n+3)(n+1)} \sim \frac{7}{4}$	$\frac{(19n+18)(n-1)n}{2(3n+1)(2n+3)(n+2)} \sim \frac{19}{12}$
$\frac{\partial^2}{(\partial u)^2} K(z, u)\Big|_{u=1}$	$\frac{2t^3}{(1-t)^5}$	$\frac{2(2-t)t^3}{(1-t)^6}$
$[x^n]\frac{\partial^2}{(\partial u)^2} K(z, u)\Big|_{u=1}$	$2\binom{3n+2}{n-3} - 6\binom{3n+1}{n-4}$	$4\binom{3n+3}{n-3} - 14\binom{3n+2}{n-4} + 6\binom{3n+1}{n-5}$
Variance	$\frac{v_1(3n+1)(n-1)n}{4(2n+5)(2n+3)^2(n+2)(n+1)^2}$ $\sim \frac{45}{16}$	$\frac{v_2(n-1)n}{4(3n+1)^2(2n+5)(2n+3)^2(n+3)(n+2)^2}$ $\sim \frac{389}{144}$

4.4 Valleys on the x-Axis

We now consider valleys that lie on the x-axis. Keeping the two types of valleys discussed in the previous subsection, a valley of type (1) contributes one return, and a valley of type (2) contributes two returns.

(1) and (2)

For valleys on the x-axis of type (1), using the variable u to count them we obtain generating function equations

$$S(z, u) = z^3 + uz^3 T(z, 1)S(z, u) + z^3 S(z, u) + z^3 S(z, 1) + z^3 S(z, u)S(z, 1),$$

$$T(z, u) = 1 + uz^3 T(z, 1)(T(z, u) - 1) + z^3 T(z, 1) + z^3 T(z, u) + z^3 S(z, 1)T(z, u).$$

Let $v_1 = 30n^3 + 43n^2 + 154n + 288$ and $v_2 = 778n^6 + 3953n^5 + 11212n^4 + 24373n^3 + 30064n^2 + 16260n + 2160$, then we obtain the results in Table 7.

Table 8 Results for valleys on the x-axis of type (2)

$K(z,u)$	$S(z,u)$	$T(z,u)$
$\left.\frac{\partial}{\partial u}K(z,u)\right\|_{u=1}$	$\frac{t^2}{(1-t)^2}$	$\frac{(2-t)t^2}{(1-t)^3}$
$[x^n]\frac{\partial}{\partial u}K(z,u)\Big\|_{u=1}$	$\binom{3n}{n-2}-3\binom{3n-1}{n-3}$	$2\binom{3n+1}{n-2}-7\binom{3n}{n-3}+3\binom{3n-1}{n-4}$
Mean	$\frac{n-1}{n+1}\sim 1$	$\frac{(11n+6)(n-1)}{2(3n+1)(2n+3)}\sim\frac{11}{12}$
$\left.\frac{\partial^2}{(\partial u)^2}K(z,u)\right\|_{u=1}$	$\frac{2t^3}{(1-t)^3}$	$\frac{2(2-t)t^3}{(1-t)^4}$
$[x^n]\frac{\partial^2}{(\partial u)^2}K(z,u)\Big\|_{u=1}$	$2\binom{3n}{n-3}-6\binom{3n-1}{n-4}$	$4\binom{3n+1}{n-3}-14\binom{3n}{n-4}+6\binom{3n-1}{n-5}$
Variance	$\frac{(3n+1)(n-1)n}{(2n+3)(n+1)^2}\sim\frac{3}{2}$	$\frac{(203n^3+437n^2+268n+12)(n-1)n}{4(3n+1)^2(2n+3)^2(n+2)}\sim\frac{203}{144}$

In a similar manner to that of Sect. 4.1.2, we find that the limiting probability generating function for valleys of type (1) on the x-axis are given by $\frac{4(u+3)}{(7-3u)^2}$ for S-Motzkin paths, and $\frac{4(u+11)}{3(7-3u)}$ for T-Motzkin paths (both of length $3n$).

For valleys on the x-axis of type (2), again using u to count the number of valleys, we obtain the generating function equations

$$S(z,u)=z^3+z^3T(z,1)S(z,u)+uz^3S(z,u)+z^3S(z,1)+uz^3S(z,u)S(z,1),$$

$$T(z,u)=1+z^3T(z,u)T(z,1)+z^3+uz^3(T(z,u)-1)+uz^3S(z,1)(T(z,u)-1)$$
$$+z^3S(z,1).$$

From these the results in Table 8 are obtained.

The limiting probability generating functions for the number of valleys of type (2) on the x-axis is given by $\frac{4}{(3-u)^2}$ for S-Motzkin paths and $\frac{13-u}{3(3-u)^2}$ for T-Motzkin paths (both of length $3n$).

5 Identities

In Sects. 4.2 and 4.3 the coefficients of the generating functions used to find the variance were greatly simplified by using derivatives (compared to extracting coefficients using Cauchy's integral formula). An example of this simplification is given: Using Cauchy's integral formula as in Sect. 4.1.1 we obtain the coefficients

$$[x^n]\frac{2t^2(1-2t)}{(1-3t)^3(1-t)}=2\sum_{k\geq 0}(k+1)3^k\left[\binom{3n-k-1}{n-k-2}-2\binom{3n-k-2}{n-k-3}\right].$$

On the other hand, the generating function can be expressed as a derivative, and by using the formula $\frac{t^3}{1-3t} = \sum_{n\geq 3} \binom{3n-3}{n-3} x^n$ we find that

$$[x^n] \frac{2t^2(1-2t)}{(1-3t)^3(1-t)} = [x^n] \frac{2}{3} \cdot \frac{d}{dx} \frac{t^3}{1-3t} = \binom{3n-1}{n-2} n.$$

It follows that

$$2\sum_{k\geq 0}(k+1)3^k\left[\binom{3n-k-1}{n-k-2} - 2\binom{3n-k-2}{n-k-3}\right] = \binom{3n-1}{n-2} n,$$

which is a special case of the more general identity

$$2\sum_{k\geq j}3^k(k+1)\left[\binom{3n-k-1}{n-k-2} - 2\binom{3n-k-2}{n-k-3}\right] = \binom{3n-j-1}{n-j-2}(n+j)3^j.$$

This and other beautiful identities such as

$$2\sum_{k\geq 0}3^k(k+2i)\binom{3n-k+i-4}{n-k-i-1} = \binom{3n+i-3}{n-i}(n-i)$$

and

$$2\sum_{k\geq 0}3^k(k+2i+1)\binom{3n-k+i-2}{n-k-i-1} = \binom{3n+i-1}{n-i}(n-i).$$

can be proved directly by induction. A table of the simplifications used to calculate variances in Sect. 4 is given in Table 9.

Table 9 Generating functions and their coefficients

Generating function	In terms of derivatives	Power series expansion
$\frac{2t^2(1-5t+8t^2-3t^3)}{(1-3t)^3(1-t)}$	$-\frac{2}{3}\frac{d}{dx}\frac{t^3}{1-3t} + 2x\frac{d}{dx}\frac{t^2}{1-3t}$	$\sum_{n\geq 3}\binom{3n-2}{n-3}\frac{n(n+3)}{(n-2)}x^n$
$\frac{2t^2(1-2t)}{(1-3t)^3(1-t)}$	$\frac{2}{3}\frac{d}{dx}\frac{t^3}{1-3t}$	$\sum_{n\geq 2}\binom{3n-1}{n-2}nx^n$
$\frac{2t^3(1-2t)(1-t)}{(1-3t)^3}$	$\frac{2}{3}x\frac{d}{dx}\frac{t^3}{1-3t}$	$\sum_{n\geq 3}\binom{3n-3}{n-3}\frac{2n}{3}x^n$
$\frac{2t^2(1-3t+3t^2)(1-t)}{(1-3t)^3}$	$x\frac{d}{dx}\frac{t^2}{1-3t} - x\frac{d}{dx}\frac{t^3}{1-3t}$	$\sum_{n\geq 2}\binom{3n-3}{n-2}nx^n$
$\frac{2t^3(1-t-3t^2)}{(1-3t)^3(1-t)}$	$\frac{1}{2}\frac{d}{dx}\frac{t^5}{1-3t} + \frac{1}{2}\frac{d}{dx}\frac{t^4}{1-3t}$	$\sum_{n\geq 3}\binom{3n-2}{n-3}(n-1)x^n$
$\frac{2t^3(2-t-9t^2)}{(1-3t)^3(1-t)^2}$	$\frac{1}{x}\frac{d}{dx}\left(\frac{4}{5}\frac{t^5}{1-3t} + \frac{3}{5}\frac{t^6}{1-3t} - \frac{t^7}{1-3t}\right)$	$\sum_{n\geq 2}\binom{3n-1}{n-2}\frac{(n-1)(n-2)}{n+1}x^n$
$\frac{2t^3(2-3t-3t^2)}{(1-3t)^3}$	$-\frac{2}{5}\frac{d}{dx}\frac{t^6}{1-3t} - \frac{1}{5}\frac{d}{dx}\frac{t^5}{1-3t} + \frac{d}{dx}\frac{t^4}{1-3t}$	$\sum_{n\geq 3}2\binom{3n-3}{n-3}(n-1)x^n$

References

1. Deutsch, E.: Dyck path enumeration. Discrete Math. **204**(1–3), 167–202 (1999)
2. Deutsch, E., Feretić, S., Noy, M.: Diagonally convex directed polyominoes and even trees: a bijection and related issues. Discrete Math. **256**(3), 645–654 (2002)
3. Drmota, M.: Systems of functional equations. Random Struct. Algorithms – Special issue: average-case analysis of algorithms **10**, 103–124 (1997)
4. Flajolet, P., Sedgewick, R.: Analytic Combinatorics. Cambridge University Press, Cambridge (2009)
5. Gu, N.S.S., Li, N.Y., Mansour, T.: 2-binary trees: bijections and related issues. Discrete Math. **308**, 1209–1221 (2008)
6. Knuth, D.: Donald Knuth's 20th Annual Christmas Tree Lecture: (3/2)-ary Trees. https://youtu.be/P4AaGQIo0HY, December (2014)
7. Mansour, T.: Counting peaks at height k in a Dyck path. J. Integer Seq. **5**, Article 02.1.1 (2002)
8. Panholzer, A., Prodinger, H.: Bijections for ternary trees and non-crossing trees. Discrete Math. **250**(1–3), 181–195 (2002)
9. Pergola, E., Pinzani, R., Rinaldi, S., Sulanke, R.A.: A bijective approach to the area of generalized Motzkin paths. Adv. Appl. Math. **28**(3–4), 580–591 (2002)
10. Petrov, F., Vershik, A.: International Mathematics Competition: Day 2 Problem 8. http://www.imc-math.org.uk/imc2018/imc2018-day2-questions.pdf (2018). Last accessed 26 Jan 2019
11. Prodinger, H.: A simple bijection between a subclass of 2-binary trees and ternary trees. Discrete Math. **309**, 959–961 (2009)
12. Prodinger, H., Wagner, S.: Minimal and maximal plateau lengths in Motzkin paths. In: AofA 2007, DMTCS Proceedings, AH 2007, pp. 353–362 (2007)
13. Sloane, N.J.A.: The On-line Encyclopedia of Integer Sequences. Published electronically at https://oeis.org (2019)

A Theorem to Reduce Certain Modular Form Relations Modulo Primes

Cristian-Silviu Radu

Dedicated to my advisor and friend Peter Paule on the occasion of his 60th birthday

1 Introduction

Let p be a prime. Let $\mathcal{A}_k^1(N)$ be the set of meromorphic modular forms of weight k for the group $\Gamma_1(N)$ and $\mathcal{A}^1(N) := \oplus_{k=-\infty}^{\infty} \mathcal{A}_k^1(N)$. Let $f_j(\tau) = \sum_{n=m_j}^{\infty} a_j(n)q^n \in \mathcal{A}^1(N)$, $j = 0, \ldots, \nu$ and $q = e^{2\pi i \tau}$ such that the $a_j(n)$ are integers. The main result of this paper is that

$$f_0(\tau) + q f_1(\tau) + q^2 f_2(\tau) + \cdots + q^\nu f_\nu(\tau) \equiv 0 \pmod{p} \tag{1}$$

iff

$$f_0(\tau) \equiv f_1(\tau) \equiv f_2(\tau) \equiv \cdots \equiv f_\nu(\tau) \equiv 0 \pmod{p}. \tag{2}$$

This result is also important from an algorithmic point of view because if we want to design an algorithm to prove relations like (1) we see that we only need to prove congruences modulo p between meromorphic modular forms. For this situation

The research was funded by the Austrian Science Fund (FWF), W1214-N15, project DK6 and by grant P2016-N18. The research was supported by the strategic program "Innovatives 2010 plus" by the Upper Austrian Government.

C.-S. Radu (✉)
Johannes Kepler University Linz, Research Institute for Symbolic Computation (RISC), Linz, Austria
e-mail: silviu.radu@risc.jku.at

© Springer Nature Switzerland AG 2020
V. Pillwein, C. Schneider (eds.), *Algorithmic Combinatorics: Enumerative Combinatorics, Special Functions and Computer Algebra*, Texts & Monographs in Symbolic Computation, https://doi.org/10.1007/978-3-030-44559-1_16

there are well-known proving tools like Sturm's theorem, etc. In this regard there is a lot of theory developed which allows automatization of proving such relations.

The organization of this paper is as follows. In Sect. 2 we introduce basic definitions and notions. In Sect. 3 the main result of this paper is proven, namely the implication $(1) \Rightarrow (2)$.

2 Basic Notions and Definitions

Let

$$M_2(\mathbb{Z}) := \left\{ \begin{pmatrix} a & b \\ c & d \end{pmatrix} : a, b, c, d \in \mathbb{Z}, ad - bc > 0 \right\}$$

and

$$SL_2(\mathbb{Z}) := \left\{ \begin{pmatrix} a & b \\ c & d \end{pmatrix} \in M_2(\mathbb{Z}), ad - bc = 1 \right\}.$$

For N a positive integer let

$$\Gamma_0(N) := \left\{ \begin{pmatrix} a & b \\ c & d \end{pmatrix} \in SL_2(\mathbb{Z}) : c \equiv 0 \ (\mathrm{mod}\ N) \right\},$$

$$\Gamma_1(N) := \left\{ \begin{pmatrix} a & b \\ c & d \end{pmatrix} \in \Gamma_0(N) : a \equiv d \equiv 1 \ (\mathrm{mod}\ N) \right\},$$

and

$$\Gamma(N) := \left\{ \begin{pmatrix} a & b \\ c & d \end{pmatrix} \in \Gamma_1(N) : b \equiv 0 \ (\mathrm{mod}\ N) \right\}.$$

Let

$$\mathbb{H} := \{ \tau \in \mathbb{C} : \mathrm{Im}(\tau) > 0 \}.$$

If f, g are meromorphic functions on $\mathbb{H}$ and $f(\tau) = g(\tau)$ for all values $\tau \in \mathbb{H}$ where f, g are defined, we simply write $f(\tau) = g(\tau)$ and omit to write where τ lives. There will be no confusion because we will always use the symbol τ for a generic $\tau \in \mathbb{H}$. For special values we will use τ with a subscript for example τ_0, $\tau_1, \ldots$, etc. Since the symbol τ is always used for generic $\tau \in \mathbb{H}$ we will often write $f(\tau)$ for the function f and for specializations of f at a point we use for the point the symbol τ_j for $j \in \mathbb{N}$. That is, $f(\tau_j)$ is the value of f at the point τ_j.

For $k \in \mathbb{Z}$, f meromorphic on $\mathbb{H}$ and $\gamma = \begin{pmatrix} a & b \\ c & d \end{pmatrix} \in M_2(\mathbb{Z})$ we define

$$(f|_k \gamma)(\tau) := (ad - bc)^{k/2}(c\tau + d)^{-k} f\left(\frac{a\tau + b}{c\tau + d}\right).$$

Then for $\gamma_1, \gamma_2 \in M_2(\mathbb{Z})$:

$$f|_k \gamma_1 |_k \gamma_2 = f|_k \gamma_1 \gamma_2.$$

A good reference for properties like this e.g. is [3].

Let N be a positive integer and k an integer. Let Γ be a subgroup of $SL_2(\mathbb{Z})$ such that $\Gamma(N) \subseteq \Gamma$. A *meromorphic modular form of weight k for Γ* is a meromorphic function f on $\mathbb{H}$ such that:

(i) for all $\gamma \in \Gamma$, $f|_k \gamma = f$;
(ii) for all $\gamma \in SL_2(\mathbb{Z})$, $(f|_k \gamma)(\tau)$ admits a Laurent expansion in powers of $e^{\frac{2\pi i \tau}{N}}$ with finite principal part.

We denote the set of meromorphic modular forms of weight k for Γ by $\mathscr{A}_k(\Gamma)$.

A *weak modular form of weight k for Γ* is a meromorphic modular form of weight k for Γ which is holomorphic on $\mathbb{H}$. We denote the set of weak modular forms of weight k for Γ by $M_k^!(\Gamma)$.

A *modular form of weight k for Γ* is a weak modular form of weight k for Γ such that $(f|_k \gamma)(\tau)$ admits a Laurent expansion in powers of $e^{\frac{2\pi i \tau}{N}}$ with principal part 0. We denote the set of modular forms of weight k for Γ by $M_k(\Gamma)$.

Remark 2.1 Let $T := \begin{pmatrix} 1 & 1 \\ 0 & 1 \end{pmatrix}$. We note that if f is a meromorphic modular form of weight k for $\Gamma_1(N)$, then since $T \in \Gamma_1(N)$ and because of *(i)* we have

$$(f|_k T)(\tau) = f(\tau \mid 1) = f(\tau).$$

Because of *(ii)*, there exist $m \in \mathbb{Z}$ and $a(n) \in \mathbb{C}$, $n \geq m$, such that

$$f(\tau) = \sum_{n=m}^{\infty} a(n) e^{\frac{2\pi i n \tau}{N}} \quad \text{and consequently} \quad f(\tau + 1) = \sum_{n=m}^{\infty} a(n) e^{\frac{2\pi i n}{N}} e^{\frac{2\pi i n \tau}{N}}.$$

In particular $f(\tau + 1) = f(\tau)$ implies that $a(n)e^{\frac{2\pi i n}{N}} = a(n)$ which is only possible iff $a(n) = 0$ unless $N | n$. This implies that there exist $m' \in \mathbb{Z}$ and $b(n) \in \mathbb{C}$, $n \geq m'$, such that

$$f(\tau) = \sum_{n=m'}^{\infty} b(n) q^n \tag{3}$$

where here and in the following

$$q = q(\tau) := e^{2\pi i \tau}.$$

Note that in the sum (3) q should be understood as $q(\tau)$.

Note 2.2 When $f \in \mathcal{A}_k(\Gamma_1(N))$ for convenience we will write

$$f(\tau) = \sum_{n=-\infty}^{\infty} a(n)q^n \qquad (4)$$

although because of *(ii)*, there exists an integer m such that $a(n) = 0$ for all $n < m$.

For simplicity we define

$$\mathcal{A}_k^!(N) := \mathcal{A}_k(\Gamma_1(N)).$$

As we observed in Remark 2.1, if $f \in \mathcal{A}_k^!(N)$ then $f(\tau) = \sum_{n=-\infty}^{\infty} a(n)q^n$. Let R be a subring of $\mathbb{C}$. If $a(n) \in R$ for all $n \in \mathbb{Z}$, we say that $f \in \mathcal{A}_k^!(N, R)$.

Similarly if $f \in \mathcal{A}_k(\Gamma(N))$, then $f(\tau) = \sum_{-\infty}^{\infty} b(n)q_N$ with $q_N := e^{\frac{2\pi i \tau}{N}}$. If $b(n) \in R$ (for R as above), for all $n \in \mathbb{Z}$, then we say that $f \in \mathcal{A}_k(\Gamma(N), R)$. Analogously we define $M_k^!(\Gamma, R)$ and $M_k(\Gamma, R)$ for an arbitrary subgroup $\Gamma \subseteq SL_2(\mathbb{Z})$.

3   Main Result

The goal of this section is to prove Theorem 3.18 which says that for given $f_j(\tau) \in \mathcal{A}^!(N) = \cup_{k=-\infty}^{\infty} \mathcal{A}_k^!(N)$ for $j = 0, \ldots, \nu$, we have

$$f_0(\tau) + q f_1(\tau) + \cdots + q^\nu f_\nu(\tau) \equiv 0 \pmod{p}$$

iff

$$f_0(\tau) \equiv f_1(\tau) \equiv \cdots \equiv f_\nu(\tau) \equiv 0 \pmod{p}.$$

We will need a couple of results for proving this and, for the sake of logical transparence we explain here shortly how they depend on each other. Lemmas 3.1 and 3.2 are used for proving Lemma 3.3. Lemma 3.8 does not depend on any lemma proven in this paper. One of the crucial results of this section is Theorem 3.10, which is proven by using Lemmas 3.8, 3.2, 3.3, and Deligne and Rapoport's result Lemma 3.9. Theorem 3.10 says that for a given positive integer N, a prime p, and

$\Phi \in \mathcal{A}_k^1(N, \mathbb{Z}_{(p)})$ with

$$\Phi(\tau) = \sum_{n=-\infty}^{\infty} b(n)q^n,$$

we have for any given prime ℓ and integers a and t with $\gcd(a, \ell N) = 1$:

$$\forall_{n \in \mathbb{Z}} \, b(\ell n + t) \equiv 0 \pmod{p} \Rightarrow \forall_{n \in \mathbb{Z}} \, b(\ell n + a^2 t) \equiv 0 \pmod{p}.$$

Here for p a prime

$$\mathbb{Z}_{(p)} := \{a/b \,|\, a, b \in \mathbb{Z}, p \nmid b\}.$$

Theorem 3.10 is only used to prove Theorem 3.11 which simply says that the q-expansion of a meromorphic modular form with integer coefficients cannot be congruent modulo p to a polynomial in q, unless this polynomial is a constant. As we will see this is the key tool needed in every intermediate result until we arrive at the proof of Theorem 3.18. A very simple but crucial ingredient needed for the induction proof of Theorem 3.18 is Lemma 3.13. For the proof of Lemma 3.13 one only needs Lemma 3.1. Lemma 3.14 is just a simple result needed when one divides two meromorphic modular forms modulo p. Theorem 3.15 is a weaker version of Theorem 3.18 which is based on Lemmas 3.14, 3.13 and Theorem 3.11. One obtains Corollary 3.16 from Theorem 3.15 which is used to prove Lemma 3.17. Finally by using Lemmas 3.17, 3.14 and 3.13 one proves Theorem 3.18.

Lemma 3.1 *Let m, N be positive integers with $m|N$ and k an integer. For $\lambda \in \mathbb{Z}$ let $M_{\lambda,m} := \begin{pmatrix} 1 & \lambda \\ 0 & m \end{pmatrix}$. Let $\Phi \in \mathcal{A}_k^1(N)$ and $\gamma = \begin{pmatrix} a & b \\ c & d \end{pmatrix} \in \Gamma_0(N)$. Then $\Phi|_k M_{\lambda,m}\gamma = (\Phi|_k\gamma)|_k M_{bd+\lambda d^2, m}$.*

Proof The statement is equivalent to proving.

$$\Phi|_k M_{\lambda,m}\gamma M_{bd+\lambda d^2,m}^{-1}\gamma^{-1} = \Phi.$$

We have that

$$M_{\lambda,m}\gamma M_{bd+\lambda d^2,m}^{-1}\gamma^{-1} = \begin{pmatrix} a + \lambda c & -\frac{(bd+\lambda d^2)(a+\lambda c)+b+\lambda d}{m} \\ mc & (-bd + \lambda d^2)c + d \end{pmatrix} \gamma^{-1} \in \Gamma_1(N)$$

because $c \equiv 0 \pmod{m}$ and therefore $ad \equiv 1 \pmod{m}$ which implies that

$$-(bd + \lambda d^2)(a + \lambda c) + b + \lambda d \equiv -b - \lambda d + b + \lambda d \equiv 0 \pmod{m}.$$

$\square$

Lemma 3.2 *Let m be a positive integer and t an integer. Let Φ be meromorphic on $\mathbb{H}$ and $\Phi(\tau) = \sum_{n=-\infty}^{\infty} a(n)q^n$. Then*

$$\frac{1}{m} \sum_{\lambda=0}^{m-1} e^{-\frac{2\pi i \lambda t}{m}} \Phi\left(\frac{\tau + \lambda}{m}\right) = q^{\frac{t}{m}} \sum_{n=-\infty}^{\infty} a(mn + t)q^n.$$

Proof

$$\frac{1}{m} \sum_{\lambda=0}^{m-1} e^{-\frac{2\pi i \lambda t}{m}} \Phi\left(\frac{\tau + \lambda}{m}\right)$$

$$= \frac{1}{m} \sum_{\lambda=0}^{m-1} e^{-\frac{2\pi i \lambda t}{m}} \sum_{n=-\infty}^{\infty} a(n)e^{2\pi i n \frac{\tau+\lambda}{m}}$$

$$= \frac{1}{m} \sum_{n=-\infty}^{\infty} a(n)e^{\frac{2\pi i n \tau}{m}} \sum_{\lambda=0}^{m-1} e^{\frac{2\pi i \lambda(n-t)}{m}}$$

$$= e^{\frac{2\pi i t \tau}{m}} \sum_{n=-\infty}^{\infty} a(mn + t)e^{2\pi i n \tau}.$$

$\square$

Lemma 3.3 *Let m, N be positive integers and t an integer. Let $\gamma = \begin{pmatrix} a & b \\ c & d \end{pmatrix} \in \Gamma_0(mN) \cap \Gamma_1(N)$ and $\Phi \in \mathcal{A}_k^1(N)$ with $\Phi(\tau) = \sum_{n=-\infty}^{\infty} a(n)q^n$. Then*

$$q^{\frac{t}{m}} \sum_{n=-\infty}^{\infty} a(mn + t)q^n |_k \gamma = q^{\frac{a^2 t}{m}} e^{\frac{2\pi i b a t}{m}} \sum_{n=-\infty}^{\infty} a(mn + a^2 t)q^n.$$

Proof We have

$$q^{\frac{t}{m}} \sum_{n=-\infty}^{\infty} a(mn + t)q^n |_k \gamma = \frac{1}{m} \sum_{\lambda=0}^{m-1} e^{-\frac{2\pi i \lambda t}{m}} m^{k/2} (\Phi|_k M_{\lambda,m})|_k \gamma$$

because of Lemma 3.2

$$= \frac{1}{m} \sum_{\lambda=0}^{m-1} e^{-\frac{2\pi i \lambda t}{m}} m^{k/2} (\Phi|_k \gamma)|_k M_{\lambda d^2 + bd, m}$$

because of Lemma 3.1

$$= \frac{1}{m} \sum_{\lambda=0}^{m-1} e^{-\frac{2\pi i \lambda t}{m}} m^{k/2} \Phi|_k M_{\lambda d^2 + bd, m}$$

because of $\Phi \in \mathcal{A}_k^1(N)$

$$= \frac{1}{m} e^{\frac{2\pi i b a t}{m}} \sum_{\lambda'=0}^{m-1} e^{-\frac{2\pi i a^2 \lambda' t}{m}} \Phi\left(\frac{\tau + \lambda'}{m}\right)$$

by using the substitution $\lambda' \equiv \lambda d^2 + bd \pmod{m}$, with inverse $\lambda \equiv a^2 \lambda' - ba$ $\pmod{m}$.

$$= q^{\frac{a^2 t}{m}} e^{\frac{2\pi i b a t}{m}} \sum_{n=-\infty}^{\infty} a(mn + a^2 t) q^n$$

because of Lemma 3.2.

$\square$

Definition 3.4 We define $\eta : \mathbb{H} \to \mathbb{C}$ by

$$\eta(\tau) = e^{\frac{\pi i \tau}{12}} \prod_{n=1}^{\infty} (1 - q^n)$$

and

$$\Delta := \eta^{24}.$$

Remark 3.5 By [5, Th. 1.64] we find that $(\eta(24\tau))^2 \in M_1(\Gamma_0(576)) \subseteq M_1(\Gamma_1(576))$ and $\Delta \in M_{12}(\mathrm{SL}_2(\mathbb{Z}))$.

Definition 3.6 We denote by $j(\tau)$ the classical modular invariant.

Remark 3.7 Note that $j(\tau) \in M_0^1(\mathrm{SL}_2(\mathbb{Z}), \mathbb{Z})$. Furthermore, $j(\tau) = q^{-1} + \ldots$.

For meromorphic functions f, g on $\mathbb{H}$ which additionally have Laurent expansions in q with coefficients in $\mathbb{Z}_{(p)}$ for some prime p, that is $f(\tau) = \sum_{n=-\infty}^{\infty} a(n) q^n$ and $g(\tau) = \sum_{n=-\infty}^{\infty} b(n) q^n$, with $a(n), b(n) \in \mathbb{Z}_{(p)}$ for all $n \in \mathbb{Z}$, we write

$$f(\tau) \equiv g(\tau) \pmod{p}$$

iff $\frac{a(n)-b(n)}{p} \in \mathbb{Z}_{(p)}$ for all $n \in \mathbb{Z}$.

Lemma 3.8 *Let* p *be a prime. Let* $f \in \mathcal{A}_k^!(N, \mathbb{Z}_{(p)})$. *Then there exist* $g \in M_k^!(\Gamma_1(N)), \mathbb{Z})$ *and a monic* $p(X) \in \mathbb{Z}[X]$ *such that*

$$\frac{g(\tau)}{p(j(\tau))} \equiv f(\tau) \pmod{p}.$$

Proof Assume that the $f(\tau)$ has n poles in the fundamental domain of $\Gamma_1(N)$ counted with multiplicity. Let $\tau_1, \ldots, \tau_n$ be the poles of $f(\tau)$. Then

$$G(\tau) := f(\tau) \prod_{j=1}^{n} (j(\tau) - j(\tau_j)) \tag{5}$$

has no poles in $\mathbb{H}$, that is $G \in M_k^!(\Gamma_1(N))$. Furthermore, there exists an $u \in \mathbb{Z}$ such that $\Delta(\tau)^u G(\tau) \in M_{k+12u}(\Gamma_1(N))$. From [7, Th. 3.52] we know that there exist

$$b_1, \ldots, b_s \in M_{k+12u}(\Gamma_1(N), \mathbb{Z})$$

such that

$$M_{k+12u}(\Gamma_1(N)) = \{c_1 b_1(\tau) + \cdots + c_s b_s(\tau), c_1, \ldots, c_s \in \mathbb{C}\}.$$

In particular there exist $c_1, \ldots, c_s \in \mathbb{C}$ such that

$$\Delta(\tau)^u G(\tau) = c_1 b_1(\tau) + \cdots + c_s b_s(\tau)$$

or equivalently

$$G(\tau) = c_1 \frac{b_1(\tau)}{\Delta(\tau)^u} + \cdots + c_s \frac{b_s(\tau)}{\Delta(\tau)^u}. \tag{6}$$

Let

$$p(X) := \prod_{j=1}^{n} (X - j(\tau_j)) = X^n + a_{n-1} X^{n-1} + \cdots + a_1 X + a_0. \tag{7}$$

Let $a_n := 1$ and V be the vector space over $\mathbb{Q}$ generated by

$$\{c_1, \ldots, c_s\} \cup \{a_0, \ldots, a_n\}.$$

Let $r_1, \ldots, r_m$ be a basis of V over $\mathbb{Q}$. Then for $i = 1, \ldots, m$

$$a_i = d_1^{(i)} r_1 + \cdots + d_m^{(i)} r_m$$

for some rational numbers $d_k^{(i)}$, $k = 1, \ldots, m$. Then by (7)

$$p(X) = \sum_{i=0}^{n} X^i a_i = \sum_{i=0}^{n} X^i \sum_{j=1}^{m} d_j^{(i)} r_j = \sum_{j=1}^{m} r_j \underbrace{\sum_{i=0}^{n} d_j^{(i)} X^i}_{=d_j p_j(X)}.$$

where $p_1(X), \ldots, p_m(X) \in \mathbb{Z}[X]$ and $d_1, \ldots, d_m \in \mathbb{Q}$ are chosen such that $p_1(X), \ldots, p_m(X)$ are primitive in the sense of Gauss. Hence

$$p(X) = r_1 d_1 p_1(X) + r_2 d_2 p_2(X) + \cdots + r_m d_m p_m(X). \tag{8}$$

Similarly for $i = 1, \ldots, s$

$$c_i = e_1^{(i)} r_1 + \cdots + e_m^{(i)} r_m$$

for some rational numbers $e_k^{(i)}$, $k = 1, \ldots, m$. Then by (6) we have that

$$G(\tau) = \sum_{i=1}^{s} c_i \frac{b_i(\tau)}{\Delta(\tau)^u} = \sum_{i=1}^{s} \sum_{j=1}^{m} e_j^{(i)} r_j \frac{b_i(\tau)}{\Delta(\tau)^u} = \sum_{j=1}^{m} r_j \underbrace{\sum_{i=1}^{s} e_j^{(i)} \frac{b_i(\tau)}{\Delta(\tau)^u}}_{=e_j f_j(\tau)}$$

where $f_j(\tau) = \sum_{n=-\infty}^{\infty} b_j(n) q^n$ and $e_1, \ldots, e_m \in \mathbb{Q}$ are chosen such that

$$\nexists_\ell \text{ prime} \forall_{n \in \mathbb{Z}} \ell | b_j(n). \tag{9}$$

Hence

$$G(\tau) = e_1 r_1 f_1(\tau) + \cdots + e_m r_m f_m(\tau) \tag{10}$$

Note that $f_i(\tau) = \frac{1}{e_i} \sum_{i=1}^{s} e_j^{(i)} \frac{b_i(\tau)}{\Delta(\tau)^u} \in M_k^!(\Gamma_1(N), \mathbb{Z})$.
In particular (5), (8) and (10) implies

$$e_1 r_1 f_1(\tau) + \cdots + e_m r_m f_m(\tau)$$
$$= \{r_1 d_1 p_1(j(\tau)) + r_2 d_2 p_2(j(\tau)) + \cdots + r_m d_m p_m(j(\tau))\} f(\tau). \tag{11}$$

Since $r_1, \ldots, r_m$ is a basis, (11) implies that

$$e_i f_i(\tau) = d_i f(\tau) p_i(j(\tau)), \quad i = 1, \ldots, m.$$

In particular, writing $\frac{e_1}{d_1} = \frac{a}{b}$ with $a, b \in \mathbb{Z}$ and $\gcd(a, b) = 1$ we obtain

$$\frac{a f_1(\tau)}{b p_1(j(\tau))} = f(\tau).$$

This implies that

$$\frac{a f_1(\tau)}{b} = f(\tau) p_1(j(\tau))$$

Since the coefficients in the q-expansion of $f(\tau) p_1(j(\tau))$ are in $\mathbb{Z}_{(p)}$ and because of (9) it follows that $p \nmid b$. Let b' be an integer such that $bb' \equiv 1 \pmod{p}$ and define

$$g(\tau) := ab' f_1(\tau).$$

Then $g(\tau) \in M_k^!(\Gamma_1(N), \mathbb{Z})$. In particular

$$g(\tau) \equiv f(\tau) p_1(j(\tau)) \pmod{p}.$$

Next we observe that there exists $c \in \{1, \ldots, p-1\}$ and monic $r(X) \in \mathbb{Z}[X]$ such that $p_1(X) \equiv c r(X) \pmod{p}$, since $p_1(X)$ is primitive. This implies that

$$g(\tau) \equiv c f(\tau) r(j(\tau)) \pmod{p}$$

or equivalently

$$\frac{c' g(\tau)}{r(j(\tau))} \equiv f(\tau) \pmod{p},$$

where c' is an integer such that $cc' \equiv 1 \pmod{p}$. $\qquad\square$

As a simple consequence of [1, VII, Cor. 3.12] we have:

Lemma 3.9 *Let k, N be positive integers and $f \in M_k(\Gamma(N), \mathbb{Z}[\xi])$ where $\xi := e^{\frac{2\pi i}{N}}$. Then for all $\gamma \in \Gamma_0(N)$, $f|_k \gamma \in M_k(\Gamma(N), \mathbb{Z}[\xi])$.*

Theorem 3.10 *Let ℓ, p be primes. Let N be a positive integer and t an integer. Let $\gamma = \begin{pmatrix} a & b \\ c & d \end{pmatrix} \in \Gamma_0(\ell N) \cap \Gamma_1(N)$. Let $\Phi \in \mathcal{A}_k^!(N, \mathbb{Z}_{(p)})$ and $\sum_{n=-\infty}^{\infty} b(n) q^n := \Phi(\tau)$. Then*

$$\sum_{n=-\infty}^{\infty} b(\ell n + t) q^n \equiv 0 \pmod{p} \Rightarrow \sum_{n=-\infty}^{\infty} b(\ell n + a^2 t) q^n \equiv 0 \pmod{p}.$$

Proof By Lemma 3.8 there exists $f \in M_k^!(\Gamma_1(N), \mathbb{Z})$ and a monic $p(X) \in \mathbb{Z}[X]$ such that

$$\Phi(\tau) \equiv \frac{f(\tau)}{p(j(\tau))} \pmod{p}.$$

Let

$$\sum_{n=-\infty}^{\infty} a(n)q^n := \frac{f(\tau)}{p(j(\tau))}.$$

We define

$$F(\tau) := p(j(\ell\tau)) \prod_{\lambda=0}^{\ell-1} p\left(j\left(\frac{\tau+\lambda}{\ell}\right)\right). \tag{12}$$

If the degree of $p(X)$ is n, we observe directly from (12) that the q-expansion of $F(\tau)$ has the form

$$F(\tau) = q^{-n(\ell+1)} + O(q^{-n(\ell+1)+1}), \tag{13}$$

and we need later in the proof that the coefficient of $q^{-n(\ell+1)}$ is 1. Let

$$Q(X) := (X - j(\ell\tau)) \prod_{\lambda=0}^{\ell-1} \left(X - j\left(\frac{\tau+\lambda}{\ell}\right)\right), \tag{14}$$

and define $e_n(\tau)$ by the relation

$$Q(X) = X^{\ell+1} + \sum_{n-1}^{\ell} e_n(\tau)X^n.$$

By [6, §4, Th. 16], $Q(X) \in \mathbb{Z}[j(\tau)][X]$. We observe from (14) that $e_n(\tau) = E_n(Y_0(\tau), Y_1(\tau), \ldots, Y_\ell(\tau))$ where

$$(Y_0(\tau), Y_1(\tau), \ldots, Y_\ell(\tau)) := \left(j(\ell\tau), j\left(\frac{\tau}{\ell}\right), \ldots, j\left(\frac{\tau+\ell-1}{\ell}\right)\right)$$

and $E_n(X_0, X_1, \ldots, X_\ell) \in \mathbb{Z}[X_0, X_1, \ldots, X_\ell]$ are the elementary symmetric polynomials. Furthermore, $F(j(\tau)) = f(Y_0(\tau), Y_1(\tau), \ldots, Y_\ell(\tau))$ where $f(X_0, X_1, \ldots, X_\ell) \in \mathbb{Z}[X_0, X_1, \ldots, X_\ell]$ is a symmetric polynomial and since every integer symmetric polynomial in $X_0, \ldots, X_\ell$ is an integer polynomial in the elementary symmetric functions $E_1, \ldots, E_n$ by [2, p. 20, (2.4)], it follows that

$$f(X_0, \ldots, X_n) = h(E_0(X_0, \ldots, X_n), \ldots, E_n(X_0, \ldots, X_n))$$

for some $h(X_0, \ldots, X_n) \in \mathbb{Z}[X_0, \ldots, X_n]$. This implies that

$$F(\tau) = h(e_0(\tau), \ldots, e_n(\tau)) \in \mathbb{Z}[j(\tau)]$$

or, in other words, there exists $r(X) \in \mathbb{Z}[X]$ such that $F(\tau) = r(j(\tau))$ and because of (13) it follows that $r(X)$ is monic.

Next we see that

$$G(\tau) := r(j(\tau))^\ell \Big(q^{\frac{t}{\ell}} \sum_{n=-\infty}^\infty \frac{a(\ell n + t)}{p} q^n \Big)^\ell \in \mathcal{A}_{k\ell}^1(N, \mathbb{Z})$$

because of Lemma 3.3. Furthermore, by Lemma 3.2,

$$r(j(\tau)) \times q^{\frac{t}{\ell}} \sum_{n=-\infty}^\infty \frac{a(\ell n + t)}{p} q^n = F(\tau) \times \frac{1}{p^2} \sum_{\lambda=0}^{\ell-1} e^{-\frac{2\pi i \lambda t}{\ell}} \frac{f\left(\frac{\tau+\lambda}{\ell}\right)}{p\left(j\left(\frac{\tau+\lambda}{\ell}\right)\right)}$$

$$= p(j(\ell\tau)) \times \frac{1}{\ell^2} \sum_{\lambda=0}^{\ell-1} e^{-\frac{2\pi i \lambda t}{\ell}} f\left(\frac{\tau+\lambda}{\ell}\right) \prod_{\alpha \neq \lambda} p\left(j\left(\frac{\tau+\alpha}{\ell}\right)\right),$$

which is holomorphic on $\mathbb{H}$ because $j(\tau)$ and $f(\tau)$ are holomorphic on $\mathbb{H}$. This implies that $G(\tau)$ is holomorphic on $\mathbb{H}$ so that $G(\tau) \in M_k^1(\Gamma_1(N), \mathbb{Z})$.

Because of $\Delta \in M_{12}(\mathrm{SL}_2(\mathbb{Z}))$, we also have $\Delta \in M_{12}(\Gamma_1(N))$ and $\Delta|_{12}\gamma$ is a q-series with positive order for all $\gamma \in \mathrm{SL}_2(\mathbb{Z})$. Because of this there exists a positive integer i such that

$$r(j(\tau))^\ell \Delta(\tau)^i \Big(q^{\frac{t}{\ell}} \sum_{n=-\infty}^\infty \frac{a(\ell n + t)}{p} q^n \Big)^\ell \in M_{12i+k}(\Gamma_1(N), \mathbb{Z}).$$

By Lemma 3.3:

$$r(j(\tau))^\ell \Delta(\tau)^i \Big(q^{\frac{t}{\ell}} \sum_{n=-\infty}^\infty \frac{a(\ell n + t)}{p} q^n \Big)^\ell |_{k+12i} \gamma$$

$$= r(j(\tau))^\ell \Delta(\tau)^i \Big(q^{\frac{a^2 t}{\ell}} \sum_{n=-\infty}^\infty \frac{a(\ell n + a^2 t)}{p} q^n \Big)^\ell.$$

Then by Lemma 3.9 the q-series

$$r(j(\tau))^\ell \Delta(\tau)^i \Big(q^{\frac{a^2 t}{\ell}} \sum_{n=-\infty}^\infty \frac{a(\ell n + a^2 t)}{p} q^n \Big)^\ell,$$

has integer coefficients. This implies that

$$\sum_{n=-\infty}^{\infty} a(\ell n + a^2 t) \equiv 0 \ (\text{mod } p),$$

finishing the proof. □

Theorem 3.11 *Let p be a prime. Let N be a positive integer and let $f \in \mathcal{A}_k^1(N, \mathbb{Z}_{(p)})$. Assume that for some $r(q) \in \mathbb{Z}[q, q^{-1}]$ we have*

$$f(\tau) = \sum_{n=-\infty}^{\infty} a(n)q^n \equiv r(q) \ (\text{mod } p).$$

Then $f(\tau) \equiv a(0) \ (\text{mod } p)$.

Proof Let r and u be such that

$$r(q) \equiv a(r)q^r + a(r-1)q^{r-1} + \cdots + a(u+1)q^{u+1} + a(u)q^u.$$

Let $t \neq 0$ be such that $a(t) \not\equiv 0 \ (\text{mod } p)$ and let $v := r + 1$ if $r \neq -1$ or $v := 1$ if $r := -1$. Let $a, b, c, d \in \mathbb{Z}$ and ℓ a prime such that

$$\ell > r + 2 - u, \tag{15}$$

$$a^2 v \equiv t \ (\text{mod } \ell), \tag{16}$$

$$a \equiv 1 \ (\text{mod } N), \tag{17}$$

$$c \equiv 0 \ (\text{mod } \ell N), \tag{18}$$

$$ad - bc = 1. \tag{19}$$

Then $\gamma = \begin{pmatrix} a & b \\ c & d \end{pmatrix} \in \Gamma_0(\ell N) \cap \Gamma_1(N)$. Furthermore, $\sum_{n=-\infty}^{\infty} a(\ell n + v) \equiv 0 \ (\text{mod } p)$. Then because of Theorem 3.10,

$$\sum_{n=-\infty}^{\infty} a(\ell n + a^2 v)q^n = \sum_{n=-\infty}^{\infty} a(\ell n + t)q^n \equiv 0 \ (\text{mod } p).$$

This is false because $a(t) \not\equiv 0 \ (\text{mod } p)$. It is left to show that there exist a, b, c, d, ℓ satisfying (15)–(19). Let $vt = 2^s m$ where m is odd. By standard properties of the Legendre symbol we obtain for any prime $\ell \neq 2$:

$$\left(\frac{vt}{\ell}\right) = \left(\frac{2^s m}{\ell}\right) = \left(\frac{2}{\ell}\right)^s \left(\frac{m}{\ell}\right) = (-1)^{s\frac{\ell^2-1}{8}}(-1)^{\frac{m-1}{2}\frac{\ell-1}{2}}\left(\frac{\ell}{m}\right).$$

Assuming that $\ell \equiv 1 \pmod 8$ we obtain

$$\left(\frac{vt}{\ell}\right) = \left(\frac{\ell}{m}\right).$$

Assuming further that $\ell \equiv 1 \pmod m$ we obtain that

$$\left(\frac{vt}{\ell}\right) = 1. \tag{20}$$

We have proven that vt is a square modulo ℓ for all primes $\ell \equiv 1 \pmod{8m}$. By Dirichlet's theorem there are infinitely many such primes ℓ. In particular there exists a prime ℓ with $\ell \nmid N$ and such that (15) is satisfied. We fix such an ℓ. Then $vt \equiv x^2 \pmod{\ell}$ because of (20). Let $a \in \mathbb{Z}$ such that

$$a \equiv xv^{-1} \pmod{\ell} \quad \text{and}$$

$$a \equiv 1 \pmod{N}.$$

Such an a clearly exists because of the Chinese remainder theorem. In particular for this a, (16)–(17) are satisfied. Set $c := \ell N$. Then we can find integer b, d such that $ad - bc = 1$ because $\gcd(a, c) = 1$. Hence we have constructed a, b, c, d and ℓ with the desired properties.   □

Definition 3.12 Let d be a positive integer. For f meromorphic on $\mathbb{H}$ we define $U_d(f)$ and $V_d(f)$ meromorphic on $\mathbb{H}$ by

$$U_d(f)(\tau) := \frac{1}{d} \sum_{\lambda=0}^{d-1} f\left(\frac{\tau + \lambda}{d}\right)$$

and $V_d(f)(\tau) := f(d\tau)$.

Lemma 3.13 *Let k and t be integers and m a positive integer and $f(\tau) \in \mathcal{A}_k^1(N)$. Then*

$$V_m U_m(q^t f)(\tau) = q^t G(\tau),$$

where $G \in \mathcal{A}_k^1(Nm^2)$.

Proof

$$U_m(q^t f)(\tau) = \frac{1}{m} \sum_{\lambda=0}^{m-1} e^{\frac{2\pi i t(\tau + \lambda)}{m}} f\left(\frac{\tau + \lambda}{m}\right)$$

$$= e^{\frac{2\pi i t \tau}{m}} \frac{1}{m} \sum_{\lambda=0}^{m-1} e^{\frac{2\pi i t \lambda}{m}} f\left(\frac{\tau + \lambda}{m}\right)$$

$$= q^{t/m} g(\tau).$$

If $f \in \mathcal{A}_k(\Gamma_1(N))$ then $f|M_{\lambda,m} \in \mathcal{A}_k(\Gamma_1(N) \cap \Gamma(m))$ because of Lemma 3.1. In particular $g(\tau) \in \mathcal{A}_k(\Gamma_1(N) \cap \Gamma(m))$. Consequently, $G := V_m g \in \mathcal{A}_k(\Gamma_1(Nm^2))$, because for $\begin{pmatrix} a & b \\ c & d \end{pmatrix} \in \Gamma_1(Nm^2)$ we have $\begin{pmatrix} a & bm \\ c/m & d \end{pmatrix} \in \Gamma_1(N) \cap \Gamma(m)$ which implies

$$V_m g|_k \begin{pmatrix} a & b \\ c & d \end{pmatrix} = V_m \left(g|_k \begin{pmatrix} a & bm \\ c/m & d \end{pmatrix} \right) = V_m g.$$

$\square$

Lemma 3.14 *Let p be a prime and $a(\tau) \in \mathcal{A}_k^1(576N, \mathbb{Z}_{(p)})$. Let*

$$a(\tau) = pb_s q^s + pb_{s+1} q^{s+1} + \cdots + pb_{s+m-1} q^{s+m-1} + \sum_{n \geq s+m} b_n q^n, \quad p \nmid b_{s+m}.$$

Then there exists $\tilde{a}(\tau) = \sum_{n \geq s+m} c_n q^n \in \mathcal{A}_k^1(576N, \mathbb{Z}_{(p)})$ such that $\tilde{a}(\tau) \equiv a(\tau)$ (mod p).

Proof Let

$$a_1(\tau) := a(\tau) - pb_s j(\tau)^{-s+2k} (\eta(24\tau))^{2k}.$$

Then

$$a_1(\tau) = pb_{s+1}^{(1)} q^{s+1} + \cdots + pb_{s+m-1}^{(1)} q^{s+m-1} + \sum_{n \geq s+m} b_n^{(1)} q^n.$$

Let

$$a_2(\tau) := a_1(\tau) - pb_{s+1}^{(1)} j(\tau)^{-(s+1)+2k} (\eta(24\tau))^{2k}.$$

Then

$$a_2(\tau) = pb_{s+2}^{(2)} q^{s+2} + \cdots + pb_{s+m-1}^{(2)} q^{s+m-1} + \sum_{n \geq s+m} b_n^{(2)} q^n.$$

Define analogously $a_3(\tau), \ldots, a_m(\tau)$. Then $\tilde{a}(\tau) := a_m(\tau)$ satisfies $a(\tau) \equiv \tilde{a}(\tau)$ (mod p). By Remarks 3.7 and 3.5 it follows that $\tilde{a}(\tau) \in \mathcal{A}_k^1(576N, \mathbb{Z}_{(p)})$. $\square$

Theorem 3.15 *Let N be a positive integer. Let $\phi_0(\tau), \ldots, \phi_n(\tau) \in \mathcal{A}_0^1(N, \mathbb{Z}_{(p)})$. Assume that*

$$\phi_n(\tau) q^n + \phi_{n-1}(\tau) q^{n-1} + \cdots + \phi_0(\tau) \equiv 0 \pmod{p}.$$

Then $\phi_0(\tau) = \phi_1(\tau) = \cdots = \phi_n(\tau) \equiv 0 \pmod{p}$.

Proof We proceed by induction on the degree with respect to q of the left hand side of the relation. We assume now that the theorem is valid for any relation of degree less than n. We assume that $\phi_0(\tau) \not\equiv 0$ because otherwise we may divide the relation by q and by the induction hypothesis we are finished. So assume $\phi_0(\tau) \not\equiv 0$ (mod p). We can also assume that $\phi_n(\tau) \not\equiv 0$ (mod p) because otherwise again we are finished by induction.

By Lemma 3.14 there exists $\tilde{\phi}_0 \in \mathcal{A}_0^1(576N, \mathbb{Z}_{(p)})$ such that $\tilde{\phi}_0(\tau) = b_s q^s + O(q^{s+1})$, $p \nmid b_s$ and $\tilde{\phi}_0(\tau) \equiv \phi_0(\tau)$ (mod p). Then $\frac{\phi_\kappa(\tau)}{\tilde{\phi}_0(\tau)} \in \mathcal{A}_0^1(576N, \mathbb{Z}_{(p)})$, for $\kappa \in 0, \ldots, n$. Let $\kappa > 0$ be minimal such that $\frac{\phi_\kappa(\tau)}{\tilde{\phi}_0(\tau)} \not\equiv 0$ (mod p). We divide the relation by $q^\kappa \tilde{\phi}_0(\tau)$ and obtain:

$$\frac{\phi_n(\tau)}{\tilde{\phi}_0(\tau)} q^{n-\kappa} + \frac{\phi_{n-1}(\tau)}{\tilde{\phi}_0(\tau)} q^{n-2-\kappa} + \cdots + \frac{\phi_\kappa(\tau)}{\tilde{\phi}_0(\tau)} + q^{-\kappa} \equiv 0 \pmod{p}. \tag{21}$$

Let

$$\frac{\phi_\kappa(\tau)}{\tilde{\phi}_0(\tau)} = \sum_{j=-\infty}^{\infty} a(j) q^j.$$

By Theorem 3.11 there exists a minimal $d > \kappa$ such that $a(d) \not\equiv 0$ (mod p) or $\sum_{n=m}^{\infty} a(n) q^n \equiv a(0)$ (mod p). Let

$$s := \begin{cases} \kappa + 1 & \text{if } \sum a(n) q^n \equiv a(0) \pmod{p}, \\ d & \text{otherwise.} \end{cases}$$

Then applying the operator $V_s U_s$ to the relation (21) yields

$$b_n(\tau) q^{n-\kappa} + b_{n-1}(\tau) q^{n-\kappa-1} + \cdots + b_\kappa(\tau) \equiv 0 \pmod{p}, \tag{22}$$

where for $i = \kappa, \ldots, n$:

$$b_i(\tau) := q^{\kappa-i} V_s U_s \left(q^{i-\kappa} \frac{\phi_i(\tau)}{\tilde{\phi}_0(\tau)} \right).$$

Note that since $s > \kappa$ we have $V_s U_s(q^{-\kappa}) = 0$.

In particular by Lemma 3.13 $b_i(\tau) \in \mathcal{A}_k^1(576Ns^2, \mathbb{Z}_{(p)})$. Next note that

$$b_\kappa(\tau) = V_s U_s\left(\sum a(n) q^n\right) = \sum a(ns) q^{ns}$$

and hence $b_\kappa(\tau) \equiv a(0)$ (mod p) if $\sum a(n) q^n \equiv a(0)$ (mod p) or $b_\kappa(\tau)$ contains the term $q^d a(d) \not\equiv 0$ (mod p), in any case $b_\kappa(\tau) \not\equiv 0$ (mod p).

However by the induction hypothesis $b_n(\tau) \equiv \cdots b_\kappa(\tau) \equiv 0 \pmod{p}$. This contradicts $b_\kappa(\tau) \not\equiv 0 \pmod{p}$ hence we have $\phi_n(\tau) \equiv \cdots \equiv \phi_0(\tau) \equiv 0 \pmod{p}$.

□

Corollary 3.16 *Let p be a prime. Let $a_0(\tau), a_1(\tau), \ldots, a_n(\tau) \in M_0^!(\Gamma_1(N), \mathbb{Z}_{(p)})$. Let $r(q) \in \mathbb{Z}[q, q^{-1}]$ be non-constant modulo p. Assume that*

$$a_n(\tau)r(q)^n + a_{n-1}(\tau)r(q)^{n-1} + \cdots + a_0(\tau) \equiv 0 \pmod{p}.$$

Then $a_0(\tau) \equiv a_1(\tau) \equiv \cdots \equiv a_n(\tau) \equiv 0 \pmod{p}$.

Proof We proceed by induction on n. Assume that $r(q)$ has positive degree and let d be its degree. Then there exist $b_{dn-1}(\tau), \ldots, b_0(\tau) \in M_0^!(\Gamma_1(N), \mathbb{Z}_{(p)})$ such that

$$a_n(\tau)q^{dn} + b_{dn-1}(\tau)q^{dn-1} + \cdots + b_0(\tau) \equiv 0 \pmod{p}$$

Then by Theorem 3.15 we have $a_n(\tau) \equiv 0 \pmod{p}$. By induction we are finished.

Next assume that $r(q)$ has negative degree and let $-d$ be its low-degree. Then there exist $b_{-dn+1}(\tau), \ldots, b_0(\tau) \in M_0^!(\Gamma_1(N), \mathbb{Z}_{(p)})$ such that

$$a_n(\tau)q^{-dn} + b_{-dn+1}(\tau) + \cdots + b_0(\tau) \equiv 0 \pmod{p}.$$

After multiplication of both sides by q^{dn}, we obtain by Theorem 3.15 that $a_n(\tau) \equiv 0 \pmod{p}$. By induction we are finished.

□

Lemma 3.17 *Let N be a positive integer and p a prime. For $i = 1, \ldots, n$ let $\Phi_i \in \mathcal{A}_{k_i}^1(N, \mathbb{Z}_{(p)})$, $k_i \in \mathbb{Z}$. Let $r(q) \in \mathbb{Z}[q, q^{-1}]$ be such that*

$$\Phi_1(\tau) + \Phi_2(\tau) + \cdots + \Phi_n(\tau) \equiv r(q) \pmod{p}. \tag{23}$$

Then $r(q) \equiv j \pmod{p}$ for some $j \in \mathbb{Z}$.

Proof Let

$$\nu := \frac{12(p-1)}{\gcd(p^2 - 1, 24)}.$$

Let $b_i^{(0)}(\tau) := \sum_{\substack{1 \le j \le n \\ k_j \equiv i \pmod{\nu}}} \Phi_j(\tau)\left(\frac{\eta(\tau)^p}{\eta(p\tau)}\right)^{\frac{2(i-k_j)}{p-1}}$. Then $\Phi_1(\tau) + \cdots + \Phi_n(\tau) \equiv$

$b_0^{(0)}(\tau) + b_1^{(0)}(\tau) + \cdots + b_{\nu-1}^{(0)}(\tau) \pmod{p}$ because of $\left(\frac{\eta(\tau)^p}{\eta(p\tau)}\right)^{\frac{2\nu}{p-1}} \equiv 1 \pmod{p}$.

In particular $b_i^{(0)}(\tau) \in \mathcal{A}_i^1(576pN)$ because by [5, Th. 1.64], $\left(\frac{\eta(\tau)^p}{\eta(p\tau)}\right)^{\frac{2\nu}{p-1}} \in A_\nu(\Gamma_0(p)) \subseteq A_\nu^1(p)$. This shows in particular that any sum $\Phi_1'(\tau) + \cdots + \Phi_{n'}'(\tau)$ with $\Phi_j'(\tau) \in \mathcal{A}_j^1(576pN)$ can be written as $b_0'(\tau) + \cdots + b_\nu'(\tau)$ with $b_i'(\tau) \in$

$\mathcal{A}_j^1(576pN)$. Taking both sides of (23) to the power of k for $k = 1, \ldots, v$ and applying this rewriting to the left hand side we obtain the following system:

$$r(q) \equiv b_0^{(0)}(\tau) + b_1^{(0)}(\tau) + \cdots + b_{v-1}^{(0)}(\tau),$$

$$r(q)^2 \equiv b_0^{(1)}(\tau) + b_1^{(1)}(\tau) + \cdots + b_{v-1}^{(1)}(\tau),$$

$$\vdots \quad \vdots \quad \vdots$$

$$r(q)^v \equiv b_0^{(v-1)}(\tau) + b_1^{(v-1)}(\tau) + \cdots + b_{v-1}^{(v-1)}(\tau),$$

for some $b_i^{(j)} \in \mathcal{A}_i^1(N)$. Let $T(\tau) := \eta(24\tau)^2$. By Remark 3.5, $T \in \mathcal{A}_1^1(576pN, \mathbb{Z}_{(p)})$. We define

$$A := \begin{pmatrix} \frac{b_0^{(0)}(\tau)}{T(\tau)^0} & \frac{b_1^{(0)}(\tau)}{T(\tau)} & \cdots & \frac{b_{v-1}^{(0)}(\tau)}{T(\tau)^{v-1}} \\ \frac{b_0^{(1)}(\tau)}{T(\tau)^0} & \frac{b_1^{(1)}(\tau)}{T(\tau)^1} & \cdots & \frac{b_{v-1}^{(1)}(\tau)}{T(\tau)^{v-1}} \\ \frac{b_0^{(2)}(\tau)}{T(\tau)^0} & \frac{b_1^{(2)}(\tau)}{T(\tau)^1} & \cdots & \frac{b_{v-1}^{(2)}(\tau)}{T(\tau)^{v-1}} \\ \vdots & \vdots & \vdots & \vdots \\ \frac{b_0^{(v-1)}(\tau)}{T(\tau)^0} & \frac{b_1^{(v-1)}(\tau)}{T(\tau)^1} & \cdots & \frac{b_{v-1}^{(v-1)}(\tau)}{T(\tau)^{v-1}} \end{pmatrix}.$$

Then

$$\begin{pmatrix} r(q) \\ r(q)^2 \\ \vdots \\ r(q)^v \end{pmatrix} \equiv A \begin{pmatrix} T(\tau)^0 \\ T(\tau)^1 \\ \vdots \\ T(\tau)^{v-1} \end{pmatrix} \pmod{p}. \tag{24}$$

Note that the entries of A are in $\mathcal{A}_0^1(576pN)$. If A is not invertible modulo p, then there exists modular functions $x_1(\tau), x_2(\tau), \ldots, x_v(\tau)$ not all identically zero such that

$$(x_1, x_2, \ldots, x_v)A \equiv (0, 0, \ldots, 0)$$

which together with (24) implies that

$$x_1(\tau)r(q) + x_2(\tau)r(q)^2 + \cdots + x_v(\tau)r(q)^v \equiv 0 \pmod{p}.$$

which is impossible by Corollary 3.16 unless $r(q)$ is constant modulo p.

If A is invertible modulo p, then

$$\text{adj}(A) \begin{pmatrix} r(q) \\ r(q)^2 \\ \vdots \\ r(q)^\nu \end{pmatrix} = \det(A) \begin{pmatrix} T(\tau)^0 \\ T(\tau)^1 \\ \vdots \\ T(\tau)^{\nu-1} \end{pmatrix}.$$

where $\text{adj}(A)$ is the adjoint of A and $\det(A) \not\equiv 0 \pmod{p}$. In particular, since A is invertible modulo p it follows that the first row of $\text{adj}(A)$ contains at least one entry which is nonzero modulo p. This leads to a relation of the form

$$r(q)a_1(\tau) + r(q)^2 a_2(\tau) + \cdots + r(q)^\nu a_\nu(\tau) \equiv \det(A) T(\tau)^0 = \det(A) \pmod{p}$$

which is impossible because of Corollary 3.16 unless $r(q)$ is constant modulo p.

$\square$

Theorem 3.18 *Let p be a prime. Let N be a positive integer. Let $S_j \subset \mathbb{Z}$ be finite for $0 \leq j \leq m$. Assume that we have a relation of the form*

$$\sum_{0 \leq j \leq m} q^j \sum_{i \in S_j} \Phi_{i,j}(\tau) \equiv 0 \pmod{p},$$

where $\Phi_{i,j} \in \mathcal{A}_i^1(N, \mathbb{Z}_{(p)})$ and $\Phi_{i,j}(\tau) \not\equiv 0 \pmod{p}$, for $0 \leq j \leq m$ and $i \in S_j$. Then $\sum_{i \in S_j} \Phi_{i,j}(\tau) \equiv 0 \pmod{p}$ for $j \in \{0, \ldots, m\}$.

The proof of this theorem follows similar steps as the proof of Lemma 3.15, therefore in this proof we will not repeat certain minor arguments.

Proof We proceed using induction on the length $|S_0| + |S_1| + \cdots + |S_m|$ of the relation. Assume that the statement hold for all relations of length less than M and we wish to prove it for a relation of length M.

Therefore assume that the length of the relation is M. Assume that the theorem is false. Without loss of generality we may assume that $\sum_{i \in S_0} \Phi_{i,0}(\tau) \not\equiv 0 \pmod{p}$ because in case not we divide the relation by an appropriate power of q to make it into the desired form. Then there exists a minimal $\kappa > 0$ such that $\sum_{i \in S_\kappa} \Phi_{i,\kappa} \not\equiv 0 \pmod{p}$.

Take $I \in S_0$, then by assumption $\Phi_{I,0}(\tau) \not\equiv 0 \pmod{p}$. By Lemma 3.14 there exists $\tilde{\Phi}_{I,0} \in \mathcal{A}_I^1(576N, \mathbb{Z}_{(p)})$ such that $\tilde{\Phi}_{I,0}(\tau) \equiv \Phi_{I,0}(\tau)$ and

$$\tilde{\Phi}_{I,0}(\tau) = b_r q^r + O(q^{r+1}), \quad p \nmid b_r.$$

Divide the relation by $\tilde{\Phi}_{I,0}(\tau) q^\kappa$. We obtain the relation

$$\sum_{\kappa \leq j \leq M} q^{j-\kappa} \sum_{i \in S_j} \frac{\Phi_{i,j}(\tau)}{\tilde{\Phi}_{I,0}(\tau)} + q^{-\kappa} + \sum_{i \in S_0, i \neq I} q^{-\kappa} \frac{\Phi_{i,0}(\tau)}{\tilde{\Phi}_{I,0}(\tau)} \equiv 0 \pmod{p}. \quad (25)$$

Let $\sum a(n)q^n = \sum_{i \in S_\kappa} \frac{\Phi_{i,\kappa}(\tau)}{\tilde{\Phi}_{I,0}(\tau)}$. By Lemma 3.17 there exists a minimal integer $d > \kappa$ such that $a(n) \not\equiv 0 \pmod{p}$ or $\sum a(n)q^n \equiv a(0) \pmod{p}$. Let

$$
s := \begin{cases} \kappa + 1 & \text{if } \sum a(n)q^n \equiv a(0) \pmod{p}, \\ d & \text{otherwise.} \end{cases}
$$

Then applying $V_s U_s$ to the relation (25) and defining for $j \in \{0\} \cup \{\kappa, \ldots, m\}$ and $i \in S_j$:

$$
B_{i-I,j}(\tau) := q^{-j+\kappa} V_s U_s \left(q^{j-\kappa} \frac{\Phi_{i,j}(\tau)}{\tilde{\Phi}_{I,0}(\tau)} \right)
$$

yields a relation of the form

$$
\sum_{\kappa \le j \le M} q^{j-\kappa} \sum_{i \in S_j} B_{i-I,j}(\tau) + \sum_{i \in S_0, i \ne I} B_{i-I,0}(\tau) \equiv 0 \pmod{p} \tag{26}
$$

and $\sum_{i \in S_\kappa} B_{i-I,\kappa}(\tau) \not\equiv 0 \pmod{p}$ by construction. Multiplying the above relation by q^κ we obtain

$$
\sum_{\kappa \le j \le M} q^j \sum_{i \in S_j} B_{i-I,j}(\tau) + \sum_{i \in S_0, i \ne I} B_{i-I,0}(\tau) \equiv 0 \pmod{p} \tag{27}
$$

Note that $B_{i,j}(\tau) \in \mathcal{A}_i^1(576Ns^2)$ because of Lemma 3.13. Thus we obtain a new relation with length $< M$ which implies by induction that $\sum_{i \in S_j} B_{i-I,j}(\tau) \equiv 0 \pmod{p}$ for all $j \in \{0, \ldots, m\}$ in particular also for $j = \kappa$ which is a contradiction. $\square$

4   An Immediate Consequence

Corollary 4.1 *Assume $p_0(X), \ldots, p_n(X) \in \mathbb{Z}[q]$ and $B(\tau) \in \mathcal{A}^1(N)$, if*

$$
p_0(q) + p_1(q)B(\tau) + \cdots + p_n(q)B(\tau)^n \equiv 0 \mod p,
$$

then $B(\tau) \equiv c \pmod{p}$, for some $c \in \mathbb{Z}$.

Proof Assume $B(\tau) = \sum_{n=m}^{\infty} b(n)q^n$, and let $C(\tau) := B(\tau) - b(0)$. Then

$$
\tilde{p}_0(q) + \tilde{p}_1(q)C(\tau) + \cdots + \tilde{p}_n(q)C(\tau)^n \equiv 0 \mod p,
$$

for some $\tilde{p}_i(X) \in \mathbb{Z}[X]$. We can rewrite this identity as

$$r_0(C(\tau) + qr_1(C(\tau)) + \cdots + r_u(C(\tau))q^u \equiv 0 \ (\mathrm{mod} \ p)$$

with $r_i(X) \in \mathbb{Z}[X]$ not all 0 modulo p. Take one j such $r_j(X) \not\equiv 0 \ (\mathrm{mod} \ p)$. Then by our main theorem $r_j(C(\tau)) \equiv 0 \ (\mathrm{mod} \ p)$. This is possible only if $C(\tau) \equiv c$ $(\mathrm{mod} \ p)$ for some $c \in \mathbb{Z}$.   □

We will use this result in an upcoming paper to prove the even case of Subbarao's conjecture [8] already proven by Ono in [4].

5 Conclusion

The conclusion of this paper is that relations of the general form in the abstract can be reduced to much simpler relations, therefore it is not very likely that one would find in the literature such general relations, at least we are not aware of any. This paper can serve as a proof of nonexistence of nontrivial relations of such general form. Here by nontrivial we mean such that are not composed by simpler relations, that is irreducible in some sense. This paper also answers a question by Peter Paule communicated to the author in a private discussion. For this reason it is published with the occasion of his 60th birthday.

References

1. Deligne, P., Rapoport, M.: Les Schémas de Modules de Courbes Elliptiques. In: Modular Functions of One Variable, II. (Proceedings of the International Summer School, Univ. Antwerp, Antwerp 1972). Lecture Notes in Mathematics, vol. 349, pp. 143–316. Springer, Berlin (1973)
2. MacDonald, I.G.: Symmetric Functions and Hall Polynomials. Oxford University Press, Oxford (1995)
3. Ogg, A.: Modular Forms and Dirichlet Series. W. A. Benjamin, Inc., New York/Amsterdam (1969)
4. Ono, K.: Parity of the partition function in arithmetic progressions. J. Reine Angew. Math. **472**, 1–15 (1996)
5. Ono, K.: The Web of Modularity: Arithmetic of the Coefficients of Modular Forms and q-Series. CBMS Regional Conference Series in Mathematics, vol. 102. Published for the Conference Board of the Mathematical Sciences, Washington, DC (2004)
6. Schoeneberg, B.: Elliptic Modular Functions. Springer, Berlin (1974)
7. Shimura, G.: Introduction to the Arithmetic Theory of Automorphic Functions. Princeton University Press, Princeton (1971)
8. Subbarao, M.V.: Some remarks on the partition function. Am. Math. Month. **73**, 851–854 (1966)

Trying to Solve a Linear System for Strict Partitions in 'Closed Form'

Volker Strehl

For Peter@60, and to the memory of Alain Lascoux

1 Introduction

1.1 History and Motivation

The problem treated and partially solved in this article has its roots in work that I have been involved in since almost 10 years. At the 63th meeting of the *Séminaire Lotharingien de Combinatoire* Christian Krattenthaler acquainted me with Arvind Ayyer, who at that time was working as a postdoc with the physicist Kirone Mallick at Saclay (France). Together they were investigating a particular combinatorial model in statistical physics: an asymmetric exclusion process, which is a continuous-time Markov process on a finite number of sites in linear order (like the familiar TASEP model). In addition to moving particles to empty sites 'to their right', the possibility of annihilation between neighboring particles is a particular feature. They had managed to compute the partition functions using the technique of transfer matrices, from which many interesting probabilistic properties of the model could be computed. One aspect, however, was left as an open problem in their article [1]: determining the eigenvalues of the generator matrices of the process, for which the transfer matrices act as intertwining matrices. They had a very precise (and surprisingly simple!) conjecture for the characteristic polynomial, but they— and a number of colleagues they had been asking for help—were unable to prove it. Christian Krattenthaler knew that I have an affinity for eigenvalue problems, and thus he encouraged Arvind Ayyer to tell me about his conjecture. I was lucky to rather quickly find a proof using a technique that I was familiar with due to my

V. Strehl (✉)
Department Informatik, Friedrich-Alexander Universität Erlangen-Nürnberg, Erlangen, Germany
e-mail: Volker.Strehl@fau.de

© Springer Nature Switzerland AG 2020
V. Pillwein, C. Schneider (eds.), *Algorithmic Combinatorics: Enumerative Combinatorics, Special Functions and Computer Algebra*, Texts & Monographs in Symbolic Computation, https://doi.org/10.1007/978-3-030-44559-1_17

interest in quantum computing: consider the Hadamard conjugates of the generator matrices, i.e., consider the problem in an other system of coordinates, obtained by a particular orthogonal transform. The surprise was that (up to tricky an additional permutation conjugation) the generator matrices do not change a lot, but in the new coordinates they become triangular. This approach to solving the eigenvalue problem has been reported in our joint FPSAC article [2].

Since this "Hadamard trick" had proved so successful, I augmented the exclusion-annihilation-model with parameters (one free parameter per site) in a way that made the same proof technique still work. In this generalized model I was able to give a concise algebraic description of the generator and transfer matrices, and thus obtain the partition functions and eigenvalues.

Over the years I have presented the results in several seminar and colloquium talks, but only recently I have started writing down all the details—see the article [3] written in parallel to the present one. It appeared to me that the "Hadamard trick" also suggests a very different approach to the partition functions, which avoids the rather laborious work with the transfer matrices. This different approach passes by a problem that is interesting by itself—and that is what the present article is about. It is a problem about an infinite linear system of equations, with rational functions as coefficients, obtained from the Hadamard-transformed fully parametrized model, and for which the knowledge of properties of the solution would easily lead to the partition functions. In particular, the components of the (unique) solution are rational functions in the site parameters of the physical model, and the knowledge of their denominators is of high interest. But even partially solving the linear system in question is not an easy task. It appered to me that the mentioned rational functions of the solution, though they themselves aren't symmetric functions in most cases (which seems natural because in the original model there is no symmetry in the site parameters), they are nonetheless closely related to symmetric functions. With lots of data at hand, I contacted Alain Lascoux, the grand-master of symmetric functions. He got interested in the problem, started his own computations, made valuable (and sometimes cryptic) suggestions on how to express these rational functions in terms of 'known' functions, and how one might try to prove this. We had just agreed to jointly work on the problem—when in 2013 Alain suddenly died, which stalled my enthusiasm for quite some time. Anyway, the work presented here owes a lot to Alain and my imagination says that he would have appreciated the outcome. I gratefully dedicate this work to his memory.

1.2   Outline

In this article I refrain from outlining the physical model that motivated the work exposed here. See [1] for the original model and [2] for the solution of the eigenvalue conjecture using what I call the "Hadamard trick". The present paper is technically completely self-contained. A detailed explanation of why the solution of the linear system studied here is of interest for obtaining the partition functions for the general fully parametrized model is contained in [3].

The original model, and the work on transfer matrices, partition functions and eigenvalues, uses matrices indexed by binary vectors of fixed length n, which is the number of sites. Binary vectors can be interpreted as strict partitions, i.e. partitions of integers with distinct parts, in an obvious way. Since the solution of the problem, i.e., solving a linear system indexed by binary state vectors, has to do with symmetric functions, Schur functions in particular, it is only natural to set up the stage in terms of partitions and tableaux from the beginning. The variables of the symmetric and related functions are essentially the site variables of the physical model: for a model with n sites there are $n + 1$ relevant variables $x_0, x_1, \ldots, x_n$. Throughout the article the number n will carry this meaning.

In Sect. 2 the scenario involving (strict) partitions and (shifted) tableaux and a valuation in terms of rational functions will be presented. After the terminological generalities of Sect. 2.1 the valuation problem, which runs over the lattice S of strict partitions, will be spelled out in Sect. 2.2, see Eq. (2) and Fig. 10, showing the first 15 equations of the (infinite) system. To each strict partition λ there is attached a rational function $[\![\lambda]\!]$, and these functions are related through the covering relation of the lattice S. For a number of small strict partitions λ the function $[\![\lambda]\!]$ is given explicitly, and properties of these solutions are observed. An explicit general solution for the particular 2-part partitions of the type $\lambda = (n, 1)$ (Sect. 2.2.5) indicates the direction in which journey goes: symmetric functions, and Schur functions in particular.

In Sect. 3 a main technical tool from the field of symmetric functions and its relatives, see, e.g., Alain Lascoux' view in [4], will enter the stage: *divided differences*. Their importance for the main problem becomes strikingly apparent in Lemma 2, which I call the *Main Lemma*: it relates the covering relation of the lattice S with the application of divided differences to the rational functions $[\![\lambda]\!]$ of the solution for the partitions $\lambda \in S$ involved. This way one can determine $[\![\lambda]\!]$ for all staircase partitions $\lambda = (n, n - 1, n - 2, \ldots, 2, 1) = \Delta_n$ (Sect. 3.5), and one can say what the denominators of the $[\![\lambda]\!]$ are in general (Sect. 3.6). It is precisely this aspect which is of particular interest for the application to the physical model, because these denominators are precisely the partition functions.

Nice as these results look, actually computing values $[\![\lambda]\!]$ for 'general' λ is tedious even for a computer, because of the rapid expression swell. The mention of 'closed form' in the title of this paper hints at the problem of finding compact expressions in terms of "known" functions. To say it right away: I don't have a complete solution for this problem—far from that, but at least for two important subclasses of partitions of S, the so-called *join-irreducibles* and the partitions consisting of two (distinct) parts I have been able to express the $[\![\lambda]\!]$ in terms of Schur functions, where usually a pair of two alphabets is required. In Sect. 4 the necessary terminology about Schur functions is introduced. Actually, not much is needed—everything done here relies on generating functions and the Jacobi-Trudi type description of the Schur polynomials, taken as a definition, see any text on symmetric functions like [4], or [5] or [6]. After the necessary tools have been prepared in Sect. 4, the general results for the cases just mentioned are proved in Sect. 5.1 for the join-irreducibles and in Sect. 5.2 for the two-part partitions. The proofs given here are technically elementary, but somewhat tedious.

2 Partitions, Tableaux, and a Rational Valuation Problem

In the first part of this section the relevant definitions and notations for partitions in general, for strict partitions in particular, and for the join-irreducibles in the lattice of strict partitions are given. The second part presents the statement of main problem addressed in this article, and together with computed data it gives an outlook of what can be expected.

2.1 The Lattice S of Strict Partitions

2.1.1 Basic Definitions and Notation

As usual, we write partitions (of nonnegative integers) as vectors $\lambda = (\lambda_1, \lambda_2, \ldots, \lambda_k)$ of positive integers, with $\lambda_1 \geq \lambda_2 \geq \cdots \geq \lambda_k > 0$. If there is no danger of ambiguity, the shorter sequence notation $\lambda_1 \lambda_2 \ldots \lambda_k$ will be used. The usual way of visualizing partitions is via (Ferrers) diagrams, like in Figs. 1 and 2 and in the left part of Fig. 3. These are arrangements of unit boxes with coordinates (i, j), with $1 \leq j \leq \lambda_i$ for $1 \leq i \leq k$, i.e., with k rows, numbered from bottom to top, and with λ_i boxes in the i-th row ($1 \leq i \leq k$). The number of parts is denoted by $|\lambda| = k$, the number partitioned or *size* of λ is $\|\lambda\| = \lambda_1 + \lambda_2 + \cdots + \lambda_k$. In the case $k = 0$ we would have the empty sequence $\emptyset$ as the unique partition of 0. For a positive integer ℓ we say that ℓ occurs in λ, denoted by $\ell \in \lambda$, if $\lambda_j = \ell$ for some j with $1 \leq j \leq k$. The *conjugate* $\widetilde{\lambda} = \widetilde{\lambda}_1 \widetilde{\lambda}_2 \ldots \widetilde{\lambda}_\ell$ of a partition $\lambda = \lambda_1 \lambda_2 \ldots \lambda_k$ is the partition obtained from λ by interchanging the role of rows and columns, i.e., $\ell = \lambda_1$, $\widetilde{\lambda}_1 = k$, and $\widetilde{\lambda}_j$ is the number of $\lambda_i \geq j$. Obviously, the conjugation mapping $\lambda \to \widetilde{\lambda}$ is an involution.

The set $\mathcal{P}$ of all partitions is a distributive lattice under the ordering by inclusion of diagrams, with the size $\|\lambda\|$ as its rank function, i.e.,

$$\lambda = \lambda_1 \lambda_2 \ldots \lambda_k \leq \mu_1 \mu_2 \ldots \mu_\ell = \mu \iff k \leq \ell \text{ and } \lambda_i \leq \mu_i \ (1 \leq i \leq k).$$

The covering relation of $\mathcal{P}$, verbally described as "adding one box in a legal position", is given by

$$\lambda = \lambda_1 \ldots \lambda_k \lessdot \lambda + \varepsilon^i \text{ for } i = 1 \text{ or for some } 1 < i \leq k \text{ with } \lambda_{i-1} > \lambda_i,$$

$$\text{or } \lambda = \lambda_1 \ldots \lambda_k \lessdot \lambda + \varepsilon^{k+1} = \lambda_1 \ldots \lambda_k 1,$$

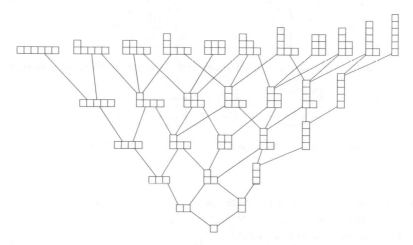

Fig. 1 Lower end (ranks between 1 and 6) of the lattice $\mathcal{P}$ of partitions

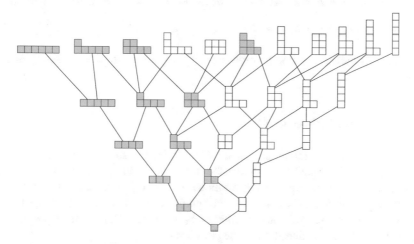

Fig. 2 Lower end (ranks between 1 and 6) of the lattice $\mathcal{S}$ of strict partitions (colored) as sublattice of the lattice $\mathcal{P}$ of all partitions

where the $\boldsymbol{\varepsilon}^i$ are the "unit vectors" $\boldsymbol{\varepsilon}^i = 0^{i-1}.1.0^{k-i}$ (of length k) for $1 \leq i \leq k$, and $\boldsymbol{\varepsilon}^{k+1} = 0^k.1$ (of length $k+1$, with the understanding that $\lambda_{k+1} = 0$). The lower part of the lattice $\mathcal{P}$ is visualized in Fig. 1.

Our concern will be mostly with *strict* partitions, i.e., partitions with *distinct* parts ($\lambda_1 > \lambda_2 > \cdots > \lambda_k > 0$), which form the sublattice $\mathcal{S}$ of $\mathcal{P}$.

When dealing with $\mathcal{S}$ it is often convenient to visualize strict partitions by *shifted* diagrams: take the standard diagram of $\boldsymbol{\lambda} = \lambda_1\lambda_2\ldots\lambda_k$ and push the boxes of the i-th row $i-1$ units to the right ($1 \leq i \leq k$), starting with the bottom row and going

Fig. 3 The strict partition $\lambda = 7641$ displayed as diagram (left) and as shifted diagram (right)

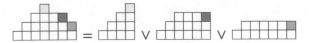

Fig. 4 The strict partition $\lambda = 7641$ as join of the join-irreducibles $<4, 3> = 4321$, $<6, 2> = 654$ and $<7, 1> = 76$

upward. The coordinates of the boxes are now (i, j) for $i \leq j < i + \lambda_i$, where $1 \leq i \leq k$. This transition is illustrated in Fig. 3.

The covering relation of S reads

$$\lambda = \lambda_1 \ldots \lambda_k \lessdot \lambda + \varepsilon^i \quad \text{for } i = 1 \quad \text{or for some } 1 < i \leq k \text{ with } \lambda_{i-1} > \lambda_i + 1,$$

or $\lambda = \lambda_1 \ldots \lambda_k \lessdot \lambda + \varepsilon^{k+1} = \lambda_1 \ldots \lambda_k 1$ if $\lambda_k \geq 2$.

Visually this means "adding a single box in a legal position for obtaining again a shifted diagram".

2.1.2 Join-Irreducibles in S

The join-irreducible elements of a distributive lattice are the 'backbone' of the lattice, as each element of the lattice is the supremum of the elements of an antichain in the poset of join-irreducibles in a unique way. The join-irreducible elements of $\mathcal{P}$ are easily identified as the 'rectangular' partitions $(r, r, \ldots, r)$—these will not be used in the sequel. The join-irreducibles of S, however, will play an important role. Again, they are easily identified as the partitions

$$<n, k> = (n, n - 1, \ldots, n - k) \quad (0 \leq k < n).$$

Figure 4 shows an example of how strict partitions are represented as joins of join-irreducibles in S.

The lower part of the lattice S, with the join-irreducible elements marked by framing them, is reproduced in Fig. 5, and the lower end of the poset of join-irreducibles is displayed in Fig. 6.

$$321 \longrightarrow 421 \longrightarrow 431 \longrightarrow 432 \longrightarrow 4321$$

$$\uparrow \qquad \uparrow \qquad \uparrow$$

$$32 \longrightarrow 42 \longrightarrow 43$$

$$\uparrow \qquad \uparrow$$

$$21 \longrightarrow 31 \longrightarrow 41$$

$$\uparrow \qquad \uparrow \qquad \uparrow$$

$$\emptyset \longrightarrow 1 \longrightarrow 2 \longrightarrow 3 \longrightarrow 4$$

Fig. 5 The lower end of the lattice S with join-irreducible elements marked by framing

$$<4, 3>$$

$$\uparrow$$

$$<3, 2> \longrightarrow <4, 2>$$

$$\uparrow \qquad \uparrow$$

$$<2, 1> \longrightarrow <3, 1> \longrightarrow <4, 1>$$

$$\uparrow \qquad \uparrow \qquad \uparrow$$

$$<1, 0> \longrightarrow <2, 0> \longrightarrow <3, 0> \longrightarrow <4, 0>$$

Fig. 6 The lower end of the poset of join-irreducibles of S

			12		
		7	11		
	3	5	10	13	
1	2	4	6	8	9

Fig. 7 A shifted standard tableau of shape $\lambda = 6421$

2.2 Shifted Tableaux and a Valuation Problem

2.2.1 Shifted Standard Tableaux

In analogy to the well known concept of *standard (Young) tableaux* for general partitions one can define *shifted standard tableaux* (sst) for the case of strict partitions. For a strict partition λ of size $\|\lambda\| = n$ a *shifted standard tableau* of shape λ is a filling t of the n boxes of the shifted diagram representing λ with the integers $\{1, 2, \ldots, n\}$ such that the numbers filled in increase strictly along rows, i.e., $t(i, j) < t(i, j + 1)$, and along columns, i.e., $t(i, j) < t(i + 1, j)$. An example of an sst for $\lambda = 6421$ is given in Fig. 7. The set of all shifted standard tableaux of a given shape λ is denoted by $sST(\lambda)$. An example is displayed in Fig. 8.

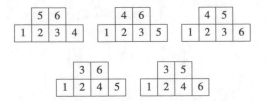

Fig. 8 The set $sST(42)$ of shifted standard tableaux of shape $\lambda = 42$

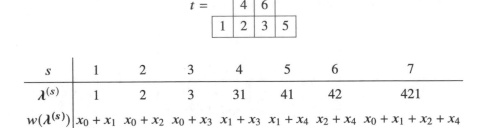

Fig. 9 A shifted tableau t of shape $\lambda = (4, 2, 1)$, and its representation as a nested sequence of strict partitions, together with their weights

A shifted standard tableau of shape λ can be seen a covering sequence of strict partitions:

$$t : \emptyset \lessdot \lambda^{(1)} \lessdot \lambda^{(2)} \lessdot \lambda^{(3)} \lessdot \cdots \lessdot \lambda^{(s)} \lessdot \lambda^{(s+1)} \lessdot \cdots \lessdot \lambda^{\|\lambda\|} = \lambda, \qquad (1)$$

by taking as $\lambda^{(s)}$ the diagram consisting of the boxes filled with the numbers $\{1, 2, \ldots, s\}$, see Fig. 9 for an example.

2.2.2 A Valuation Problem for Shifted Standard Tableaux

In the sequel we will be considering polynomials and rational functions in variables taken from a finite subset of $X = \{x_0, x_1, x_2, \ldots\}$. Generally, for integers $0 \leq a \leq b$ the subset $\{x_a, x_{a+1}, x_{a+2}, \ldots, x_{b-1}, x_b\} \subset X$ of the variables will be denoted by $X_{a,b}$. The subset $X_{a,a} = \{x_a\}$ will be identified with x_a.

For each strict partition $\lambda = \lambda_1 \lambda_2 \ldots \lambda_k \in S$ we define its *weight* as the linear polynomial

$$w(\lambda) = x_{\lambda_1} + x_{\lambda_2} + \cdots + x_{\lambda_k}(+x_0),$$

where the additional term x_0 is taken or not so as to make the total number of summands even. As a shorthand, we occasionally write $x_{ijk\ldots}$ in place of $x_i + x_j + x_k + \ldots$.

For a shifted tableau t of shape λ, written as a nested sequence of strict partitions, as in Eq. (1), we define its *total weight* as

$$w(t) = \prod_{1 \leq s \leq \|\lambda\|} w(\lambda^{(s)}).$$

For the shifted tableau of Fig. 9 the total weight is

$$w(t) = (x_0 + x_1)(x_0 + x_2)(x_0 + x_3)(x_1 + x_3)(x_1 + x_4)(x_2 + x_4)(x_0 + x_1 + x_2 + x_4)$$

$$= (x_{01})(x_{02})(x_{03})(x_{13})(x_{14})(x_{24})(x_{0124}).$$

I admit that at this point the use of the variable x_0 is not well motivated. It comes from the investigation of the model for the asymmetric exclusion process mentioned in the Introduction. For the purpose of this article, x_0 could be set to 0 without loosing much, but keeping it makes the formulas more homogeneous, so I will continue to do so.

Here is now the central problem for the remainder of this article:

$$\text{For } \lambda \in S \text{ compute} \quad [\![\lambda]\!] = \sum \left\{ \frac{1}{w(t)} \, ; \, t \in sST(\lambda) \right\}. \tag{2}$$

For a strict partition λ, written as $\lambda_1 \lambda_2 \ldots \lambda_k$ or as $(\lambda_1, \lambda_2, \ldots, \lambda_k)$, the expression $[\![\lambda]\!]$ represents a rational function in the variables $x_0, x_1, x_2, \ldots, x_{\lambda_1}$. The notation $[\![\lambda_1 \lambda_2 \ldots \lambda_k]\!]$ will usually be used, sometimes with separating commas.

A task equivalent to (2) is: solve—as explicitly as possible—the linear system that runs over the lattice S of strict partitions with its covering relation $\lessdot$:

$$[\![\lambda]\!] = \frac{1}{w(\lambda)} \sum_{\mu \lessdot \lambda} [\![\mu]\!], \quad \text{with } [\![\emptyset]\!] = 1 \quad (\lambda \in S). \tag{3}$$

In Fig. 10 the first 15 equations of this infinite system are displayed.
The first equation with three terms on the right hand side would be

$$[\![531]\!] = [\![431]\!] + [\![521]\!] + [\![53]\!].$$

Conceptually, hardly anything could be simpler: this is a (very sparse) triangular system—but writing down the solution by backward substitution requires repeated division, where the divisors are sums (of even length) of the variables, so that the $[\![\lambda]\!]$ are rational functions in the variables $x_0, x_1, x_2, \ldots$, and it is easy to see that the numerator and denominator polynomials are homogeneous. If we define for any rational function f/g the *rank* of f/g as the difference between denominator degree and numerator degree $\deg g - \deg f$, i.e., the negative of the degree as defined in

$$(x_0 + x_1)\,[\![1]\!] = [\![0]\!]$$

$$(x_0 + x_2)\,[\![2]\!] = \qquad [\![1]\!]$$

$$(x_1 + x_2)\,[\![21]\!] = \qquad\qquad [\![2]\!]$$

$$(x_0 + x_3)\,[\![3]\!] = \qquad\qquad [\![2]\!]$$

$$(x_1 + x_3)\,[\![31]\!] = \qquad\qquad [\![21]\!] + [\![3]\!]$$

$$(x_2 + x_3)\,[\![32]\!] = \qquad\qquad\qquad [\![31]\!]$$

$$(x_0 + x_1 + x_2 + x_3)\,[\![321]\!] = \qquad\qquad\qquad [\![32]\!]$$

$$(x_0 + x_4)\,[\![4]\!] = \qquad\qquad\qquad [\![3]\!]$$

$$(x_1 + x_4)\,[\![41]\!] = \qquad\qquad\qquad [\![31]\!] + [\![4]\!]$$

$$(x_2 + x_4)\,[\![42]\!] = \qquad\qquad\qquad [\![32]\!] + [\![41]\!]$$

$$(x_0 + x_1 + x_2 + x_4)\,[\![421]\!] = \qquad\qquad\qquad [\![321]\!] + [\![42]\!]$$

$$(x_3 + x_4)[\![43]\!] = \qquad\qquad\qquad\qquad [\![42]\!]$$

$$(x_0 + x_1 + x_3 + x_4)[\![431]\!] = \qquad\qquad\qquad [\![421]\!] + [\![43]\!]$$

$$(x_0 + x_2 + x_3 + x_4)[\![432]\!] = \qquad\qquad\qquad\qquad [\![431]\!]$$

$$(x_1 + x_2 + x_3 + x_4)[\![4321]\!] = \qquad\qquad\qquad\qquad [\![432]\!]$$

Fig. 10 The first 15 equation of the linear system

Sect. 3.2, then it is simply checked that the rank of $[\![\lambda]\!]$ equals the size of the partition λ, as defined above.

2.2.3   Some Computed Data

The explicit values of the solution $[\![\lambda]\!]$ for very small λ are given in Fig. 11.

So far everything looks pretty simple, except for the following observation: from the definition we have

$$[\![321]\!] = \frac{1}{x_0 + x_1 + x_2 + x_3}\,[\![32]\!],$$

but the quadrinomial $x_0 + x_1 + x_2 + x_3$ from the division is cancelled, because it occurs in $[\![32]\!]$ as numerator. This matching numerator has been 'created' in the previous step

$$[\![31]\!] = \frac{1}{x_1 + x_3}\,([\![21]\!] + [\![3]\!]).$$

This is apparently not an accident. If one continues the evaluation one never finds multinomials of four or more terms in the denominator. The denominators are

$$[\![1]\!] = \frac{1}{x_0 + x_1}$$

$$[\![2]\!] = \frac{1}{(x_0 + x_1)(x_0 + x_2)}$$

$$[\![21]\!] = \frac{1}{(x_0 + x_1)(x_0 + x_2)(x_1 + x_2)}$$

$$[\![3]\!] = \frac{1}{(x_0 + x_1)(x_0 + x_2)(x_0 + x_3)}$$

$$[\![31]\!] = \frac{x_0 + x_1 + x_2 + x_3}{(x_0 + x_1)(x_0 + x_2)(x_0 + x_3)(x_1 + x_2)(x_1 + x_3)}$$

$$[\![32]\!] = \frac{x_0 + x_1 + x_2 + x_3}{(x_0 + x_1)(x_0 + x_2)(x_0 + x_3)(x_1 + x_2)(x_1 + x_3)(x_2 + x_3)}$$

$$[\![321]\!] = \frac{1}{(x_0 + x_1)(x_0 + x_2)(x_0 + x_3)(x_1 + x_2)(x_1 + x_3)(x_2 + x_3)}$$

Fig. 11 First values of the solution

always products of binomials—this will be proved later, see Sect. 3.6—but at this point this looks like magic.

Continuing the evaluation one finds that the expressions soon become 'unwieldy', e.g.,

$$[\![42]\!] = \frac{1}{x_2 + x_4}([\![32]\!] + [\![41]\!])$$

is a rational function with numerator

$$x_1 x_0^2 + x_2 x_0^2 + x_3 x_0^2 + x_4 x_0^2 + x_1^2 x_0 + x_2^2 x_0 + x_3^2 x_0 + x_4^2 x_0$$

$$+ 2x_1 x_2 x_0 + 2x_1 x_3 x_0 + 2x_2 x_3 x_0 + 2x_1 x_4 x_0 + 2x_2 x_4 x_0 + 2x_3 x_4 x_0 + x_1 x_2^2$$

$$+ x_1 x_3^2 + x_2 x_3^2 + x_1 x_4^2 + x_2 x_4^2 + x_3 x_4^2 + x_1^2 x_2 + x_1^2 x_3 + x_2^2 x_3$$

$$+ 2x_1 x_2 x_3 + x_1^2 x_4 + x_2^2 x_4 + x_3^2 x_4 + 2x_1 x_2 x_4 + 2x_1 x_3 x_4 + 2x_2 x_3 x_4$$

and denominator

$$(x_0+x_1)(x_0+x_2)(x_1+x_2)(x_0+x_3)(x_1+x_3)(x_2+x_3)(x_0+x_4)(x_1+x_4)(x_2+x_4) \,.$$

The numerator can be 'simplified' into

$$(x_1 + x_2 + x_3 + x_4) x_0^2 + (x_1 + x_2 + x_3 + x_4)^2 x_0 + x_1 (x_2 + x_3 + x_4)^2$$

$$+ (x_2 + x_3)(x_2 + x_4)(x_3 + x_4) + x_1^2 (x_2 + x_3 + x_4),$$

but this still looks mysterious. If we look at

$$[\![531]\!] = \frac{1}{x_0 + x_1 + x_3 + x_5}([\![431]\!] + [\![521]\!] + [\![53]\!]),$$

we get (using $x_{ijk...}$ again as an abbreviation for $x_i + x_j + x_k \cdots$) as numerator (as simplified by Mathematica)

$$x_5^2 (x_{234})^3 + x_5^3 (x_{234})^2 + (x_{23})^2 x_4 (x_{24}) (x_{34}) + (x_{23}) (x_{24}) (x_{34}) (x_{234}) x_5$$

$$+ x_1 \left((x_{234}) x_5^3 + 2 (x_{234})^2 x_5^2 + \left(x_2^3 + 4 (x_{34}) x_2^2 + 4 (x_{34})^2 x_2 \right.\right.$$

$$+ (x_{34}) \left(x_3^2 + 3x_4 x_3 + x_4^2 \right) \Big) x_5 + (x_{23}) (x_{24}) (x_{34}) (x_{234})$$

$$+ x_1^2 \left((x_{345}) x_2^2 + (x_{345})^2 x_2 + (x_{34}) (x_{35}) (x_{45}) \right)$$

$$+ x_0^2 \left((x_{2345}) x_1^2 + (x_{2345})^2 x_1 + (x_{234}) x_5^2 + (x_{23}) (x_{24}) (x_{34}) + (x_{234})^2 x_5 \right) \Big)$$

$$+ x_0 (x_{2345}) \left((x_{2345}) x_1^2 + (x_{2345})^2 x_1 + (x_{234}) x_5^2 + (x_{23}) (x_{24}) (x_{34}) + (x_{234})^2 x_5 \right) \Big)$$

and as denominator a product of binomials

$$\prod_{0 \le i \le 3} \prod_{i < j \le 5} (x_i + x_j).$$

Again, the combination of the three rational functions $[\![431]\!]$, $[\![521]\!]$, and $[\![53]\!]$ (not shown here, due to their size) with hard to digest expressions in their numerators somehow produces the factor $x_0 + x_1 + x_3 + x_5$ in the numerator, that finally gets cancelled.

2.2.4 About Denominators

As far as the denominators of the $[\![\lambda]\!]$ are concerned, let us define for $n > k \ge 0$ what we call the *standard denominator polynomials*:

$$q_{n,k}(x_0, x_1, x_2, \ldots) = \prod_{0 \le i \le k} \prod_{i < j \le n} (x_i + x_j) = \prod_{(j,i) \le (n,k)} (x_i + x_j), \qquad (4)$$

where the second way of writing the product implicitly refers to the poset of join-irreducibles of the lattice S of strict partitions. Indeed, the polynomials $q_{n,k}$ are in 1-1-correspondence with the join irreducibles in the lattice S, see Sect. 2.1.2.

Note that $q_{n,k}(x_0, x_1, x_2, \ldots)$ is a polynomial that is separately symmetric in the two sets of variables $X_{0,k} = \{x_0, x_1, \ldots, x_k\}$ and $X_{k+1,n} = \{x_{k+1}, x_{k+2}, \ldots, x_n\}$. The case $k = n - 1$ is special, however, because $q_{n,n-1}$ is symmetric in all variables

$X_{0,n} = \{x_0, x_1, \ldots, x_n\}$. The degree of $q_{n,k}$ is

$$\sum_{i=0}^{k}(n-i) = (k+1)n - \binom{k+1}{2} = \frac{(k+1)(2n-k)}{2}.$$

Looking back to the data given above, we can state that in these examples the $q_{n,k}$ appear as the denominator polynomials of $[\![\lambda]\!]$ precisely if $\lambda = (n, k, \ldots)$, i.e., the denominator of the function $[\![\lambda]\!]$ depends only on the two largest parts of λ. This is true in general, as will be shown in Sect. 3.6, Corollary 2.

The case of partitions consisting of one part is very simple, see the next subsection.

2.2.5 Outlook: Evaluating $[\![n, 1]\!]$ in General

As an encouraging experience, we will now compute the rational functions $[\![\lambda]\!]$ for the very special case of 2-part partitions of type $\lambda = (n, 1)$ for all $n \geq 2$. Even this task is not completely trivial, and a bit of intuition is needed to guess the 'good' form of the solution. The final statement makes use of the elementary and homogeneous symmetric functions over distinct alphabets (sets of variables), which will be amalgamated later (Sect. 4.3) in the notion of Schur functions over a pair of alphabets.

To start with, note that for partitions $\lambda = (n)$ consisting of one part the situation is trivial. Because of

$$[\![n]\!] = \frac{1}{x_0 + x_n}[\![n-1]\!] \quad \text{for} \quad n > 0,$$

we have

$$[\![n]\!] = \frac{1}{\prod_{1 \leq i \leq n}(x_0 + x_i)} = \frac{1}{q_{n,0}}. \tag{5}$$

But already the computation of $[\![n, 1]\!]$ in general is not obvious. One gets as first values (with the usual abbreviations)

$$[\![21]\!] = \frac{1}{x_{12}}[\![2]\!] \qquad = \frac{1}{x_{01}x_{02}x_{12}} = \frac{1}{q_{2,1}},$$

$$[\![31]\!] = \frac{1}{x_{13}}([\![21]\!] + [\![3]\!]) = \frac{x_{0123}}{x_{01}x_{02}x_{03}x_{12}x_{13}} = \frac{x_{03} + x_{12}}{q_{3,1}},$$

$$[\![41]\!] = \frac{1}{x_{14}}([\![31]\!] + [\![4]\!]) = \ldots = \frac{x_{03}x_{04} + x_{12}x_{04} + x_{12}x_{13}}{q_{4,1}},$$

$$[\![51]\!] = \frac{1}{x_{15}}([\![41]\!] + [\![5]\!]) = \ldots = \frac{x_{05}x_{04}x_{03} + x_{05}x_{04}x_{12} + x_{05}x_{13}x_{12} + x_{14}x_{13}x_{12}}{q_{5,1}}.$$

The way of writing the numerators on the right was not obtained by machine simplification, but by human inspection and trial. It suggests that in general

$$[\![n, 1]\!] = \frac{\sum_\lambda \prod_{i=1}^{n-2}(x_{\lambda_i} + x_{\lambda_i'})}{q_{n,1}}, \tag{6}$$

where $\sum_\lambda$ runs over all $\lambda \in S$ with

$$\underline{\tau}^{(n)} = (n-1, n-2, \ldots, 3, 2) \leq \lambda \leq \overline{\tau}^{(n)} = (n, n-1, \ldots 4, 3) \quad \text{and} \quad \lambda' = \overline{\tau}^{(n)} - \lambda.$$

Here λ' is not a strict partition, but a vector of type $0^\ell 1^{n-2-\ell}$.

To illustrate this in the case $n = 5$: the relevant λ with $\underline{\tau}^{(5)} = 432 \leq \lambda \leq \overline{\tau}^{(5)} = 543$ are 543 (with $\lambda' = 000$), 542 (with $\lambda' = 001$), 532 (with $\lambda' = 011$), and 432 (with $\lambda' = 111$).

The claim (6) is routinely verified by induction:

$$[\![n+1, 1]\!] = \frac{1}{x_1 + x_{n+1}}([\![n, 1]\!] + [\![n+1]\!])$$

$$= \frac{1}{x_1 + x_{n+1}} \left(\frac{\sum_\lambda \prod_{i=1}^{n-2}(x_{\lambda_i} + x_{\lambda_i'})}{q_{n,1}} + \frac{1}{q_{n+1,0}} \right)$$

$$= \frac{1}{q_{n+1,1}} \frac{1}{(x_1 + x_{n+1})} \times$$

$$\times \left(\sum_\lambda \prod_{i=1}^{n-2}(x_{\lambda_i} + x_{\lambda_i'}) \cdot (x_0 + x_{n+1})(x_1 + x_{n+1}) + \prod_{1<j\leq n+1}(x_1 + x_j) \right)$$

$$= \frac{1}{q_{n+1,1}} \left(\sum_\lambda \prod_{i=1}^{n-2}(x_{\lambda_i} + x_{\lambda_i'}) \cdot (x_0 + x_{n+1}) + \prod_{1<j\leq n}(x_1 + x_j) \right).$$

The expression in parentheses is precisely the corresponding $\sum_\mu$-summation for $n + 1$, running over strict partitions μ with $\underline{\tau}^{(n+1)} \leq \mu \leq \overline{\tau}^{n+1}$. The $\sum_\lambda$-part covers all $\mu = (n + 1, \lambda)$, whereas the last term gives the contribution from $\mu = (n, n - 1, \ldots, 3, 2)$.

This form of $[\![n, 1]\!]$ does not look particularly attractive, but it is possible to rewrite it in a much neater way in terms of symmetric functions. Indeed,

$$[\![n, 1]\!] = \frac{1}{q_{n,1}} \sum_{k=0}^{n-2} h_k(X_{0,1}) \cdot e_{n-2-k}(X_{2,n}),$$

where the $h_k(A)$ resp. $e_\ell(B)$ denote the homogeneous resp. elementary symmetric functions over the alphabets A resp. B—see Sect.4 for notation.

Once one has made this guess, it is a routine matter to verify that indeed both sides of

$$\sum_{\lambda} \prod_{i=1}^{n-2} (x_{\lambda_i} + x_{\lambda_i'}) = \sum_{k=0}^{n-2} h_k(X_{0,1}) \cdot e_{n-2-k}(X_{2,n}) \tag{7}$$

contain the same monomials. As an illustration for $n = 4$:

$$h_2(X_{0,1}) + h_1(X_{0,1})e_1(X_{2,4}) + e_2(X_{2,4})$$
$$= x_1^2 + (x_0 + x_1 + x_2 + x_3)(x_0 + x_4) + x_1(x_2 + x_3) + x_2 x_3$$
$$= (x_0 + x_3)(x_0 + x_4) + (x_2 + x_3)(x_0 + x_4) + (x_1 + x_2)(x_2 + x_3).$$

The term on the right hand side of (7) can also be written as a Schur polynomial over a pair alphabets, viz.

$$\sum_{k=0}^{n-2} h_k(X_{0,1}) \cdot e_{n-2-k}(X_{2,n}) = S_{n-2}(X_{0,1}|X_{2,n}).$$

This view is interesting because the numerator polynomials $q_{n,1}$, and the $q_{n,m}$ in general, can be written as Schur polynomials over a pair of alphabets:

$$q_{n,m} = S_{\widetilde{<n,m>}}(X_{0,m}|X_{m+1,n}) = S_{<n,m>}(X_{m+1,n}|X_{0,m}), \tag{8}$$

where $<n, m> = (n, n-1, \ldots, n-m)$ is join-irreducible, and where $\widetilde{<n, m>}$ is the conjugate partition of $<n, m>$. See Sect. 4.3 for the definition of Schur functions over a pair of alphabets.

3 Divided Differences

The main technical tool for computing $[\![\lambda]\!]$ is a standard one when computing in the area of symmetric functions and its vicinity: *divided differences*. It turns out that there is an intimate connection between the lattice structure of S and the nature of divided differences as symmetrizing operators.

3.1 Definitions and Properties

We consider functions (here polynomials and rational functions only) in the variables $X = \{x_0, x_1, \ldots\}$. For $r \geq 0$ let $\sigma_r : X \to X$ denote the transposition of variables $x_r \leftrightarrow x_{r+1}$. For a function $f = f(X)$ we write f^{σ_r} in place of the composition $f \circ \sigma_r$.

If $f(x_0, x_1, \ldots)$ is any function, then we define the divided difference of f w.r.t. σ_r, denoted by $f\partial_r$, as the function given by

$$(f\partial_r)(X) = \frac{f^{\sigma_r}(X) - f(X)}{x_r - x_{r+1}}.$$

We adopt here Lascoux' habit to write the operator ∂_r on the right of the function it acts on. Thus iterations ('cascades') of divided differences like $\partial_b \partial_{b-1} \ldots \partial_a$ mean that ∂_b is executed first and ∂_a is executed last.

Note that our definition differs from the usual one, where one takes $\frac{f^{\sigma_r} - f}{x_{r+1} - x_r}$ instead, by the sign. The reason for this trivial change is to avoid carrying minus-signs around when taking iterated divided differences.

A function f is *symmetric* w.r.t. σ_r, if $f^{\sigma_r} = f$, and this is precisely the case if $f\partial_r = 0$.

As an operator on functions, ∂_r has the usual properties, like linearity, product rule, quotient rule (see below, and generally refer to [4]). The relations among the ∂_r for different r that are derived from the (Coxeter-) relations for the transpositions σ_r are:

$$\partial_r \partial_r = 0 \qquad\qquad \text{if } r \geq 0,$$
$$\partial_r \partial_s = \partial_s \partial_r \qquad\qquad \text{if } |r - s| \geq 2,$$
$$\partial_r \partial_{r+1} \partial_r, = \partial_{r+1} \partial_r \partial_{r+1} \qquad\qquad \text{if } r \geq 0.$$

Well known rules for computing with divided differences are the product rule,

$$(f \cdot g)\partial_r = f^{\sigma_r} \cdot g\partial_r + f\partial_r \cdot g$$
$$= f\partial_r \cdot g^{\sigma_r} + f \cdot g\partial_r$$

and the rules for reciprocals and quotients

$$f^{-1}\partial_r = -\frac{1}{f \cdot f^{\sigma_r}} \cdot (f\partial_r),$$

$$(f/g)\,\partial_r = \frac{1}{g \cdot g^{\sigma_r}}(g \cdot f\partial_r - f \cdot g\partial_r)$$

$$= \frac{1}{g \cdot g^{\sigma_r}}(g^{\sigma_r} \cdot f\partial_r - f^{\sigma_r} \cdot g\partial_r).$$

3.2 Divided Differences for Rational Functions

Define the *degree* of a rational function f/g, where f and g are polynomials, not necessarily coprime, as $\deg(f/g) = \deg(f) - \deg(g)$. Divided differences change the degree of rational functions

$$\deg((f/g)\partial_r) = \begin{cases} \deg(f/g) - 1 & \text{if } (f/g)\partial_r \neq 0, \\ \text{undefined} & \text{if } (f/g)\partial_r = 0. \end{cases}$$

For a single polynomial q we set

$$\tilde{q} = \frac{q}{\gcd(q, q^{\sigma_r})} = \frac{\text{lcm}(q, q^{\sigma_r})}{q^{\sigma_r}}.$$

Then

$$\left(\frac{1}{q}\right)\partial_r = \frac{1}{x_r - x_{r+1}} \cdot \left(\frac{1}{q^{\sigma_r}} - \frac{1}{q}\right)$$

$$= \frac{1}{\text{lcm}(q, q^{\sigma_r})} \cdot \frac{\tilde{q}^{\sigma_r} - \tilde{q}}{x_r - x_{r+1}}$$

$$= \frac{1}{\text{lcm}(q, q^{\sigma_r})} \cdot \tilde{q}\partial_r.$$

Similarly one gets for polynomials p and q

$$\left(\frac{p}{q}\right)\partial_r = \frac{1}{\text{lcm}(q, q^{\sigma_r})} \left(\tilde{q}^{\sigma_r} \cdot p\partial_r + p \cdot \tilde{q}\partial_r\right).$$

3.3 Standard Denominator Polynomials

Recall from Sect. 2.2.4 and Eq. (4) the notion of standard denominator polynomials, $q_{n,k}(X) = \prod_{\substack{0 \leq i \leq k \\ i < j \leq n}} (x_i + x_j)$, and their properties. We will now compute the divided differences

$$\frac{1}{q_{n,k}}\partial_r.$$

From the remark about symmetry in Sect. 2.2.4 it follows that the only interesting cases are $r = k < n-1$ and $r = n$. In all other cases the divided difference vanishes because then $q_{n,k}$ is symmetric w.r.t. $\sigma_r = x_r \leftrightarrow x_{r+1}$.

- In the case $r = n$, it follows from the geometric picture of the join-irreducibles that

$$\gcd(q_{n,k}, q_{n,k}^{\sigma_n}) = q_{n-1,k}, \qquad \text{lcm}(q_{n,k}, q_{n,k}^{\sigma_n}) = q_{n+1,k}.$$

In the case $k = 0$ the term $q_{n,-1}$ has to be taken as 1. Then

$$\widetilde{q}_{n,k} = \frac{q}{\gcd(q_{n,k}, q_{n,k}^{\sigma_n})} = \prod_{0 \le i \le k} (x_i + x_n)$$

and

$$\frac{1}{q_{n,k}} \partial_n = \frac{1}{q_{n+1,k}} \cdot \widetilde{q}_{n,k} \partial_n.$$

– In the case $r = k < n - 1$ we have

$$q_{n,k} = q_{n,k-1} \cdot (x_k + x_{k+1}) \cdot \prod_{j > k} (x_k + x_j),$$

$$q_{n,k}^{\sigma_k} = q_{n,k-1} \cdot (x_k + x_{k+1}) \cdot \prod_{j > k} (x_{k+1} + x_j),$$

and so from the 'geometry' of the join-irreducibles it follows that

$$\gcd(q_{n,k}, q_{n,k}^{\sigma_k}) = (x_k + x_{k+1}) q_{n,k-1}, \qquad \mathrm{lcm}(q_{n,k}, q_{n,k}^{\sigma_k}) = q_{n,k+1}.$$

Then

$$\widetilde{q}_{n,k} = \frac{q}{\gcd(q_{n,k}, q_{n,k}^{\sigma_n})} = \prod_{k+2 \le j \le n} (x_k + x_j)$$

and

$$\frac{1}{q_{n,k}} \partial_k = \frac{1}{q_{n,k+1}} \cdot \widetilde{q}_{n,k} \partial_k.$$

For later reference we state the following consequences:

Lemma 1 *For any polynomial p and $0 \le k < n$ we have*

$$\left[\frac{p}{q_{n,k}} \right] \partial_r = \begin{cases} \dfrac{f}{q_{n+1,k}} & \text{if } r = n, \\[2mm] \dfrac{g}{q_{n,k+1}} & \text{if } r = k < n - 1, \\[2mm] \dfrac{1}{q_{n,k}} \cdot p \partial_r & \text{otherwise,} \end{cases}$$

where f and g are polynomials.

3.4 The Main Lemma

In this section we will prove the main technical result that relates the valuations $[\![\lambda]\!]$ and divided differences. For convenience, we set $[\![\lambda]\!] = 0$ if λ is a partition that is not strict, i.e., $\lambda \in \mathcal{P} \setminus \mathcal{S}$. In what follows, the $\boldsymbol{\varepsilon}^j$ are the "unit vectors" of Sect. 2.1.1.

Lemma 2 (Main Lemma) *Let* $\lambda = \lambda_1 \lambda_2 \ldots \lambda_k \in \mathcal{S}$ *and* $\ell \geq 1$. *Then*

$$[\![\lambda]\!] \, \partial_\ell = \begin{cases} [\![\lambda + \boldsymbol{\varepsilon}^j]\!] & \text{if } \ell \in \lambda, \ell = \lambda_j, \\ 0 & \text{if } \ell \notin \lambda. \end{cases}$$

The first alternative includes the case where $[\![\lambda]\!] \, \partial_\ell = 0$, *when* $\lambda + \boldsymbol{\varepsilon}^j \notin \mathcal{S}$.

Proof To start with, consider the action of a divided difference ∂_ℓ on the defining expression

$$[\![\lambda]\!] = \frac{1}{x_\lambda} \sum_{1 \leq i \leq k} [\![\lambda - \boldsymbol{\varepsilon}^i]\!],$$

viz.

$$[\![\lambda]\!] \, \partial_\ell = \frac{1}{x_\lambda^{\sigma_\ell}} \cdot \sum_{1 \leq i \leq k} [\![\lambda - \boldsymbol{\varepsilon}^i]\!] \, \partial_\ell + \frac{1}{x_\lambda} \partial_\ell \cdot x_\lambda \cdot [\![\lambda]\!].$$

Various situations can occur, depending on whether ℓ and $\ell + 1$ belong or don't belong to λ. Thus the proof has to deal with four distinct cases.

(i) $\boxed{\ell = \lambda_j \in \lambda, \ell + 1 \notin \lambda}$

Here $\lambda' = \lambda + \boldsymbol{\varepsilon}^j \in \mathcal{S}$, x_λ contains x_ℓ as a summand, but not $x_{\ell+1}$, and thus $x_\lambda^{\sigma_\ell} = x_\lambda - x_\ell + x_{\ell+1} = x_{\lambda'}$. Therefore

$$\frac{1}{x_\lambda} \partial_\ell = (\frac{1}{x_\lambda} - \frac{1}{x_\lambda^{\sigma_\ell}})/(x_{\ell+1} - x_\ell) = \frac{1}{x_\lambda} \frac{1}{x_\lambda^{\sigma_\ell}} = \frac{1}{x_\lambda} \frac{1}{x_{\lambda'}}$$

and

$$[\![\lambda]\!] \, \partial_\ell = \frac{1}{x_{\lambda'}} \left(\sum_{1 \leq i \leq k} [\![\lambda - \boldsymbol{\varepsilon}^i]\!] \partial_\ell + [\![\lambda]\!] \right).$$

By induction we may use

$$[\![\lambda - \boldsymbol{\varepsilon}^i]\!] \partial_\ell = [\![\lambda - \boldsymbol{\varepsilon}^i + \boldsymbol{\varepsilon}^j]\!]$$

in all cases where $i \neq j$ without problems. But the case $i = j$ has to be considered separately. We have to show that $[\![\lambda - \varepsilon^j]\!]\partial_\ell = 0$, because then we would get

$$\frac{1}{x_{\lambda'}} \left(\sum_{1 \leq i \leq k} [\![\lambda - \varepsilon^i]\!]\partial_\ell + [\![\lambda]\!] \right) = \frac{1}{x_{\lambda'}} \left(\sum_{\substack{1 \leq i \leq k \\ i \neq j}} [\![\lambda - \varepsilon^i + \varepsilon^j]\!] + [\![\lambda]\!] \right),$$

which, by definition of the valuation, equals $[\![\lambda']\!] = [\![\lambda + \varepsilon^j]\!]$.

Now for $[\![\lambda - \varepsilon^j]\!]\partial_\ell$ there are two possibilities: either $\lambda - \varepsilon^j \notin S$ (i.e., $\lambda_j - 1 = \lambda_{j+1}$), then $[\![\lambda - \varepsilon^j]\!] = 0$ anyway; or $\lambda - \varepsilon^j \in S$, then neither $\lambda_j = \ell$ nor $\lambda_j + 1 = \ell + 1$ belong to $\lambda - \varepsilon^j$ and we have $[\![\lambda - \varepsilon^j]\!]\partial_\ell = 0$, by induction making use of case (iii).

(ii) $\boxed{\ell \notin \lambda, \ell + 1 = \lambda_j \in \lambda}$

In this situation, contrary to case (i), we have, with $\lambda' = \lambda - \varepsilon^j$ and $x_{\lambda'} = x_\lambda - x_{\ell+1} + x_\ell$,

$$\frac{1}{x_\lambda}\partial_\ell = \left(\frac{1}{x_\lambda^{\sigma_\ell}} - \frac{1}{x_\lambda} \right) / (x_{\ell+1} - x_\ell) = -\frac{1}{x_\lambda}\frac{1}{x_\lambda^{\sigma_\ell}} = -\frac{1}{x_\lambda}\frac{1}{x_{\lambda'}}.$$

Hence

$$[\![\lambda]\!]\,\partial_\ell = \frac{1}{x_{\lambda'}} \left(\sum_{1 \leq i \leq k} [\![\lambda - \varepsilon^i]\!]\partial_\ell - [\![\lambda]\!] \right).$$

Now $\ell \notin \lambda - \varepsilon^i$ for all $i \neq j$, so that $[\![\lambda - \varepsilon^i]\!]\partial_\ell = 0$ by induction. The only exception is the case $i = j$, which by induction gives

$$[\![\lambda - \varepsilon^j]\!]\partial_\ell = [\![\lambda - \varepsilon^j + \varepsilon^j]\!] = [\![\lambda]\!],$$

so that in the expression for $[\![\lambda]\!]\,\partial_\ell$ only the two terms $[\![\lambda]\!]$ and $-[\![\lambda]\!]$ survive, but cancelling each other.

(iii) $\boxed{\ell \notin \lambda, \ell + 1 \notin \lambda}$

Now x_λ is symmetric w.r.t. $\sigma_\ell : x_\ell \leftrightarrow x_{\ell+1}$, so that

$$[\![\lambda]\!]\,\partial_\ell = \frac{1}{x_\lambda} \cdot \sum_{1 \leq i \leq k} [\![\lambda - \varepsilon^i]\!]\partial_\ell.$$

Assume $\lambda_j > \ell + 1 > \ell > \lambda_{j+1}$. There are two possible cases:

– If $\lambda_j - 1 > \ell + 1$, the both ℓ and $\ell + 1$ do not occur in $\lambda - \varepsilon^j$, nor in any other $\lambda - \varepsilon^i$, so, by induction, $[\![\lambda - \varepsilon^i]\!]\partial_\ell = 0$ for all i.

- If, however, $\lambda_j = \ell + 2 > \ell + 1 > \ell > \lambda_{j+1}$, then for $\lambda' = \lambda - \varepsilon^j$ we have $\lambda'_j = \ell + 1 > \ell > \lambda_{j+1}$ and $[\![\lambda - \varepsilon^j]\!]\partial_\ell = 0$ by induction for case (ii), whereas $[\![\lambda - \varepsilon^i]\!]\partial_\ell = 0$ for all $i \neq j$ as before.

(iv) $\boxed{\ell = \lambda_j \in \lambda, \ell + 1 = \lambda_{j-1} \in \lambda}$

Then $\lambda + \varepsilon^j \notin S$ and one has to show that $[\![\lambda]\!]\partial_\ell = 0$. Since both x_ℓ and $x_{\ell+1}$ belong to λ, x_λ is symmetric w.r.t. σ_ℓ and the expression for $[\![\lambda]\!]\partial_\ell$ reduces to

$$[\![\lambda]\!]\partial_\ell = \frac{1}{x_\lambda} \cdot \sum_{1 \leq i \leq k} [\![\lambda - \varepsilon^i]\!]\partial_\ell.$$

The following situations occur:

- If $i < j - 1$ or $i > j$, then either $\lambda' = \lambda - \varepsilon^i \notin S$, or $\ell = \lambda'_j$, $\ell + 1 = \lambda'_{j-1}$ and $[\![\lambda']\!]\partial_\ell = 0$ follows by induction.
- If $i = j - 1$, $\lambda' = \lambda - \varepsilon^{j-1}$, we have $\lambda'_{j-1} = \lambda_{j-1} - 1 = \ell$ and $\lambda'_j = \lambda_j = \ell$, so that λ' is not strict and already $[\![\lambda']\!] = 0$.
- If $i = j$, $\lambda' = \lambda - \varepsilon^j$, we have $\lambda'_{j-1} = \lambda_{j-1} = \lambda_j + 1 = \ell + 1$ and $\lambda'_j = \lambda_j - 1 = \ell - 1$, so that λ' does not contain $\ell = \lambda_j$ and $[\![\lambda']\!]\partial_\ell = 0$ by induction from case (iii).

All summands vanish and thus $[\![\lambda]\!]\partial_\ell = 0$. $\square$

Remark The situation of augmenting a strict partition $\lambda = \lambda_1\lambda_2 \ldots \lambda_k$ with $\lambda_k \geq 2$ by adding a new $(k + 1)$-st part $\lambda_{k+1} = 1$ is not covered by the lemma. We state without proof that

$$[\![\lambda_1\lambda_2 \ldots \lambda_k]\!]\partial_0 = \begin{cases} [\![\lambda_1\lambda_2 \ldots \lambda_k 1]\!] & \text{if } k \text{ is odd and } \lambda_k \geq 2, \\ 0 & \text{if } k \text{ is even or } \lambda_k = 1. \end{cases}$$

Lemma 2 suggests a method for computing $[\![\lambda]\!]$ using divided differences: let $\lambda = \lambda_1\lambda_2 \ldots \lambda_k \in S$ then:

- Start with the staircase partition

$$\Delta_k = (k, k - 1, k - 2, \ldots, 1) = <k, k - 1>,$$

see Sect. 3.5, where $[\![\Delta_k]\!]$ is given.
- Then (in terms if shifted diagrams) extend the first row of Δ_k from k boxes to λ_1 boxes, which amounts to applying the cascade $\partial_k \partial_{k+1} \ldots \partial_{\lambda_1-1}$ to $[\![\Delta_k]\!]$, which gives $[\![(\lambda_1, k - 1, k - 2, \ldots, 1)]\!]$.
- Then extend the second row from $k - 1$ boxes to λ_2 boxes, which means applying the cascade $\partial_{k-1}\partial_k \ldots \partial_{\lambda_2-1}$ to the function $[\![(\lambda_1, k - 1, k - 2, \ldots, 1)]\!]$, which gives $[\![(\lambda_1, \lambda_2, k - 2, \ldots, 1)]\!]$.

– Continue this way row by row until extending the top row of the shifted diagram, which means applying the cascade $\partial_1 \partial_2 \ldots \partial_{\lambda_k - 1}$ to the result from extending the previous row, which gives $[\![\lambda_1, \lambda_2, \ldots, \lambda_k]\!]$.
– In all, we get

$$[\![\lambda]\!] = [\![\Delta_k]\!] \partial_k \partial_{k+1} \ldots \partial_{\lambda_1 - 1} \, \partial_{k-1} \partial_k \ldots \partial_{\lambda_2 - 1} \cdots \partial_1 \partial_2 \ldots \partial_{\lambda_k - 1}.$$

3.5 Application: Staircase Partitions

In this subsection we will use divided differences to compute $[\![\lambda]\!]$ for the staircase partitions $\Delta_n = (n, n-1, n-2, \ldots, 2, 1) = \,<n, n-1>$ and the truncated staircase partitions $\Delta'_n = (n, n-1, n-2, \ldots, 3, 2) = \,<n, n-2>$. From what has just been stated, cascades of divided differences will be used, without computing explicitly the intermediate expressions.

Theorem 1 *For all $n \geq 1$,*

$$[\![\Delta'_n]\!] = \frac{w(\Delta_n)}{q_{n,n-1}} = \frac{(x_0+)x_1 + x_2 + \ldots + x_n}{\prod_{0 \leq i < j \leq n}(x_i + x_j)}, \tag{9}$$

$$[\![\Delta_n]\!] = \frac{1}{q_{n,n-1}} = \frac{1}{\prod_{0 \leq i < j \leq n}(x_i + x_j)}. \tag{10}$$

Proof From the fact that in the lattice $\mathcal{S}$ the staircase partition Δ_n covers only the truncated staircase partition Δ'_n one has

$$[\![\Delta_n]\!] = \frac{1}{w(\Delta_n)}[\![\Delta'_n]\!],$$

and thus (10) follows immediately from (9).

We will now use the cascade operation

$$[\![\Delta'_{n+1}]\!] = [\![\Delta_n]\!] \partial_n \partial_{n-1} \ldots \partial_1.$$

Writing down the claim explicitly gives

$$\frac{w(\Delta_{n+1})}{\prod_{0 \leq i < j \leq n+1}(x_i + x_j)} = \left[\frac{1}{\prod_{0 \leq i < j \leq n}(x_i + x_j)} \right] \partial_n \partial_{n-1} \ldots \partial_1$$

$$= \left[\frac{\prod_{0 \leq i \leq n}(x_i + x_{n+1})}{\prod_{0 \leq i < j \leq n+1}(x_i + x_j)} \right] \partial_n \partial_{n-1} \ldots \partial_1.$$

Now $\prod_{0 \le i < j \le n+1}(x_i + x_j)$ is symmetric w.r.t. all σ_r $(1 \le r \le n)$, so that this term cancels on both sides and we are left with the task to prove the equivalent assertion about polynomials:

$$w(\Delta_{n+1}) = \left[\prod_{0 \le i \le n} (x_i + x_{n+1}) \right] \partial_n \partial_{n-1} \ldots \partial_1.$$

This will be shown in the proposition that follows. □

Denote $Q_n = \prod_{0 \le i < n}(x_i + x_n)$. In terms of elementary and homogeneous symmetric functions this can be written as

$$Q_n = \sum_{k=0}^{n} e_k(X_{0,n-1}) h_{n-k}(X_{n,n}).$$

In the terminology of symmetric functions this is a Schur polynomial of degree n over two sets $X_{0,n-1} = \{x_0, x_1, \ldots, x_{n-1}\}$ and $X_{n,n} = \{x_n\}$ of variables, denoted as $S_n(x_n | X_{0,n-1})$—see Sect. 4.3 for the notation.

Proposition 1 *For all $n \ge 1$,*

$$Q_{n+1} \partial_n \partial_{n-1} \ldots \partial_1 = \begin{cases} x_0 + x_1 + \ldots + x_{n+1} & \text{if } n \text{ is even,} \\ x_1 + x_2 + \cdots + x_{n+1} & \text{if } n \text{ is odd.} \end{cases} \tag{11}$$

Proof One way to prove Eq. (11) goes by working through the following sequence of statements:

– The left hand side of (11) is a homogeneous polynomial of degree 1.
– The left hand side of (11) is symmetric in $X_{0,n+1}$, resp. $X_{1,n+1}$.
  (this can be done by using the braid relations for the divided differences).
– The coefficient of x_{n+1} in the left hand side is 1 (which is not so obvious).

An alternative way proceeds follows:

– Apply ∂_n to Q_{n+1} to obtain

$$Q_{n+1} \partial_n = \underbrace{\sum_{k=0}^{n} e_k(X_{0,n}) h_{n-k}(X_{n,n+1})}_{Q_n^\ell} - \underbrace{\sum_{k=0}^{n} e_k(X_{0,n-1}) h_{n-k}(X_{n,n})}_{Q_n^r}.$$

– As for Q_n^r, this is nothing but Q_n. We can use induction to obtain

$$(Q_n^r) \partial_{n-1} \ldots \partial_1 = (x_0 +) x_1 + \ldots + x_n.$$

– As for Q_n^ℓ, note that the factors $e_k(X_{0,n})$ are symmetric in the variables $X_{0,n}$ and thus behave like scalars w.r.t. $\partial_n, \partial_{n-1}, \ldots, \partial_1$. Hence

$$(Q_n^\ell)\partial_{n-1}\partial_{n-2}\ldots\partial_1 = \sum_{k=0}^{n} e_k(X_{0,n}) \cdot h_{n-k}(X_{n,n+1})\partial_{n-1}\ldots\partial_1$$

$$= \sum_{k=0}^{n} e_k(X_{0,n}) \cdot h_{1-k}(X_{1,n+1})$$

$$= e_0(X_{0,n})h_1(X_{1,n+1}) + e_1(X_{0,n})h_0(X_{1,n+1})$$

$$= h_1(X_{1,n+1}) + e_1(X_{0,n})$$

$$= x_0 + 2x_1 + 2x_2 + \cdots + 2x_n + x_{n+1}.$$

Thus

$$Q_{n+1}\partial_n \ldots \partial_1 = \begin{cases} x_0 + x_1 + \cdots + x_{n+1} & \text{if } Q_n\partial_{n-1} \ldots \partial_n = x_1 + x_2 + \cdots + x_n, \\ x_1 + x_2 + \cdots + x_{n+1} & \text{if } Q_n\partial_{n-1} \ldots \partial_n = x_0 + x_1 + \cdots + x_n. \end{cases}$$

$\square$

3.6 Application: Denominators

Now it can be shown that the rational functions $[\![\lambda]\!]$ can be written by using the standard denominator polynomials $q_{n,m}$ of Sects. 2.2.4 and 3.3 as denominators, showing that the denominator of $[\![\lambda]\!]$ depends only on the two largest parts of λ. The proof of Theorem 2 does not show immediately that this representation is indeed reduced, i.e. numerator and denominator polynomials have no common factor. A further short argument is needed to make this clear.

Theorem 2 *For strict partitions $\lambda \in S$, the rational functions $[\![\lambda]\!]$ can be written as*

$$[\![\lambda]\!] = \begin{cases} \dfrac{1}{q_{n,0}} & \text{if } \lambda = (n), \\[2ex] \dfrac{p_\lambda}{q_{n,m}} & \text{if } \lambda = (n, m, \ldots), \end{cases}$$

where p_λ is a homogeneous polynomial in the variables $X_{0,n}$.

Proof The one-part case $\lambda = (n)$ has been mentioned in Sect. 2.2.5, see Eq. (5).

Let now $\lambda = (\lambda_1, \lambda_2, \ldots, \lambda_k) = (n, m, r, \ldots)$ be a strict partition with $k \geq 2$ parts. We look at the construction of λ as shown in Sect. 3.4, starting with the staircase partition Δ_k and applying divided differencing cascades:

$$[\![\lambda]\!] = [\![\Delta_k]\!] \underbrace{\partial_k \partial_{k+1} \ldots \partial_{n-1}}_{\partial_{k\ldots n-1}} \underbrace{\partial_{k-1} \partial_k \ldots \partial_{m-1}}_{\partial_{k-1\ldots m-1}} \underbrace{\partial_{k-2} \partial_{k-1} \ldots \partial_{r-1}}_{\partial_{k-2\ldots r-1}} \ldots$$

We know from Theorem 1, Eq. (10), that

$$[\![\Delta_k]\!] = \frac{1}{q_{k,k-1}}.$$

Applying the cascade $\partial_{k\ldots n-1}$ to it means applying $n-1-k$ times the first part of Lemma 1, which gives

$$[\![\Delta_k]\!]\partial_{k\ldots n-1} = \frac{P(n,k-1,k-2,\ldots,1)}{q_{n,k-1}}.$$

Applying now the cascade $\partial_{k-1\ldots m-1}$ to this gives, see the second part of Lemma 1,

$$[\![\Delta_k]\!]\partial_{k\ldots n-1}\partial_{k-1\ldots m-1} = \frac{P(n,m,k-2,\ldots,1)}{q_{n,m}}.$$

If $k = 2$ then we are finished. Otherwise assume that there is a third part $\lambda_3 = r$. When applying the cascade $\partial_{k-2\ldots r-1}$ to the last expression note that the denominator polynomial $q_{n,m} = \prod_{0 \leq i \leq m, i < j \leq n}(x_i + x_j)$ is symmetric w.r.t. all σ_ℓ with $k-2 \leq \ell \leq r-1$ because $r < m$. Thus the denominator behaves like a scalar when acting with $\partial_{k-2\ldots r-1}$ and thus

$$[\![\Delta_k]\!]\partial_{k\ldots n-1}\partial_{k-1\ldots m-1}\partial_{k-2\ldots r-1} = \frac{1}{q_{n,m}} \cdot \left[P(n,m,k-2,\ldots,1)\right]\partial_{k-2\ldots r-1}$$

$$= \frac{P(n,m,k-2,\ldots,1)\,\partial_{k-2\ldots r-1}}{q_{n,m}} = \frac{P(n,m,r,\ldots)}{q_{n,m}}.$$

If $k = 3$ we are done, the same argument can be played repeatedly for any $k > 3$.   □

For a partition $\lambda \in S$ let $\mathrm{sdp}(\lambda)$ denote the standard denominator polynomial of λ, i.e., $\mathrm{sdp}(\lambda) = q_{n,m}$ if $\lambda = (n, m, \ldots)$ (or $= q_{n,0}$ if $\lambda = (n)$). In view of the relation of the standard denominator polynomials to the join-irreducibles of S on can draw the immediate consequence.

Corollary 1 *For strict partitions* $\lambda, \mu \in S$, *if* $\lambda \subseteq \mu$ *then* $\mathrm{sdp}(\lambda) | \mathrm{sdp}(\mu)$. *In particular,*

$$\mathrm{sdp}(\Delta_n) = q_{n,n-1} = \prod_{0 \leq i < j \leq n} (x_i + x_j)$$

is the least common multiple for all denominators coming from the $[\![\lambda]\!]$ *with* $\lambda \subseteq \Delta_n$.

The statement about the lcm-property of the $\text{sdp}(\Delta_n)$ is of major interest for the physical model that triggered the investigations presented here. More details are given in [3].

The statement about the denominators of the $[\![\lambda]\!]$ can be made even more precise. What is observed from the examples given Sects. 2.2.3 and 2.2.4 is indeed true in general.

Corollary 2 *For strict partitions* $\lambda = (n, m, \ldots) \in S$ *the standard denominator polynomial* $\text{sdp}(\lambda) = q_{n,m}$ *is the true denominator of* $[\![\lambda]\!]$, *when written as a reduced fraction.*

Proof Consider the possible one-step extensions leading from $\lambda = (n, m, \ldots)$ to $\lambda' = \lambda + \varepsilon^k$ i.e., by "adding one box in a legal position", and what this does to the denominators. The new binomials $x_i + x_j$ with $i < j$ that may appear in the denominator are (see Lemma 1) those where for

- $k = 1: 0 \le i \le m$ and $j = n + 1$;
- $k = 2: i = m + 1$ and $m + 1 < j \le n$ (observe that $m < n - 1$ in this case);
- $k > 2$: none new terms.

Imagine that one extends λ' further and that immediately or later one of the new binomials just created, $x_{i_0} + x_{j_0}$, say, disappears due to cancellation with the numerator polynomials. Due to the same argument just made for the appearance of this 'lost' binomial $x_{i_0} + x_{j_0}$, this term cannot reappear at any later stage of extension. But since $\lambda' \subseteq \Delta_N$ for N big enough ($N = n + 1$ will do), and the fact that by Theorem 1 the binomial $x_{i_0} + x_{j_0}$ definitely belongs to the denominator of $[\![\Delta_N]\!]$, it cannot have disappeared before. No binomials in the denominators of the $[\![\lambda]\!]$ ever 'get lost' in the extension process. □

4　Schur Functions and Variants

For two families of strict partitions the answer to the problem of saying how the rational function $[\![\lambda]\!]$ can be expressed in terms of 'known' functions will be given in this article. It turns out that symmetric functions, and Schur functions in particular, nicely do the job. The present section contains just the necessary definitions and facts required here. As stated in the introduction, the reader who is less familiar with this concepts is invited to consult standard texts like [4–6] for more information. The essential tool for working here with Schur functions is what is known as Jacobi-Trudi determinants. This is taken for the definitions, other combinatorial or algebraic definitions of Schur functions won't be used here and are therefore not introduced.

4.1 Elementary and Homogeneous Symmetric Functions

For any alphabet A (or set of variables, usually finite, and for us always of type $A = X_{a,b} = \{x_a, x_{a+1}, \ldots, x_n\}$, as before) the elementary symmetric functions $e_n(A)$ and the homogeneous (complete) symmetric functions $h_n(A)$ are defined by their generating functions[1]

$$e(A) = \sum_{n \geq 0} e_n(A) = \prod_{\alpha \in A} (1 + \alpha),$$

$$h(A) = \sum_{n \geq 0} h_n(A) = \frac{1}{\prod_{\alpha \in A} (1 - \alpha)},$$

where $e_n(A)$ resp. $h_n(A)$ are the homogeneous parts of degree n if the products on the right hand side are expanded 'as usual'. The definition implies that $e_n(A) = 0$ if $n \geq \sharp A$, that $e_0(A) = 1 = h_1(A)$, and that $e_n(A) = 0 = h_n(A)$ for $n < 0$.

Both families can be used to define (multiplicative) bases of the vector space of symmetric functions, parametrized by partitions, where the basis elements are the products $e_\lambda(A) = e_{\lambda_1} e_{\lambda_2} \cdots e_{\lambda_k}$ resp. $h_\lambda(A) = h_{\lambda_1} h_{\lambda_2} \cdots h_{\lambda_k}$ for partitions $\lambda = \lambda_1 \lambda_2 \ldots \lambda_k$. These products will not be used in the sequel. But the $e_n(A)$ and $h_n(A)$ can be used to defined another, non multiplicative basis of the space of symmetric functions, the *Schur functions*, undeniably the most interesting and most important family in the realm of classical symmetric functions. There are several equivalent ways, of algebraic and of combinatorial character, to define Schur functions. For the purpose of this work I will restrict the formalism to the definition (and use) of the so-called Jacobi-Trudi determinants.

4.2 Schur Functions Over a Single Alphabet

The particular Schur functions $S_n(A)$, where n is a single integer, and where A is an alphabet (usually finite), are nothing but the homogeneous symmetric functions: $S_n(A) = h_n(A)$. In particular, $S_n(A) = 0$ for $n < 0$.

---

[1] Often one writes $e(A; t) = \sum_{n \geq 0} e_n(A) t^n = \prod_{\alpha \in A} (1 + \alpha t)$, etc., using an explicit counting variable t for the degree. This is not really necessary, so t is left implicit, except for a situation in and after Eq. (23) in Sect. 5.1, where it improves readability. Note that using 'negative alphabets' $e(-A)$ mean the same using $e(A; -t)$, etc.

For any partition $\lambda = (\lambda_1, \lambda_2, \ldots, \lambda_k)$ and any alphabet A the Schur function $S_\lambda(A)$ is defined as

$$S_\lambda(A) = \det\left[S_{\lambda_{k-j}-i+j}(A)\right]_{0 \le i, j < k}$$

$$= \det\begin{bmatrix} S_{\lambda_k} & S_{\lambda_{k-1}+1} & \cdots & S_{\lambda_1+k-1} \\ S_{\lambda_k-1} & S_{\lambda_{k-1}} & \cdots & S_{\lambda_1+k-2} \\ \vdots & \vdots & \ddots & \vdots \\ S_{\lambda_k-k+1} & S_{\lambda_{k-1}-k+2} & \cdots & S_{\lambda_1} \end{bmatrix}_A, \qquad (12)$$

where the subscript A to the matrix indicates that all the Schur functions appearing in the matrix have to be taken over the alphabet A. Note that the parts of λ appear as indices in increasing order along the main diagonal of the defining matrix, and that the indices decrease by 1 along each column.

The entries of the matrix that defines $S_\lambda(A)$ are all homogeneous symmetric functions. One may as well define Schur functions using the elementary symmetric functions. Indeed,

$$S_{\tilde{\lambda}}(A) = \det\left[e_{\lambda_{k-j}-i+j}(A)\right]_{0 \le i, j < k}$$

$$= \det\begin{bmatrix} e_{\lambda_k} & e_{\lambda_{k-1}+1} & \cdots & e_{\lambda_1+k-1} \\ e_{\lambda_k-1} & e_{\lambda_{k-1}} & \cdots & e_{\lambda_1+k-2} \\ \vdots & \vdots & \ddots & \vdots \\ e_{\lambda_k-k+1} & e_{\lambda_{k-1}-k+2} & \cdots & e_{\lambda_1} \end{bmatrix}_A, \qquad (13)$$

where $\tilde{\lambda}$ denotes the conjugate partition of λ.

4.3 Schur Functions Over a Pair of Alphabets

If A and B are two alphabets (usually finite and disjoint), for any integer n the Schur function $S_n(A|B)$ is defined as a convolution of homogeneous and elementary symmetric functions:

$$S_n(A|B) = \sum_{0 \le k \le n} h_k(A) \cdot e_{n-k}(B).$$

In particular, $S_n(A|\emptyset) = h_n(A)$ and $S_n(\emptyset|B) = e_n(B)$. The generating function of the $S_n(A|B)$ is

$$S(A|B) = \sum_{n \ge 0} S_n(A|B) = \frac{\prod_{\beta \in B}(1 + \beta)}{\prod_{\alpha \in A}(1 - \alpha)} = h(A) \cdot e(B).$$

For any partition $\lambda = \lambda_1 \lambda_2 \ldots \lambda_k$ the Schur function $S_\lambda(A|B)$ is defined by a Jacobi-Trudi determinant:

$$S_\lambda(A|B) = \det \left[S_{\lambda_k - j - i + j}(A|B) \right]_{0 \leq i, j < k},$$

i.e., by the same determinant as for $S_n(A)$, but with the alphabet A replaced by the pair $(A|B)$ of alphabets.

The following way of writing Schur functions $S_n(A|B)$ over a pair of alphabets in terms of functions over a single alphabet will be useful later.

Proposition 2 *For integer n and any pair $(A|B)$ of alphabets one has*

$$S_n(A|B) = \sum_k (-1)^k e_k(B^2) \cdot S_{n-2k}(A + B).$$

In this statement, the term $e_k(B^2)$ denotes the k-th elementary symmetric function over an alphabet which consists of the squares of the elements of B. The sum $A + B$ is nothing but the union of the two alphabets A and B. Furthermore, if A is any alphabet, then $-A$ is an alphabet which consists of the negatives of the elements of A. Thus

$$e(-A) = \prod_{\alpha \in A}(1 - \alpha) \quad \text{and} \quad h(-A) = \frac{1}{\prod_{\alpha \in A}(1 + \alpha)}.$$

The close relationship between elementary and homogeneous symmetric functions is thus concisely expressed as reciprocity between the generating series

$$e(-A) \cdot h(A) = 1.$$

Proof of Proposition 2 This follows from expanding both sides of the generating function identity:

$$S(A|B) = \sum_{n \geq 0} S_n(A|B) = \frac{\prod_{\beta \in B}(1 + \beta)}{\prod_{\alpha \in A}(1 - \alpha)}$$

$$= \frac{\prod_{\beta \in B}(1 - \beta^2)}{\prod_{\alpha \in A}(1 - \alpha) \prod_{\beta \in B}(1 - \beta)}$$

$$= \prod_{\beta \in B}(1 - \beta^2) \cdot \sum_{n \geq 0} S_n(A + B)$$

$$= e(-B^2) \cdot S(A + B).$$

$\square$

As we will often deal with staircase partitions, it is good to know the following property of their Schur functions over two alphabets. It shows that in this particular case the Schur functions over a pair of alphabets may be considered over their union as a single alphabet.

Proposition 3 *For staircase partitions* Δ_n, *let* $X = A \uplus B = A' \uplus B'$ *be two decompositions of the alphabet* X. *Then*

$$S_{\Delta_n}(A|B) = S_{\Delta_n}(A'|B') = S_{\Delta_n}(X).$$

Proof Let $\gamma \in A$, $A' = A \setminus \gamma$ and $B' = B \cup \{\gamma\}$. Then

$$S(A'|B') = \frac{\prod_{\beta \in B'}(1 + \beta)}{\prod_{\alpha \in A'}(1 - \alpha)} = (1 - \gamma^2)\frac{\prod_{\beta \in B}(1 + \beta)}{\prod_{\alpha \in A}(1 - \alpha)} = (1 - \gamma^2)S(A|B),$$

so that

$$S_n(A'|B') = S_n(A|B) - \gamma^2 S_{n-2}(A|B).$$

Writing down the Jacobi-Trudi determinant defining $S_{\Delta_n}(A'|B')$ and using this property one obtains the matrix defining $S_{\Delta_n}(A|B)$ by elementary column operations.   □

It has been mentioned before, see Eq. (8), that all standard denominator polynomials $q_{n,m}$ are indeed Schur functions (over two alphabets, which in the case of S_{Δ_n} reduces to one, if desired). A proof of this can be given by using the so-called Pieri rules for the Schur functions.

4.4 Block Schur Functions

Only a special case of this construct is needed here. For a partition λ, an integer m and two alphabets A, C (not necessarily disjoint) we define

$$S_{\lambda;m}(A; C) = \det \begin{bmatrix} S_{\lambda_k} & S_{\lambda_{k-1}+1} & \cdots & S_{\lambda_1+k-1} & S_{m+k} \\ S_{\lambda_k-1} & S_{\lambda_{k-1}} & \cdots & S_{\lambda_1+k-2} & S_{m+k-1} \\ \vdots & \vdots & \ddots & \vdots & \vdots \\ S_{\lambda_k-k+1} & S_{\lambda_{k-1}-k+2} & \cdots & S_{\lambda_1} & S_{m+1} \\ S_{\lambda_k-k} & S_{\lambda_{k-1}-k+1} & \cdots & S_{\lambda_1-1} & S_m \end{bmatrix}_{A; C}.$$

The matrix is like the usual matrix for the Schur function of a partition $\lambda = \lambda_0\lambda_1 \ldots \lambda_k$ over the alphabet A, in which the last column (the one which would refer to λ_0) is replaced by $[S_{m+k}, S_{m+k-1}, \ldots, S_{m+1}, S_m]^\top$ over the alphabet C.

This construction can be extended to cover the situation where the alphabet A is replaced by a pair of alphabets $(A|B)$, and C is replaced by a pair $(C|D)$ of alphabets.

Another extension, which will be used below, is by adjoining two columns (and extending the matrix by two rows accordingly)

$$S_{\lambda;m,n}(A; C) = \det \begin{bmatrix} S_{\lambda_k} & S_{\lambda_{k-1}+1} & \cdots & S_{\lambda_1+k-1} & S_{m+k} & S_{n+k+1} \\ S_{\lambda_k-1} & S_{\lambda_{k-1}} & \cdots & S_{\lambda_1+k-2} & S_{m+k-1} & S_{n+k} \\ \vdots & \vdots & \ddots & \vdots & \vdots & \vdots \\ S_{\lambda_k-k+1} & S_{\lambda_{k-1}-k+2} & \cdots & S_{\lambda_1} & S_{m+1} & S_{n+2} \\ S_{\lambda_k-k} & S_{\lambda_{k-1}-k+1} & \cdots & S_{\lambda_1-1} & S_m & S_{n+1} \\ S_{\lambda_k-k-1} & S_{\lambda_{k-1}-k} & \cdots & S_{\lambda_1-2} & S_{m-1} & S_n \end{bmatrix}_{A;\, C}.$$

Now the last two columns are to be taken over the alphabet C.

The next proposition illustrates the use of these constructions.

Proposition 4 *For any partition* λ *and alphabets* A, B *and a letter* x *not belonging to* $A \cup B$ *one has*

$$S_{\lambda;0}(A + x|B; x) = S_\lambda(A|B), \tag{14}$$

$$S_{\lambda;0}(A|B; -x) = S_\lambda(A + x|B). \tag{15}$$

Proof From the generating functions

$$S(A + x|B) = \frac{\prod_{\beta \in B}(1 + \beta)}{(1 - x) \prod_{\alpha \in A}(1 - \alpha)} = \frac{1}{1 - x} S(A|B),$$

which is

$$e(-x) \cdot S(A + x|B) = S(A|B),$$

or

$$S_m(A + x|B) - x\, S_{m-1}(A + x|B) = S_m(A|B).$$

Now use row operations to turn the matrix defining $S_{\lambda;0}(A + x|B)$ (the last column of which is $\left[\ldots, x^3, x^2, x, 1\right]^\top$ into

$$\left[\begin{array}{c|c} \boxed{S_\lambda(A|B)} & \begin{matrix} 0 \\ \vdots \\ 0 \end{matrix} \\ \hline * \cdots * & 1 \end{array}\right],$$

where $\boxed{S_\lambda(A|B)}$ stands for the matrix defining $S_\lambda(A|B)$. This proves the first identity.

The second one is proved in a similar way by using

$$S(A|B + x) = (1 + x) \cdot S(A|B) = e(x) \cdot S(A|B). \qquad \square$$

The extensions for separating two elements from an alphabet are similar and are given next.

Proposition 5 *For any partition* λ *and alphabets* A, B *and letters* x, y *not belonging to* $A \cup B$ *one has*

$$S_{\lambda;0,0}(A + x + y|B; x + y) = S_\lambda(A|B), \qquad (16)$$

$$S_{\lambda;0,0}(A|B; -x - y) = S_\lambda(A|B + x + y). \qquad (17)$$

The proofs are very similar to those for Proposition 2, now based on

$$e(x + y) \cdot S(A + x + y|B) = S(A|B),$$

$$S(A|B + x + y) = e(-x - y) \cdot S(A|B).$$

5 Results

There are two families of strict partitions for which a complete answer can be given for the problem of expressing $[\![\lambda]\!]$ in 'closed form': join irreducible strict partitions and strict partitions consisting of two parts. The notion of 'closed form' means: in terms of symmetric functions, and Schur functions in particular.

5.1 Evaluating $[\![\lambda]\!]$ for the Join-Irreducibles in S

In this section we will prove a 'closed-form' evaluation of $[\![\lambda]\!]$ for the case of join-irreducible strict partitions λ.

Theorem 3 *For the join-irreducible elements* $\lambda = <n, k> = (n, n - 1, \ldots, n - k)$ *of the lattice* S *of strict partitions, where* $0 \le k < n$, *one has*

$$[\![\lambda]\!] = \frac{S_{\Delta_{n-k-1}}(X_{k+1 \bmod 2, n})}{S_{\Delta_n}(X_{0,n})}. \qquad (18)$$

Special cases that we have seen (Sect. 2.2.5 and Theorem 1) are:

- the one-part partitions $<n, 0> = (n)$

$$[\![<n, 0>]\!] = [\![n]\!] = \frac{S_{\Delta_{n-1}}(X_{1,n})}{S_{\Delta_n}(X_{0,n})} = \frac{\prod_{1 \leq i < j \leq n}(x_i + x_j)}{\prod_{0 \leq i < j \leq n}(x_i + x_j)} = \frac{1}{\prod_{1 \leq j \leq n}(x_0 + x_j)};$$

- the truncated staircases $\Delta'_n = <n, n-2> = (n, n-1, \ldots, 2)$

$$[\![\Delta'_n]\!] = \begin{cases} \dfrac{S_{\Delta_1}(X_{1,n})}{S_{\Delta_n}(X_{0,n})} = \dfrac{x_1 + x_2 + \cdots + x_n}{\prod_{0 \leq i < j \leq n}(x_i + x_j)} & \text{if } n \text{ is even,} \\[3ex] \dfrac{S_{\Delta_1}(X_{0,n})}{S_{\Delta_n}(X_{0,n})} = \dfrac{x_0 + x_1 + x_2 + \cdots + x_n}{\prod_{0 \leq i < j \leq n}(x_i + x_j)} & \text{if } n \text{ is odd;} \end{cases}$$

- the staircases $\Delta_n = <n, n-1> = (n, n-1, \ldots, 1)$

$$[\![\Delta_n]\!] = \begin{cases} \dfrac{S_{\Delta_0}(X_{0,n})}{S_{\Delta_n}(X_{0,n})} = \dfrac{1}{\prod_{0 \leq i < j \leq n}(x_i + x_j)} & \text{if } n \text{ is even,} \\[3ex] \dfrac{S_{\Delta_0}(X_{1,n})}{S_{\Delta_n}(X_{0,n})} = \dfrac{1}{\prod_{0 \leq i < j \leq n}(x_i + x_j)} & \text{if } n \text{ is odd.} \end{cases}$$

Proof The proof of the theorem is quite intricate and requires repeated rewriting of expressions involving symmetric functions over varying alphabets. It proceeds by induction over n for fixed k, which is the number of parts of each of the partitions involved in this process. The induction basis is $\Delta_k = <k, k-1>$, for which the result has been established in Theorem 1. In the induction step the join-irreducible strict partition $<n, k-1> = (n, n-1, \ldots, n-k+1)$ is extended in k divided difference steps to the join-irreducible strict partition $<n+1, k-1> = (n+1, n, \ldots, n+2-k)$, which amounts to applying the difference cascade $\partial_n \partial_{n-1} \ldots \partial_{n-k+1}$ to $[\![<n, k-1>]\!]$:

$$[\![<n, k-1>]\!] \xrightarrow{\partial_n \partial_{n-1} \ldots \partial_{n-k+1}} [\![<n+1, k-1>]\!], \tag{19}$$

via

$$\begin{aligned} <n, k-1> &= (n, n-1, n-2 \ldots, n-k+1) \\ &<_n (n+1, n-1, n-2 \ldots, n-k+1) \\ &<_{n-1} (n+1, n, n-2, \ldots, n-k+1) \\ &\vdots \qquad \vdots \\ &<_{n-k+1} (n+1, n, n-1, \ldots, n-k+2) \\ &= <n+1, k-1>, \end{aligned}$$

as displayed in Fig. 12.

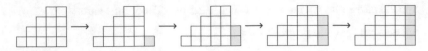

Fig. 12 Showing $<n, k-1>=<5, 3>$ $\longrightarrow$ $<6, 3>=<n+1, k-1>$

By plugging in the asserted expressions from (18) on both sides, the induction step (19) requires the verification of the identities

$$\frac{S_{\Delta_{n-k}}(X_{\varepsilon,n})}{S_{\Delta_n}(X_{0,n})} \partial_n \partial_{n-1} \cdots \partial_{n-k+1} = \frac{S_{\Delta_{n-k+1}}(X_{\varepsilon,n+1})}{S_{\Delta_{n+1}}(X_{0,n+1})} \qquad \text{with } \varepsilon = k \bmod 2.$$

Now $S_{\Delta_{n+1}}(X_{0,n+1})$ is symmetric w.r.t. all the variables $\{x_0, x_1, \ldots, x_{n+1}\}$, hence behaves like a scalar w.r.t. the cascade $\partial_n \partial_{n-1} \ldots \partial_{n-k+1}$. Multiplying both sides by $S_{\Delta_{n+1}}(X_{0,n+1})$ yields the equivalent claim

$$\left[\frac{S_{\Delta_{n+1}}(X_{0,n+1})}{S_{\Delta_n}(X_{0,n})} S_{\Delta_{n-k}}(X_{\varepsilon,n}) \right] \partial_n \partial_{n-1} \ldots \partial_{n-k+1} = S_{\Delta_{n-k+1}}(X_{\varepsilon,n+1}),$$

which in view of

$$\frac{S_{\Delta_{n+1}}(X_{0,n+1})}{S_{\Delta_n}(X_{0,n})} = \frac{\prod_{0 \le i < j \le n+1}(x_i + x_j)}{\prod_{0 \le i < j \le n}(x_i + x_j)} = \prod_{0 \le i \le n}(x_i + x_{n+1}) = S_{n+1}(x_{n+1}|X_{0,n})$$

can be written as a polynomial identity

$$\left[S_{n+1}(x_{n+1}|X_{0,n}) \cdot S_{\Delta_{n-k}}(X_{\varepsilon,n}) \right] \partial_n \partial_{n-1} \ldots \partial_{n-k+1} = S_{\Delta_{n+1-k}}(X'_{\varepsilon,n+1}). \qquad (20)$$

For better readability the notation will be changed a bit. In place of n we will write b, in place of $n - k$ write a, so that k gets replaced $b - a$. The alphabet $X_{0,a}$ will be denoted by A, for the alphabet $X_{a+1,b}$ resp. $X_{a+1,b+1}$ we write B resp. B', so that $A + B$ means the same as $X_{0,b}$ and $A + B'$ stands for $X_{0,b+1}$.

The goal is now to evaluate the left-hand side of (20), viz.

$$\left[S_{b+1}(x_{b+1}|A + B) \cdot S_{\Delta_a}(A + B) \right] \partial_b \partial_{b-1} \ldots \partial_{a+1}. \qquad (21)$$

This is presented in a sequence of steps.

(i) The inconvenience of (21) is that the divided differences ∂_b etc. act on both factors in the bracket. It is therefore plausible to try to pass to an equivalent expression which avoids that. We claim that

$$S_{b+1}(x_{b+1}|A + B) \cdot S_{\Delta_a}(A + B) = S_{\Delta_a;b+1}(A + B + x_{b+1}; x_{b+1}|A + B).$$

For the proof of this, note that the last column of the matrix for the block Schur function $S_{\Delta_a;b+1}(A + B + x_{b+1}; x_{b+1}|A + B)$ is

$$[S_{b+1+m}(x_{b+1}|A + B)]_{a \geq m \geq 0}^\top.$$

Since $e_k(A + B) = 0$ if $k > \sharp(A + B) = b + 1$, we have for the terms in question

$$S_{b+1+m}(x_{b+1}|A + B) = x_{b+1}^m \sum_{j=0}^{b+1} x_{b+1}^{b+1-j} e_j(A + B)$$

$$= x_{b+1}^m \cdot S_{b+1}(x_{b+1}|A + B),$$

and therefore, taking Proposition 2 (with A resp. $A + B$ replaced by B resp. $\emptyset$, and x by x_{b+1}) into account, we get

$$S_{\Delta_a;b+1}(A + B + x_{b+1}; x_{b+1}|A + B)$$

$$= S_{\Delta_a;0}(A + B + x_{b+1}; x_{b+1}) \cdot S_{b+1}(x_{b+1}|A + B)$$

$$= S_{\Delta_a}(A + B) \cdot S_{b+1}(x_{b+1}|A + B).$$

(ii) In view of Proposition 3 we can rewrite

$$S_{\Delta_a;b+1}(A + B + x_{b+1}; x_{b+1}|A + B)$$

by replacing the $S_{b+1+m}(x_{b+1}|A+B)$ in the last column of the corresponding matrix by sums over products of $e_\ell(A^2 + B^2)$ and $S_{\Delta_a;b+1-2\ell}(A + B + x_{b+1})$ terms. This gives

$$S_{\Delta_a;b+1}(A+B+x_{b+1}; x_{b+1}|A+B) = \sum_{\ell \geq 0}(-1)^\ell e_\ell(A^2+B^2) \cdot S_{\Delta_a;b+1-2\ell}(A+B'),$$

where $A + B' = A + B + x_{b+1}$.

(iii) The expression to be evaluated is now

$$\left[\sum_{\ell \geq 0}(-1)^\ell e_\ell(A^2 + B^2) \cdot S_{\Delta_a;b+1-2\ell}(A + B')\right] \partial_b \partial_{b-1} \ldots \partial_{a+1}.$$

Here the terms $S_{\Delta_a;b+1-2\ell}(A + B')$ are symmetric in the variables of $A + B'$, hence they behave as scalars w.r.t. the difference cascade $\partial_b \partial_{b-1} \cdots \partial_{a+1}$. Now the difference cascade acts only as for

$$e_\ell(A^2 + B^2)\partial_b \partial_{b-1} \cdots \partial_{a+1}.$$

(iv) The required result is

$$e_\ell(A^2 + B^2)\partial_b\partial_{b-1}\cdots\partial_{a+1} = \sum_{j\geq 0}(-1)^j e_{\ell-b+a+j}(A^2)e_{b-a-2j}(B'),$$

which can be demonstrated by induction.

(v) The expression to be evaluated has turned into

$$\sum_\ell (-1)^\ell S_{\Delta_a;b+1-2\ell}(A + B')\sum_j(-1)^j e_{\ell-b+a+j}(A^2)e_{b-a-2j}(B').$$

Putting $m = 2\ell - b + a$ and replacing $j \mapsto j - m + \ell$, hence

$$\ell - b + a + j \mapsto j, \qquad\qquad b - a - 2j \mapsto m - 2j,$$
$$j + \ell \mapsto a - b + j, \qquad b + 1 - 2\ell \mapsto a + 1 + m,$$

gives

$$\sum_{m\equiv_2 b-a} S_{\Delta_a;a+1-m}(A + B')\sum_j(-1)^j e_j(A^2)e_{m-2j}(B'),$$

or

$$\sum_j(-1)^j e_j(A^2)\sum_{m\equiv_2 b-a} S_{\Delta_a;a+1-m}(A + B')e_{m-2j}(B'). \tag{22}$$

(vi) Consider now the sum (22) for the position (r, a) (where rows and columns are indexed from 0 to a) in the matrix used to define $S_{\Delta_a;a+1-m}(A + B')$. This is

$$(*_r) \qquad \sum_j(-1)^j e_j(A^2)\sum_{m\equiv_2 b-a} h_{2a+1-r-m}(A + B')e_{m-2j}(B').$$

Put $N_r = 2a + 1 - r$. Then this can be written as

$$[t^{N_r}]\prod_{\alpha\in A}(1 - \alpha^2 t^2)\prod_{\alpha\in A}\frac{1}{1 - \alpha t}\prod_{\beta\in B'}\frac{1}{1 - \beta t}\left[\prod_{\beta\in B'}(1 + \beta t)\right]', \tag{23}$$

where $[t^N](\ldots)$ means "the coefficient of t^N (=homogeneous part of degree N) in $(\ldots)$, and where in $[\ldots]'$ only the terms of degree $\equiv b - a$ mod 2 have to be taken. Equivalently,

$$[t^{N_r}] \prod_{a \in A} (1 + \alpha t) \prod_{\beta \in B'} \frac{1}{1 - \beta t} \cdot \frac{1}{2} \left\{ \prod_{\beta \in B'} (1 + \beta t) \pm \prod_{\beta \in B'} (1 - \beta t) \right\}$$

$$= [t^{N_r}] \prod_{a \in A} (1 + \alpha t) \prod_{\beta \in B'} (1 + \beta t) \cdot \frac{1}{2} \left\{ \prod_{\beta \in B'} \frac{1}{1 - \beta t} \pm \prod_{\beta \in B'} \frac{1}{1 + \beta t} \right\}$$

$$= [t^{N_r}] e(A + B', t) \cdot \frac{1}{2} (h(B', t) \pm h(B', -t)).$$

(vii) This turns the problem into the evaluation of

$$\sum_{\substack{0 \le k \le 2a+1 \\ k \equiv 2b-a}} h_k(B') \cdot \det \begin{bmatrix} e_1 & e_3 & \cdots & e_{2a-1} & e_{2a+1-k} \\ e_0 & e_2 & \cdots & e_{2a-2} & e_{2a-k} \\ \vdots & \vdots & \ddots & \vdots & \vdots \\ \cdots\cdots\cdots & e_a & e_{a+2-k} \\ \cdots\cdots\cdots & e_{a-1} & e_{a+1-k} \end{bmatrix}_{A + B'}. \tag{24}$$

It has to be shown that this expression evaluates to $S_{\Delta_{a+1}}(A + B')$. There are two different cases to consider, according to the parity of $b - a$.

- If $b - a$ is even, then the k in the sum of (24) runs over the even numbers $0, 2, 4, \ldots, 2a$. In all cases, except $k = 0$, the last column of the respective matrix is identical to one of the earlier columns, hence the determinant vanishes. In the case $k = 0$ the matrix is precisely the matrix that defines $S_{\Delta_{a+1}}(\emptyset | A + B')$, which in view of Proposition 4 is the same as $S_{\Delta_{a+1}}(A + B')$.

- If $b - a$ is odd, then the k in the sum of (24) runs over the odd numbers $1, 3, \ldots, 2a + 1$. Here the last column of the respective matrix is never identical to one of the earlier columns. But the matrix nevertheless defines a Schur function—but since the index of the last entry e_{a+1-k} is (except for the case $k = 1$) strictly less than the index of the next to last diagonal element e_a, the last column has to be moved to the left in order to re-establish the condition of (weak) growth of the diagonal elements as required for the definition of Schur functions.

  An example will make this point clear. The case $a = 3$ is displayed in Fig. 13. Note that since the determinants are expressed in terms of elementary symmetric functions, the Schur functions that appear are are not indexed by the partitions corresponding to the indices of the diagonal elements, but rather by their conjugates—see Eq. (13).

(viii) We continue with the treatment in the case $a \not\equiv b \mod 2$. It should be clear from the example given in Fig. 13 how this part of the procedure works in

$$\sum_{k\in\{1,3,5,7\}} h_k(B') \cdot \det \begin{bmatrix} e_1 & e_3 & e_5 & e_{7-k} \\ e_0 & e_2 & e_4 & e_{6-k} \\ 0 & e_1 & e_3 & e_{5-k} \\ 0 & e_0 & e_2 & e_{4-k} \end{bmatrix}_{A+B'}$$

$$= h_1(B') \cdot \det \begin{bmatrix} e_1 & e_3 & e_5 & e_6 \\ e_0 & e_2 & e_4 & e_5 \\ 0 & e_1 & e_3 & e_4 \\ 0 & e_0 & e_2 & e_3 \end{bmatrix}_{A+B'} + h_3(B') \cdot \det \begin{bmatrix} e_1 & e_3 & e_5 & e_4 \\ e_0 & e_2 & e_4 & e_3 \\ 0 & e_1 & e_3 & e_2 \\ 0 & e_0 & e_2 & e_0 \end{bmatrix}_{A+B'}$$

$$+ h_5(B') \cdot \det \begin{bmatrix} e_1 & e_3 & e_5 & e_2 \\ e_0 & e_2 & e_4 & e_1 \\ 0 & e_1 & e_3 & e_0 \\ 0 & e_0 & e_2 & 0 \end{bmatrix}_{A+B'} + h_7(B') \cdot \det \begin{bmatrix} e_1 & e_3 & e_5 & e_0 \\ e_0 & e_2 & e_4 & 0 \\ 0 & e_1 & e_3 & 0 \\ 0 & e_0 & e_2 & 0 \end{bmatrix}_{A+B'}$$

$$= h_1(B') \cdot \det \begin{bmatrix} e_1 & e_3 & e_5 & e_6 \\ e_0 & e_2 & e_4 & e_5 \\ 0 & e_1 & e_3 & e_4 \\ 0 & e_0 & e_2 & e_3 \end{bmatrix}_{A+B'} - h_3(B') \cdot \det \begin{bmatrix} e_1 & e_3 & e_4 & e_5 \\ e_0 & e_2 & e_3 & e_4 \\ 0 & e_1 & e_2 & e_3 \\ 0 & e_0 & e_1 & e_2 \end{bmatrix}_{A+B'}$$

$$+ h_5(B') \cdot \det \begin{bmatrix} e_1 & e_2 & e_3 & e_5 \\ e_0 & e_1 & e_2 & e_4 \\ 0 & e_0 & e_1 & e_3 \\ 0 & 0 & e_0 & e_2 \end{bmatrix}_{A+B'} - h_7(B') \cdot \det \begin{bmatrix} e_0 & e_1 & e_3 & e_5 \\ 0 & e_0 & e_2 & e_4 \\ 0 & 0 & e_1 & e_3 \\ 0 & 0 & e_0 & e_2 \end{bmatrix}_{A+B'}$$

$$= h_1(B')S_{\overline{3321}}(\emptyset|A+B') - h_3(B')S_{\overline{2221}}(\emptyset|A+B')$$
$$+ h_5(B')S_{\overline{2111}}(\emptyset|A+B') - h_7(B')S_{\overline{21}}(\emptyset|A+B')$$

$$= h_1(B')S_{432}(A+B') - h_3(B')S_{43}(A+B')$$
$$+ h_5(B')S_{41}(A+B') - h_7(B')S_{21}(A+B')$$

$$= \begin{bmatrix} h_1 & h_3 & h_5 & h_7 \end{bmatrix}_{B'} \cdot \begin{bmatrix} S_{432} \\ -S_{43} \\ S_{41} \\ -S_{21} \end{bmatrix}_{A+B'}$$

Fig. 13 Evaluation steps in the case $b - a$ odd, here $a = 3$

general. One obtains

$$\sum_{0 \leq \ell \leq a} (-1)^{\ell} h_{2\ell+1}(B') \cdot S_{\lambda_a^{(\ell)}}(A + B')$$

$$= [h_1 \ h_3 \ \ldots \ h_{2a+1}]_{B'} \cdot \begin{bmatrix} S_{\lambda_a^{(0)}} \\ S_{\lambda_a^{(1)}} \\ \vdots \\ S_{\lambda_a^{(a)}} \end{bmatrix}_{A + B'}, \qquad (25)$$

where for $a \geq 1$ the sequence of partitions $\lambda_a^{(0)}, \lambda_a^{(1)}, \ldots, \lambda_a^{(a)}$ (starting with $\lambda_a^{(0)} = (a + 1, a, \ldots, 3, 2)$ and then subtracting 2 from the last, the next to last, etc. parts) given in vector notation

$$\lambda_a^{(j)} = (a + 1, a, a - 1, \ldots, j + 3, j + 2, j - 1, j - 2, \ldots, 2, 1, 0).$$

The next two steps will also be illustrated in the case $a = 3$, which is sufficiently general to show how the ideas work.

(ix) Note that in the expressions showing up in (25) the h- and the S_{λ}-factors have different alphabets attached. This can easily be cured.

It follows from the duality between elementary and homogeneous symmetric functions, namely

$$e(-A) \cdot h(A + B') = h(B'),$$

that

$$[e_0 \ -e_1 \ e_2 \ -e_3]_A \begin{bmatrix} h_1 & h_3 & h_5 & h_7 \\ h_0 & h_2 & h_4 & h_6 \\ 0 & h_1 & h_3 & h_5 \\ 0 & h_0 & h_2 & h_4 \end{bmatrix}_{A + B'} - [h_1 \ h_3 \ h_5 \ h_7]_{B'}.$$

(x) Using this fact in (25) brings up the problem of computing

$$\begin{bmatrix} h_1 & h_3 & h_5 & h_7 \\ h_0 & h_2 & h_4 & h_6 \\ 0 & h_1 & h_3 & h_5 \\ 0 & h_0 & h_2 & h_4 \end{bmatrix} \begin{bmatrix} S_{432} \\ -S_{43} \\ S_{41} \\ -S_{21} \end{bmatrix} \qquad (26)$$

over an arbitrary alphabet. This could be done directly by an argument that uses Pieri's rule for computing the product of Schur functions and homogeneous symmetric functions as a sum of Schur functions, by referring to the diagrams of the partitions involved. Here we content ourselves with simple application of Cramer's rule, which shows that the unique solution of the linear system

$$
\begin{bmatrix}
h_1 & h_3 & h_5 & h_7 \\
h_0 & h_2 & h_4 & h_6 \\
0 & h_1 & h_3 & h_5 \\
0 & h_0 & h_2 & h_4
\end{bmatrix}
\begin{bmatrix}
y_0 \\
y_1 \\
y_2 \\
y_3
\end{bmatrix}
=
\begin{bmatrix}
S_{4321} \\
0 \\
0 \\
0
\end{bmatrix}
$$

is given (over any alphabet) by $\begin{bmatrix} y_0 & y_1 & y_2 & y_3 \end{bmatrix} = \begin{bmatrix} S_{432} & -S_{43} & S_{41} & -S_{21} \end{bmatrix}$. Indeed,

$$
y_0 = \frac{1}{S_{4321}}
\begin{bmatrix}
S_{4321} & h_3 & h_5 & h_7 \\
0 & h_2 & h_4 & h_6 \\
0 & h_1 & h_3 & h_5 \\
0 & h_0 & h_2 & h_4
\end{bmatrix}
=
\begin{bmatrix}
h_2 & h_4 & h_6 \\
h_1 & h_3 & h_5 \\
h_0 & h_2 & h_4
\end{bmatrix}
= S_{432},
$$

$$
y_1 = \frac{1}{S_{4321}}
\begin{bmatrix}
h_1 & S_{4321} & h_5 & h_7 \\
h_0 & 0 & h_4 & h_6 \\
0 & 0 & h_3 & h_5 \\
0 & 0 & h_2 & h_4
\end{bmatrix}
= -
\begin{bmatrix}
h_3 & h_5 \\
h_2 & h_4
\end{bmatrix}
= -S_{43}, \quad \text{etc.}
$$

Finally we get

$$
\begin{bmatrix} h_1 & h_3 & h_5 & h_7 \end{bmatrix}_{B'}
\begin{bmatrix}
S_{432} \\
-S_{43} \\
S_{41} \\
-S_{21}
\end{bmatrix}_{A+B'}
=
\begin{bmatrix} e_0 & -e_1 & e_2 & -e_3 \end{bmatrix}_A
\begin{bmatrix}
h_1 & h_3 & h_5 & h_7 \\
h_0 & h_2 & h_4 & h_6 \\
0 & h_1 & h_3 & h_5 \\
0 & h_0 & h_2 & h_4
\end{bmatrix}_{A+B'}
\begin{bmatrix}
S_{432} \\
-S_{43} \\
S_{41} \\
-S_{21}
\end{bmatrix}_{A+B'}
$$

$$
=
\begin{bmatrix} e_0 & -e_1 & e_2 & -e_3 \end{bmatrix}_A
\begin{bmatrix}
S_{4321} \\
0 \\
0 \\
0
\end{bmatrix}_{A+B'}
= S_{\Delta_4}(A+B').
$$

This way it is shown that also in the second case ($a \not\equiv b \mod 2$) the expression in (24) evaluates to $S_{\Delta_{a+1}}(A + B')$. This concludes the proof of Theorem 3. □

5.2 Evaluating $[\![\lambda]\!]$ for the Two-Part Partitions in S

In this last part the result of the evaluation of $[\![\lambda]\!]$ for the class of strict partitions with two parts will be stated and proved. There are two ways of obtaining results, depending on which of the two parts will be extended. Figure 14 shows an example

Fig. 14 Showing the extensions $(n, m) = (5, 2) \rightarrow (6, 2) = (n + 1, m)$ and $(n, m) = (5, 2) \rightarrow (5, 3) = (n, m + 1)$

both possibilities: (n, m) is always extendable to $(n + 1, m)$, but extendable to $(n, m + 1)$ only if $m + 1 < n$. See Eq. (28) for the view of divided difference.

Theorem 4 *For two-part strict partitions (n, m) with $0 < m < n$:*

$$[\![n, m]\!] = \frac{S_{<n-2,m-1>}(X_{0,m}|X_{m+1,n})}{S_{\widetilde{<n,m>}}(X_{0,m}|X_{m+1,n})}. \tag{27}$$

Recall that the denominator

$$S_{\widetilde{<n,m>}}(X_{0,m}|X_{m+1,n}) = \prod_{\substack{0 \leq i < j \leq n \\ i \leq m}} (x_i + x_j) = q_{n,m}$$

is the standard denominator for strict partitions with largest part n and second largest part m.

Note that Theorems 3 and 4 both cover the case of two-part strict partitions $(n, n - 1) = <n, 1>$, and that in these cases the expressions in (18) and (27) agree indeed.

The proof of Theorem 4 will not be given in full detail, sometimes exemplary data (with an obvious way of generalizing) will suffice. A few preliminary indications are in order. The basic facts are, of course,

$$[\![n + 1, m]\!] = [\![n, m]\!] \, \partial_n \qquad \text{and} \qquad [\![n, m + 1]\!] = [\![n, m]\!] \, \partial_m. \tag{28}$$

By plugging in the representation asserted in (27) and clearing denominators (respecting the compatiblity with ∂_n resp. ∂_m) one is left with the task to prove the *polynomial* relations, i.e., for ∂_n with $1 \leq m < n$:

$$\left[S_{m+1}(x_{n+1}|X_{0,m}) \cdot S_{<n-2,m-1>}(X_{0,m}|X_{m+1,n}) \right] \partial_n$$
$$= S_{<n-1,m-1>}(X_{0,m}|X_{m+1,n+1}), \tag{29}$$

and for ∂_m with $1 \leq m \leq n - 2$:

$$\left[S_{n-m-1}(x_{m+1}|X_{m+2,n}) \cdot S_{<n-2,m-1>}(X_{0,m}|X_{m+1,n}) \right] \partial_m$$
$$= S_{<n-2,m>}(X_{0,m+1}|X_{m+2,n}). \tag{30}$$

Note a particular feature of these identities, which shows that they are not plain obvious: in (29) the right-hand side is symmetric w.r.t. the variables $X_{m+1,n+1}$, but for the expression on the left-hand side this is not clear, because the symmetrizing operator ∂_n creates a symmetry between x_n and x_{n+1} (or else annihilates the expression, if the symmetry was already there), but usually symmetries between x_n and other variables like x_{n-1} are destroyed. A similar observation can be made w.r.t. the terms in (30). A way to resolve this for (29) is to expand both sides over the triple of alphabets $(X_{0,m}, X_{m+1,n-1}, X_{n,n+1})$, and similarly for (30).

Proof of (29) As mentioned, rather than writing out the proof in full detail, I will go through it using a generic example—it should become obvious that and how the procedure generalizes.

Let us consider the case $n = 5, m = 3$, so that (29) becomes

$$\left[S_4(x_6|X_{0,3}) \cdot S_{<3,2>}(X_{0,3}|X_{4,5}) \right] \partial_5 = S_{<4,2>}(X_{0,3}|X_{4,6}),$$

where $<3, 2>$ is the partition $(3, 2, 1)$ and $<4, 2>$ is the partition $(4, 3, 2)$.

The expression in the brackets of the left hand side may be written as

$$\sum_{i=0}^{4} h_i(x_6) e_{4-i}(X_{0,3}) \cdot S_{<3,2>;0}(X_{0,3}|x_4; x_5),$$

where the second identity in Proposition 4 has been used to rewrite the defining matrix for the Schur function $S_{<3,2>}(X_{0,3}|X_{4,5})$. Now use

$$h_i(x_6) = h_i(X_{5,6}) - x_5 h_{i-1}(X_{5,6}) \qquad \text{(first term for each } i\text{)},$$

$$h_i(x_5) = h_i(X_{5,6}) - x_6 h_{i-1}(X_{5,6}) \qquad \text{(last col. of matrix for } S_{<3,2>;0}\text{)},$$

to rewrite this as

$$\sum_{i=0}^{4} e_{4-i}(X_{0,3}) \cdot (h_i(X_{5,6}) - x_5 h_{i-1}(X_{5,6})) \cdot (D_0 - x_6 D_{-1}), \tag{31}$$

where

$$D_0 = S_{<3,2>;0}(X_{0,3}|x_4; X_{5,6}) = \det \begin{bmatrix} S_1 & S_3 & S_5 & -h_3 \\ S_0 & S_2 & S_4 & h_2 \\ 0 & S_1 & S_3 & -h_1 \\ 0 & S_0 & S_2 & h_0 \end{bmatrix}_{(X_{0,3}|x_4); X_{5,6}},$$

where the S_i are taken over the alphabet $(X_{0,3}, x_4)$ and the h_j are taken over the alphabet $X_{5,6}$, and where

$$D_{-1} = S_{<3,2>;-1}(X_{0,3}|x_4; X_{5,6}) = \det \begin{bmatrix} S_1 & S_3 & S_5 & h_2 \\ S_0 & S_2 & S_4 & -h_1 \\ 0 & S_1 & S_3 & h_0 \\ 0 & S_0 & S_2 & -h_{-1} \end{bmatrix}_{(X_{0,3}|x_4); X_{5,6}}.$$

Now, expand (31) into

$$\sum_{i=0}^{4} e_{4-i} \cdot (h_i D_0 - x_5 h_{i-1} D_0 - x_6 h_i D_{-1} + x_5 x_6 h_{i-1} D_{-1}).$$

Note that the first and the last terms of the expression in parentheses are symmetric w.r.t. $\sigma_5 : x_5 \leftrightarrow x_6$ because D_0 and D_{-1} and the h_i are. Hence they will disappear under application of ∂_5 and we obtain

$$\left[\sum_{i=0}^{4} h_i(x_6) e_{4-i}(X_{0,3}) \cdot S_{<3,2>;0}(X_{0,3}|x_4; x_5) \right] \partial_5$$

$$= \sum_{i=0}^{4} e_{4-i} [-x_5 \partial_5 \cdot h_{i-1} D_0 - x_6 \partial_5 \cdot h_i D_{-1}] = \sum_{i=0}^{4} e_{4-i} [h_{i-1} D_0 - h_i D_{i-1}]$$

$$= S_3(X_{5,6}|X_{0,3}) \cdot D_1 - S_4(X_{5,6}|X_{0,3}) \cdot D_{-1}.$$

Consequently, the left hand side of (29) can be written in determinantal form as

$$\det \begin{bmatrix} 0 & 0 & 0 & -S_3' & S_4' \\ S_1 & S_3 & S_5 & h_2 & -h_3 \\ S_0 & S_2 & S_4 & -h_1 & h_2 \\ 0 & S_1 & S_3 & h_0 & -h_1 \\ 0 & S_0 & S_2 & 0 & h_0 \end{bmatrix},$$

where all the S_i are taken over the alphabet $(X_{0,3}|x_4)$, the S_i' are taken over $(X_{5,6}|X_{0,3})$, and all the h_i are taken over the alphabet $X_{5,6}$.

Now look at the right hand side of (29) and rewrite it according to the second identity in Proposition 5 as

$$S_{<4,3>}(X_{0,3}|X_{4,6}) = \det \begin{bmatrix} S_2 & S_4 & S_6 & -h_3 & h_4 \\ S_1 & S_3 & S_5 & h_2 & -h_3 \\ S_0 & S_2 & S_4 & -h_1 & h_2 \\ 0 & S_1 & S_3 & h_0 & -h_1 \\ 0 & S_0 & S_2 & 0 & h_0 \end{bmatrix},$$

with the same alphabets as just stated. Hence we are left with the task of showing

$$\det \begin{bmatrix} 0 & 0 & 0 & -S'_3 & S'_4 \\ S_1 & S_3 & S_5 & h_2 & -h_3 \\ S_0 & S_2 & S_4 & -h_1 & h_2 \\ 0 & S_1 & S_3 & h_0 & -h_1 \\ 0 & S_0 & S_2 & 0 & h_0 \end{bmatrix} = \det \begin{bmatrix} S_2 & S_4 & S_6 & -h_3 & h_4 \\ S_1 & S_3 & S_5 & h_2 & -h_3 \\ S_0 & S_2 & S_4 & -h_1 & h_2 \\ 0 & S_1 & S_3 & h_0 & -h_1 \\ 0 & S_0 & S_2 & 0 & h_0 \end{bmatrix}.$$

But this is easily explained. Multiply the matrix on the right hand side by the row vector

$$\left[e_4(-X_{0,3}),\ e_3(-X_{0,3}),\ e_2(-X_{0,3}),\ e_1(-X_{0,3}),\ e_0(-X_{0,3}) \right],$$

and then replace the first row by the resulting vector, which as a row operation does not change the determinant. Then the first three entries vanish, because

$$e(-X_{0,3}) \cdot \frac{1+x_4}{\prod_{i=0}^{3}(1-x_i)} = 1 + x_4$$

has no terms of degree ≥ 1. As for the last two entries, just note that they are coefficients (for the correct degree and with the correct sign) of

$$e(-X_{0,3}) \cdot \frac{1}{(1-x_5)(1-x_6)}.$$

This finishes the example proof of (29).   □

Proof of (30) Similar to the general formula (see Proposition 2)

$$S_n(A|B) = \sum_k (-1)^k e_k(B^2) S_{n-2k}(A + B),$$

one has in the particular case of shifting just one element from right to left in the alphabets

$$S_n(A + a|B + b) = S_n(A + a + b|B) - b^2 S_{n-2}(A + a + b|B),$$

and hence for the divided difference $\partial_{a,b}$ related to $\sigma_{a,b} : a \leftrightarrow b$

$$S_n(A + a|B + b)\, \partial_{a,b} = -(a + b) S_{n-2}(A + a + b|B),$$

since, except for the b^2, everything else is symmetric w.r.t. $\sigma_{a,b}$.

These two facts can be used to write down divided differences of Schur functions of the type $S_\lambda(A + a|B + b)\partial_{a,b}$ in general. We will consider here only the particular situation of join-irreducible partitions $<n, m>$, for which the result is easy to state.

Let now $1 \leq m \leq n$ and further denote $X_{0,m-1} = A$, $x_m = n$, $x_{m+1} = b$, $X_{m+2,n+2} = B$. Then the $m \times m$-matrix defining $S_{<n,m-1>}(A + a | B + b)$ for the join-irreducible partition $<n, m-1> = (n, n-1, \ldots, n-m+1)$ is

$$
\begin{bmatrix}
S_{n-m+1} & S_{n-m+3} & \cdots & S_{n+m-1} \\
S_{n-m} & S_{n-m+2} & \cdots & S_{n+m-2} \\
\vdots & \vdots & \ddots & \vdots \\
S_{n-2m+2} & S_{n-2m+4} & \cdots & S_n
\end{bmatrix}_{(A+a|B+b)}
=
$$

$$
\begin{bmatrix}
S_{n-m+1} - b^2 S_{n-m-1} & S_{n-m+3} - b^2 S_{n-m+1} & \cdots & S_{n+m-1} - b^2 S_{n+m-3} \\
S_{n-m} - b^2 S_{n-m-2} & S_{n-m+2} - b^2 S_{n-m} & \cdots & S_{n-m-2} - b^2 S_{n+m-4} \\
\vdots & \vdots & \ddots & \vdots \\
S_{n-2m+2} - b^2 S_{n-2m} & S_{n-2m+4} - b^2 S_{n-2m+2} & \cdots & S_n - b^2 S_{n-2}
\end{bmatrix}_{(A+a+b|B)}.
$$

The fact that the S-terms without b^2-factor in each column appear with a b^2-factor in the subsequent column can be used to considerably simplify the writing of the determinant of this matrix. The result that one gets is in terms of a $(m+1) \times (m+1)$-matrix

$$S_{<n,m-1>}(A+a|B+b) =$$

$$
\det
\begin{bmatrix}
1 & b^2 & b^4 & \cdots & b^{2m} \\
S_{n-m-1} & S_{n-m+1} & S_{n-m+3} & \cdots & S_{n+m-1} \\
S_{n-m-2} & S_{n-m} & S_{n-m+2} & \cdots & S_{n+m-2} \\
\vdots & \vdots & \vdots & \ddots & \vdots \\
S_{n-2m} & S_{n-2m+2} & S_{n-2m+4} & \cdots & S_n
\end{bmatrix}_{(A+a+b|B)} . \tag{32}
$$

This allows to write the divided difference $S_{<n,m-1>}(A + a | B + b)\partial_{a,b}$ in concise form, because the $\partial_{a,b}$ acts only on the terms of the first row.

$$S_{<n,m-1>}(A + a | B + b)\partial_{a,b} =$$

$$
(a+b) \cdot \det
\begin{bmatrix}
0 & h_0(a^2, b^2) & h_1(a^2, b^2) & \cdots & h_{m-1}(a^2, b^2) \\
S_{n-m-1} & S_{n-m+1} & S_{n-m+3} & \cdots & S_{n+m-1} \\
S_{n-m-2} & S_{n-m} & S_{n-m+2} & \cdots & S_{n+m-2} \\
\vdots & \vdots & \vdots & \ddots & \vdots \\
S_{n-2m} & S_{n-2m+2} & S_{n-2m+4} & \cdots & S_n
\end{bmatrix}_{(A+a+b|B)}
$$

$$\tag{33}$$

What needs to be shown, is

$$\left[S_{n-m+1}(b|B) \cdot S_{<n,m-1>}(A+a|B+b)\right]\partial_{a,b} = S_{<n,m>}(A+a+b|B).$$

We have

$$\left[S_{n-m+1}(b|B) \cdot S_{<n,m-1>}(A+a|B+b)\right]\partial_{a,b}$$

$$= S_{n-m+1}(a|B) \cdot \left[S_{<n,m-1>}(A+a|B+b)\right]\partial_{a,b}$$

$$+ \left[S_{n-m+1}(b|B)\right]\partial_{a,b} \cdot S_{<n,m-1>}(A+a|B+b)$$

$$= S_{n-m+1}(a|B) \cdot (a+b) \cdot D_{-1} + S_{n-m}(a+b|B) \times D_0$$

$$= S_{n-m+2}(a|B+b) \cdot D_{-1} + S_{n-m}(a+b|B) \times D_0$$

$$= \left[S_{n-m+2}(a+b|B) - b^2 S_{n-m}(a+b|B)\right] \times D_{-1} + S_{n-m}(a+b|B) \times D_0,$$

where D_0 resp. D_{-1} are the determinants showing up in (32) resp. (33). The elements in positions k (for $0 \le k \le m$) in the top row of the corresponding matrix are

$$\left(S_{n-m+2}(a+b|B) - b^2 S_{n-m}(a+b|B)\right) h_{k-1}(a^2, b^2) + S_{n-m}(a+b|B)h_k(b^2).$$

These are is easily seen to be the same as $S_{n-m+2k}(a+b|B)$.

For the verification the alphabet B plays no role here, it suffices to check the identity

$$(h_{n+1}(a+b) - b^2 h_{n-1}(a+b)) \cdot h_{k-1}(a^2+b^2) + h_{n-1}(a+b)h_k(b^2) = h_{n-1+2k}(a+b),$$

or, equivalently (and exhibiting the symmetry),

$$h_k(a+b)h_\ell(a^2+b^2) - a^2 b^2 h_{k-2}(a+b)h_{\ell-1}(a^2+b^2) = h_{k+2\ell}(a+b).$$

What we have shown up to this point is

$$\left[S_{n-m+1}(b|B) \cdot S_{<n,m-1>}(A+a|B+b)\right]\partial_{a,b} = \det \begin{bmatrix} S'_{n-m} & S'_{n-m+2} & \cdots & S'_{n+m} \\ S_{n-m-1} & S_{n-m+1} & \cdots & S_{n+m-1} \\ S_{n-m-2} & S_{n-m} & \cdots & S_{n+m-2} \\ \vdots & \vdots & \ddots & \vdots \\ S_{n-2m} & S_{n-2m+2} & \cdots & S_n \end{bmatrix},$$

where the unprimed S-functions are to be taken over the alphabet $(A+a+b|B)$ and the primed S-functions are to be taken over the alphabet $(a+b|B)$, whereas our goal was to arrive at

$$S_{<n,m>}(A + a + b|B) = \det \begin{bmatrix} S_{n-m} & S_{n-m+2} & \cdots & S_{n+m} \\ S_{n-m-1} & S_{n-m+1} & \cdots & S_{n+m-1} \\ S_{n-m-2} & S_{n-m} & \cdots & S_{n+m-2} \\ \vdots & \vdots & \ddots & \vdots \\ S_{n-2m} & S_{n-2m+2} & \cdots & S_n \end{bmatrix}.$$

But the last two determinants are easily seen to be equal by executing successive row operations that are based on

$$S_n(A|B) = \sum_{k \geq 0} (-1)^k S_{n-k}(A + X|B) e_k(X).$$

This terminates the proof of (30). □

Both of (29) and (30) independently yield the validity of Theorem 4 for strict two-part partitions.

6 Conclusion and Outlook

In this article I have investigated a simple linear system with rational function coefficients, indexed by the elements of the lattice of strict partitions. The interest in this particular system comes from a model in combinatorial physics, an asymmetric exclusion process with annihilation. Explaining the precise relation between the the problem of determining the partition functions of that model and the work presented here has been deferred to the article [3]—but it can be stated that the partition function for the n-site model in question appears as the denominator of the solution $[\![\Delta_n]\!]$ attached to the staircase partition Δ_n. It is therefore important to know what $[\![\Delta_n]\!]$ is—see Eq. (10) in Theorem 1, and that the denominator is indeed the least common multiple of the denominators of all the solutions $[\![\lambda]\!]$ that appear for partitions $\lambda \subseteq \Delta_n$, as stated in Corollaries 1 and 2.

The behavior of the denominators – only products of binomials appear as factors, despite the denominators in the definition of the system—provoked me to look into the numerators of the solutions as well, which is way more complicated. It leads into expressions that can be concisely presented in terms of Schur functions as 'closed forms'. A general solution is still far away, but I have been able to give a precise answer for two interesting families of strict partitions:

- the join-irreducible elements of the lattice S of strict partitions (Theorem 3);
- the strict-partitions consisting of two parts (Theorem 4).

In both cases, as before in the investigation of the denominators, the main technical tool, i.e., the use of divided differences for the iterative construction of solution components $\lambda \subseteq \Delta_n$ along the covering relation of the lattice $\mathcal{S}$—see the Main Lemma 2—plays a decisive role.

On the 'to-do' side, I have already hinted at exposing the precise relevance of the results obtained here for the model of combinatorial physics that inspired the investigations presented here. Furthermore, I think that determining 'closed-form' solutions for other classes of partitions will be very difficult. Experimental experience suggests that Schur functions over more than two alphabets and with blocks will show up.

References

1. Ayyer, A., Mallick, K.: Exact results for an asymmetric annihilation process with open boundaries. J. Phys. A **43**(4), 045003, 22 pp. (2010) . MR2578722
2. Ayyer, A., Strehl, V.: The spectrum of an asymmetric annihilation process. In: 22nd International Conference on Formal Power Series and Algebraic Combinatorics (FPSAC 2010). Discrete Mathematics & Theoretical Computer Science Proceedings, AN. Association of Discrete Mathematics & Theoretical Computer Science, Nancy (2010), pp. 461–472. MR2673858
3. Strehl, V.: The fully parametrized asymmetric exclusion process with annihilation. Seminaire Lotharingien de Combinatoire **B81a**, 35 pp. (2020)
4. Lascoux, A.: Symmetric functions and combinatorial operators on polynomials. In: CBMS Regional Conference Series in Mathematics, vol. 99, xii+268 pp. American Mathematical Society, Providence, RI (2003). ISBN: 0-8218-2871-1. MR2017492
5. Macdonald, I.G.: Symmetric Functions and Hall Polynomials. Oxford Mathematical Monographs, viii+180 pp. The Clarendon Press, Oxford University Press, New York (1979). ISBN: 0-19-853530-9. MR0553598
6. Stanley, R.P.: Enumerative Combinatorics, vol. 2. Cambridge Studies in Advanced Mathematics, vol. 62, xii+581 pp. Cambridge University Press, Cambridge (1999). ISBN: 0-521-56069-1; 0-521-78987-7. MR1676282

Untying the Gordian Knot via Experimental Mathematics

Yukun Yao and Doron Zeilberger

This article is dedicated to Peter Paule, one of the great pioneers of experimental mathematics and symbolic computation. In particular, it is greatly inspired by his masterpiece, co-authored with Manuel Kauers, 'The Concrete Tetrahedron' [3], where a whole chapter is dedicated to our favorite ansatz, the C-finite ansatz

1 Introduction

Once upon a time there was a knot that no one could untangle, it was so complicated. Then came Alexander the Great and, in one second, *cut* it with his sword.

Analogously, many mathematical problems are very hard, and the current party line is that in order for it be considered solved, the solution, or answer, should be given a logical, rigorous, *deductive* proof.

Suppose that you want to answer the following question:

Find a closed-form formula, as an expression in n, for the real part of the n-th complex root of the Riemann zeta function, $\zeta(s)$.

Let's call this quantity $a(n)$. Then you compute these real numbers, and find out that $a(n) = \frac{1}{2}$ for $n \leq 1000$. Later you are told by Andrew Odlyzko that $a(n) = \frac{1}{2}$ for all $1 \leq n \leq 10^{10}$. Can you conclude that $a(n) = \frac{1}{2}$ for *all n*? We would, but, at this time of writing, there is no way to deduce it rigorously, so it remains an open problem. It is very possible that one day it will turn out that $a(n)$ (the real part of the *n*-th complex root of $\zeta(s)$) belongs to a certain *ansatz*, and that checking it for the first N_0 cases implies its truth in general, but this remains to be seen.

There are also frameworks, e.g. *Pisot sequences* (see [5, 10]), where the *inductive* approach fails miserably.

Y. Yao (✉) · D. Zeilberger
Rutgers University, New Brunswick, NJ, USA
e-mail: yao@math.rutgers.edu

© Springer Nature Switzerland AG 2020
V. Pillwein, C. Schneider (eds.), *Algorithmic Combinatorics: Enumerative Combinatorics, Special Functions and Computer Algebra*, Texts & Monographs in Symbolic Computation, https://doi.org/10.1007/978-3-030-44559-1_18

On the other hand, in order to (rigorously) prove that $1^3 + 2^3 + 3^3 + \cdots + n^3 = (n(n + 1)/2)^2$, for *every* positive integer n, it suffices to check it for the five special cases $0 \leq n \leq 4$, since both sides are polynomials of **degree** 4, hence the difference is a polynomial of degree ≤ 4, given by five 'degrees of freedom'.

This is an example of what is called the 'N_0 principle'. In the case of a polynomial identity (like this one), N_0 is simply the degree plus one.

But our favorite *ansatz* is the C-finite ansatz. A sequence of numbers $\{a(n)\}$ $(0 \leq n < \infty)$ is C-finite if it satisfies a *linear recurrence equation with constant coefficients*. For example the Fibonacci sequence that satisfies $F(n) - F(n - 1) - F(n - 2) = 0$ for $n \geq 2$.

The C-finite ansatz is beautifully described in chapter 4 of the masterpiece 'The Concrete Tetrahedron' [3], by Manuel Kauers and Peter Paule, and discussed at length in [9].

Here the 'N_0 principle' also holds (see [8]), i.e. by looking at the 'big picture' one can determine *a priori*, a positive integer, often not that large, such that checking that $a(n) = b(n)$ for $1 \leq n \leq N_0$ implies that $a(n) = b(n)$ for all $n > 0$.

A sequence $\{a(n)\}_{n=0}^{\infty}$ is C-finite if and only if its (ordinary) *generating function* $f(t) := \sum_{n=0}^{\infty} a(n) t^n$ is a **rational function** of t, i.e. $f(t) = P(t)/Q(t)$ for some *polynomials* $P(t)$ and $Q(t)$. For example, famously, the generating function of the Fibonacci sequence is $t/(1 - t - t^2)$.

Phrased in terms of generating functions, the C-finite ansatz is the subject of chapter 4 of yet another masterpiece, Richard Stanley's 'Enumerative Combinatorics' (volume 1) [6]. There it is shown, using the 'transfer matrix method' (that originated in physics), that in many combinatorial situations, where there are finitely many states, one is guaranteed, *a priori*, that the generating function is rational.

Alas, finding this transfer matrix, at each specific case, is not easy! The human has to first figure out the set of states, and then using human ingenuity, figure out how they interact.

A better way is to automate it. Let the computer do the research, and using 'symbolic dynamical programming', the computer, automatically, finds the set of states, and constructs, *all by itself* (without any human pre-processing) the set of states and the transfer matrix. But this may not be so efficient for two reasons. First, at the very end, one has to invert a matrix with *symbolic* entries, hence compute symbolic determinants, that is time-consuming. Second, setting up the 'infra-structure' and writing a program that would enable the computer to do 'machine-learning' can be very daunting.

In this article, we will describe two *case studies* where, by 'general nonsense', we know that the generating functions are rational, and it is easy to bound the degree of the denominator (alias the order of the recurrence satisfied by the sequence). Hence a simple-minded, *empirical*, approach of computing the first few terms and then 'fitting' a recurrence (equivalently rational function) is possible.

The first case-study concerns counting spanning trees in families of grid-graphs, studied by Paul Raff [4], and F.J. Faase [2]. In their research, the human first analyzes the intricate combinatorics, manually sets up the transfer matrix, and only at the end lets a computer-algebra system evaluate the symbolic determinant.

Our key observation, that enabled us to 'cut the Gordian knot' is that the terms of the studied sequences are expressible as *numerical* determinants. Since computing numerical determinants is so fast, it is easy to compute sufficiently many terms, and then fit the data into a rational function. Since we easily have an upper bound for the degree of the denominator of the rational function, everything is rigorous.

The second case-study is computing generating functions for sequences of determinants of 'almost diagonal Toeplitz matrices'. Here, in addition to the 'naive' approach of cranking enough data and then fitting it into a rational function, we also describe the 'symbolic dynamical programming method', that surprisingly is faster for the range of examples that we considered. But we believe that for sufficiently large cases, the naive approach will eventually be more efficient, since the 'deductive' approach works equally well for the analogous problem of finding the sequence of permanents of these almost diagonal Toeplitz matrices, for which the naive approach will soon be intractable.

This article may be viewed as a *tutorial*, hence we include lots of implementation details, and Maple code. We hope that it will inspire readers (and their computers!) to apply it in other situations.

2 Accompanying Maple Packages

This article is accompanied by three Maple packages, GFMatrix.txt, JointConductance.txt, and SpanningTrees.txt, all available from the http://sites.math.rutgers.edu/~zeilberg/mamarim/mamarimhtml/gordian.html. In that page there are also links to numerous sample input and output files.

3 The Human Approach to Enumerating Spanning Trees of Grid Graphs

In order to illustrate the advantage of "keeping it simple", we will review the human approach to the enumeration task that we will later redo using the 'Gordian knot' way. While the human approach is definitely interesting for its own sake, it is rather painful.

Our goal is to enumerate the number of spanning trees in certain families of graphs, notably grid graphs and their generalizations. Let's examine Paul Raff's interesting approach described in his paper *Spanning Trees in Grid Graph* [4]. Raff's approach was inspired by the pioneering work of F.J. Faase [2].

The goal is to find generating functions that enumerate spanning trees in grid graphs and the product of an arbitrary graph and a path or a cycle.

Grid graphs have two parameters, let's call them k and n. For a $k \times n$ grid graph, let's think of k as *fixed* while n is the discrete input variable of interest.

Definition The $k \times n$ grid graph $G_k(n)$ is the following graph given in terms of its vertex set V and edge set E:

$$V = \{v_{ij} | 1 \leq i \leq k, 1 \leq j \leq n\},$$

$$S = \{\{v_{ij}, v_{i'j'}\} | |i - i'| + |j - j'| = 1\}.$$

The main idea in the human approach is to consider the collection of set-partitions of $[k] = \{1, 2, \ldots, k\}$ and figure out the transition when we extend a $k \times n$ grid graph to a $k \times (n + 1)$ one.

Let $\mathcal{B}_k$ be the collection of all set-partitions of $[k]$. $B_k = |\mathcal{B}_k|$ is called the k-th Bell number. Famously, the exponential generating function of B_k, namely $\sum_{k=0}^{\infty} \frac{B_k}{k!} t^k$, equals $e^{e^t - 1}$.

A lexicographic ordering on $\mathcal{B}_k$ is defined as follows:

Definition Given two partitions P_1 and P_2 of $[k]$, for $i \in [k]$, let X_i be the block of P_1 containing i and Y_i be the block of P_2 containing i. Let j be the minimum number such that $X_i \neq Y_i$. Then $P_1 < P_2$ iff

1. $|P_1| < |P_2|$ or
2. $|P_1| = |P_2|$ and $X_j \prec Y_j$ where $\prec$ denotes the normal lexicographic order.

For example, here is the ordering for $k = 3$:

$$\mathcal{B}_3 = \{\{\{1, 2, 3\}\}, \{\{1\}, \{2, 3\}, \{\{1, 2\}, \{3\}\}, \{\{1, 3\}, \{2\}\}, \{\{1\}, \{2\}, \{3\}\}\} \quad .$$

For simplicity, we can rewrite it as follows:

$$\mathcal{B}_3 = \{123, 1/23, 12/3, 13/2, 1/2/3\}.$$

Definition Given a spanning forest F of $G_k(n)$, the partition induced by F is obtained from the equivalence relation

$$i \sim j \iff v_{n,i}, v_{n,j} \text{ are in the same component of } F.$$

For example, the partition induced by any spanning tree of $G_k(n)$ is $123 \ldots k$ because by definition, in a spanning tree, all $v_{n,i}, 1 \leq i \leq k$ are in the same component. For the other extreme, where every component only consists of one vertex, the corresponding set-partition is $1/2/3/ \ldots / k - 1/k$ because no two $v_{n,i}, v_{n,j}$ are in the same component for $1 \leq i < j \leq k$.

Definition Given a spanning forest F of $G_k(n)$ and a set-partition P of $[k]$, we say that F is consistent with P if:

1. The number of trees in F is precisely $|P|$.
2. P is the partition induced by F.

Let E_n be the set of edges $E(G_k(n)) \backslash E(G_k(n-1))$, then E_n has $2k-1$ members.

Given a forest F of $G_k(n-1)$ and some subset $X \subseteq E_n$, we can combine them to get a forest of $G_k(n)$ as follows. We just need to know how many subsets of E_n can transfer a forest consistent with some partition to a forest consistent with another partition. This leads to the following definition:

Definition Given two partitions P_1 and P_2 in $\mathcal{B}_k$, a subset $X \subseteq E_n$ transfers from P_1 to P_2 if a forest consistent with P_1 becomes a forest consistent with P_2 after the addition of X. In this case, we write $X \diamond P_1 = P_2$.

With the above definitions, it is natural to define a $B_k \times B_k$ transfer matrix A_k by the following:

$$A_k(i, j) = |\{A \subseteq E_{n+1} | A \diamond P_j = P_i\}|.$$

Let's look at the $k = 2$ case as an example. We have

$$\mathcal{B}_2 = \{12, 1/2\}, E_{n+1} = \{\{v_{1,n}, v_{1,n+1}\}, \{v_{2,n}, v_{2,n+1}\}, \{v_{1,n+1}, v_{2,n+1}\}\}.$$

For simplicity, let's call the edges in E_{n+1} e_1, e_2, e_3. Then to transfer the set-partition $P_1 = 12$ to itself, we have the following three ways: $\{e_1, e_2\}, \{e_1, e_3\}, \{e_2, e_3\}$. In order to transfer the partition $P_2 = 1/2$ into P_1, we only have one way, namely: $\{e_1, e_2, e_3\}$. Similarly, there are two ways to transfer P_1 to P_2 and one way to transfer P_2 to itself Hence the transfer matrix is the following 2×2 matrix:

$$A = \begin{bmatrix} 3 & 1 \\ 2 & 1 \end{bmatrix}.$$

Let $T_1(n), T_2(n)$ be the number of forests of $G_k(n)$ which are consistent with the partitions P_1 and P_2, respectively. Let

$$v_n = \begin{bmatrix} T_1(n) \\ T_2(n) \end{bmatrix},$$

then

$$v_n = Av_{n-1}.$$

The characteristic polynomial of A is

$$\chi_\lambda(A) = \lambda^2 - 4\lambda + 1.$$

By the Cayley-Hamilton Theorem, A satisfies

$$A^2 - 4A + 1 = 0.$$

Hence the recurrence relation for $T_1(n)$ is

$$T_1(n) = 4T_1(n-1) - T_1(n-2),$$

the sequence is

$$\{1, 4, 15, 56, 209, 780, 2911, 10864, 40545, 151316, \ldots\} \quad \text{(OEIS A001353)}$$

and the generating function is

$$\frac{x}{1 - 4x + x^2}.$$

Similarly, for the $k = 3$ case, the transfer matrix

$$A_3 = \begin{bmatrix} 8 & 3 & 3 & 4 & 1 \\ 4 & 3 & 2 & 2 & 1 \\ 4 & 2 & 3 & 2 & 1 \\ 1 & 0 & 0 & 1 & 0 \\ 3 & 2 & 2 & 2 & 1 \end{bmatrix}.$$

The transfer matrix method can be generalized to general graphs of the form $G \times P_n$, especially cylinder graphs.

As one can see, we had to think very hard. First we had to establish a 'canonical' ordering over set-partitions, then define the consistence between partitions and forests, then look for the transfer matrix and finally worry about initial conditions.

Rather than think so hard, let's compute sufficiently many terms of the enumeration sequence, and try to guess a linear recurrence equation with constant coefficients, that would be provable *a posteriori* just because we know that *there exists* a transfer matrix without worrying about finding it explicitly. But how do we generate sufficiently many terms? Luckily, we can use the celebrated **Matrix Tree Theorem**.

4   The Matrix Tree Theorem

Matrix Tree Theorem If $A = (a_{ij})$ is the adjacency matrix of an arbitrary graph G, then the number of spanning trees is equal to the determinant of any co-factor of the Laplacian matrix L of G, where

$$L = \begin{bmatrix} a_{12} + \cdots + a_{1n} & -a_{12} & \cdots & -a_{1,n} \\ -a_{21} & a_{21} + \cdots + a_{2n} & \cdots & -a_{2,n} \\ \vdots & \vdots & \ddots & \vdots \\ -a_{n1} & -a_{n2} & \cdots & a_{n1} + \cdots + a_{n,n-1} \end{bmatrix}.$$

For instance, taking the (n, n) co-factor, we have that the number of spanning trees of G equals

$$\begin{vmatrix} a_{12} + \cdots + a_{1n} & -a_{12} & \cdots & -a_{1,n-1} \\ -a_{21} & a_{21} + \cdots + a_{2n} & \cdots & -a_{2,n-1} \\ \vdots & \vdots & \ddots & \vdots \\ -a_{n-1,1} & -a_{n-1,2} & \cdots & a_{n-1,1} + \cdots + a_{n-1,n} \end{vmatrix}.$$

Since computing determinants for numeric matrices is very fast, we can find the generating functions for the number of spanning trees in grid graphs and more generalized graphs by experimental methods, using the C-finite ansatz.

5 The GuessRec Maple Procedure

Our engine is the Maple procedure GuessRec(L) that resides in the Maple packages accompanying this article. We used the 'vanilla', straightforward, linear algebra approach for guessing, using *undetermined coefficients*. A more efficient way is via the celebrated Berlekamp-Massey algorithm [7]. Since the guessing part is not the *bottle-neck* of our approach (it is rather the data-generation part), we preferred to keep it simple.

Naturally, we need to collect enough data. The input is the data (given as a list) and the output is a conjectured recurrence relation derived from that data.

Procedure GuessRec(L) inputs a list, L, and attempts to output a linear recurrence equation with constant coefficients satisfied by the list. It is based on procedure GuessRec1(L,d) that looks for such a recurrence of order d.

The output of GuessRec1(L,d) consists of the the list of initial d values ('initial conditions') and the recurrence equation represented as a list. For instance, if the input is $L = [1, 1, 1, 1, 1, 1]$ and $d = 1$, then the output will be $[[1], [1]]$; if the input is $L = [1, 4, 15, 56, 209, 780, 2911, 10864, 40545, 151316]$ as the $k = 2$ case for grid graphs and $d = 2$, then the output will be $[[1, 4], [4, -1]]$. This means that our sequence satisfies the recurrence $a(n) = 4a(n - 1) - a(n - 2)$, subject to the initial conditions $a(0) = 1, a(1) = 4$.

Here is the Maple code:

```
GuessRec1:=proc(L,d) local eq,var,a,i,n:
if nops(L)<=2*d+2 then
  print('The list must be of size >=', 2*d+3 ):
  RETURN(FAIL) :
fi:
var:={seq(a[i],i=1..d)}:
eq:={seq(L[n]-add(a[i]*L[n-i],i=1..d),n=d+1..nops(L))}:
var:=solve(eq,var) :
if var=NULL then
  RETURN(FAIL) :
```

```
else
  RETURN([[op(1..d,L)],[seq(subs(var,a[i]),i=1..d)]]):
fi:
end:
```

The idea is that having a long enough list L ($|L| > 2d + 2$) of data, we use the data after the d-th one to discover whether there exists a linear recurrence relation, the first d data points being the initial condition. With the unknowns $a_1, a_2, \ldots, a_d$, we have a linear systems of no less than $d + 3$ equations. If there is a solution, it is extremely likely that the recurrence relation holds in general. The first list of length d in the output constitutes the list of initial conditions while the second list, R, codes the linear recurrence, where $[R[1], \ldots R[d]]$ stands for the following recurrence:

$$L[n] = \sum_{i=1}^{d} R[i]L[n - i].$$

Here is the Maple procedure GuessRec (L):

```
GuessRec:=proc(L) local gu,d:
for d from 1 to trunc(nops(L)/2)-2 do
  gu:=GuessRec1(L,d):
  if gu<>FAIL then
    RETURN(gu):
fi:
od:
FAIL:
end:
```

This procedure inputs a sequence L and tries to guess a recurrence equation with constant coefficients satisfying it. It returns the initial values and the recurrence equation as a pair of lists. Since the length of L is limited, the maximum degree of the recurrence cannot be more than $\lfloor |L|/2 - 2 \rfloor$. With this procedure, we just need to input $L = [1, 4, 15, 56, 209, 780, 2911, 10864, 40545, 151316]$ to get the recurrence (and initial conditions) $[[1, 4], [4, -1]]$.

Once the recurrence relation, let's call it S, is discovered, procedure CtoR (S, t) finds the generating function for the sequence. Here is the Maple code:

```
CtoR:=proc(S,t) local D1,i,N1,L1,f,f1,L:
if not (type(S,list) and nops(S)=2 and type(S[1],list)
  and type(S[2],list) and nops(S[1])=nops(S[2])
  and type(t, symbol) ) then
  print('Bad input'):
  RETURN(FAIL):
fi:
D1:=1-add(S[2][i]*t**i,i=1..nops(S[2])):
N1:=add(S[1][i]*t**(i-1),i=1..nops(S[1])):
L1:=expand(D1*N1):
L1:=add(coeff(L1,t,i)*t**i,i=0..nops(S[1])-1):
f:=L1/D1:
```

```
L:=degree(D1,t)+10:
f1:=taylor(f,t=0,L+1):
if expand([seq(coeff(f1,t,i),i=0..L)])
                              <>expand(SeqFromRec(S,L+1))
then
   print([seq(coeff(f1,t,i),i=0..L)],SeqFromRec(S,L+1)):
   RETURN(FAIL):
else
   RETURN(f):
fi:
end:
```

Procedure SeqFromRec used above (see the package) simply generates many terms using the recurrence.

Procedure CtoR(S,t) outputs the rational function in t, whose coefficients are the members of the C-finite sequence S. For example:

$$\text{CtoR}([[1, 1], [1, 1]], t) = \frac{1}{-t^2 - t + 1}.$$

Briefly, the idea is that the denominator of the rational function can be easily determined by the recurrence relation and we use the initial condition to find the starting terms of the generating function, then multiply it by the denominator, yielding the numerator.

6   Application of GuessRec for Enumerating Spanning Trees of Grid Graphs and $G \times P_n$

With the powerful procedures GuessRec and CtoR, we are able to find generating functions for the number of spanning trees of generalized graphs of the form $G \times P_n$. We will illustrate the application of GuessRec to finding the generating function for the number of spanning trees in grid graphs.

First, using procedure GridMN(k,n), we get the $k \times n$ grid graph.

Then, procedure SpFn uses the Matrix Tree Theorem to evaluate the determinant of the co-factor of the Laplacian matrix of the grid graph which is the number of spanning trees in this particular graph. For a fixed k, we need to generate a sufficiently long list of data for the number of spanning trees in $G_k(n), n \in [l(k), u(k)]$. The lower bound $l(k)$ can't be too small since the first several terms are the initial condition; the upper bound $u(k)$ can't be too small as well since we need sufficient data to obtain the recurrence relation. Notice that there is a symmetry for the recurrence relation, and to take advantage of this fact, we modified GuessRec to get the more efficient GuessSymRec (requiring less data). Once the recurrence relation, and the initial conditions, are given, applying CtoR(S,t) will give the desirable generating function, that, of course, is a rational function of t. All the above is incorporated in procedure GFGridKN(k,t) which inputs a positive integer k and a symbol t, and outputs the generating function whose coefficient of t^n is the number of spanning trees in $G_k(n)$, i.e. if we let $s(k, n)$ be the number of spanning trees in $G_k(n)$, the generating function

$$F_k(t) = \sum_{n=0}^{\infty} s(k, n) t^n.$$

We now list the generating functions $F_k(t)$ for $1 \leq k \leq 7$: except for $k = 7$, these were already found by Raff [4] and Faase [2], but it is reassuring that, using our new approach, we got the same output. The case $k = 7$ seems to be new.

Theorem 1 *The generating function for the number of spanning trees in $G_1(n)$ is:*

$$F_1(t) = \frac{t}{1 - t}.$$

Theorem 2 *The generating function for the number of spanning trees in $G_2(n)$ is:*

$$F_2 = \frac{t}{t^2 - 4t + 1}.$$

Theorem 3 *The generating function for the number of spanning trees in $G_3(n)$ is:*

$$F_3 = \frac{-t^3 + t}{t^4 - 15t^3 + 32t^2 - 15t + 1}.$$

Theorem 4 *The generating function for the number of spanning trees in $G_4(n)$ is:*

$$F_4 = \frac{t^7 - 49t^5 + 112t^4 - 49t^3 + t}{t^8 - 56t^7 + 672t^6 - 2632t^5 + 4094t^4 - 2632t^3 + 672t^2 - 56t + 1}.$$

For $5 \leq k \leq 7$, since the formulas are too long, we present their numerators and denominators separately.

Theorem 5 *The generating function for the number of spanning trees in $G_5(n)$ is:*

$$F_5 = \frac{N_5}{D_5}$$

where

$$N_5 = -t^{15} + 1440t^{13} - 26752t^{12} + 185889t^{11} - 574750t^{10} + 708928t^9$$
$$-708928t^7 + 574750t^6 - 185889t^5 + 26752t^4 - 1440t^3 + t,$$
$$D_5 = t^{16} - 209t^{15} + 11936t^{14} - 274208t^{13} + 3112032t^{12} - 19456019t^{11}$$
$$+70651107t^{10} - 152325888t^9 + 196664896t^8 - 152325888t^7$$

$$+70651107\,t^6 - 19456019\,t^5 + 3112032\,t^4 - 274208\,t^3$$
$$+11936\,t^2 - 209\,t + 1.$$

Theorem 6 *The generating function for the number of spanning trees in $G_6(n)$ is:*

$$F_6 = \frac{N_6}{D_6}$$

where

$$N_6 = t^{31} - 33359\,t^{29} + 3642600\,t^{28} - 173371343\,t^{27} + 4540320720\,t^{26}$$
$$-70164186331\,t^{25} + 634164906960\,t^{24} - 2844883304348\,t^{23}$$
$$-1842793012320\,t^{22} + 104844096982372\,t^{21} - 678752492380560\,t^{20}$$
$$+2471590551535210\,t^{19} - 5926092273213840\,t^{18} + 9869538714631398\,t^{17}$$
$$-11674018886109840\,t^{16} + 9869538714631398\,t^{15}$$
$$-5926092273213840\,t^{14} + 2471590551535210\,t^{13}$$
$$-678752492380560\,t^{12} + 104844096982372\,t^{11} - 1842793012320\,t^{10}$$
$$-2844883304348\,t^9 + 634164906960\,t^8 - 70164186331\,t^7$$
$$+4540320720\,t^6 - 173371343\,t^5 + 3642600\,t^4 - 33359\,t^3 + t,$$

$$D_6 = t^{32} - 780\,t^{31} + 194881\,t^{30} - 22377420\,t^{29} + 1419219792\,t^{28}$$
$$-55284715980\,t^{27} + 1410775106597\,t^{26} - 24574215822780\,t^{25}$$
$$+300429297446885\,t^{24} - 2629946465331120\,t^{23} + 16741727755133760\,t^{22}$$
$$-78475174345180080\,t^{21} + 273689714665707178\,t^{20}$$
$$-716370537293731320\,t^{19} + 1417056251105102122\,t^{18}$$
$$-2129255507292156360\,t^{17} + 2437932520099475424\,t^{16}$$
$$-2129255507292156360\,t^{15} + 1417056251105102122\,t^{14}$$
$$-716370537293731320\,t^{13} + 273689714665707178\,t^{12}$$
$$-78475174345180080\,t^{11} + 16741727755133760\,t^{10}$$
$$-2629946465331120\,t^9 + 300429297446885\,t^8 - 24574215822780\,t^7$$
$$+1410775106597\,t^6 - 55284715980\,t^5$$
$$+1419219792\,t^4 - 22377420\,t^3 + 194881\,t^2 - 780\,t + 1.$$

Theorem 7 *The generating function for the number of spanning trees in $G_7(n)$ is:*

$$F_7 = \frac{N_7}{D_7}$$

where

$$N_7 = -t^{47} - 142\,t^{46} + 661245\,t^{45} - 279917500\,t^{44} + 53184503243\,t^{43}$$
$$-5570891154842\,t^{42} + 341638600598298\,t^{41} - 11886702497030032\,t^{40}$$
$$+164458937576610742\,t^{39} + 4371158470492451828\,t^{38}$$
$$-288737344956855301342\,t^{37} + 7736513993329973661368\,t^{36}$$
$$-131582338768322853956994\,t^{35} + 1573202877300834187134466\,t^{34}$$
$$-13805721749199518460916737\,t^{33} + 90975567796174070740787232\,t^{32}$$
$$-455915282590547643587452175\,t^{31} + 1747901867578637315747826286\,t^{30}$$
$$-5126323837327170557921412877\,t^{29} + 11416779122947828869806142972\,t^{28}$$
$$-18924703166237080216745900796\,t^{27} + 22194247945745188489023284104\,t^{26}$$
$$-15563815847174688069871470516\,t^{25} + 15563815847174688069871470516\,t^{23}$$
$$-22194247945745188489023284104\,t^{22} + 18924703166237080216745900796\,t^{21}$$
$$-11416779122947828869806142972\,t^{20} + 5126323837327170557921412877\,t^{19}$$
$$-1747901867578637315747826286\,t^{18} + 455915282590547643587452175\,t^{17}$$
$$-90975567796174070740787232\,t^{16} + 13805721749199518460916737\,t^{15}$$
$$-1573202877300834187134466\,t^{14} + 131582338768322853956994\,t^{13}$$
$$-7736513993329973661368\,t^{12} + 288737344956855301342\,t^{11}$$
$$-4371158470492451828\,t^{10} - 164458937576610742\,t^{9}$$
$$+11886702497030032\,t^{8} - 341638600598298\,t^{7} + 5570891154842\,t^{6}$$
$$-53184503243\,t^{5} + 279917500\,t^{4} - 661245\,t^{3} + 142\,t^{2} + t,$$

$$D_7 = t^{48} - 2769\,t^{47} + 2630641\,t^{46} - 1195782497\,t^{45} + 305993127089\,t^{44}$$
$$-48551559344145\,t^{43} + 5083730101530753\,t^{42} - 366971376492201338\,t^{41}$$
$$+18871718211768417242\,t^{40} - 709234610141846974874\,t^{39}$$
$$+19874722637854592209338\,t^{38} - 422023241997789381263002\,t^{37}$$
$$+6880098547452856483997402\,t^{36} - 87057778313447181201990522\,t^{35}$$
$$+862879164715733847737203343\,t^{34} - 6750900711491569851736413311\,t^{33}$$
$$+41958615314622858303912597215\,t^{32} - 208258356862493902206466194607\,t^{31}$$
$$+828959040281722890327985220255\,t^{30} - 2654944041424536277948746010303\,t^{29}$$
$$+6859440538554030239641036025103\,t^{28} - 14324708604336971207868317957868\,t^{27}$$
$$+24214587194571650834572683444012\,t^{26} - 33166490975387358866518005011884\,t^{25}$$

$$+368308503833758374810960263578684\,t^{24} - 3316649097538735886651800501118844\,t^{23}$$

$$+242145871945716508345726834440124\,t^{22} - 1432470860433697120786831795786844\,t^{21}$$

$$+6859440538554030239641036025103\,t^{20} - 26549440414245362779487460103034\,t^{19}$$

$$+82895904028172289032798522025554\,t^{18} - 208258356862493902206466194607\,t^{17}$$

$$+4195861531462285830391259721554\,t^{16} - 67509007114915698517364133114\,t^{15}$$

$$+86287916471573384773720334334\,t^{14} - 8705777831344718120199052244\,t^{13}$$

$$+6880098547452856483997402\,t^{12} - 42202324199778938126300244\,t^{11}$$

$$+19874722637854592209338\,t^{10} - 7092346101418469748744\,t^{9}$$

$$+18871718211768417242\,t^{8} - 369713764922013384\,t^{7} + 5083730101530753\,t^{6}$$

$$-48551559344145\,t^{5} + 305993127089\,t^{4} - 1195782497\,t^{3} + 2630641\,t^{2} - 2769t + 1.$$

Note that, surprisingly, the degree of the denominator of $F_7(t)$ is 48 rather than the expected 64 since the first six generating functions' denominator have degree 2^{k-1}, $1 \le k \le 6$. With a larger computer, one should be able to compute F_k for larger k, using this experimental approach.

Generally, for an arbitrary graph G, we consider the number of spanning trees in $G \times P_n$. With the same methodology, a list of data can be obtained empirically with which a generating function follows.

7 Joint Resistance

The original motivation for the Matrix Tree Theorem, first discovered by Kirchhoff (of Kirchhoff's laws fame) came from the desire to efficiently compute joint resistances in an electrical network.

Suppose one is interested in the joint resistance in an electric network in the form of a grid graph between two diagonal vertices $[1, 1]$ and $[k, n]$. We assume that each edge has resistance 1 Ohm. To obtain it, all we need is, in addition for the number of spanning trees (that's the numerator), the number of spanning forests $SF_k(n)$ of the graph $G_k(n)$ that have exactly two components, each component containing exactly one of the members of the pair $\{[1, 1], [k, n]\}$ (this is the denominator). The joint resistance is just the ratio.

In principle, we can apply the same method to obtain the generating function S_k. Empirically, we found that the denominator of S_k is always the square of the denominator of F_k times another polynomial C_k. Once the denominator is known, we can find the numerator in the same way as above. So our focus is to find C_k. The procedure DenomSFKN(k,t) in the Maple package JointConductance .txt, calculates C_k. For $2 \le k \le 4$, we have

$$C_2 = t - 1,$$

$$C_3 = t^4 - 8t^3 + 17t^2 - 8t + 1,$$

$$C_4 = t^{12} - 46t^{11} + 770t^{10} - 6062t^9 + 24579t^8 - 55388t^7$$
$$+ 72324t^6 - 55388t^5 + 24579t^4 - 6062t^3 + 770t^2 - 46t + 1.$$

Remark By looking at the output of our Maple package, we conjectured that $R(k, n)$, the resistance between vertex $[1, 1]$ and vertex $[k, n]$ in the $k \times n$ grid graph, $G_k(n)$, where each edge is a resistor of 1 Ohm, is asymptotically n/k, for any fixed k, as $n \to \infty$. We proved it rigorously for $k \leq 6$, and we wondered whether there is a human-generated "electric proof". Naturally we emailed Peter Doyle, the co-author of the delightful masterpiece [1], who quickly came up with the following argument.

Making the horizontal resistors into almost resistance-less gold wires gives the lower bound $R(k, n) \geq (n - 1)/k$ since it is a parallel circuit of k resistors of $n - 1$ Ohms. For an upper bound of the same order, put 1 Ampere in at $[1,1]$ and out at $[k, n]$, routing $1/k$ Ampere up each of the k verticals. The energy dissipation is $k(n - 1)/k^2 + C(k) = (n - 1)/k + C(k)$, where the constant $C(k)$ is the energy dissipated along the top and bottom resistors. Specifically, $C(k) = 2(1 - 1/k)^2 + (1 - 2/k)^2 + \cdots + (1/k)^2)$. So $(n - 1)/k \leq R(k, n) \leq (n - 1)/k + C(k)$.

We thank Peter Doyle for his kind permission to reproduce this *electrifying* argument.

8   The Statistic of the Number of Vertical Edges in Spanning Trees of Grid Graphs

Often in enumerative combinatorics, the class of interest has natural 'statistics', like height, weight, and IQ for humans. Recall that the *naive counting* is

$$|A| := \sum_{a \in A} 1,$$

getting a **number**. Define:

$$|A|_x := \sum_{a \in A} x^{f(a)},$$

where $f := A \to \mathbb{Z}$ is the statistic of interest. To go from the weighted enumeration (a certain Laurent polynomial) to straight enumeration, one simply plugs-in $x = 1$, i.e. $|A|_1 = |A|$.

The *scaled* random variable is defined as follows. Let $E(f)$ and $Var(f)$ be the *expectation* and *variance*, respectively, of the statistic f defined on A, and define the *scaled* random variable, for $a \in A$, by

$$X(a) := \frac{f(a) - E(f)}{\sqrt{Var(f)}}.$$

In this section, we are interested in the statistic 'number of vertical edges', defined on spanning trees of grid graphs. For given k and n, let, as above, $G_k(n)$ denote the $k \times n$ grid-graph. Let $\mathcal{F}_{k,n}$ be its set of spanning trees. If the weight is 1, then $\sum_{f \in \mathcal{F}_{k,n}} 1 = |\mathcal{F}_{k,n}|$ is the naive counting. Now let's define a natural statistic $ver(T)$ = the number of vertical edges in the spanning tree T, and the weight $w(T) = v^{ver(T)}$, then the weighted counting follows:

$$Ver_{k,n}(v) = \sum_{T \in \mathcal{F}_{k,n}} w(T)$$

where $\mathcal{F}_{k,n}$ is the set of spanning trees of $G_k(n)$.

We define the bivariate generating function

$$g_k(v, t) = \sum_{n=0}^{\infty} Ver_{k,n} t^n.$$

More generally, with our Maple package GFMatrix.txt, and procedure VerGF, we are able to obtain the bivariate generating function for an arbitrary graph of the form $G \times P_n$. The procedure VerGF takes inputs G (an arbitrary graph), N (an integer determining how many data we use to find the recurrence relation) and two symbols v and t.

The main tool for computing VerGF is still the Matrix Tree Theorem and GuessRec. But we need to modify the Laplacian matrix for the graph. Instead of letting $a_{ij} = -1$ for $i \neq j$ and $\{i, j\} \in E(G \times P_n)$, we should consider whether the edge $\{i, j\}$ is a vertical edge. If so, we let $a_{i,j} = -v, a_{j,i} = -v$. The diagonal elements which are $(-1) \times$ (the sum of the rest entries on the same row) should change accordingly. The following theorems are for grid graphs when $2 \leq k \leq 4$ while $k = 1$ is a trivial case because there are no vertical edges.

Theorem 8 *The bivariate generating function for the weighted counting according to the number of vertical edges of spanning trees in $G_2(n)$ is:*

$$g_2(v, t) = \frac{vt}{1 - (2v + 2)t + t^2}.$$

Theorem 9 *The bivariate generating function for the weighted counting according to the number of vertical edges vertical edges of spanning trees in $G_3(n)$ is:*

$$g_3(v, t) = \frac{-t^3 v^2 + v^2 t}{1 - (3v^2 + 8v + 4)t - (-10v^2 - 16v - 6)t^2 - (3v^2 + 8v + 4)t^3 + t^4}.$$

Theorem 10 *The bivariate generating function for the weighted counting according to the number of vertical edges of spanning trees in $G_4(n)$ is:*

$$g_4(v, t) = \frac{numer(g_4)}{denom(g_4)}$$

where

$$numer(g_4) = v^3 t + \left(-16 v^5 - 24 v^4 - 9 v^3\right) t^3 + \left(8 v^6 + 40 v^5 + 48 v^4 + 16 v^3\right) t^4$$
$$+ \left(-16 v^5 - 24 v^4 - 9 v^3\right) t^5 + v^3 t^7$$

and

$$denom(g_4) = 1 - \left(4 v^3 + 20 v^2 + 24 v + 8\right) t$$
$$- \left(-52 v^4 - 192 v^3 - 256 v^2 - 144 v - 28\right) t^2$$
$$- \left(64 v^5 + 416 v^4 + 892 v^3 + 844 v^2 + 360 v + 56\right) t^3$$
$$- \left(-16 v^6 - 160 v^5 - 744 v^4 - 1408 v^3 - 1216 v^2 - 480 v - 70\right) t^4$$
$$- \left(64 v^5 + 416 v^4 + 892 v^3 + 844 v^2 + 360 v + 56\right) t^5$$
$$- \left(-52 v^4 - 192 v^3 - 256 v^2 - 144 v - 28\right) t^6$$
$$- \left(4 v^3 + 20 v^2 + 24 v + 8\right) t^7 + t^8.$$

With the Maple package `BiVariateMoms.txt` and its `Story` procedure from http://sites.math.rutgers.edu/~zeilberg/tokhniot/BiVariateMoms.txt, the expectation, variance and higher moments can be easily analyzed. We calculated up to the 4th moment for $G_2(n)$. For $k = 3, 4$, you can find the output files from http://sites.math.rutgers.edu/~yao/OutputStatisticVerticalk=3.txt; http://sites.math.rutgers.edu/~yao/OutputStatisticVerticalk=4.txt.

Theorem 11 *The moments of the statistic: the number of vertical edges in the spanning trees of $G_2(n)$ are as follows:*
Let b be the largest positive root of the polynomial equation

$$b^2 - 4b + 1 = 0$$

whose floating-point approximation is 3.732050808, then the size of the n-th family (i.e. straight enumeration) is very close to

$$\frac{b^{n+1}}{-2 + 4b}.$$

The average of the statistics is, asymptotically

$$\frac{1}{3} + \frac{1}{3}\frac{(-1+2b)n}{b}.$$

The variance of the statistics is, asymptotically

$$-\frac{1}{9} + \frac{1}{9}\frac{(7b-2)n}{-1+4b}.$$

The skewness of the statistics is, asymptotically

$$\frac{780b - 209}{(4053b - 1086)n^3 + (-7020b + 1881)n^2 + (4053b - 1086)n - 780b + 209}.$$

The kurtosis of the statistics is, asymptotically

$$3\frac{(32592b - 8733)n^2 + (-56451b + 15126)n + 21728b - 5822}{(32592b - 8733)n^2 + (-37634b + 10084)n + 10864b - 2911}.$$

9 Application of the C-finite Ansatz to Computing Generating Functions of Determinants (and Permanents) of Almost-Diagonal Toeplitz Matrices

So far, we have seen applications of the C-finite ansatz methodology for automatically computing generating functions for enumerating spanning trees/forests for certain infinite families of graphs.

The second case study is completely different, and in a sense more general, since the former framework may be subsumed in this new context.

Definition Diagonal matrices A are square matrices in which the entries outside the main diagonal are 0, i.e. $a_{ij} = 0$ if $i \neq j$.

Definition An almost-diagonal Toeplitz matrix A is a square matrices in which $a_{i,j} = 0$ if $j - i \geq k_1$ or $i - j \geq k_2$ for some fixed positive integers k_1, k_2 and $\forall i_1, j_1, i_2, j_2$, if $i_1 - j_1 = i_2 - j_2$, then $a_{i_1 j_1} = a_{i_2 j_2}$.

For simplicity, we use the notation $L = [n, $ [the first k_1 entries in the first row], [the first k_2 entries in the first column]] to denote the $n \times n$ matrix with these specifications. Note that this notation already contains all information we need to reconstruct this matrix. For example, [6, [1,2,3], [1,4]] is the matrix

$$\begin{bmatrix} 1 & 2 & 3 & 0 & 0 & 0 \\ 4 & 1 & 2 & 3 & 0 & 0 \\ 0 & 4 & 1 & 2 & 3 & 0 \\ 0 & 0 & 4 & 1 & 2 & 3 \\ 0 & 0 & 0 & 4 & 1 & 2 \\ 0 & 0 & 0 & 0 & 4 & 1 \end{bmatrix}.$$

The following is the Maple procedure `DiagMatrixL` (in our Maple package `GFMatrix.txt`), which inputs such a list L and outputs the corresponding matrix.

```
DiagMatrixL:=proc(L) local n, r1, c1,p,q,S,M,i:
n:=L[1]:
r1:=L[2]:
c1:=L[3]:
p:=nops(r1)-1:
q:=nops(c1)-1:
if r1[1] <> c1[1] then
  return fail:
fi:
S:=[0$(n-1-q), seq(c1[q-i+1],i=0..q-1), op(r1),
  0$(n-1-p)]:
M:=[0$n]:
for i from 1 to n do
  M[i]:=[op(max(0,n-1-q)+q+2-i..max(0,n-1-q)+q+1
  +n-i,S)]:
od:
return M:
end:
```

For this matrix, $k_1 = 3$ and $k_2 = 2$. Let k_1, k_2 be fixed and M_1, M_2 be two lists of numbers or symbols of length k_1 and k_2 respectively, A_k is the almost-diagonal Toeplitz matrix represented by the list $L_k = [k, M_1, M_2]$. Note that the first elements in the lists M_1 and M_2 must be identical.

Having fixed two lists M_1 of length k_1 and M_2 of length k_2, (where $M_1[1] = M_2[1]$), it is of interest to derive *automatically*, the generating function (that is always a rational function for reasons that will soon become clear), $\sum_{k=0}^{\infty} a_k t^k$, where a_k denotes the determinant of the $k \times k$ almost-diagonal Toeplitz matrix whose first row starts with M_1, and first column starts with M_2. Analogously, it is also of interest to do the analogous problem when the determinant is replaced by the permanent.

Here is the Maple procedure `GFfamilyDet` which takes inputs (i) A: a name of a Maple procedure that inputs an integer n and outputs an $n \times n$ matrix according to some rule, e.g., the almost-diagonal Toeplitz matrices, (ii) a variable name t, (iii) two integers m and n which are the lower and upper bounds of the sequence of determinants we consider. It outputs a rational function in t, say $R(t)$, which is the generating function of the sequence.

```
GFfamilyDet:=proc(A,t,m,n) local i,rec,GF,B,gu,Denom,
   L,Numer:
L:=[seq(det(A(i)),i=1..n)]:
rec:=GuessRec([op(m..n,L)])[2]:
gu:=solve(B-1-add(t**i*rec[i]*B,i=1..nops(rec)),  B):
Denom:=denom(subs(gu,B)):
Numer:=Denom*(1+add(L[i]*t**i,  i=1..n)):
Numer:=add(coeff(Numer,t,i)*t**i,  i=0..degree
   (Denom,t)):
Numer/Denom:
end:
```

Similarly we have procedure `GFfamilyPer` for the permanent. Let's look at an example. The following is a sample procedure which considers the family of almost diagonal Toeplitz matrices which the first row $[2, 3]$ and the first column $[2, 4, 5]$.

```
SampleB:=proc(n) local L,M:
L:=[n,  [2,3],  [2,4,5]]:
M:=DiagMatrixL(L):
end:
```

Then `GFfamilyDet(SampleB, t, 10, 50)` will return the generating function

$$-\frac{1}{45\,t^3 - 12\,t^2 + 2\,t - 1}.$$

It turns out, that for this problem, the more 'conceptual' approach of setting up a transfer matrix also works well. But don't worry, the computer can do the 'research' all by itself, with only a minimum amount of human pre-processing.

We will now describe this more conceptual approach, that may be called *symbolic dynamical programming*, where the computer sets up, *automatically*, a finite-state scheme, by *dynamically* discovering the set of states, and automatically figures out the transfer matrix.

10   The Transfer Matrix Method for Almost-Diagonal Toeplitz Matrices

Recall from Linear Algebra 101, the

Cofactor Expansion   Let $|A|$ denote the determinant of an $n \times n$ matrix A, then

$$|A| = \sum_{j=1}^{n} (-1)^{i+j} a_{ij} M_{ij}, \quad \forall i \in [n],$$

where M_{ij} is the (i, j)-minor.

We'd like to consider the Cofactor Expansion for almost-diagonal Toeplitz matrices along the first row. For simplicity, we assume while $a_{i,j} = 0$ if $j-i \geq k_1$ or $i - j \geq k_2$ for some fixed positive integers k_1, k_2, and if $-k_2 < j_1 - i_1 < j_2 - i_2 < k_1$, then $a_{i_1 j_1} \neq a_{i_2 j_2}$. Under this assumption, for any minors we obtain through recursive Cofactor Expansion along the first row, the dimension, the first row and the first column should provide enough information to reconstruct the matrix.

For an almost-diagonal Toeplitz matrix represented by $L =$[Dimension, [the first k_1 entries in the first row], [the first k_2 entries in the first column]], any minor can be represented by [Dimension, [entries in the first row up to the last nonzero entry], [entries in the first column up to the last nonzero entry]].

Our goal in this section is the same as the last one, to get a generating function for the determinant or permanent of almost-diagonal Toeplitz matrices A_k with dimension k. Once we have those almost-diagonal Toeplitz matrices, the first step is to do a one-step expansion as follows:

```
ExpandMatrixL:=proc(L,L1)
local n,R,C,dim,R1,C1,i,r,S,candidate,newrow,newcol,
  gu,mu,
  temp,p,q,j:
n:=L[1]:
R:=L[2]:
C:=L[3]:
p:=nops(R)-1:
q:=nops(C)-1:
dim:=L1[1]:
R1:=L1[2]:
C1:=L1[3]:
if R1=[] or C1=[] then
  return :
elif R[1]<>C[1] or R1[1]<>C1[1] or dim>n then
  return fail:
else
S:={}:
gu:=[0$(n-1-q), seq(C[q-i+1],i=0..q-1),
  op(R), 0$(n-1-p)]:
candidate:=[0$nops(R1),R1[-1]]:
```

```
for i from 1 to nops(R1) do
  mu:=R1[i]:
for j from n-q to nops(gu) do
    if gu[j]=mu then
      candidate[i]:=gu[j-1]:
    fi:
  od:
od:
for i from n-q to nops(gu) do
  if gu[i] = R1[2] then
    temp:=i:
    break:
  fi:
od:
for i from 1 to nops(R1) do
  if i = 1 then
    mu:=[R1[i]*(-1)**(i+1),
      [dim-1,[op(i+1..nops(candidate), candidate)],
      [seq(gu[temp-i],i=1..temp-n+q)]]]:
    S:=S union mu:
  else
    mu:=[R1[i]*(-1)**(i+1), [dim-1, [op(1..i-1,
      candidate),
      op(i+1..nops(candidate), candidate)],
      [op(2..nops(C1), C1)]]]:
    S:=S union mu:
  fi:
od:
  return S:
fi:
end:
```

The ExpandMatrixL procedure inputs a data structure $L = $ [Dimension, first_row=[ ], first_col=[ ]] as the matrix we start and the other data structure $L1$ as the current minor we have, expands $L1$ along its first row and outputs a list of [multiplicity, data structure].

We would like to generate all the "children" of an almost-diagonal Toeplitz matrix regardless of the dimension, i.e., two lists L represent the same child as long as their first_rows and first_columns are the same, respectively. The set of "children" is the scheme of the almost diagonal Toeplitz matrices in this case.

The following is the Maple procedure ChildrenMatrixL which inputs a data structure L and outputs the set of its "children" under Cofactor Expansion along the first row:

```
ChildrenMatrixL:=proc(L) local S,t,T,dim,U,u,s:
dim:=L[1]:
S:={[op(2..3,L)]}:
T:={seq([op(2..3,t[2])],t in ExpandMatrixL(L,L))}:
while T minus S <> {} do
  U:=T minus S:
  S:=S union T:
  T:={}:
```

```
   for u in U do
   T:=T union {seq([op(2..3,t[2])],t
     in ExpandMatrixL(L,[dim,op(u)]))}:
od:
od:
for s in S do
   if s[1]=[] or s[2]=[] then
     S:=S minus {s}:
   fi:
od:
S:
end:
```

After we have the scheme S, by the Cofactor Expansion of any element in the scheme, a system of algebraic equations follows. For children in S, it's convenient to let the almost-diagonal Toeplitz matrix be the first one C_1 and for the rest, any arbitrary ordering will do. For example, if after Cofactor Expansion for C_1, c_2 "copies" of C_2 and c_3 "copies" of C_3 are obtained, then the equation will be

$$C_1 = 1 + c_2 t C_2 + c_3 t C_3.$$

However, if the above equation is for $C_i, i \neq 1$, i.e. C_i is not the almost-diagonal Toeplitz matrix itself, then the equation will be slightly different:

$$C_i = c_2 t C_2 + c_3 t C_3.$$

Here t is a symbol as we assume the generating function is a rational function of t.

Here is the Maple code that implements how we get the generating function for the determinant of a family of almost-diagonal Toeplitz matrices by solving a system of algebraic equations:

```
GFMatrixL:=proc(L,t)  local S,dim,var,eq,n,A,i,result,
  gu,mu:
dim:=L[1]:
S:=ChildrenMatrixL(L):
S:=[[op(2..3,L)], op(S minus {[op(2..3,L)]})]:
n:=nops(S):
var:={seq(A[i],i=1..n)}:
eq:={}:
for i from 1 to 1 do
  result:=ExpandMatrixL(L,[dim,op(S[i])]):
  for gu in result do
    if gu[2][2]=[] or gu[2][3]=[] then
      result:=result minus {gu}:
    fi:
  od:
  eq:=eq union {A[i] - 1
    - add(gu[1]*t*A[CountRank(S, [op(2..3, gu[2])])],
    gu in result)}:
od:
```

```
for i from 2 to n do
  result:=ExpandMatrixL(L,[dim,op(S[i])]):
  for gu in result do
  if gu[2][2]=[] or gu[2][3]=[] then
    result:=result minus gu:
  fi:
od:
eq:=eq union {A[i]
  - add(gu[1]*t*A[CountRank(S, [op(2..3, gu[2])])]),
    gu in result)}:
od:
gu:=solve(eq, var)[1]:
subs(gu, A[1]):
end:
```

`GFMatrixL([20, [2, 3], [2, 4, 5]], t)` returns

$$-\frac{1}{45\,t^3 - 12\,t^2 + 2\,t - 1}.$$

Compared to empirical approach, the 'symbolic dynamical programming' method is faster and more efficient for the moderate-size examples that we tried out. However, as the lists will grow larger, it is likely that the former method will win out, since with this non-guessing approach, it is equally fast to get generating functions for determinants and permanents, and as we all know, permanents are hard.

The advantage of the present method is that it is more appealing to humans, and does not require any 'meta-level' act of faith. However, both methods are very versatile and are great experimental approaches for enumerative combinatorics problems. We hope that our readers will find other applications.

11 Summary

Rather than trying to tackle each enumeration problem, one at a time, using ad hoc human ingenuity each time, building up an intricate transfer matrix, and only using the computer at the end as a symbolic calculator, it is a much better use of our beloved silicon servants (soon to become our masters!) to replace 'thinking' by 'meta-thinking', i.e. develop experimental mathematics methods that can handle many different types of problems. In the two case studies discussed here, everything was made rigorous, but if one can make semi-rigorous and even non-rigorous discoveries, as long as they are *interesting*, one should not be hung up on rigorous proofs. In other words, if you can find a rigorous justification (like in these two case studies) that's nice, but if you can't, that's also nice!

Acknowledgements Many thanks are due to a very careful referee that pointed out many minor, but annoying errors, that we hope corrected. Also thanks to Peter Doyle for permission to include his elegant electric argument.

References

1. Doyle, P., Snell, L.: Random Walks and Electrical Networks. Carus Mathematical Monographs, vol. 22. Mathematical Association of America, Washington (1984)
2. Faase, F.J.: On the number of specific spanning subgraphs of the graphs $g \times p_n$. Ars Combinatorica **49**, 129–154 (1998)
3. Kauers, M., Paule, P.: The Concrete Tetrahedron. Springer, New York (2011)
4. Raff, P.: Spanning Trees in Grid Graph. https://arxiv.org/abs/0809.2551
5. Shalosh, B., Ekhad, N.J.A.: Sloane and Doron Zeilberger. Automated proof (or disproof) of linear recurrences satisfied by pisot sequences. Personal Journal of Shalosh B. Ekhad and Doron Zeilberger. Available from http://www.math.rutgers.edu/~zeilberg/mamarim/mamarimhtml/pisot.html
6. Stanley, R.: Enumerative Combinatorics, vol. 1, 1st edn. Wadsworth & Brooks/Cole, Belmont, CA (1986). 2nd edn. Cambridge University Press, Cambridge (2011)
7. Wikipedia contributors. "Berlekamp-Massey algorithm." Wikipedia, The Free Encyclopedia. Wikipedia, The Free Encyclopedia, 26 November 2018. Web. 7 Jan. 201
8. Zeilberger, D.: An enquiry concerning human (and computer!) [mathematical] understanding. In: Calude, C.S. (ed.) Randomness & Complexity, from Leibniz to Chaitin, pp. 383–410. World Scientific, Singapore (2007). Available from http://www.math.rutgers.edu/~zeilberg/mamarim/mamarimhtml/enquiry.html
9. Zeilberger, D.: The C-finite ansatz. Ramanujan J. **31**, 23–32 (2013). Available from http://www.math.rutgers.edu/~zeilberg/mamarim/mamarimhtml/cfinite.html
10. Zeilberger, D.: Why the cautionary tales supplied by Richard Guy's strong law of small numbers should not be overstated. Personal J. Shalosh B. Ekhad and Doron Zeilberger. Available from http://www.math.rutgers.edu/~zeilberg/mamarim/mamarimhtml/small.html

Printed in the United States
by Baker & Taylor Publisher Services